Statistics and Computing

Series Editors:
J. Chambers
W. Eddy
W. Härdle
S. Sheather
L. Tierney

Statistics and Computing

José C. Pinheiro
Douglas M. Bates

Mixed-Effects Models in S and S-PLUS

With 172 Illustrations

 Springer

José C. Pinheiro
Department of Biostatistics
Novartis Pharmaceuticals
One Health Plaza
East Hanover, NJ 07936-1080
USA
jose.pinheiro@pharma.novartis.com

Douglas M. Bates
Department of Statistics
University of Wisconsin
Madison, WI 53706-1685
USA
bates@stat.wisc.edu

Series Editors:

J. Chambers
 Bell Labs,
 Lucent Technologies
600 Mountain Ave.
Murray Hill, NJ 07974
USA

W. Eddy
Department of Statistics
Carnegie Mellon University
Pittsburgh, PA 15213
USA

W. Härdle
Institut für Statistik
 und Ökonometrie
Humboldt-Universität
 zu Berlin
Spandauer Str. 1
D-10178 Berlin
Germany

S. Sheather
Australian Graduate School
 of Management
University of New South Wales
Sydney NSW 2052
Australia

L. Tierney
School of Statistics
University of Minnesota
Vincent Hall
Minneapolis, MN 55455
USA

Library of Congress Cataloging-in-Publication Data
Pinheiro, José C.
    Mixed-effects models in S and S-PLUS/José C. Pinheiro, Douglas M. Bates
        p. cm.—(Statistics and computing)
    Includes bibliographical references and index.
    ISBN 978-1-4419-0317-4 (soft cover)
    I. Bates, Douglas M. II. Title. III. Series.
    QA76.73.Sl5P56   2000
    005.13′3—dc21   99-053566

Printed on acid-free paper.

Printed in the United States of America.

Springer Verlag is a part of *Springer Science + Business Media*

springeronline.com

To Elisa and Laura

To Mary Ellen, Barbara, and Michael

Preface

Mixed-effects models provide a flexible and powerful tool for the analysis of grouped data, which arise in many areas as diverse as agriculture, biology, economics, manufacturing, and geophysics. Examples of grouped data include longitudinal data, repeated measures, blocked designs, and multilevel data. The increasing popularity of mixed-effects models is explained by the flexibility they offer in modeling the within-group correlation often present in grouped data, by the handling of balanced and unbalanced data in a unified framework, and by the availability of reliable and efficient software for fitting them.

This book provides an overview of the theory and application of linear and nonlinear mixed-effects models in the analysis of grouped data. A unified model-building strategy for both linear and nonlinear models is presented and applied to the analysis of over 20 real datasets from a wide variety of areas, including pharmacokinetics, agriculture, and manufacturing. A strong emphasis is placed on the use of graphical displays at the various phases of the model-building process, starting with exploratory plots of the data and concluding with diagnostic plots to assess the adequacy of a fitted model. Over 170 figures are included in the book.

The class of mixed-effects models considered in this book assumes that both the random effects and the errors follow Gaussian distributions. These models are intended for grouped data in which the response variable is (at least approximately) continuous. This covers a large number of practical applications of mixed-effects models, but does not include, for example, generalized linear mixed-effects models (Diggle, Liang and Zeger, 1994).

The balanced mix of real data examples, modeling software, and theory makes this book a useful reference for practitioners who use, or intend to use, mixed-effects models in their data analyses. It can also be used as a text for a one-semester graduate-level applied course in mixed-effects models. Researchers in statistical computing will also find this book appealing for its presentation of novel and efficient computational methods for fitting linear and nonlinear mixed-effects models.

The nlme library we developed for analyzing mixed-effects models in implementations of the S language, including S-PLUS and R, provides the underlying software for implementing the methods presented in the text, being described and illustrated in detail throughout the book. All analyses included in the book were produced using version 3.1 of nlme with S-PLUS 3.4 running on an Iris 5.4 Unix platform. Because of platform dependencies, the analysis results may be expected to vary slightly with different computers or operating systems and with different implementations of S. Furthermore, the current version of the nlme library for R does not support the same range of graphics presentations as does the S-PLUS version. The latest version of nlme and further information on the NLME project can be obtained at

http://nlme.stat.wisc.edu or

http://cm.bell-labs.com/stat/NLME.

Errata and updates of the material in the book will be made available on-line at the same sites.

The book is divided into parts. Part I, comprising five chapters, is dedicated to the linear mixed-effects (LME) model and Part II, comprising three chapters, covers the nonlinear mixed-effects (NLME) model. Chapter 1 gives an overview of LME models, introducing some examples of grouped data and the type of analyses that applies to them. The theory and computational methods for LME models are the topics of Chapter 2. Chapter 3 describes the structure of grouped data and the many facilities available in the nlme library to display and summarize such data. The model-building approach we propose is described and illustrated in detail in the context of LME models in Chapter 4. Extensions of the basic LME model to include variance functions and correlation structures for the within-group errors are considered in Chapter 5. The second part of the book follows an organization similar to the first. Chapter 6 provides an overview of NLME models and some of the analysis tools available for them in nlme. The theory and computational methods for NLME models are described in Chapter 7. The final chapter is dedicated to model building in the context of NLME models and to illustrating in detail the nonlinear modeling facilities available in the nlme library.

Even though the material covered in the book is, for the most part, self-contained, we assume that the reader has some familiarity with linear regression models, say at the level of Draper and Smith (1998). Although enough theory is covered in the text to understand the strengths and weaknesses of mixed-effects models, we emphasize the applied aspects of these. Readers who desire to learn in more detail the theory of mixed-effects are referred to the excellent book by Davidian and Giltinan (1995). Some knowledge of the S language is definitely desireable, but not a pre-requisite for following the material in the book. For those who are new to, or less familiar with S, we suggest using in conjunction with this book the, by now, *classic* reference Venables and Ripley (1999), which provides an overview of S and an introduction to a wide variety of statistical models to be used with S.

The authors may be contacted via electronic mail at

`jcp@research.bell-labs.com`

`bates@stat.wisc.edu`

and would appreciate being informed of typos, errors, and improvements to the contents of this book.

Typographical Conventions:

The S language objects and commands referenced throughout the book are printed in a monospaced typewriter font `like this`, while the S classes are printed in sans-serif font like this. The standard prompt > is used for S commands and the prompt + is used to indicate continuation lines.

To save space, some of the S output has been edited. Omission of complete lines are usually indicated by

. . .

but some blank lines have been removed without indication. The S output was generated using the options settings

```
> options( width = 68, digits = 5 )
```

The default settings are for 80 and 7, respectively.

Acknowledgements:

This book would not exist without the help and encouragement of many people with whom we have interacted over the years. We are grateful to the people who have read and commented on earlier versions of the book; their many useful suggestions have had a direct impact on the current organization of the book. Our thanks also go to the many people who have tested and made suggestions on the nlme library, in particular the beta testers for the current version of the software. It would not be possible to name all these people here, but in particular we would like to thank John Chambers, Yonghua Chen, Bill Cleveland, Saikat DebRoy, Ramón Días-Uriarte, David James, Diane Lambert, Renaud Lancelot, David Lansky, Brian Ripley, Elisa Santos, Duncan Temple Lang, Silvia Vega, and Bill

Venables. Finally, we would like to thank our editor, John Kimmel, for his continuous encouragement and support.

José C. Pinheiro
Douglas M. Bates
March 2000

Contents

Part I

Linear Mixed-Effects Models

1
Linear Mixed-Effects Models: Basic Concepts and Examples

Many common statistical models can be expressed as linear models that incorporate both *fixed effects*, which are parameters associated with an entire population or with certain repeatable levels of experimental factors, and *random effects*, which are associated with individual experimental units drawn at random from a population. A model with both fixed effects and random effects is called a *mixed-effects* model.

Mixed-effects models are primarily used to describe relationships between a response variable and some covariates in data that are grouped according to one or more classification factors. Examples of such *grouped data* include *longitudinal data, repeated measures data, multilevel data,* and *block designs.* By associating common random effects to observations sharing the same level of a classification factor, mixed-effects models flexibly represent the covariance structure induced by the grouping of the data.

In this chapter we present an overview of linear mixed-effects (LME) models, introducing their basic concepts through the analysis of several real-data examples, starting from simple models and gradually moving to more complex models. Although the S code to fit these models is shown, the purpose here is to present the motivation for using LME models to analyze grouped data and not to concentrate on the software for fitting and displaying the models. This chapter serves as an appetizer for the material covered in later chapters: the theoretical and computational methods for LME models described in Chapter 2 and the linear mixed-effects modeling facilities available in the nlme library, covered in detail in Chapter 4.

The examples described in this chapter also serve to illustrate the breadth of applications of linear mixed-effects models.

1.1   A Simple Example of Random Effects

The data shown in Figure 1.1 are from an experiment in nondestructive testing for longitudinal stress in railway rails cited in Devore (2000, Example 10.10, p. 427). Six rails were chosen at random and tested three times each by measuring the time it took for a certain type of ultrasonic wave to travel the length of the rail. The only experimental setting that changes between the observations is the rail. We say these observations are arranged in a *one-way classification* because they are classified according to a single characteristic—the rail on which the observation was made. These data are described in greater detail in Appendix A.26.

The quantities the engineers were interested in estimating from this experiment are the average travel time for a "typical" rail (the *expected travel time*), the variation in average travel times among rails (the *between-rail variability*), and the variation in the observed travel times for a single rail (the *within-rail variability*). We can see from Figure 1.1 that there is considerable variability in the mean travel time for the different rails. Overall the between-rail variability is much greater than the within-rail variability.

The data on the rails experiment are given in an object called `Rail` that is available with the nlme library. Giving the name `Rail` by itself to the S interpreter will result in the data being displayed.

```
> Rail
Grouped Data: travel ~ 1 | Rail
   Rail travel
  1    1    55
  2    1    53
  3    1    54
 . . .
 17    6    85
 18    6    83
```

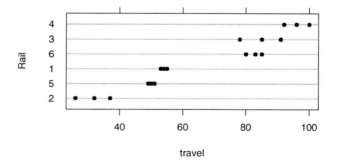

FIGURE 1.1. Travel time in nanoseconds for ultrasonic head-waves in a sample of six railroad rails. The times shown are the result of subtracting 36,100 nanoseconds from the original observation.

As would be expected, the structure of the data is quite simple—each row corresponds to one observation for which the rail and the travel time are recorded. The names of the variables in the data frame are `Rail` and `travel`. There is also a formula, `travel ~ 1 | Rail`, associated with the data. This formula is discussed in Chapter 3, where we describe structures available in the nlme library for representing grouped data.

Data from a one-way classification like the rails example can be analyzed either with a fixed-effects model or with a random-effects model. The distinction between the two models is according to whether we wish to make inferences about those particular levels of the classification factor that were used in the experiment or to make inferences about the population from which these levels were drawn. In the latter case the "levels" usually correspond to different subjects or different plots or different experimental units of some sort.

To illustrate the importance of accounting for the classification factor when modeling grouped data such as the rails example, we initially ignore the grouping structure of the data and assume the simple model

$$y_{ij} = \beta + \epsilon_{ij}, \quad i = 1, \ldots, M, \quad j = 1, \ldots, n_i, \tag{1.1}$$

where y_{ij} is the observed travel time for observation j on rail i, β is the mean travel time across the population of rails being sampled, and the ϵ_{ij} are independent $\mathcal{N}(0, \sigma^2)$ error terms. The number of rails is M and the number of observations on rail i is n_i. In this case $M = 6$ and $n_1 = n_2 = \cdots = n_6 = 3$. The total number of observations is $N = \sum_{i=1}^{M} n_i = 18$.

The `lm` function is used to fit the single-mean model (1.1) in S. Its first argument is a formula describing the model and its second argument is a data frame containing the variables named in the model formula.

```
> fm1Rail.lm <- lm( travel ~ 1, data = Rail )
> fm1Rail.lm
Call:
lm(formula = travel ~ 1, data = Rail)

Coefficients:
 (Intercept)
        66.5

Degrees of freedom: 18 total; 17 residual
Residual standard error: 23.645
```

As is typical with S, we do not produce output directly from the fitting process. Instead we store the fitted model as an object called `fm1Rail.lm` then cause this object to be displayed. It contains the parameter estimates $\widehat{\beta} = 66.5$ and $\widehat{\sigma} = 23.645$.

The boxplots of the residuals from the `fm1Rail.lm` fit by rail number, displayed in Figure 1.2, illustrate the fundamental problem with ignoring

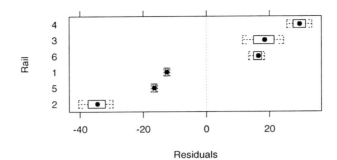

FIGURE 1.2. Boxplots of residuals by rail number for the lm fit of the single-mean model (1.1) to the data from the rail experiment.

the classification factor when modeling grouped data: the "group effects" are incorporated into the residuals (which, in this case, have identical signs for each rail), leading to an inflated estimate of the within-rail variability.

The "rail effects" indicated in Figure 1.2 may be incorporated into the model for the travel times by allowing the mean of each rail to be represented by a separate parameter. This *fixed-effects* model for the one-way classification is written

$$y_{ij} = \beta_i + \epsilon_{ij}, \quad i = 1, \dots, M, \quad j = 1, \dots, n_i, \tag{1.2}$$

where the β_i represents the mean travel time of rail i and, as in (1.1), the errors ϵ_{ij} are assumed to be independently distributed as $\mathcal{N}(0, \sigma^2)$. We can again use lm to fit (1.2).

```
> fm2Rail.lm <- lm( travel ~ Rail - 1, data = Rail )
> fm2Rail.lm
Call:
lm(formula = travel ~ Rail - 1, data = Rail)

Coefficients:
  Rail2 Rail5 Rail1  Rail6  Rail3 Rail4
 31.667    50    54 82.667 84.667    96

Degrees of freedom: 18 total; 12 residual
Residual standard error: 4.0208
```

A -1 is used in the model formula to prevent the default inclusion of an intercept term in the model. As expected, there is considerable variation in the estimated mean travel times per rail. The residual standard error obtained for the fixed-effects model (1.2), $\hat{\sigma} = 4.0208$, is about one-sixth of the corresponding estimate obtained for the single-mean model (1.1), indicating that the fm2Rail.lm model has successfully accounted for the rail effects. This is better illustrated by the boxplots of the fm2Rail.lm residuals

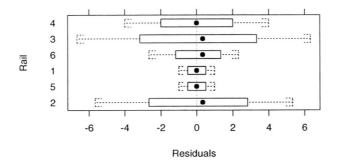

FIGURE 1.3. Boxplots of residuals by rail number for the `lm` fit of the fixed-effects model (1.2) to the data from the rail experiment.

by rail number, shown in Figure 1.3. The residuals are now centered around zero and have considerably smaller magnitudes than those in Figure 1.2.

Even though the fixed-effects model (1.2) accounts for the rail effects, it does not provide a useful representation of the rails data. Its basic problem is that it only models the specific sample of rails used in the experiment, while the main interest is in the population of rails from which the sample was drawn. In particular, `fm2Rail.lm` does not provide an estimate of the between-rail variability, which is one of the central quantities of interest in the rails experiment. Another drawback of this fixed-effects model is that the number of parameters in the model increases linearly with the number of rails.

A *random-effects* model circumvents these problems by treating the rail effects as random variations around a population mean. The following re-parameterization of model (1.2) helps motivate the random-effects model for the rails data. We write

$$y_{ij} = \bar{\beta} + \left(\beta_i - \bar{\beta}\right) + \epsilon_{ij}, \tag{1.3}$$

where $\bar{\beta} = \sum_{i=1}^{6} \beta_i/6$ represents the average travel time for the rails in the experiment. The random-effects model replaces $\bar{\beta}$ by the mean travel time over the *population of rails* and replaces the deviations $\beta_i - \bar{\beta}$ by random variables whose distribution is to be estimated.

A random-effects model for the one-way classification used in the rails experiment is written

$$y_{ij} = \beta + b_i + \epsilon_{ij}, \tag{1.4}$$

where β is the mean travel time across the population of rails being sampled, b_i is a random variable representing the deviation from the population mean of the mean travel time for the ith rail, and ϵ_{ij} is a random variable representing the deviation in travel time for observation j on rail i from the mean travel time for rail i.

To complete the statistical model, we must specify the distribution of the random variables $b_i, i = 1, \ldots, M$ and $\epsilon_{ij}, i = 1, \ldots, M; j = 1, \ldots, n_i$. We begin by modeling both of these as independent, constant variance, normally distributed random variables with mean zero. The variances are denoted σ_b^2 for the b_i, or "between-rail" variability, and σ^2 for the ϵ_{ij}, or "within-rail" variability. That is,

$$b_i \sim \mathcal{N}(0, \sigma_b^2), \quad \epsilon_{ij} \sim \mathcal{N}(0, \sigma^2). \tag{1.5}$$

This model may be modified if it does not seem appropriate. As described in Chapter 4, we encourage using graphical and numerical diagnostic tools to assess the validity of the model and to suggest ways in which it could be modified. To start, however, we will use this simple model.

This model with two sources of random variation, b_i and ϵ_{ij}, is sometimes called a *hierarchical* model (Lindley and Smith, 1972; Bryk and Raudenbush, 1992) or a *multilevel* model (Goldstein, 1995). The b_i are called *random* effects because they are associated with the particular experimental units—rails in this case—that are selected at random from the population of interest. They are *effects* because they represent a deviation from an overall mean. That is, the "effect" of choosing rail i is to shift the mean travel time from β to $\beta + b_i$. Because observations made on the same rail share the same random effect b_i, they are correlated. The covariance between observations on the same rail is σ_b^2 corresponding to a correlation of $\sigma_b^2 / (\sigma_b^2 + \sigma^2)$.

The parameters of the statistical model created by combining (1.4) and (1.5) are β, σ_b^2, and σ^2. Note that the number of parameters will always be three, irrespective of the number of rails in the experiment. Although the random effects, $b_i, i = 1, \ldots, M$ may behave like parameters, formally they are just another level of random variation in the model so we do not "estimate" them as such. We will, however, form predictions $\widehat{b}_i$ of the values of these random variables, given the data we observed.

1.1.1   Fitting the Random-Effects Model With lme

The lme function from the nlme library for S can be used to fit linear mixed-effects models, using either *maximum likelihood* (ML) or *restricted maximum likelihood* (REML). These estimation methods for the parameters in LME models are described in detail in §2.2.

A typical call to lme is similar to a call to lm. As in lm, the first two arguments to lme, fixed and data, give the model for the expected response (the fixed-effects part of the model) and the object containing the data to which the model should be fit. The third argument, random, is a one-sided formula describing the random effects and the grouping structure for the model. Another important argument is method. Specifying method = "ML" produces maximum likelihood fits while method = "REML", the default, produces restricted maximum likelihood fits.

Many variations in the specifications of linear mixed-effects models for lme are possible, as shown later in this and other chapters. Details of all the possible arguments and their forms are given in Appendix B.

We obtain the restricted maximum likelihood fit of the model given by (1.4) and (1.5) to the Rail data with

```
> fm1Rail.lme <- lme(travel ~ 1, data = Rail, random = ~ 1 | Rail)
```

The first argument indicates that the response is travel and that there is a single fixed effect, the intercept. The second argument indicates that the data will be found in the object named Rail. The third argument indicates that there is a single random effect for each group and that the grouping is given by the variable Rail. Note that there is a variable or column Rail within the data frame that is also named Rail. Because no estimation method is specified, the default, "REML", is used.

We can query the fitted lme object, fm1Rail.lme, using different accessor functions, also described in detail in Appendix B. One of the most useful of these is the summary function

```
> summary( fm1Rail.lme )
Linear mixed-effects model fit by REML
 Data: Rail
       AIC    BIC   logLik
  128.18 130.68 -61.089

Random effects:
 Formula:  ~ 1 | Rail
         (Intercept) Residual
StdDev:       24.805   4.0208

Fixed effects: travel ~ 1
            Value Std.Error DF t-value p-value
(Intercept)  66.5    10.171 12  6.5382  <.0001

Standardized Within-Group Residuals:
      Min        Q1       Med       Q3      Max
  -1.6188 -0.28218  0.035693  0.21956 1.6144

Number of Observations: 18
Number of Groups: 6
```

We see that the REML estimates for the parameters have been calculated as

$$\widehat{\beta} = 66.5, \quad \widehat{\sigma}_b = 24.805, \quad \widehat{\sigma} = 4.0208,$$

corresponding to a log-restricted-likelihood of -61.089. The estimated mean travel time $\widehat{\beta}$ is identical to the estimated intercept in the fm1Rail.lm fit, and the estimated within-rail standard deviation $\hat{\sigma}$ is identical to the residual standard error from `fm2Rail.lm`. This will not occur in general; it is a consequence of the `Rail` data being a *balanced* one-way classification that has the same number of observations on each rail. We also note that the estimated between-rail standard deviation $\hat{\sigma}_b$ is similar to the residual standard error from the `fm1Rail.lm` fit.

The output of the `summary` function includes the values of the *Akaike Information Criterion* (*AIC*) (Sakamoto, Ishiguro and Kitagawa, 1986) and the *Bayesian Information Criterion* (*BIC*) (Schwarz, 1978), which is also sometimes called *Schwarz's Bayesian Criterion* (*SBC*). These are model comparison criteria evaluated as

$$AIC = -2\log\mathrm{Lik} + 2n_{par},$$
$$BIC = -2\log\mathrm{Lik} + n_{par}\log(N),$$

where n_{par} denotes the number of parameters in the model and N the total number of observations used to fit the model. Under these definitions, "smaller is better." That is, if we are using AIC to compare two or more models for the same data, we prefer the model with the lowest AIC. Similarly, when using BIC we prefer the model with the lowest BIC.

To examine the maximum likelihood estimates we would call `lme` with the same arguments as for `fm1Rail.lme` except for `method = "ML"`. A convenient way of fitting such alternative models is to use the `update` function where we only need to specify the arguments that are different from those in the earlier fit.

```
> fm1Rail.lmeML <- update( fm1Rail.lme, method = "ML" )
> summary( fm1Rail.lmeML )
Linear mixed-effects model fit by maximum likelihood
 Data: Rail
      AIC    BIC  logLik
   134.56 137.23 -64.28

Random effects:
 Formula:  ~ 1 | Rail
        (Intercept) Residual
StdDev:      22.624   4.0208

Fixed effects: travel ~ 1
             Value Std.Error DF t-value p-value
(Intercept)   66.5     9.554 12  6.9604  <.0001

Standardized Within-Group Residuals:
    Min       Q1      Med      Q3     Max
 -1.611 -0.28887 0.034542 0.21373 1.6222

Number of Observations: 18
Number of Groups: 6
```

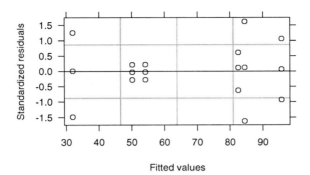

FIGURE 1.4. Standardized residuals versus the fitted values for the REML fit of a random-effects model to the data from the rail experiment.

Notice that the ML estimate of σ is 4.0208, the same as the REML estimate. Equality of the ML and REML estimates of σ occurs for this simple model, but will not occur in general. The ML estimate of σ_b, 22.624, is smaller than the REML estimate, 24.805. Finally the ML estimate of β, 66.5, is the same as the REML estimate. Again, exact equality of the ML and REML estimates of the fixed effects need not occur in more complex models, but it is commonplace for them to be nearly identical.

1.1.2 Assessing the Fitted Model

The fitted model can, and should, be examined using graphical and numerical summaries. One graphical summary that should be examined routinely is a plot of the residuals versus the fitted responses from the model. This plot is used to assess the assumption of constant variance of the ϵ_{ij}. Because this plot is a common diagnostic, it is the default `plot` method for a fitted `lme` model. That is, it is produced by the simple call

```
> plot( fm1Rail.lme )                    # produces Figure 1.4
```

The standardized residuals, shown on the vertical axis in Figure 1.4, are the raw residuals, $e_{ij} = y_{ij} - \widehat{\beta} - \widehat{b}_i$, divided by the estimated standard deviation, $\widehat{\sigma}$, of the ϵ_{ij}.

In this plot we are looking for a systematic increase (or, less commonly, a systematic decrease) in the variance of the ϵ_{ij} as the level of the response increases. If this is present, the residuals on the right-hand side of the plot will have a greater vertical spread than those on the left, forming a horizontal "wedge-shaped" pattern. Such a pattern is not evident in Figure 1.4.

With more complicated models there are other diagnostic plots that we may want to examine, as discussed in Chapter 4.

We should also examine numerical summaries of the model. A basic summary is a set of confidence intervals on the parameters, β, σ and σ_b, as produced by the intervals function.

```
> intervals( fm1Rail.lme )
Approximate 95% confidence intervals

 Fixed effects:
              lower est.  upper
 (Intercept) 44.339 66.5 88.661

 Random Effects:
  Level: Rail
                     lower   est.  upper
 sd((Intercept)) 13.274 24.805 46.354

 Within-group standard error:
 lower   est.  upper
 2.695 4.0208 5.9988
```

We can see that there is considerable imprecision in the estimates of all three of these parameters.

Another numerical summary, used to assess the significance of terms in the fixed-effects part of the model, is produced by the anova function

```
> anova( fm1Rail.lme )
            numDF denDF F-value p-value
 (Intercept)     1    12  42.748  <.0001
```

In this case, the fixed-effects model is so simple that the analysis of variance is trivial. The hypothesis being tested here is $\beta = 0$. The p-value, which is that probability of observing data as unusual as these or even more so when β actually is 0, is so small as to rule out this possibility. Regardless of the p-value, the hypothesis $\beta = 0$ is of no practical interest here because the data have been shifted by subtracting 36,100 nanoseconds from each measurement.

1.2   A Randomized Block Design

In the railway rails example of the last section, the observations were classified according to one characteristic only—the rail on which the observation was made. In other experiments we may have more than one classification factor for each observation. A *randomized block design* is a type of experiment in which there are two classification factors: an *experimental* factor for which we use fixed effects and a *blocking* factor for which we use random effects.

The data shown in Figure 1.5 and available as the object ergoStool in

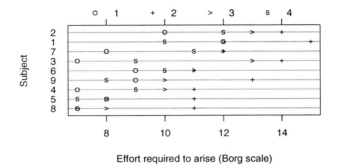

Effort required to arise (Borg scale)

FIGURE 1.5. Effort required (Borg scale) to arise from a stool for nine different subjects each using four different types of stools. Different symbols, shown in the key at the top of the plot, are used for the different types of stools.

the nlme library are from an ergometrics experiment that has a randomized block design. The experimenters recorded the effort required by each of nine different subjects to arise from each of four types of stools. We want to compare these four particular types of stools so we use fixed effects for the Type factor. The nine different subjects represent a sample from the population about which we wish to make inferences so we use random effects to model the Subject factor.

From Figure 1.5 it appears that there are systematic differences between stool types on this measurement. For example, the T2 stool type required the greatest effort from each subject while the T1 stool type was consistently one of the low effort types. The subjects also exhibited variability in their scoring of the effort, but we would expect this. We say that Subject to be a *blocking factor* because it represents a known source of variability in the experiment. Type is said to be an *experimental factor* because the purpose of the experiment is to determine if there are systematic differences in the level of effort to arise from the different types of stools.

We can visually compare the magnitude of the effects of the Type and Subject factors using a "design plot"

```
> plot.design( ergoStool )                    # produces Figure 1.6
```

This plot is produced by averaging the responses at each level of each factor and plotting these averages. We see that the variability associated with the Type factor is comparable to the variability associated with the Subject factor. We also see that the average effort according to stool type is in the order T1 ≤ T4 ≤ T3 ≤ T2.

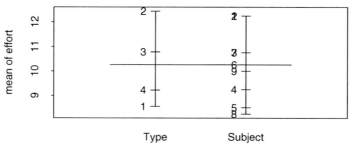

FIGURE 1.6. Design plot for the data in the stool ergometric experiment. The mean of the response (**effort**) is plotted for each level of each of the factors **Type** and **Subject**.

1.2.1   Choosing Contrasts for Fixed-Effects Terms

A model with fixed effects β_j for the **Type** factor and random effects b_i for the **Subject** factor could be written

$$y_{ij} = \beta_j + b_i + \epsilon_{ij}, \quad i = 1, \ldots, 9, \quad j = 1, \ldots, 4,$$
$$b_i \sim \mathcal{N}(0, \sigma_b^2), \quad \epsilon_{ij} \sim \mathcal{N}(0, \sigma^2),$$

$$(1.6)$$

or, equivalently,

$$\boldsymbol{y}_i = \boldsymbol{X}_i\boldsymbol{\beta} + \boldsymbol{Z}_i b_i + \boldsymbol{\epsilon}_i, \quad i = 1, \ldots, 9,$$
$$b_i \sim \mathcal{N}(0, \sigma_b^2), \quad \boldsymbol{\epsilon}_i \sim \mathcal{N}(\boldsymbol{0}, \sigma^2 \boldsymbol{I}),$$

where, for $i = 1, \ldots, 9,$

$$\boldsymbol{y}_i = \begin{bmatrix} y_{i1} \\ y_{i2} \\ y_{i3} \\ y_{i4} \end{bmatrix}, \quad \boldsymbol{X}_i = \begin{bmatrix} 1 & 0 & 0 & 0 \\ 0 & 1 & 0 & 0 \\ 0 & 0 & 1 & 0 \\ 0 & 0 & 0 & 1 \end{bmatrix}, \quad \boldsymbol{Z}_i = \boldsymbol{1} = \begin{bmatrix} 1 \\ 1 \\ 1 \\ 1 \end{bmatrix}, \quad \boldsymbol{\epsilon}_i = \begin{bmatrix} \epsilon_{i1} \\ \epsilon_{i2} \\ \epsilon_{i3} \\ \epsilon_{i4} \end{bmatrix}.$$

This form of fixed-effects matrix $\boldsymbol{X}_i$ is sometimes called the *cell means* form because the jth component of $\boldsymbol{\beta}$ represents what would be the mean effort to arise from the jth type of stool if the whole population were tested.

These β_j have a simple interpretation, but are not convenient to use when assessing differences between stool types. To make it easier to assess these differences we use an alternative form of the $\boldsymbol{X}_i$ matrices with one column representing some "overall mean" or reference level and three columns representing changes between the types of stools. The three columns representing the changes are called the *contrasts*. There are several different choices available for these contrasts (Venables and Ripley, 1999, §6.2). In S-PLUS, the default choice for unordered factors, such as the **Type** factor, is the *Helmert* contrasts

```
> contrasts( ergoStool$Type )
  [,1] [,2] [,3]
1  -1   -1   -1
2   1   -1   -1
3   0    2   -1
4   0    0    3
```

(In R the default contrasts for an unordered factor are the "treatment" contrasts, which are described below.)

The X_i matrices for a given set of contrasts can be displayed with the model.matrix function. To save space we show the X_1 matrix only.

```
> ergoStool1 <- ergoStool[ ergoStool$Subject == "1", ]
> model.matrix( effort ~ Type, ergoStool1 ) # X matrix for Subject 1
  (Intercept) Type1 Type2 Type3
1           1    -1    -1    -1
2           1     1    -1    -1
3           1     0     2    -1
4           1     0     0     3
```

Using the Helmert contrasts shown above, the components of β represent:

- β_1—Mean level of effort for the four stool types.

- β_2—Difference between T2 and T1.

- β_3—Twice the difference between T3 and the average of T1 and T2.

- β_4—Three times the difference between T4 and the average of T1, T2, and T3.

Fitting the model in this form with lme produces

```
> fm1Stool <-
+   lme(effort ~ Type, data = ergoStool, random = ~ 1 | Subject)
> summary( fm1Stool )
 . . .
Random effects:
 Formula:  ~ 1 | Subject
         (Intercept) Residual
StdDev:       1.3325   1.1003

Fixed effects: effort ~ Type
             Value Std.Error DF t-value p-value
(Intercept) 10.250   0.48052 24  21.331  <.0001
      Type1  1.944   0.25934 24   7.498  <.0001
      Type2  0.093   0.14973 24   0.618  0.5421
      Type3 -0.343   0.10588 24  -3.236  0.0035
 Correlation:
   (Intr)  T1     T2
T1  0
```

```
T2  0      0
T3  0      0     0
```

```
Standardized Within-Group Residuals:
    Min        Q1      Med     Q3      Max
  -1.802 -0.64317 0.057831 0.701 1.6314
```

```
Number of Observations: 36
Number of Groups: 9
```

By convention, the coefficient corresponding to the first column in the X_i, which is the column of 1's, is called the *intercept*. The name originated with models like the analysis of covariance model of §1.4 where a straight-line model for each group is written in terms of its slope and its intercept with the y-axis. In those cases, this parameter is the y-intercept. For the model considered here, the parameter labelled (Intercept) is the estimate of mean effort for all four types of stools across the population. The other three parameters, labelled Type1, Type2, and Type3, are described above. Their individual interpretations are not as important as the collective variability among the stool types they represent. The significance of this variability, and hence the overall significance of the Type term, is assessed with the **anova** function.

```
> anova( fm1Stool )
            numDF denDF F-value p-value
(Intercept)     1    24  455.01  <.0001
       Type     3    24   22.36  <.0001
```

On some occasions we may want to switch to other contrasts that provide more meaningful parameter estimates for the experiment. For example, if stool type T1 was a "standard" stool and we wished to compare the other types to this standard type, we could use the contrasts called the *treatment* contrasts. These contrasts represent the change from the first level of the factor to each of the other levels.

One way to cause the treatment contrasts to be used is to reset the contrasts option. Its value should be a vector of two character strings. The first string is the name of the function to use for factors, such as Type, and the second is the function to use for ordered factors, which are described in §1.6.

```
> options( contrasts = c( factor = "contr.treatment",
+                         ordered = "contr.poly" ) )
> contrasts( ergoStool$Type )
  2 3 4
1 0 0 0
2 1 0 0
3 0 1 0
4 0 0 1
```

```
> fm2Stool <-
+   lme(effort ~ Type, data = ergoStool, random = ~ 1 | Subject)
> summary( fm2Stool )
Linear mixed-effects model fit by REML
 Data: ergoStool
      AIC    BIC  logLik
   133.13 141.93 -60.565

Random effects:
 Formula:   ~ 1 | Subject
         (Intercept) Residual
StdDev:      1.3325   1.1003

Fixed effects: effort ~ Type
              Value Std.Error DF t-value p-value
(Intercept) 8.5556   0.57601 24  14.853  <.0001
      Type2 3.8889   0.51868 24   7.498  <.0001
      Type3 2.2222   0.51868 24   4.284  0.0003
      Type4 0.6667   0.51868 24   1.285  0.2110
 Correlation:
        (Intr) Type2 Type3
Type2 -0.45
Type3 -0.45    0.50
Type4 -0.45    0.50  0.50
 . . .
> anova( fm2Stool )
            numDF denDF F-value p-value
(Intercept)     1    24  455.01  <.0001
       Type     3    24   22.36  <.0001
```

Although the individual parameter estimates for the `Type` factor are different between the two fits, the `anova` results are the same. The difference in the parameter estimates simply reflects the fact that different contrasts are being estimated. The similarity of the `anova` results indicates that the overall variability attributed to the `Type` factor does not change. In each case, the row labelled `Type` in the analysis of variance table represents a test of the hypothesis

$$H_0 : \beta_2 = \beta_3 = \beta_4 = 0,$$

which is equivalent to reducing model (1.6) to

$$y_i = 1\beta + Z_i b_i + \epsilon_i, \quad i = 1,\ldots,9, \quad b_i \sim \mathcal{N}(0, \sigma_b^2), \quad \epsilon_i \sim \mathcal{N}(0, \sigma^2 I).$$

This reduced model is invariant under the change in contrasts.

There is a more subtle effect of changing from one form of X_j matrix to another—the value of the REML criterion changes. As described in §2.2.5, when the model parameters are kept at a fixed value, a change in the X_j matrices results in change in the value of the restricted likelihood function.

Even though we converge to the same variance component estimates $\hat{\sigma} = 1.1003$ and $\hat{\sigma}_b = 1.3325$, the value of the estimation criterion itself changes. Because the AIC and BIC criteria are based on the REML criterion, they will also change.

As a consequence, when using REML estimation we can only use likelihood ratio tests or comparisons of AIC or BIC for models with the same fixed-effects structure and the same contrasts for any factors used in the fixed-effects structure.

We can fit the "cell means" parameterization of the model if we add the term -1 to the formula for the fixed effects. This causes the column of 1's to be removed from the model matrices $\boldsymbol{X}_j$,

```
> model.matrix( effort ~ Type - 1, ergoStool1 )
  Type1 Type2 Type3 Type4
1     1     0     0     0
2     0     1     0     0
3     0     0     1     0
4     0     0     0     1
```

and the fitted model is now expressed in terms of the mean effort for each stool type over the population

```
> fm3Stool <-
+ lme(effort ~ Type - 1, data = ergoStool, random = ~ 1 | Subject)
> summary( fm3Stool )
Linear mixed-effects model fit by REML
 Data: ergoStool
      AIC    BIC   logLik
   133.13 141.93 -60.565

Random effects:
 Formula: ~ 1 | Subject
         (Intercept) Residual
StdDev:       1.3325   1.1003

Fixed effects: effort ~ Type - 1
        Value Std.Error DF t-value p-value
Type1   8.556  0.57601 24  14.853  <.0001
Type2  12.444  0.57601 24  21.604  <.0001
Type3  10.778  0.57601 24  18.711  <.0001
Type4   9.222  0.57601 24  16.010  <.0001
 Correlation:
       Type1 Type2 Type3
Type2 0.595
Type3 0.595 0.595
Type4 0.595 0.595 0.595
```

```
Standardized Within-Group Residuals:
    Min      Q1      Med     Q3     Max
  -1.802 -0.64317 0.057831 0.701 1.6314
```

```
Number of Observations: 36
Number of Groups: 9
```

This change in the fixed-effects structure does change the anova results.

```
> anova( fm3Stool )
        numDF denDF F-value p-value
Type      4    24  130.52  <.0001
```

The hypothesis being tested by anova for this model is

$$H_0 : \beta_1 = \beta_2 = \beta_3 = \beta_4 = 0,$$

which is equivalent to reducing model (1.6) to

$$\boldsymbol{y}_i = \boldsymbol{Z}_i b_i + \boldsymbol{\epsilon}_i, \quad i = 1, \ldots, 9, \qquad b_i \sim \mathcal{N}(0, \sigma_b^2), \quad \boldsymbol{\epsilon}_i \sim \mathcal{N}(\boldsymbol{0}, \sigma^2 \boldsymbol{I}).$$

That is, the hypothesis H_0 completely eliminates the fixed-effects parameters from the model so the mean response across the population would be zero. This hypothesis is not meaningful in the context of this experiment.

To reiterate, some general principles to keep in mind regarding fixed-effects terms for factors are:

- The overall effect of the factor should be assessed with anova, not by examining the t-value's or p-value's associated with the fixed-effects parameters. The anova output does not depend on the choice of contrasts as long as the intercept term is retained in the model.

- Interpretation of the parameter estimates for a fixed-effects term depends on the contrasts being used.

- For REML estimation, likelihood-ratio tests or comparisons of AIC or BIC require the same fixed-effects structure and the same choice of contrasts in all models.

- The "cell means" parameters can be estimated by adding -1 to a model formula but this will usually make the results of anova meaningless.

1.2.2 Examining the Model

As in the rail example, we should examine the fitted model both graphically and numerically. The intervals function provides an indication of the precision of the estimates of the variance components.

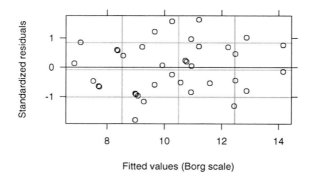

Fitted values (Borg scale)

FIGURE 1.7. Standardized residuals versus the fitted values for the REML fit of a random-effects model to the data in the ergometric experiment on types of stools.

```
> intervals( fm1Stool )
Approximate 95% confidence intervals

 Fixed effects:
                 lower       est.      upper
 (Intercept)   9.25825  10.250000  11.24175
       Type1   1.40919   1.944444    2.47970
       Type2  -0.21644   0.092593    0.40162
       Type3  -0.56111  -0.342593   -0.12408

 Random Effects:
  Level: Subject
                     lower    est.   upper
 sd((Intercept))   0.74923  1.3325  2.3697

 Within-group standard error:
   lower    est.   upper
  0.82894  1.1003  1.4605
```

We see that σ is estimated relatively precisely, whereas σ_b can vary by a factor of about 5, which is a factor of 25 if we express the estimates as variances.

The plot of the standardized residuals versus the fitted values, shown in Figure 1.7, does not indicate a violation of the assumption of constant variance for the ϵ_{ij} terms.

Figure 1.7 shows the overall behavior of the residuals relative to the fitted values. It may be more informative to examine this behavior according to the Subject factor or according to the Type factor, which we can do by providing an explicit formula to the plot method for the fitted lme model. The formula can use functions such as resid or fitted applied to the fitted model. As a shortcut, a "." appearing in the formula is interpreted as the

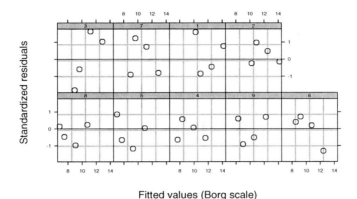

Fitted values (Borg scale)

FIGURE 1.8. Standardized residuals versus the fitted values by Subject for the REML fit of a random-effects model to the data in the ergometric experiment on types of stools.

fitted model object itself. Thus, to plot the standardized, or "Pearson," residuals versus the fitted values by Subject, we use

```
> plot( fm1Stool,                          # produces Figure 1.8
+       form = resid(., type = "p") ~ fitted(.) | Subject,
+       abline = 0 )
```

The argument abline = 0 adds a horizontal reference line at $y = 0$ to each panel. In its more general form, the value of the abline argument should be a numeric vector of length two giving the intercept and the slope of the line to be drawn on each panel. Diagnostic plots for assessing the adequacy of lme fits are discussed in detail in §4.3.

1.3 Mixed-Effects Models for Replicated, Blocked Designs

In the ergometric experiment on the types of stools, each subject tried each type of stool once. We say this design is *unreplicated* because only one observation is available at each combination of experimental conditions. In other experiments like this it may be feasible to take replicate measurements. For example, the Machines data, described in Milliken and Johnson (1992, Chapter 23) and shown in Figure 1.9, gives the productivity score for each of six randomly chosen workers tested on each of three different machine types. Each worker used each machine three times so we have three *replicates* at each set of conditions.

In Figure 1.9 we can see that there are strong indications of differences between machines and also some indications of differences between workers.

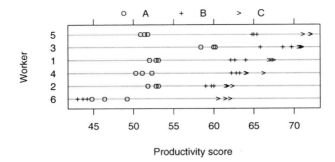

FIGURE 1.9. Productivity scores for three types of machines as used by six different workers. Scores take into account the number and the quality of components produced.

We note that there is very little variability in the productivity score for the same worker using the same machine.

As we did for the experiment on the types of stools, we will model the subject or `Worker` factor with random effects and the type or `Machine` factor with fixed effects. The replications in this experiment will allow us to assess the presence of *interactions* between worker and machine. That is, we can address the question of whether the effect of changing from one type of machine to another is different for different workers.

The comparative dotplot in Figure 1.9 allows us to see patterns across the workers and to see differences between machines within each worker. However, the possibility of interactions is not easy to assess in this plot. An alternative plot, called an *interaction plot*, shows the potential interactions more clearly. It is produced by averaging the scores for each worker on each machine, plotting these averages versus the machine type, and joining the points for each worker. The function `interaction.plot` in S creates such a plot. It is most easily called after `attach`'ing the data frame with the data so the variables in the data frame can be accessed by name.

```
> attach( Machines )  # make variables in Machines available by name
> interaction.plot( Machine, Worker, score, las = 1)   # Figure 1.10
> detach()                # undo the effect of 'attach( Machines )'
```

(The optional argument `las = 1` to `interaction.plot` alters the "label style" on the vertical axis to a more appealing form.)

If there were no interactions between machines and workers, the lines in the interaction plot would be approximately parallel. The lines in Figure 1.10 do not seem to be parallel, especially relative to the variability within the replicates that we can see in Figure 1.9. Worker 6 has an unusual pattern compared to the other workers.

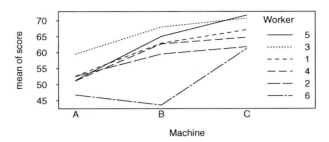

FIGURE 1.10. An interaction plot for the productivity scores for six different workers using three different machine types.

1.3.1 Fitting Random Interaction Terms

A model without interactions has the same form as the model for the ergometric experiment.

$$y_{ijk} = \beta_j + b_i + \epsilon_{ijk}, \quad i = 1, \ldots, 6, \quad j = 1, \ldots, 3, \quad k = 1, \ldots, 3,$$
$$b_i \sim \mathcal{N}(0, \sigma_b^2), \quad \epsilon_{ijk} \sim \mathcal{N}(0, \sigma^2). \tag{1.7}$$

There is a fixed effect for each type of machine and a random effect for each worker. As before, the fixed effects for the machines will be re-coded as an intercept and a set of contrasts when we fit this model as

```
> fm1Machine <-
+   lme( score ~ Machine, data = Machines, random = ~ 1 | Worker )
> fm1Machine
Linear mixed-effects model fit by REML
  Data: Machines
  Log-restricted-likelihood: -145.23
  Fixed: score ~ Machine
 (Intercept) Machine1 Machine2
       59.65   3.9833   3.3111

Random effects:
 Formula:   ~ 1 | Worker
        (Intercept) Residual
StdDev:      5.1466   3.1616

Number of Observations: 54
Number of Groups: 6
```

Because the workers represent a random sample from the population of interest, any interaction terms modeling differences between workers in changing from one machine to another will also be expressed as random effects. The model incorporating the random interaction terms, $b_{ij}, i =$

$1, \ldots, 6$, $j = 1, \ldots, 3$, is

$$y_{ijk} = \beta_j + b_i + b_{ij} + \epsilon_{ijk}, \quad i = 1, \ldots, 6, \quad j = 1, \ldots, 3, \quad k = 1, \ldots, 3,$$
$$b_i \sim \mathcal{N}(0, \sigma_1^2), \quad b_{ij} \sim \mathcal{N}(0, \sigma_2^2), \quad \epsilon_{ijk} \sim \mathcal{N}(0, \sigma^2).$$

This model has random effects at two levels: the effects b_i for the worker and the effects b_{ij} for the type of machine within each worker. In a call to lme we can express this nesting as Worker/Machine in the formula for the random effects. This expression is read as "Worker and 'Machine within Worker'". We can update the previous model with a new specification for the random effects.

```
> fm2Machine <- update( fm1Machine, random = ~ 1 | Worker/Machine )
> fm2Machine
Linear mixed-effects model fit by REML
  Data: Machines
  Log-restricted-likelihood: -109.64
  Fixed: score ~ Machine
(Intercept) Machine1 Machine2
      59.65   3.9833   3.3111

Random effects:
 Formula:   ~ 1 | Worker
         (Intercept)
StdDev:      4.7814

 Formula:   ~ 1 | Machine %in% Worker
         (Intercept) Residual
StdDev:      3.7294  0.96158

Number of Observations: 54
Number of Groups:
 Worker Machine %in% Worker
      6                  18
```

This model has produced a value of the REML criterion of -109.64, which is considerably greater than that of fm1Machine, -145.23. The anova function, when given two or more arguments representing fitted models, produces likelihood ratio tests comparing the models.

```
> anova( fm1Machine, fm2Machine )
            Model df    AIC    BIC  logLik   Test L.Ratio p-value
fm1Machine      1  5 300.46 310.12 -145.23
fm2Machine      2  6 231.27 242.86 -109.64 1 vs 2  71.191  <.0001
```

The likelihood ratio statistic comparing the more general model (fm2Machine) to the more specific model (fm2Machine) is huge and the p-value for the test is essentially zero, so we prefer fm2Machine.

The `anova` function with multiple arguments also reproduces the values of the AIC and the BIC criteria for each model. As described in §1.1.1, these criteria can be used to decide which model to prefer. Because the preference is according to "smaller is better," both these criteria show a strong preference for `fm2Machine` over `fm1Machine`.

1.3.2 Unbalanced Data

The `Machines` data are balanced in that every `Worker` is tested on every `Machine` exactly three times. Milliken and Johnson (1992) analyze this example both as balanced data and as unbalanced data. To obtain the unbalanced data, they randomly deleted ten observations, as indicated in their Table 23.1. They observe that the software they used to estimate the random effects components (SAS PROC VARCOMP) did not produce sensible maximum likelihood estimates (although the current version of this software does). The `lme` function does produce sensible maximum likelihood estimates or restricted maximum likelihood estimates from the unbalanced data.

```
>    ## delete selected rows from the Machines data
> MachinesUnbal <- Machines[ -c(2,3,6,8,9,12,19,20,27,33), ]
>    ## check that the result is indeed unbalanced
> table(MachinesUnbal$Machine, MachinesUnbal$Worker)
    6 2 4 1 3 5
A 3 2 2 1 1 3
B 3 3 3 1 2 2
C 3 3 3 3 3 3
> fm1MachinesU <- lme( score ~ Machine, data = MachinesUnbal,
+    random = ~ 1 | Worker/Machine )
> fm1MachinesU
Linear mixed-effects model fit by REML
  Data: MachinesUnbal
  Log-restricted-likelihood: -92.728
  Fixed: score ~ Machine
(Intercept) Machine1 Machine2
     59.648    3.9812   3.3123

Random effects:
 Formula:   ~ 1 | Worker
        (Intercept)
StdDev:     4.7387

 Formula:   ~ 1 | Machine %in% Worker
        (Intercept) Residual
StdDev:     3.7728    0.9332

Number of Observations: 44
```

```
Number of Groups:
 Worker Machine %in% Worker
      6                 18
> intervals( fm1MachinesU )
Approximate 95% confidence intervals

 Fixed effects:
              lower     est.   upper
(Intercept) 55.2598 59.6476 64.0353
   Machine1  1.5139  3.9812  6.4485
   Machine2  1.8940  3.3123  4.7307

 Random Effects:
  Level: Worker
                   lower    est.  upper
sd((Intercept)) 2.2162 4.7388 10.132
  Level: Machine
                   lower    est.  upper
sd((Intercept)) 2.4091 3.7728 5.9084

 Within-group standard error:
   lower    est.  upper
 0.71202 0.9332 1.2231
```

The estimates of the standard deviations and the confidence intervals
on these parameters look reasonable when compared to those from the
full data set. The techniques used in `lme` for parameter estimation do not
depend on the data being balanced.

However, for either balanced or unbalanced data we must have sufficient
information in the data to be able to estimate the variance components and
the fixed-effects parameters. We can fit a model with random interaction
effects to the `Machines` data because there are replications. If we tried to
fit a nested model to unreplicated data, such as the `ergoStool` data, it may
appear that we are successful until we examine the intervals on the variance
components.

```
> fm4Stool <- lme( effort ~ Type, ergoStool, ~ 1 | Subject/Type )
> intervals( fm4Stool )
Approximate 95% confidence intervals

 Fixed effects:
                lower       est.     upper
(Intercept)  9.25825 10.250000 11.24175
      Type1  1.40919  1.944444  2.47970
      Type2 -0.21644  0.092593  0.40162
      Type3 -0.56111 -0.342593 -0.12408
```

```
Random Effects:
  Level: Subject
                 lower    est.   upper
sd((Intercept)) 0.74952 1.3325 2.3688
  Level: Type
                 lower    est.   upper
sd((Intercept)) 0.05386 0.99958 18.551

Within-group standard error:
     lower     est.   upper
 4.3603e-07 0.45988 485050
```

Apparently the standard deviations σ_2 and σ could vary over twelve orders of magnitude!

If we write this model for these data, taking into account that each subject only tries each type of stool once, we would have

$$y_{ij} = \beta_i + b_j + b_{ij} + \epsilon_{ij}, \quad i = 1, \ldots, 3, \quad j = 1, \ldots, 6,$$
$$b_j \sim \mathcal{N}(0, \sigma_1^2), \quad b_{ij} \sim \mathcal{N}(0, \sigma_2^2), \quad \epsilon_{ij} \sim \mathcal{N}(0, \sigma^2).$$

We can see that the b_{ij} are totally confounded with the ϵ_{ij} so we cannot estimate separate standard deviations for these two random terms. In fact, the estimates reported for σ and σ_2 in this model give a combined variance that corresponds to $\hat{\sigma}^2$ from fm1Stool.

```
> (fm1Stool$sigma)^2
[1] 1.2106
> (fm4Stool$sigma)^2 + 0.79621^2
[1] 1.2107
```

The lesson here is that it is always a good idea to check the confidence intervals on the variance components after fitting a model. Having abnormally wide intervals usually indicates problems with the model definition. In particular, a model with nested interaction terms can only be fit when there are replications available in the data.

1.3.3 More General Models for the Random Interaction Effects

In the model (1.3.1), the random interaction terms all have the same variance σ_2^2. Furthermore, these random interactions are assumed to be independent of one another, even within the same subject. A more general model could treat the random interactions for each subject as a vector and allow the variance–covariance matrix for that vector to be estimated from the set of all positive-definite matrices.

To express this model we return to the matrix/vector representation used in (1.2.1). We define $\boldsymbol{y}_i$ to be the entire response vector for the ith subject,

β to be the three-dimensional vector of fixed-effects parameters for the population, b_i to be the three-dimensional vector of random effects for the ith subject, X_i to be the 9×3 fixed-effects design matrix for subject i, and Z_i to be the 9×3 random-effects design matrix for subject i. The general form of the model is then

$$y_i = X_i\beta + Z_ib_i + \epsilon_i, \quad i = 1, \ldots, 6,$$
$$b_i \sim \mathcal{N}(0, \Psi), \quad \epsilon_i \sim \mathcal{N}(0, \sigma^2 I),$$

where Ψ is a positive-definite, symmetric 3×3 matrix.

To be more specific we must define the matrices X_i and Z_i or, equivalently, define the formulae that generate these matrices as model matrices. As discussed in §1.2.1, in the fixed-effects we generally use a formula that creates X_i with a single column of 1's and two columns of contrasts. We could do the same for the Z_i but, because the random effects are assumed to have an expected value of 0 anyway, it is often more informative to use a formula such as ~ Machine - 1 that removes the intercept column.

Sample model matrices, evaluated on the Worker1's data only, are

```
> Machine1 <- Machines[ Machines$Worker == "1", ]
> model.matrix( score ~ Machine, Machine1 )        # fixed-effects X_i
     (Intercept) Machine1 Machine2
1            1        -1       -1
2            1        -1       -1
3            1        -1       -1
19           1         1       -1
20           1         1       -1
21           1         1       -1
37           1         0        2
38           1         0        2
39           1         0        2
> model.matrix( ~ Machine - 1, Machine1 )          # random-effects Z_i
     MachineA MachineB MachineC
1           1        0        0
2           1        0        0
3           1        0        0
19          0        1        0
20          0        1        0
21          0        1        0
37          0        0        1
38          0        0        1
39          0        0        1
```

The fitted model using this formulation is

```
> fm3Machine <- update( fm1Machine, random = ~Machine - 1 |Worker)
> summary( fm3Machine )
Linear mixed-effects model fit by REML
```

```
Data: Machines
     AIC    BIC   logLik
  231.89 251.21 -105.95

Random effects:
 Formula: ~ Machine - 1 | Worker
 Structure: General positive-definite
          StdDev   Corr
MachineA 4.07928 MachnA MachnB
MachineB 8.62529 0.803
MachineC 4.38948 0.623  0.771
Residual 0.96158

Fixed effects: score ~ Machine
              Value Std.Error DF t-value p-value
(Intercept) 59.650    2.1447 46  27.813  <.0001
   Machine1  3.983    1.2104 46   3.291  0.0019
   Machine2  3.311    0.5491 46   6.030  <.0001
 Correlation:
          (Intr) Machn1
Machine1  0.811
Machine2 -0.540 -0.453

Standardized Within-Group Residuals:
     Min      Q1      Med      Q3      Max
 -2.3935 -0.51378 0.026908 0.47245 2.5334

Number of Observations: 54
Number of Groups: 6
```

This model can be compared to the previous two models using the multi-argument form of `anova`.

```
> anova( fm1Machine, fm2Machine, fm3Machine )
           Model df    AIC    BIC  logLik   Test L.Ratio p-value
fm1Machine     1  5 300.46 310.12 -145.23
fm2Machine     2  6 231.27 242.86 -109.64 1 vs 2  71.191  <.0001
fm3Machine     3 10 231.89 251.21 -105.95 2 vs 3   7.376  0.1173
```

Because the p-value for the test comparing models 2 and 3 is about 12%, we would conclude that the fit `fm3Machine` is not significantly better than `fm2Machine`, taking into account the fact that `fm3Maching` requires four additional parameters in the model.

The AIC criterion is nearly the same for models 2 and 3, indicating that there is no strong preference between these models. The BIC criterion does indicate a strong preference for model 2 relative to model 3. In general BIC puts a heavier penalty than does AIC on having more parameters in the model. Because there are a total of ten parameters in model 3 compared to six parameters in model 2, the BIC criterion will tend to prefer model 2 unless model 3 provides a substantially better fit.

1.4   An Analysis of Covariance Model

Traditionally, the term *analysis of variance* has been applied to models for a continuous response as it relates to various classification factors for the observations. The model for the rails example described in §1.1

$$y_{ij} = \beta + b_i + \epsilon_{ij}, \quad i = 1, \dots, M, \quad j = 1, \dots, n_i,$$
$$b_i \sim \mathcal{N}(0, \sigma_b^2), \quad \epsilon_{ij} \sim \mathcal{N}(0, \sigma^2)$$

is an example of an analysis of variance model with random-effects terms. A *linear regression model*, such as

$$y_i = \beta_1 + \beta_2 x_i + \epsilon_i, \quad i = 1, \dots, N, \quad \epsilon_i \sim \mathcal{N}(0, \sigma^2),$$

relates a continuous response (the y_i) to one or more continuous covariates (the x_i).

The term *analysis of covariance* designates a type of model that relates a continuous response to both a classification factor and to a continuous covariate. If y_{ij} is the jth observation in the ith group of data and x_{ij} is the corresponding value of the covariate, an analysis of covariance model with a random effect for the intercept would be

$$y_{ij} = \beta_1 + b_i + \beta_2 x_{ij} + \epsilon_{ij}, \quad i = 1, \dots, M, \quad j = 1, \dots, n_i,$$
$$b_i \sim \mathcal{N}(0, \sigma_b^2), \quad \epsilon_{ij} \sim \mathcal{N}(0, \sigma^2). \tag{1.8}$$

This model combines a random-effects analysis of variance model with a linear regression model.

1.4.1   Modeling Simple Linear Growth Curves

A common application of random-effects analysis of covariance models is in modeling *growth curve* data—the results on different subjects of repeated measurements of some characteristic over time. The terms *repeated measures* and *longitudinal data* are also applied to such data.

A classic example of such data, given in Potthoff and Roy (1964), is a set of measurements of the distance from the pituitary gland to the pterygomaxillary fissure taken every two years from 8 years of age until 14 years of age on a sample of 27 children—16 males and 11 females. The data, available as the S object Orthodont and shown in Figure 1.11, were collected by orthodontists from x-rays of the children's skulls. The pituitary gland and the pterygomaxillary fissure are two easily located points on these x-rays.

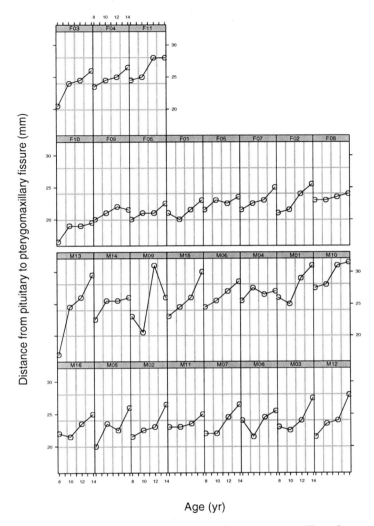

Age (yr)

FIGURE 1.11. Distance from the pituitary to the pterygomaxillary fissure versus age for a sample of 16 boys (subjects M01 to M16) and 11 girls (subjects F01 to F11). The aspect ratio for the panels has been chosen to facilitate comparison of the slope of the lines.

From Figure 1.11 it appears that there are qualitative differences between boys and girls in their growth patterns for this measurement. In Chapter 4 we will model some of these differences, but for now it is easier to restrict our modeling to the data from the female subjects only. To extract the data for the females only we first check on the names of the variables in the Orthodont object, then check for the names of the levels of the variables

Sex, then extract only those rows for which the Sex variable has the value
"Female".

```
> names( Orthodont )
[1] "distance" "age"       "Subject"  "Sex"
> levels( Orthodont$Sex )
[1] "Male"    "Female"
> OrthoFem <- Orthodont[ Orthodont$Sex == "Female", ]
```

Figure 1.11 indicates that, for most of the female subjects, the orthodontic measurement increases with age and that the growth is approximately
linear over this range of ages. It appears that the intercepts, and possibly
the slopes, of these growth curves may differ between girls. For example,
subject 10 has considerably smaller measurements than does subject 11,
and the growth rate for subjects 2 and 3 is considerably greater than that
for subjects 5 and 8.

To explore this potential linear relationship further, we fit separate linear
regression models for each girl using the lmList function.

```
> fm1OrthF.lis <- lmList( distance ~ age, data = OrthoFem )
> coef( fm1OrthF.lis )
     (Intercept)    age
F10        13.55  0.450
F09        18.10  0.275
F06        17.00  0.375
F01        17.25  0.375
F05        19.60  0.275
F08        21.45  0.175
F07        16.95  0.550
F02        14.20  0.800
F03        14.40  0.850
F04        19.65  0.475
F11        18.95  0.675
```

The function coef is a generic function (Chambers and Hastie, 1992,
Appendix A) that extracts the estimated coefficients from a fitted model
object. For an lmList object the coefficients are returned as a matrix with
one row for each of the groups of observations.

We might wish to consider whether we need to allow different slopes for
each girl. There are formal statistical tests to assess this and we will discuss
them later. For now we can proceed informally and examine individual
confidence intervals on the parameters. As we have seen, the intervals
function is used to create confidence intervals on the parameters in an
object representing a fitted model.

```
> intervals( fm1OrthF.lis )
, , (Intercept)
      lower  est.  upper
F10 10.071 13.55 17.029
```

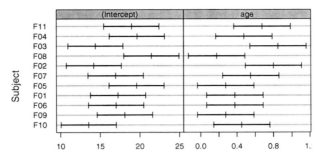

FIGURE 1.12. Comparison of 95% confidence intervals on the coefficients of simple linear regression models fitted to the orthodontic growth curve data for the female subjects.

```
F09  14.621  18.10  21.579
F06  13.521  17.00  20.479
F01  13.771  17.25  20.729
F05  16.121  19.60  23.079
F07  13.471  16.95  20.429
F02  10.721  14.20  17.679
F08  17.971  21.45  24.929
F03  10.921  14.40  17.879
F04  16.171  19.65  23.129
F11  15.471  18.95  22.429


, , age
       lower   est.   upper
F10   0.1401  0.450  0.7599
F09  -0.0349  0.275  0.5849
F06   0.0651  0.375  0.6849
F01   0.0651  0.375  0.6849
F05  -0.0349  0.275  0.5849
F07   0.2401  0.550  0.8599
F02   0.4901  0.800  1.1099
F08  -0.1349  0.175  0.4849
F03   0.5401  0.850  1.1599
F04   0.1651  0.475  0.7849
F11   0.3651  0.675  0.9849
```

As often happens, displaying the intervals as a table of numbers is not very informative. We find it much more effective to plot these intervals using

```
> plot( intervals ( fm1OrthF.lis ) )        # produces Figure 1.12
```

Figure 1.12 is of interest as much for what it does not show as for what it does show. First, consider what the figure does show. We notice that the intervals for the intercepts are all the same width, as are the intervals

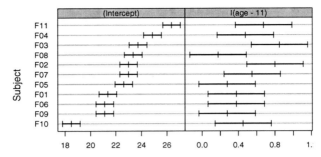

FIGURE 1.13. Comparison of 95% confidence intervals on the coefficients of simple linear regression models fitted to the centered orthodontic growth curve data for the female subjects.

for the slope with respect to age. This is a consequence of having *balanced* data; that is, all the subjects were observed the same number of times and at the same ages. We also notice that there is considerable overlap in the set of intervals for the slope with respect to age. It may be feasible to use a model with a common slope.

The surprising thing about Figure 1.12 is that it does not show the substantial differences in the intercepts that Figure 1.11 would lead us to expect. Furthermore, even though we have ordered the groups from the one with the smallest average distance (subject **F10**) to the one with the largest average distance (subject **F11**), this ordering is not reflected in the intercepts. Finally, we see that the pattern across subjects in the intervals for the intercepts is nearly a reflection of the pattern in the intervals for the slopes.

Those with experience analyzing regression models may already have guessed why this reflection of the pattern occurs. It occurs because all the data were collected between age 8 and age 14, but the intercept represents a distance at age 0. The extrapolation back to age 0 will result in a high negative correlation (about -0.98) between the estimates of the slopes and their corresponding intercept estimate.

We will remove this correlation if we center the data. In this case, we would fit the distance as a linear function of **age - 11** so the two coefficients being estimated are the distance at 11 years of age and the slope or growth rate. If we fit this revised model and plot the confidence intervals

```
> fm2OrthF.lis <- update( fm1OrthF.lis, distance ~ I( age - 11 ) )
> plot( intervals( fm2OrthF.lis ) )          # produces Figure 1.13
```

then these intervals (Figure 1.13) show the expected trend in the (**Intercept**) term, which now represents the fitted distance at 11 years.

To continue with the analysis of these data we could fit a regression model to the centered data with a common growth rate but separate intercepts for each girl. Before doing that we should consider what we could infer from such a model. We could use such a model to make inferences about the growth rate for this sample of girls. Also, we could make inferences about the expected distance for each girl at 11 years of age. Using combinations of the parameters we could make inferences about the expected distance for each of these girls at other ages. The key point is that we are in some ways restricting ourselves to the distances that we have or could observe on these particular girls

By fitting a mixed-effects model to these data we allow ourselves to make inferences about the fixed effects, which represent average characteristics of the population represented by these subjects, and the variability amongst subjects. A call to lme to fit linear growth curves with common slopes but randomly distributed shifts to the girls' orthodontic data is

```
> fm1OrthF <-
+    lme( distance ~ age, data = OrthoFem, random = ~ 1 | Subject )
> summary( fm1OrthF )
Linear mixed-effects model fit by REML
 Data: OrthoFem
      AIC     BIC   logLik
  149.22  156.17  -70.609

Random effects:
 Formula:  ~ 1 | Subject
         (Intercept) Residual
StdDev:       2.0685  0.78003

Fixed effects: distance ~ age
              Value Std.Error DF t-value p-value
(Intercept) 17.373   0.85874 32  20.230  <.0001
        age  0.480   0.05259 32   9.119  <.0001
 Correlation:
    (Intr)
age -0.674

Standardized Within-Group Residuals:
     Min       Q1      Med      Q3     Max
 -2.2736 -0.70902 0.17282 0.41221 1.6325

Number of Observations: 44
Number of Groups: 11
```

We could also fit a model with the formula distance ~ I(age - 11) but, because of the requirement of a common slope, for model building purposes the properties of the centered model are essentially equivalent to the uncentered model. Using the uncentered model makes it easier to compare with other models described below.

The model being fit would be expressed in matrix notation as

$$y_i = X_i\beta + Z_i b_i + \epsilon_i, \quad i = 1, \ldots, 11,$$
$$b_i \sim \mathcal{N}(0, \Psi), \quad \epsilon_i \sim \mathcal{N}(0, \sigma^2 I),$$

with matrices

$$X_1 = \cdots = X_{11} = \begin{bmatrix} 1 & 8 \\ 1 & 10 \\ 1 & 12 \\ 1 & 14 \end{bmatrix}, \quad Z_1 = \cdots = Z_{11} = \begin{bmatrix} 1 \\ 1 \\ 1 \\ 1 \end{bmatrix}.$$

The two-dimensional fixed-effects vector β consists of the mean intercept, β_1, for the population and the common slope or growth rate, β_2. The one-dimensional random-effects vectors, $b_i, i = 1, \ldots, 11$, describe a shift in the intercept for each subject. Because there is a common growth rate, these shifts are preserved for all values of age. The matrix $\Psi = \sigma_b^2$ will be a 1×1 matrix in this case. It represents the variance of the measurements in the population at a fixed value of age.

The REML estimates for the parameters are

$$\widehat{\sigma}_b = 2.0685, \quad \widehat{\sigma} = 0.78003, \quad \widehat{\beta}_1 = 17.373, \quad \widehat{\beta}_2 = 0.480.$$

To obtain the maximum likelihood estimates we use method = "ML".

```
> fm1OrthFM <- update( fm1OrthF, method = "ML" )
> summary( fm1OrthFM )
Linear mixed-effects model fit by maximum likelihood
 Data: OrthoFem
      AIC    BIC   logLik
  146.03 153.17 -69.015

Random effects:
 Formula:  ~ 1 | Subject
         (Intercept) Residual
StdDev:       1.9699  0.76812

Fixed effects: distance ~ age
             Value Std.Error DF t-value p-value
(Intercept) 17.373   0.85063 32  20.423  <.0001
        age  0.480   0.05301 32   9.047  <.0001
 Correlation:
     (Intr)
age -0.685

Standardized Within-Group Residuals:
     Min        Q1      Med      Q3      Max
 -2.3056 -0.71924 0.17636 0.4258 1.6689
```

```
Number of Observations: 44
Number of Groups: 11
```

Notice that, to the accuracy printed here, the estimates of the fixed-effects parameters are the same for ML and REML. The ML estimates of the standard deviations, $\hat{\sigma}_b = 1.9699$ and $\hat{\sigma} = 0.76812$ are smaller than the corresponding REML estimates. This is to be expected—the REML criterion was created to compensate for the downward bias of the maximum likelihood estimates of variance components, so it should produce larger estimates.

We have made the assumption of a common slope or growth rate for all the subjects. To test this we can fit a model with random effects for both the intercept and the slope.

```
> fm2OrthF <- update( fm1OrthF, random = ~ age | Subject )
```

The predictions from this model are shown in Figure 1.14. We compare the two models with the anova function.

```
> anova( fm1OrthF, fm2OrthF )
          Model df    AIC    BIC  logLik    Test L.Ratio p-value
fm1OrthF      1  4 149.22 156.17 -70.609
fm2OrthF      2  6 149.43 159.85 -68.714 1 vs 2  3.7896  0.1503
```

Because the p-value for the second model versus the first is about 15%, we conclude that the simpler model, fm1OrthF, is adequate.

1.4.2 Predictions of the Response and the Random Effects

The derivation of predicted values for the response and for the random effects in the linear mixed-effects model is described in §2.5. We can extract the *best linear unbiased predictions* (BLUPs) of the random effects from the fitted model with the random.effects function.

```
> random.effects( fm1OrthF )
     (Intercept)
F10    -4.005329
F09    -1.470449
F06    -1.470449
F01    -1.229032
F05    -0.021947
F07     0.340179
F02     0.340179
F08     0.702304
F03     1.064430
F04     2.150807
F11     3.599309
```

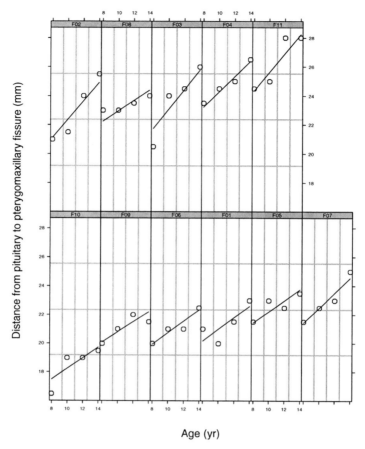

FIGURE 1.14. Original data and fitted linear relationships from a mixed-effects model for the girls' orthodontic data. This model incorporates random effects for both the slope and the intercept.

The shorter name `ranef` is a synonym for `random.effects`.

```
> ranef( fm1OrthFM )
     (Intercept)
F10   -3.995835
F09   -1.466964
F06   -1.466964
F01   -1.226119
F05   -0.021895
F07    0.339372
F02    0.339372
F08    0.700640
F03    1.061907
```

```
F04    2.145709
F11    3.590778
```

The `coefficients` function (or its shorter form `coef`) is used to extract the coefficients of the fitted lines for each subject. For the fitted model `fm1OrthF` the intercept of the fitted line for subject i is $\widehat{\beta}_1 + \widehat{b}_i$ and the slope is $\widehat{\beta}_2$.

```
> coef( fm1OrthF )
     (Intercept)     age
F10      13.367 0.47955
F09      15.902 0.47955
F06      15.902 0.47955
F01      16.144 0.47955
F05      17.351 0.47955
F07      17.713 0.47955
F02      17.713 0.47955
F08      18.075 0.47955
F03      18.437 0.47955
F04      19.524 0.47955
F11      20.972 0.47955
```

Looking back at the BLUPs of the random effects for the ML and REML fits, we can see that they are very similar. The same will be true of the coefficients for each subject and hence for the fitted lines themselves. To show this we can plot either the estimated BLUPs or the estimated coefficients. The `compareFits` function is helpful here because it allows us to put both sets of coefficients on the same panels.

```
> plot(compareFits(coef(fm1OrthF), coef(fm1OrthFM)))   # Figure 1.15
```

In Figure 1.15 each line corresponds to one subject. In the left panel the estimated intercepts from the REML fit are shown as open circles while those from the ML fit are shown as +'s. The two estimates for each subject are essentially identical. In the right panel the estimates for the coefficient with respect to `age` are shown. Because there is no random effect associated with this coefficient, the estimates do not vary between subjects. Again, the ML estimates and the REML estimates are essentially identical.

We may also want to examine the predictions for each subject from the fitted model. The `augPred` function produces predictions of the response for each of the groups over the observed range of the covariate (i.e. the range 8–14 for `age`). These predictions are augmented with the original data to produce a plot of the predicted lines for each subject superposed on the original data as in Figure 1.16.

```
> plot( augPred(fm1OrthF), aspect = "xy", grid = T )     # Fig. 1.16
```

Further diagnostic plots, such as plots of residuals versus the fitted values by subject (not shown), did not indicate any serious deficiencies in this

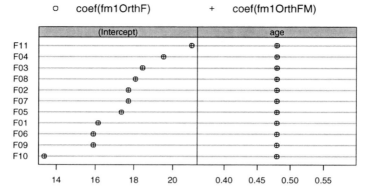

FIGURE 1.15. A comparison of the coefficients of the fitted lines for each female subject in the orthodontic example. The two sets of coefficients are from the restricted maximum likelihood fit (fm1OrthF) and the maximum likelihood fit (fm1OrthFM).

model. We will return to this data set in Chapter 4 where we will fit a combined model to the data for both the males and the females. This allows us to check for sex-related differences in the growth patterns.

1.5 Models for Nested Classification Factors

In the Machines example of §1.3 we introduced the concept of nested random effects to model an interaction between a fixed-effects factor and a random-effects factor. Nested random-effects terms are also used when we have nested classification factors.

Data from an experiment on the pixel intensity in computerized tomography (CT) scans, available as the object Pixel, are shown in Figure 1.17 and are described in Appendix A.24. The experimenters injected each of ten dogs with a dye contrast then recorded the mean pixel intensities from CT scans of the right and left lymph nodes in the axillary region on several occasions up to 21 days post injection.

Each observation is classified according to the Dog and the Side of the Dog on which it was made. The nature of the experiment is such that the left and right sides are expected to be different, but the difference is not expected to be systematic in terms of left and right. That is, for one dog the left side may have greater pixel intensities than the right, while for another dog the opposite may be true. Thus Dog and Side are considered to be nested classification factors. We will associate random-effects terms with the Dog factor, and with the Side factor nested within Dog.

Figure 1.17 indicates that the intensities generally increase then decrease over time, reaching a peak after about 10 days. There is, however, consid-

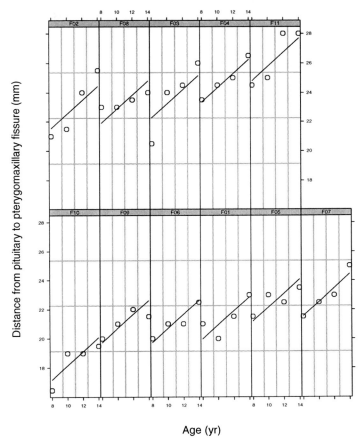

FIGURE 1.16. Original data and fitted growth curves for each female subject in the orthodontic example. The fitted curves are from a restricted maximum likelihood fit of the analysis of covariance model.

erable variability between dogs in this pattern. Within the same dog, the left and the right side generally follow the same pattern over time but often with a vertical shift between sides.

We will start with a quadratic model with respect to the day covariate so we can model the pattern of reaching a peak. We use random-effects for both the intercept and the linear term at the Dog level and a single random effect for the intercept at the Side within Dog level. This allows the overall pattern to vary between dogs in terms of the location of the peak, but not in terms of the curvature at the peak. The only difference between sides for the same dog will be a shift in the intercept.

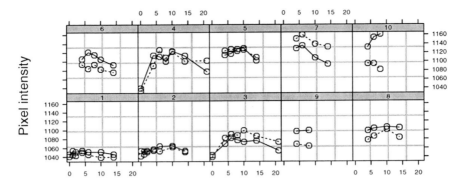

Time post injection (days)

FIGURE 1.17. Pixel intensity on CT scans over time for lymph nodes on the left and the right sides of 10 dogs.

```
> fm1Pixel <- lme( pixel ~ day + day^2, data = Pixel,
+    random = list( Dog = ~ day, Side = ~ 1 ) )
> intervals( fm1Pixel )
Approximate 95% confidence intervals

  Fixed effects:
                  lower        est.       upper
  (Intercept)  1053.0968  1073.33914  1093.5814
          day     4.3797     6.12960     7.8795
     I(day^2)    -0.4349    -0.36735    -0.2998

  Random Effects:
  Level: Dog
                        lower      est.     upper
      sd((Intercept))  15.92849  28.36994  50.52918
             sd(day)    1.08085   1.84375   3.14514
  cor((Intercept),day)  -0.89452  -0.55472   0.19138
  Level: Side
                     lower    est.   upper
  sd((Intercept))  10.417  16.824  27.173

  Within-group standard error:
    lower    est.   upper
  7.6345  8.9896  10.585

> plot( augPred( fm1Pixel ) )              # produces Figure 1.18
```

If we write the pixel intensity of the jth side's on the ith dog at the kth occasion as $y_{ijk}\, i = 1, \ldots, 10; j = 1, 2; k = 1, \ldots, n_{ij}$, and the time of the

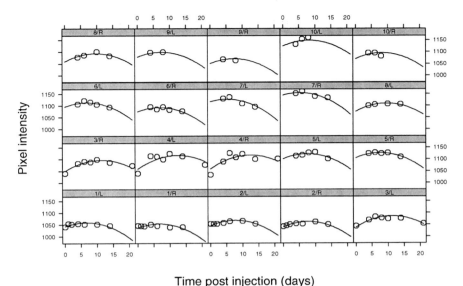

Time post injection (days)

FIGURE 1.18. The fitted curves from a quadratic model fit to the pixel intensity data. The model includes random effects in the intercept and the linear term for each dog, and a random effect in the intercept for each side of each dog.

kth scan on the ith dog as d_{ik}, the model being fit can be expressed as

$$y_{ijk} = \beta_1 + \beta_2 d_{ik} + \beta_3 d_{ik}^2 + b_{i,1} + b_{i,2} d_{ik} + b_{ij} + \epsilon_{ijk},$$
$$i = 1, \ldots, 10, \quad j = 1, 2, \quad k = 1, \ldots, n_{ij}. \tag{1.9}$$

To describe the variance and covariance terms in the model, we consider the n_{ij}-vector $\boldsymbol{y}_{ij}$ of intensities measured on side $j = 1, 2$ within dog $i = 1, \ldots, 10$. In this experiment n_{ij}, the number of observations on side j of dog i, does not depend on j but does depend on i. For example, dog 9 was scanned on only two occasions but dogs 3 and 4 were each scanned on seven occasions. The model can be represented in terms of design matrices $\boldsymbol{X}_{ij}$ for the fixed effects, $\boldsymbol{Z}_{ij}$ for the random effects for side j within dog i, and $\boldsymbol{Z}_{i,j}$ for the random effects for dog i on the measurements for side j within dog i.

Because both sides are scanned at the same times these matrices depend on i but not on j. For example, because dog 8 was scanned on days 4, 6, 10, and 14,

$$\boldsymbol{X}_{81} = \boldsymbol{X}_{82} = \begin{bmatrix} 1 & 4 & 16 \\ 1 & 6 & 36 \\ 1 & 10 & 100 \\ 1 & 14 & 196 \end{bmatrix}, \ \boldsymbol{Z}_{8,1} = \boldsymbol{Z}_{8,2} = \begin{bmatrix} 1 & 4 \\ 1 & 6 \\ 1 & 10 \\ 1 & 14 \end{bmatrix}, \ \boldsymbol{Z}_{81} = \boldsymbol{Z}_{82} = \begin{bmatrix} 1 \\ 1 \\ 1 \\ 1 \end{bmatrix}.$$

As before, $\boldsymbol{\beta}$ is the vector of fixed effects. In this case, $\boldsymbol{\beta}$ is three-dimensional. The two-dimensional vector of random effects for dog i is written $\boldsymbol{b}_i$ and the one-dimension vector of random effects for side j within dog i is written b_{ij}. Model (1.9) can then be expressed as

$$\boldsymbol{y}_{ij} = \boldsymbol{X}_{ij}\boldsymbol{\beta} + \boldsymbol{Z}_{i,j}\boldsymbol{b}_i + \boldsymbol{Z}_{ij}b_{ij} + \boldsymbol{\epsilon}_{ij},$$
$$\boldsymbol{b}_i \sim \mathcal{N}(\boldsymbol{0}, \boldsymbol{\Psi}_1), \quad b_{ij} \sim \mathcal{N}(0, \sigma_2^2), \quad \boldsymbol{\epsilon}_{ij} \sim \mathcal{N}(\boldsymbol{0}, \sigma^2 \boldsymbol{I}).$$

The parameters in this model are $\boldsymbol{\beta}$, $\boldsymbol{\Psi}_1$, σ_2^2, and σ^2. The summary and intervals functions express the estimates of the variance components as standard deviations and correlations, not as variances. We can use the VarCorr function to examine them on both the standard deviation scale and the variance scale.

```
> VarCorr( fm1Pixel )
                Variance   StdDev   Corr
     Dog = pdSymm(~ day)
(Intercept) 804.8535       28.3699 (Intr)
        day   3.3994        1.8437 -0.555
   Side = pdSymm(~ 1)
(Intercept) 283.0551       16.8242
   Residual  80.8131        8.9896
```

1.5.1   Model Building for Multilevel Models

As when modeling data with a single level of random effects, we should evaluate whether the fixed-effects structure and the random-effects structure are adequate to describe the observed data. We should also check if we have incorporated unnecessary terms in the model.

The summary table for the fixed-effects terms

```
> summary( fm1Pixel )
...
Fixed effects: pixel ~ day + day^2
              Value Std.Error DF t-value p-value
(Intercept) 1073.3    10.172  80  105.52  <.0001
        day    6.1     0.879  80    6.97  <.0001
   I(day^2)   -0.4     0.034  80  -10.82  <.0001
...
```

indicates that the quadratic term is highly significant. In a polynomial model like this we would generally retain the linear term and the intercept term if we retain the quadratic term. Thus, we will accept the fixed-effects model as it is and go on to examine the random-effects terms.

The first question to address is whether there is a need to have a random effect for each Side within each Dog. We can fit the previous model without the random effect for Side and compare the two fits with anova.

```
> fm2Pixel <- update( fm1Pixel, random = ~ day | Dog)
> anova( fm1Pixel, fm2Pixel )
          Model df    AIC    BIC  logLik   Test L.Ratio p-value
fm1Pixel      1  8 841.21 861.97 -412.61
fm2Pixel      2  7 884.52 902.69 -435.26 1 vs 2  45.309  <.0001
```

The *p*-value is extremely small indicating that the more general model, fm1Pixel, is definitely superior. The AIC and BIC values confirm this.

We can also check if the random effect for day at the Dog level is warranted. If we eliminate this term the only random effects will be a random effect for the intercept for each dog and for each side of each dog. We fit this model and compare it to fm1Pixel with

```
> fm3Pixel <- update( fm1Pixel, random = ~ 1 | Dog/Side )
> anova( fm1Pixel, fm3Pixel )
          Model df    AIC    BIC  logLik   Test L.Ratio p-value
fm1Pixel      1  8 841.21 861.97 -412.61
fm3Pixel      2  6 876.84 892.41 -432.42 1 vs 2  39.629  <.0001
```

Again, the likelihood-ratio test, and the AIC and BIC criteria, all strongly favor the more general model, fm1Pixel.

Earlier we stated that there does not appear to be a systematic difference between the left and the right sides of the dogs. For some dogs the left side produces higher pixel densities while for other dogs the right side does. We can check that this indeed is the case by adding a term for Side to the fixed effects.

```
> fm4Pixel <- update( fm1Pixel, pixel ~ day + day^2 + Side )
> summary( fm4Pixel )
...
Fixed effects: pixel ~ day + I(day^2) + Side
             Value Std.Error DF t-value p-value
(Intercept) 1073.3    10.171 80  105.53  <.0001
        day    6.1     0.879 80    6.97  <.0001
   I(day^2)   -0.4     0.034 80  -10.83  <.0001
       Side   -4.6     3.813  9   -1.21  0.2576
...
```

With a *p*-value of over 25% the fixed-effects term for Side would not be considered significant.

Finally, we would examine residual plots such as Figure 1.19 for deficiencies in the model. There are no alarming patterns in this figure.

1.6   A Split-Plot Experiment

Multiple nested levels of random effects are also used in the analysis of *split-plot* experiments such as that represented by the Oats data, shown in

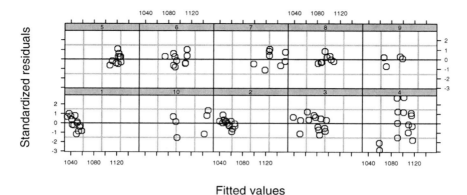

FIGURE 1.19. Standardized residuals versus fitted values by dog for a multilevel mixed-effects model fit to the pixel data..

Figure 1.20. As described in Appendix A.15, the treatment structure in this experiment was a 3 × 4 full factorial, with three varieties of oats and four nitrogen concentrations. The term *full factorial* means that every variety was used with every nitrogen concentration.

The agricultural plots for this experiment were grouped into six blocks, each with three plots. Each plot was subdivided into four subplots. The varieties were randomly assigned to the plots within each block. The nitrogen concentrations were randomly assigned to the subplots within each plot.

Physically, there are three levels of grouping of the experimental units: block, plot, and subplot. Because the treatments are randomly assigned at each level of grouping, we may be tempted to associate random effects with each level. However, because there is only one yield recorded for each subplot we cannot do this as we would saturate the model with random effects. We use a random intercept at each of the block and the whole plot levels.

Generally, we begin modeling a split-plot experiment using fixed effects for each of the experimental factors and for their interaction. For this experiment the `nitro` factor is recorded as a numeric variable. If we wish to allow general patterns in the dependencies of `yield` on `nitro` we should "coerce" it to a factor using, say, `factor(nitro)`. In this particular example, there is a natural ordering of the levels of nitrogen applied so it makes sense to coerce `nitro` to an *ordered factor* using `ordered(nitro)`. As the name implies, an ordered factor is a factor for which there is a natural ordering of the levels. One consequence of using an ordered factor instead of a factor is that the default contrasts for an ordered factor are orthogonal polynomial contrasts. The first contrast estimates the linear trend, the second estimates the quadratic effect orthogonal to the linear term, and so on.

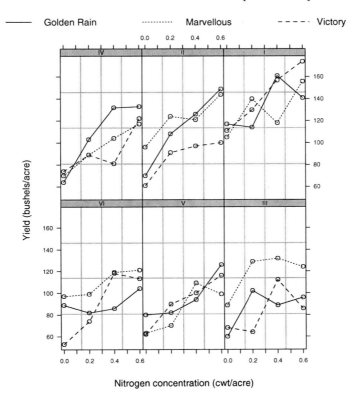

FIGURE 1.20. Yield in bushels/acre of three different varieties of oats at four different concentrations of nitrogen (hundred weight/acre). The experimental units were arranged into six blocks, each with three whole-plots subdivided into four subplots. One variety of oats was used in each whole-plot with all four concentrations of nitrogen, one concentration in each of the four subplots. The panels correspond to the blocks.

The model with fixed effects for both experimental factors and for their interaction and with random effects for both the Block factor and the Variety (whole-plot) factor is fit with

```
> fm1Oats <- lme( yield ~ ordered(nitro) * Variety, data = Oats,
+    random = ~ 1 | Block/Variety )
> anova( fm1Oats )
```

	numDF	denDF	F-value	p-value
(Intercept)	1	45	245.15	<.0001
ordered(nitro)	3	45	37.69	<.0001
Variety	2	10	1.49	0.2724
ordered(nitro):Variety	6	45	0.30	0.9322

The anova results indicate that nitro is a significant factor, but that neither Variety nor the interaction between Variety and nitro are significant.

If we drop the interaction term and refit, we obtain essentially the same results for the two main effects, Variety and nitro.

```
> fm2Oats <- update( fm1Oats, yield ~ ordered(nitro) + Variety )
> anova( fm2Oats )
                numDF denDF F-value p-value
    (Intercept)     1    51  245.14  <.0001

ordered(nitro)      3    51   41.05  <.0001
       Variety      2    10    1.49  0.2724
> summary( fm2Oats )
...
Random effects:
 Formula:   ~ 1 | Block
         (Intercept)
StdDev:       14.645

 Formula:   ~ 1 | Variety %in% Block
         (Intercept) Residual
StdDev:       10.473    12.75

Fixed effects: yield ~ ordered(nitro) + Variety
                 Value Std.Error DF t-value p-value
    (Intercept) 103.97    6.6406 51  15.657  <.0001
ordered(nitro).L 32.94    3.0052 51  10.963  <.0001
ordered(nitro).Q -5.17    3.0052 51  -1.719  0.0916
ordered(nitro).C -0.45    3.0052 51  -0.149  0.8823
        Variety1  2.65    3.5395 10   0.748  0.4720
        Variety2 -3.17    2.0435 10  -1.553  0.1515
...
```

In this model there is a random effect for Variety %in% Block as well as a fixed effect for Variety. These terms model different characteristics of the response. The random effects term, as a nested random effect, is allowing for different intercepts at the level of plots within blocks. The fact that each plot is planted with one variety means that we can use the Variety factor to indicate the plot as long as we have Variety nested within Block. As seen in Figure 1.20 the yields in one of the plots within a block may be greater than those on another plot in the same block for all levels of nitro. For example, in block III the plot that was planted with the Marvellous variety had greater yields than the other two plots at each level of nitro. The random effect at the level of Variety %in% Block allows shifts like this that may be related to the fertility of the soil in that plot, for example.

On the other hand, the fixed-effects term for Variety is used to model a systematic difference in the yields that would be due to the variety of oats planted in the plot. There do not appear to be such systematic differences. For example, even though the plot planted with the Marvellous variety is the highest yielding plot in block III, the Marvellous plot is one of the lowest yielding in block V.

Because the fixed effect for `Variety` and the random effect for `Variety %in% Block` are modeling different types of behavior, it makes sense to remove the fixed effect while retaining the random effect

```
> fm3Oats <- update( fm1Oats, yield ~ ordered( nitro ) )
> summary( fm3Oats )
...
Random effects:
 Formula:   ~ 1 | Block
        (Intercept)
StdDev:      14.506

 Formula:   ~ 1 | Variety %in% Block
        (Intercept) Residual
StdDev:      11.039     12.75

Fixed effects: yield ~ ordered(nitro)
                  Value Std.Error DF t-value p-value
     (Intercept) 103.97    6.6406 51  15.657  <.0001
ordered(nitro).L  32.94    3.0052 51  10.963  <.0001
ordered(nitro).Q  -5.17    3.0052 51  -1.719  0.0916
ordered(nitro).C  -0.45    3.0052 51  -0.149  0.8823
```

We see that the estimates for the random-effects variances and the fixed-effects for `nitro` have changed very little, if at all.

We can now examine the effect of nitrogen in more detail. We notice that the linear term, `ordered(nitro).L`, is highly significant, but the quadratic and cubic terms (`.Q` and `.C` extensions) are not. To remove the cubic and quadratic terms in the model, we simply revert to using `nitro` as a numeric variable.

```
> fm4Oats <-
+   lme( yield ~ nitro, data = Oats, random = ~ 1 | Block/Variety )
> summary( fm4Oats )
. . .
Random effects:
 Formula:   ~ 1 | Block
        (Intercept)
StdDev:      14.506
 Formula:   ~ 1 | Variety %in% Block
        (Intercept) Residual
StdDev:      11.005    12.867

Fixed effects: yield ~ nitro
            Value Std.Error DF t-value p-value
(Intercept) 81.872    6.9453 53  11.788  <.0001
      nitro 73.667    6.7815 53  10.863  <.0001
```

```
Correlation:
     (Intrc
nitro -0.293

Standardized Within-Group Residuals:
     Min       Q1       Med       Q3     Max
  -1.7438  -0.66475  0.017104  0.54299  1.803

Number of Observations: 72
Number of Groups:
 Block Variety %in% Block
     6              18
```

With VarCorr and intervals we can examine the variance components
and their confidence intervals for this model

```
> VarCorr( fm4Oats )
                  Variance StdDev
    Block = pdSymm(~ 1)
(Intercept) 210.42        14.506
  Variety = pdSymm(~ 1)
(Intercept) 121.10        11.005
   Residual 165.56        12.867
> intervals( fm4Oats )
Approximate 95% confidence intervals

 Fixed effects:
             lower    est.   upper
(Intercept) 67.942 81.872 95.803
      nitro 60.065 73.667 87.269

 Random Effects:
  Level: Block
                 lower   est.   upper
sd((Intercept)) 6.6086 14.506 31.841
  Level: Variety
                 lower   est.   upper
sd((Intercept)) 6.408 11.005 18.899

 Within-group standard error:
  lower    est.   upper
 10.637 12.867 15.565
```

We can see that the random effects at the Block and plot levels account
for a substantial amount of the variability in the response. Although the
standard deviations of these random effects are not estimated very precisely,
it does not appear reasonable that they could be zero. To check this we
would fit models without these random effects and use likelihood ratio tests
to compare them to fm4Oats. We do not show that here.

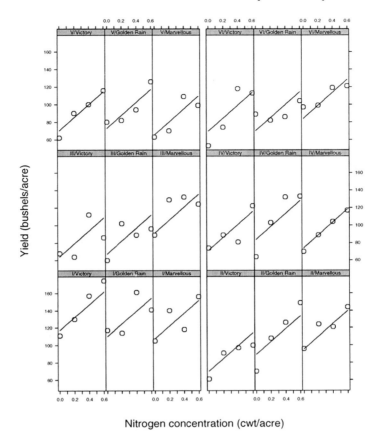

FIGURE 1.21. Observed and predicted yields in bushels/acre for three different varieties of oats at four different concentrations of nitrogen (hundred weight/acre) by block and variety. Although the model has a random effect for variety, the whole-plot factor, there is no fixed effect for variety.

The modeling of the dependence on nitrogen level by a simple linear term appears adequate. Plots of the original data and the fitted curves, obtained with

```
> plot( augPred( fm4Oats ), aspect = 2.5, layout = c(6, 3),
+         between = list( x = c(0, 0, 0.5) ) ) # produces Figure 1.21
```

do not show any systematic lack of fit. (The extra arguments in the plot call are used to enhance the appearance of the plot. They are described in §3.3.)

We could (and did) examine other common diagnostic plots to check for inadequacies in this model, but did not find any. We now have a simple, adequate model to explain the dependence of the response on both the levels of nitrogen applied, using fixed-effects terms, and the random variability in blocks and plots, using random-effects terms.

1.7 Chapter Summary

In this chapter we have presented some motivation for the use of linear mixed-effects models and an overview of their application. We showed several examples of the types of data to which these models can be applied.

The defining characteristics of mixed-effects models are that they are applied to data where the observations are grouped according to one or more levels of experimental units and that they incorporate both fixed-effects terms and random-effects terms. A fixed-effects term in a model describes the behavior of the entire population or of those units associated with repeatable levels of experimental factors. A random-effects term describes the distribution within the population of a coefficient. The "effects" in a random-effects term are associated with the individual experimental units sampled from the population.

A linear mixed-effects model is fit with the `lme` function. A preliminary list of fitted linear models by experimental unit can be obtained with the `lmList` function. Both `lmList` and `lme` fits can be examined with the accessor functions `coef`, `ranef`, `fixef`, `residuals`, and `fitted`. A summary of the fit is obtained with the `summary` function. Diagnostic plots are generated with the `plot` function or with specialized constructors such as `augPred`, `compareFits`, and `comparePred`.

The significance of fixed-effects terms is assessed with the single-argument form of the `anova` function or directly from the the `summary` function. Different forms of random-effects terms can be compared by fitting different models and comparing them with the multiple-argument form of `anova`.

The purpose of this chapter is to present the motivation for using LME models with grouped data and to set the stage for later chapters in the book, dealing with the theory and computational methods for LME models (Chapter 2) and the linear mixed-effects modeling facilities in the nlme library (Chapter 4).

Exercises

1. The `PBIB` data (Appendix A.22) are from an agricultural experiment that was laid out as a partially balanced incomplete block design. This is described in more detail in §2.4.2. The roles of the variables in these data are indicated by the names: `response`, `Treatment`, and `Block`. The structure is similar to that of the `ergoStool` data but, because there

are only four observations in each block and there are 15 levels of Treatment, each block receives only a subset of the treatments.

(a) Plot the PBIB data with plot(PBIB). Do there appear to be systematic differences between blocks?

(b) Create a design plot, like Figure 1.6 (p. 14), for the PBIB data.

(c) Fit a linear mixed-effects model, with fixed effects for Treatment and random effects for the intercept by Block, to the PBIB data. The call to lme would be like that used to fit fm1Stool in §1.2.

(d) Apply anova to the fitted model. Is the Treatment term significant? Describe the hypothesis being tested.

(e) Create a plot of the residuals versus the fitted values for this fitted model. This plot is like Figure 1.4. Does this plot indicate greater variance in the residuals at higher levels of the response?

(f) Create a plot of the standardized residuals versus the fitted values by Block. This plot is like Figure 1.8 (p. 21). Does this plot indicate systematic patterns in the residuals by Block?

We will discuss the anova results for this fitted model in more detail in §2.4.2.

2. The Oxboys data described in §3.1 and Appendix A.19 consist of the heights of 26 boys from Oxford, England, each measured on nine different occasions. The structure is similar to that of the OrthoFem data of §1.4.

(a) Plot the data (using plot(Oxboys)) and verify that a simple linear regression model gives a suitable representation of the boys' growth patterns. Do there appear to be significant differences in the individual growth patterns?

(b) Fit a simple linear regression model to height versus age using the lm function, ignoring the Subject effects. Obtain the boxplots of the residuals by Subject with bwplot(Subject ~ resid(object), Oxboys), where object should be replaced with the name of the fitted lm object. Explain the observed pattern.

(c) Use the lmList function to fit separate simple linear regression models for each Subject, using a call similar to the one used to produce fm1OrthF.lis in §1.4. Compare the boxplots of the residuals by Subject for the lmList fit (obtained with plot(object, Subject ~ resid(.)), with object replaced with the name of the lmList object) to those obtained for the lm fit. Compare also the residual standard errors from the two fits and comment.

(d) Plot the individual confidence intervals on the parameters estimated in the lmList fit and verify that both the intercept and the slope vary significantly with Subject.

(e) Use the `lme` function to fit an LME model to the data with random effects for both the intercept and the slope, using a call similar to the one used to obtain `fm1OrthF` in §1.4. Examine the boxplots of the residuals by `Subject`, comparing them to those obtained for the `lm` and `lmList` fits.

(f) Produce the plot of the standardized residuals versus fitted values (`plot(object)`) and the normal plot of the standardized residuals (`qqnorm(object)`). (In both cases `object` should be replaced with the name of the `lme` object.) Can you identify any departures from the model's assumptions?

(g) Plot the augmented predictions for the `lme` fit (obtained with `plot(augPred(object))`). Do the linear models for each subject appear adequate?

(h) Another way of assessing the linear models for each subject is to plot the residuals versus `age` by `Subject` (use `plot(object, resid(.) ~ age | Subject)`, replacing `object` with the name of the `lme` object). Several subjects have a noticeable "scooping" pattern in their residuals, indicating the need for a model with curvature.

(i) Use the `lmList` function to fit separate quadratic models for each subject. A quadratic model in `age`, as shown in `fm1Pixel` of §1.5, would be fit with `lmList(height ~ age + age^2, Oxboys)`.

(j) Examine a plot of the confidence intervals on coefficients from this second `lmList` fit. Are there indications that the coefficients differ between subjects? Are the quadratic coefficients significantly different from zero for some subjects?

(k) Fit the full mixed-effects model corresponding to the last `lmList` fit. The model will have linear and quadratic terms for `age` in the fixed-effects and the random effects. A simple way to describe this model is `lme(object)` replacing `object` with the name of the `lmList` fit.

(l) Check residual plots and numerical summaries for this `lme` model. Do there appear to be deficiencies in the fit? Do there appear to be terms in the model that could be eliminated?

3. The LME model used for the `Pixel` data in §1.5 uses random effects for the intercept and the slope at the `Dog` level and a single random effect for the intercept at the `Side` within `Dog` level. We did not discuss there how that random-effects model was chosen. The `lmList` function can be used with multilevel data to investigate which terms in an LME model require random effects.

(a) Use `lmList` to fit a separate quadratic model in `day` for each `Dog`. Print the fitted object and examine the estimated coefficients.

Can you explain the error message printed in the lmList fit? Notice that lmList was able to recover from the error and proceed to normal completion.

(b) Plot the individual confidence intervals for the coefficients in the lmList fit. Verify that only the intercept and the linear coefficient seem to vary significantly with Dog.

(c) Use the level argument to lmList to fit separate quadratic models in day for each Side within Dog (use Dog/Side as the grouping expression and set level=2). Print the fitted object using summary and explain the missing values (NA) for the standard errors of Dog 10.

(d) Plot the individual confidence intervals for the coefficients in the lmList fit by Side within Dog and verify that there is more variation among the intercepts and the linear coefficients than among the quadratic coefficients.

(e) Fit an LME model with random effects for the intercept and the linear term at both levels of grouping. Compare the resulting lme fit to the fm1Pixel object in §1.5 using anova. Which model should be the preferred?

4. The Alfalfa data described in Appendix A.1 is another example of a split-plot experiment. The structure is similar to that of the Oats data of §1.6: a 3×4 full factorial on varieties of alfalfa and date of third cutting is used with 6 blocks each subdivided into 4 plots according to a split-plot arrangement. The whole-plot treatments are given by the varieties and the subplot treatments by the date of third cutting.

(a) Plot the data (using plot(Alfalfa)). Do there appear to be cutting dates that are consistently worse/better than the others? What can you say about the block-to-block variation in the yields?

(b) Use lme to fit a two-level LME model with grouping structure Block/Variety, including a single random intercept for each level of grouping (i.e., random = ˜1 | Block/Variety). Assume a full factorial structure with main effects and interactions for the fixed effects (i.e., fixed = Yield ˜ Date * Variety). Use the *treatment* contrasts (options(contrasts = c("contr.treatment", "contr.poly"))) to get more interpretable coefficients for the fixed effects.

(c) Examine the significance of the terms in the model using anova, verifying that there are no significant differences between varieties and no significant interactions between varieties and cutting dates.

(d) Because the data are balanced, a similar ANOVA model can be fit using `aov` and the `Error` function (use `aov(Yield ~ Date * Variety + Error(Block/Variety), Alfalfa)`). Compare the results from the `aov` and `lme` fits, in particular the F-values and p-values for testing the terms in the fixed-effects model (these are obtained for the `aov` object using the `summary` function). In this case, because of the balanced structure of the data, the REML fit (obtained with `lme`) and the ANOVA fit (obtained with `aov`) are identical.

(e) Refit the LME model using fixed effects for `Date` only (a simple way to do this is to use `update(object, Yield ~ Date)`, where `object` should be replaced with name of the previous `lme` object). Print the resulting object using `summary` and investigate the differences between the cutting dates (recall that, for the *treatment* contrasts, the coefficients represent differences with respect to the cutting date labelled `None`). Can you identify a trend in the effect of cutting date on yield?

(f) Examine the plot of the residuals versus fitted values and the normal plot of the residuals. Can you identify any departures from the LME model's assumptions?

2
Theory and Computational Methods for Linear Mixed-Effects Models

In this chapter we present the theory for the linear mixed-effects model introduced in Chapter 1. A general formulation of LME models is presented and illustrated with examples. Estimation methods for LME models, based on the likelihood or the restricted likelihood of the parameters, are described, together with the computational methods used to implement them in the lme function. Asymptotic results on the distribution of the maximum likelihood estimators and the restricted maximum likelihood estimators are used to derive confidence intervals and hypotheses tests for the model's parameters.

The purpose of this chapter is to present an overview of the theoretical and computational aspects of LME models that allows the evaluation of the strengths and limitations of such models for practical applications. It is not the purpose of this chapter to present a thorough theoretical description of LME models. Such a comprehensive treatment of the theory of linear mixed-effects models can be found, for example, in Searle, Casella and McCulloch (1992) or in Vonesh and Chinchilli (1997).

Readers who are more interested in the applications of LME models and the use of the functions and methods in the nlme library to fit such models can, without loss of continuity, skip this chapter and go straight to Chapter 3. If you decide to skip this chapter at a first reading, it is recommended that you return to it (especially §2.1) at a later time to get a good understanding of the LME model formulation and its assumptions and limitations.

2.1   The LME Model Formulation

Linear mixed-effects models are mixed-effects models in which both the fixed and the random effects occur linearly in the model function. They extend linear models by incorporating random effects, which can be regarded as additional error terms, to account for correlation among observations within the same group.

In this section we present a general formulation for LME models proposed by Laird and Ware (1982). The original single-level formulation is described in §2.1.1 and its multilevel extension is described in §2.1.2.

2.1.1   Single Level of Grouping

For a single level of grouping, the linear mixed-effects model described by Laird and Ware (1982) expresses the n_i-dimensional response vector $\boldsymbol{y}_i$ for the ith group as

$$\boldsymbol{y}_i = \boldsymbol{X}_i\boldsymbol{\beta} + \boldsymbol{Z}_i\boldsymbol{b}_i + \boldsymbol{\epsilon}_i, \quad i = 1, \ldots, M,$$
$$\boldsymbol{b}_i \sim \mathcal{N}(\boldsymbol{0}, \boldsymbol{\Sigma}), \quad \boldsymbol{\epsilon}_i \sim \mathcal{N}(\boldsymbol{0}, \sigma^2\boldsymbol{I}), \tag{2.1}$$

where $\boldsymbol{\beta}$ is the p-dimensional vector of *fixed effects*, $\boldsymbol{b}_i$ is the q-dimensional vector of *random effects*, $\boldsymbol{X}_i$ (of size $n_i \times p$) and $\boldsymbol{Z}_i$ (of size $n_i \times q$) are known fixed-effects and random-effects regressor matrices, and $\boldsymbol{\epsilon}_i$ is the n_i-dimensional *within-group error* vector with a spherical Gaussian distribution. The assumption $\mathrm{Var}(\boldsymbol{\epsilon}_i) = \sigma^2\boldsymbol{I}$ can be relaxed as shown in Chapter 5, where we describe extensions that allow us to model nonconstant variances or special within-group correlation structures. The random effects $\boldsymbol{b}_i$ and the within-group errors $\boldsymbol{\epsilon}_i$ are assumed to be independent for different groups and to be independent of each other for the same group.

Because the distribution of the random effects vectors $\boldsymbol{b}_i$ is assumed to be normal (or Gaussian) with a mean of $\boldsymbol{0}$, it is completely characterized by its variance–covariance matrix $\boldsymbol{\Psi}$. This matrix must be symmetric and positive semi-definite; that is, all its eigenvalues must be non-negative. We will make the stronger assumption that it is *positive-definite* which is to say that all its eigenvalues must be strictly positive. We can make this restriction because an indefinite model can always be re-expressed as a positive-definite model of lower dimension.

The random effects $\boldsymbol{b}_i$ are defined to have a mean of $\boldsymbol{0}$ and therefore any nonzero mean for a term in the random effects must be expressed as part of the fixed-effects terms. Thus, the columns of $\boldsymbol{Z}_i$ are usually a subset of the columns of $\boldsymbol{X}_i$.

When computing with the model it is more convenient to express the variance–covariance matrix in the form of a *relative precision factor*, $\boldsymbol{\Delta}$, which is any matrix that satisfies

$$\frac{\Psi^{-1}}{1/\sigma^2} = \Delta^T \Delta.$$

If Ψ is positive-definite then such a Δ will exist, but it need not be unique. The Cholesky factor (Thisted, 1988, §3.3) of $\sigma^2\Psi^{-1}$ is one possible Δ. The matrix Δ is called a *relative precision factor* because it factors the *precision matrix*, Ψ^{-1}, of the random effects, expressed relative to the precision, $1/\sigma^2$, of the ϵ_i.

We use some of the examples in Chapter 1 to illustrate the general LME model formulation.

Railway Rails Experiment

In the case of the rails data introduced in §1.1, $M = 6$, $n_i = 3$, $i = 1, \ldots, 6$, $p = q = 1$, and the regressor matrices for the fixed and random effects are particularly simple:

$$X_i = Z_i = 1 = \begin{bmatrix} 1 \\ 1 \\ 1 \end{bmatrix}, \quad i = 1, \ldots, 6.$$

The random effects $b_i, i = 1, \ldots, 6$ are scalars; hence their variance σ_b^2 is also a scalar, as is the relative precision factor, Δ. There is only one choice for Δ (up to changes in sign) and that is

$$\Delta = \sqrt{\sigma^2/\sigma_b^2}.$$

Ergometric Experiment of Types of Stools

The data for the stools ergometric experiment of §1.2 are balanced, with $M = 6$, $n_i = 4$, $i = 1, \ldots, 6$, $p = 4$, and $q = 1$. The fixed-effects regressor matrices X_i are determined by the contrasts chosen to represent the types of stool. For the Helmert contrasts parameterization used in the fit of the fm1Stool object in §1.2.1, we have

$$X_i = \begin{bmatrix} 1 & -1 & -1 & -1 \\ 1 & 1 & -1 & -1 \\ 1 & 0 & 2 & -1 \\ 1 & 0 & 0 & 3 \end{bmatrix}, \quad i = 1, \ldots, 6.$$

The random-effects regression matrices Z_i and the relative precision factor Δ are the same as in the rails example.

Orthodontic Growth Curve in Girls

The orthodontic growth curve data for females presented in §1.4.1 are also balanced, with $M = 11$, $n_i = 4$, $i = 1, \ldots, 11$. For the LME model with random effects for both the intercept and the slope, used to fit the `fm2OrthF` object in §1.4.1, we have $p = q = 2$ and the fixed- and random-effects regressor matrices are identical and given by

$$
\boldsymbol{X}_i = \boldsymbol{Z}_i = \begin{bmatrix} 1 & 8 \\ 1 & 10 \\ 1 & 12 \\ 1 & 14 \end{bmatrix}, \quad i = 1, \ldots, 11.
$$

Any square-root of the 2×2 matrix $\sigma^2 \boldsymbol{\Psi}^{-1}$ can be used as a relative precision factor in this case.

2.1.2   A Multilevel LME Model

The Laird–Ware formulation for single-level LME models presented in §2.1.1 can be extended to multiple, nested levels of random effects. In the case of two nested levels of random effects the response vectors at the innermost level of grouping are written $\boldsymbol{y}_{ij}, i = 1, \ldots, M, j = 1, \ldots, M_i$ where M is the number of first-level groups and M_i is the number of second-level groups within first-level group i. The length of $\boldsymbol{y}_{ij}$ is n_{ij}.

The fixed-effects model matrices are $\boldsymbol{X}_{ij}, i = 1, \ldots, M, j = 1, \ldots, M_i$ of size $n_{ij} \times p$. Using first-level random effects $\boldsymbol{b}_i$ of length q_1 and second-level random effects $\boldsymbol{b}_{ij}$ of length q_2 with corresponding model matrices $\boldsymbol{Z}_{i,j}$ of size $n_i \times q_1$ and $\boldsymbol{Z}_{ij}$ of size $n_i \times q_2$, we write the model as

$$
\boldsymbol{y}_{ij} = \boldsymbol{X}_{ij} \boldsymbol{\beta} + \boldsymbol{Z}_{i,j} \boldsymbol{b}_i + \boldsymbol{Z}_{ij} \boldsymbol{b}_{ij} + \boldsymbol{\epsilon}_{ij}, \quad i = 1, \ldots, M, \quad j = 1, \ldots, M_i,
\tag{2.2}
$$

$$
\boldsymbol{b}_i \sim \mathcal{N}(\boldsymbol{0}, \boldsymbol{\Psi}_1), \quad \boldsymbol{b}_{ij} \sim \mathcal{N}(\boldsymbol{0}, \boldsymbol{\Psi}_2), \quad \boldsymbol{\epsilon}_{ij} \sim \mathcal{N}(\boldsymbol{0}, \sigma^2 \boldsymbol{I}).
$$

The level-1 random effects $\boldsymbol{b}_i$ are assumed to be independent for different i, the level-2 random effects $\boldsymbol{b}_{ij}$ are assumed to be independent for different i or j and to be independent of the level-1 random effects, and the within-group errors $\boldsymbol{\epsilon}_{ij}$ are assumed to be independent for different i or j and to be independent of the random effects.

Extensions to an arbitrary number Q of levels of random effects follow the same general pattern. For example, with $Q = 3$ the response for the kth level-3 unit within the jth level-2 unit within the ith level-1 unit is written

$$
\boldsymbol{y}_{ijk} = \boldsymbol{X}_{ijk} \boldsymbol{\beta} + \boldsymbol{Z}_{i,jk} \boldsymbol{b}_i + \boldsymbol{Z}_{ij,k} \boldsymbol{b}_{ij} + \boldsymbol{Z}_{ijk} \boldsymbol{b}_{ijk} + \boldsymbol{\epsilon}_{ijk},
$$

$$
i = 1, \ldots, M, \quad j = 1, \ldots, M_i, \quad k = 1, \ldots, M_{ij},
$$

$$
\boldsymbol{b}_i \sim \mathcal{N}(\boldsymbol{0}, \boldsymbol{\Sigma}_1), \quad \boldsymbol{b}_{ij} \sim \mathcal{N}(\boldsymbol{0}, \boldsymbol{\Sigma}_2), \quad \boldsymbol{b}_{ijk} \sim \mathcal{N}(\boldsymbol{0}, \boldsymbol{\Sigma}_3), \quad \boldsymbol{\epsilon}_{ijk} \sim \mathcal{N}(\boldsymbol{0}, \sigma^2 \boldsymbol{I}).
$$

Note that the distinction between, say, the kth horizontal section of the regressor matrix for the level-2 random effect $\boldsymbol{b}_{ij}$, written $\boldsymbol{Z}_{ij,k}$, and the

jkth horizontal section of the regressor matrix for the level-1 random effect b_i, written $Z_{i,jk}$, is the position of the comma in the subscripts.

As with a single level of random effects, we will express the variance–covariance matrices, Ψ_q, $q = 1, \ldots, Q$, in terms of relative precision factors Δ_q.

In this book, we only consider mixed-effects models with a multivariate normal (or Gaussian) distribution for the random effects and the within-group errors. Generally we assume that the variance–covariance matrix Ψ_q for the level-q random effects can be any positive-definite, symmetric matrix. In some models we will further restrict the form of Ψ_q, say by requiring that it be diagonal or that it be a multiple of the identity.

Those familiar with the *multilevel modeling* literature (Bryk and Raudenbush, 1992; Goldstein, 1995) may notice that we count "levels" differently. In that literature the model (2.1) is called a two-level model because there are two levels of random variation. Similarly, the model (2.2) is called a three-level model. We prefer the terminology from the experimental design literature and count the number of "levels" as the number of nested levels of random effects.

Split-Plot Experiment on Varieties of Oats and Nitrogen Levels

We use the split-plot experiment on the yield of three different varieties of oats measured at four different concentrations of nitrogen, described in §1.6, to illustrate the multilevel LME model formulation. The final model used in that section, corresponding to the fitted object `fm4Oats`, represents the yield y_{ijk} for the jth variety of oat at the kth nitrogen concentration N_k in the ith block as

$$y_{ijk} = \beta_0 + \beta_1 N_k + b_i + b_{ij} + \epsilon_{ijk}, \ i = 1, \ldots, 6, \ j = 1, \ldots, 3, \ k = 1, \ldots, 4,$$
$$b_i \sim \mathcal{N}\left(0, \sigma_1^2\right), \quad b_{ij} \sim \mathcal{N}\left(0, \sigma_2^2\right), \quad \epsilon_{ijk} \sim \mathcal{N}(0, \sigma^2).$$

The fixed effects are the intercept β_0 and the nitrogen slope β_1. The b_i denote the Block random effects, the b_{ij} denote the Variety within Block random effects, and the ϵ_{ijk} denote the within-group errors. This is an example of a two-level mixed-effects model, with the b_{ij} random effects *nested* within the b_i random effects.

In this example, $M = 6$, $M_i = 3$, $n_{ij} = 4$, $i = 1, \ldots, 6$ $j = 1, \ldots, 3$, $p = 2$, and $q_1 = q_2 = 1$. The regressor matrices are

$$X_{ij} = \begin{bmatrix} 1 & 0.0 \\ 1 & 0.2 \\ 1 & 0.4 \\ 1 & 0.6 \end{bmatrix}, \quad Z_{i,j} = Z_{ij} = \begin{bmatrix} 1 \\ 1 \\ 1 \\ 1 \end{bmatrix}, \quad i = 1, \ldots, 6, \quad j = 1, \ldots, 3.$$

Because all the random effects are scalars, the precision factors are uniquely defined (up to changes in sign) as

$$\Delta_1 = \sqrt{\sigma^2/\sigma_1^2} \quad \text{and} \quad \Delta_2 = \sqrt{\sigma^2/\sigma_2^2}.$$

2.2   Likelihood Estimation for LME Models

Several methods of parameter estimation have been used for linear mixed-effects models. We will concentrate on two general methods: maximum likelihood (ML) and restricted maximum likelihood (REML). Descriptions and comparisons of the various estimation methods used for LME models can be found, for example, in Searle et al. (1992) and Vonesh and Chinchilli (1997).

2.2.1   The Single-Level LME Likelihood Function

Consider first the model (2.1) that has a single level of random effects. The parameters of the model are β, σ^2, and whatever parameters determine Δ. We use θ to represent an unconstrained set of parameters that determine Δ. We will discuss parameterizations of Δ in §2.2.7—for now we will simply assume that a suitable parameterization has been chosen.

The *likelihood function* for the model (2.1) is the probability density for the data given the parameters, but regarded as a function of the parameters with the data fixed, instead of as a function of the data with the parameters fixed. That is,

$$L\left(\beta, \theta, \sigma^2 | y\right) = p(y | \beta, \theta, \sigma^2),$$

where L is the likelihood, p is a probability density, and y is the entire N-dimensional response vector, $N = \sum_{i=1}^{M} n_i$.

Because the nonobservable random effects $b_i, i = 1, \ldots, M$ are part of the model, we must integrate the conditional density of the data given the random effects with respect to the marginal density of the random effects to obtain the marginal density for the data. We can use the independence of the b_i and the ϵ_i to express this as

$$\begin{aligned}
L\left(\beta, \theta, \sigma^2 | y\right) &= \prod_{i=1}^{M} p\left(y_i | \beta, \theta, \sigma^2\right) \\
&= \prod_{i=1}^{M} \int p\left(y_i | b_i, \beta, \sigma^2\right) p\left(b_i | \theta, \sigma^2\right) db_i,
\end{aligned} \tag{2.3}$$

where the conditional density of $\boldsymbol{y}_i$ is multivariate normal

$$p\left(\boldsymbol{y}_i | \boldsymbol{b}_i, \boldsymbol{\beta}, \sigma^2\right) = \frac{\exp\left(-\|\boldsymbol{y}_i - \boldsymbol{X}_i\boldsymbol{\beta} - \boldsymbol{Z}_i\boldsymbol{b}_i\|^2 / 2\sigma^2\right)}{(2\pi\sigma^2)^{n_i/2}} \qquad (2.4)$$

and the marginal density of $\boldsymbol{b}_i$ is also multivariate normal

$$\begin{aligned} p\left(\boldsymbol{b}_i | \boldsymbol{\theta}, \sigma^2\right) &= \frac{\exp\left(-\boldsymbol{b}_i^T \boldsymbol{\Psi}^{-1} \boldsymbol{b}_i\right)}{(2\pi)^{q/2} \sqrt{|\boldsymbol{\Psi}|}} \\ &= \frac{\exp\left(-\|\boldsymbol{\Delta}\boldsymbol{b}_i\|^2 / 2\sigma^2\right)}{(2\pi\sigma^2)^{q/2} \operatorname{abs}|\boldsymbol{\Delta}|^{-1}}, \end{aligned} \qquad (2.5)$$

where $|\boldsymbol{A}|$ denotes the determinant of the matrix $\boldsymbol{A}$. Substituting (2.4) and (2.5) into (2.3) provides the likelihood as

$$L\left(\boldsymbol{\beta}, \boldsymbol{\theta}, \sigma^2 | \boldsymbol{y}\right) =$$

$$\prod_{i=1}^{M} \frac{\operatorname{abs}|\boldsymbol{\Delta}|}{(2\pi\sigma^2)^{n_i/2}} \int \frac{\exp\left[-\left(\|\boldsymbol{y}_i - \boldsymbol{X}_i\boldsymbol{\beta} - \boldsymbol{Z}_i\boldsymbol{b}_i\|^2 + \|\boldsymbol{\Delta}\boldsymbol{b}_i\|^2\right)/2\sigma^2\right]}{(2\pi\sigma^2)^{q/2}} \, d\boldsymbol{b}_i$$

$$= \prod_{i=1}^{M} \frac{\operatorname{abs}|\boldsymbol{\Delta}|}{(2\pi\sigma^2)^{n_i/2}} \int \frac{\exp\left(-\left\|\tilde{\boldsymbol{y}}_i - \tilde{\boldsymbol{X}}_i\boldsymbol{\beta} - \tilde{\boldsymbol{Z}}_i\boldsymbol{b}_i\right\|^2 / 2\sigma^2\right)}{(2\pi\sigma^2)^{q/2}} \, d\boldsymbol{b}_i, \qquad (2.6)$$

where

$$\tilde{\boldsymbol{y}}_i = \begin{bmatrix} \boldsymbol{y}_i \\ \boldsymbol{0} \end{bmatrix}, \quad \tilde{\boldsymbol{X}}_i = \begin{bmatrix} \boldsymbol{X}_i \\ \boldsymbol{0} \end{bmatrix}, \quad \tilde{\boldsymbol{Z}}_i = \begin{bmatrix} \boldsymbol{Z}_i \\ \boldsymbol{\Delta} \end{bmatrix}, \qquad (2.7)$$

are augmented data vectors and model matrices. This approach of changing the contribution of the marginal distribution of the random effects into extra rows for the response and the design matrices is called a *pseudo-data* approach because it creates the effect of the marginal distribution by adding "pseudo" observations.

The exponent in the integral of (2.6) is in the form of a squared norm or, more specifically, a residual sum-of-squares. We can determine the conditional modes of the random effects given the data, written $\widehat{\boldsymbol{b}}_i$, by minimizing this residual sum-of-squares. This is a standard least squares problem for which we could write the solution as

$$\widehat{\boldsymbol{b}}_i = \left(\tilde{\boldsymbol{Z}}_i^T \tilde{\boldsymbol{Z}}_i\right)^{-1} \tilde{\boldsymbol{Z}}_i^T \left(\tilde{\boldsymbol{y}}_i - \tilde{\boldsymbol{X}}_i\boldsymbol{\beta}\right).$$

The squared norm can then be expressed as

$$\begin{aligned} \left\|\tilde{\boldsymbol{y}}_i - \tilde{\boldsymbol{X}}_i\boldsymbol{\beta} - \tilde{\boldsymbol{Z}}_i\boldsymbol{b}_i\right\|^2 &= \left\|\tilde{\boldsymbol{y}}_i - \tilde{\boldsymbol{X}}_i\boldsymbol{\beta} - \tilde{\boldsymbol{Z}}_i\widehat{\boldsymbol{b}}_i\right\|^2 + \left\|\tilde{\boldsymbol{Z}}_i\left(\boldsymbol{b}_i - \widehat{\boldsymbol{b}}_i\right)\right\|^2 \\ &= \left\|\tilde{\boldsymbol{y}}_i - \tilde{\boldsymbol{X}}_i\boldsymbol{\beta} - \tilde{\boldsymbol{Z}}_i\widehat{\boldsymbol{b}}_i\right\|^2 + \left(\boldsymbol{b}_i - \widehat{\boldsymbol{b}}_i\right)^T \tilde{\boldsymbol{Z}}_i^T \tilde{\boldsymbol{Z}}_i \left(\boldsymbol{b}_i - \widehat{\boldsymbol{b}}_i\right). \end{aligned} \qquad (2.8)$$

The first term in (2.8) does not depend on $\boldsymbol{b}_i$ so its exponential can be factored out of the integral in (2.6). Integrating the exponential of the second term in (2.8) is equivalent, up to a constant, to integrating a multivariate normal density function. Note that

$$
\frac{\sqrt{|\tilde{\boldsymbol{Z}}_i^T \tilde{\boldsymbol{Z}}_i|}}{\sqrt{|\tilde{\boldsymbol{Z}}_i^T \tilde{\boldsymbol{Z}}_i|}} \int \frac{\exp\left[-\left(\boldsymbol{b}_i - \hat{\boldsymbol{b}}_i\right)^T \tilde{\boldsymbol{Z}}_i^T \tilde{\boldsymbol{Z}}_i \left(\boldsymbol{b}_i - \hat{\boldsymbol{b}}_i\right)/2\sigma^2\right]}{(2\pi\sigma^2)^{q/2}}\, d\boldsymbol{b}_i
$$

$$
= \frac{1}{\sqrt{|\tilde{\boldsymbol{Z}}_i^T \tilde{\boldsymbol{Z}}_i|}} \int \frac{\exp\left[-\left(\boldsymbol{b}_i - \hat{\boldsymbol{b}}_i\right)^T \tilde{\boldsymbol{Z}}_i^T \tilde{\boldsymbol{Z}}_i \left(\boldsymbol{b}_i - \hat{\boldsymbol{b}}_i\right)/2\sigma^2\right]}{(2\pi\sigma^2)^{q/2}/\sqrt{|\tilde{\boldsymbol{Z}}_i^T \tilde{\boldsymbol{Z}}_i|}}\, d\boldsymbol{b}_i
$$

$$
= \frac{1}{\sqrt{|\tilde{\boldsymbol{Z}}_i^T \tilde{\boldsymbol{Z}}_i|}} = \frac{1}{\sqrt{|\boldsymbol{Z}_i^T \boldsymbol{Z}_i + \boldsymbol{\Delta}^T \boldsymbol{\Delta}|}}. \tag{2.9}
$$

By combining (2.8) and (2.9) we can express the integral in (2.6) as

$$
\int \frac{\exp\left[-\left\|\tilde{\boldsymbol{y}}_i - \tilde{\boldsymbol{X}}_i\boldsymbol{\beta} - \tilde{\boldsymbol{Z}}_i\boldsymbol{b}_i\right\|^2/2\sigma^2\right]}{(2\pi\sigma^2)^{q/2}}\, d\boldsymbol{b}_i
$$

$$
= \frac{\exp\left(-\left\|\tilde{\boldsymbol{y}}_i - \tilde{\boldsymbol{X}}_i\boldsymbol{\beta} - \tilde{\boldsymbol{Z}}_i\hat{\boldsymbol{b}}_i\right\|^2/2\sigma^2\right)}{\sqrt{|\tilde{\boldsymbol{Z}}_i^T \tilde{\boldsymbol{Z}}_i|}}
$$

to give

$$
L\left(\boldsymbol{\beta}, \boldsymbol{\theta}, \sigma^2|\boldsymbol{y}\right)
$$

$$
= \frac{1}{(2\pi\sigma^2)^{N/2}} \exp\left(\frac{-\sum_{i=1}^{M} \left\|\tilde{\boldsymbol{y}}_i - \tilde{\boldsymbol{X}}_i\boldsymbol{\beta} - \tilde{\boldsymbol{Z}}_i\hat{\boldsymbol{b}}_i\right\|^2}{2\sigma^2}\right) \prod_{i=1}^{M} \frac{\text{abs}|\boldsymbol{\Delta}|}{\sqrt{|\tilde{\boldsymbol{Z}}_i^T \tilde{\boldsymbol{Z}}_i|}}. \tag{2.10}
$$

The expression (2.10) could be used directly in an optimization routine to calculate the maximum likelihood estimates for $\boldsymbol{\beta}$, $\boldsymbol{\theta}$, and $\boldsymbol{\sigma}^2$. However, the optimization is much simpler if we first *concentrate* or *profile* the likelihood so it is a function of $\boldsymbol{\theta}$ only. That is, we calculate the conditional estimates $\hat{\boldsymbol{\beta}}(\boldsymbol{\theta})$ and $\hat{\sigma}^2(\boldsymbol{\theta})$ as the values that maximize $L(\boldsymbol{\beta}, \boldsymbol{\theta}, \sigma^2)$ for a given $\boldsymbol{\theta}$. Notice that the parts of (2.10) involving $\boldsymbol{\beta}$ and σ^2 are identical in form to the likelihood for a linear regression model so $\hat{\boldsymbol{\beta}}(\boldsymbol{\theta})$ and $\hat{\sigma}^2(\boldsymbol{\theta})$ can be determined from standard linear regression theory.

We do need to be careful because the least squares estimates for $\boldsymbol{\beta}$ will depend on the conditional modes $\widehat{\boldsymbol{b}}_i$ and these, in turn, depend on $\boldsymbol{\beta}$. Thus, we must determine these least squares values jointly as the least squares solution to

$$\left(\widehat{\boldsymbol{b}}_1^T, \ldots, \widehat{\boldsymbol{b}}_M^T, \widehat{\boldsymbol{\beta}}^T\right)^T = \arg\min_{b_1, \ldots, b_M, \boldsymbol{\beta}} \|\boldsymbol{y}_e - \boldsymbol{X}_e(\boldsymbol{b}_1, \ldots, \boldsymbol{b}_M, \boldsymbol{\beta})^T\|^2,$$

where

$$\boldsymbol{X}_e = \begin{bmatrix} \boldsymbol{Z}_1 & \boldsymbol{0} & \cdots & \boldsymbol{0} & \boldsymbol{X}_1 \\ \boldsymbol{\Delta} & \boldsymbol{0} & \cdots & \boldsymbol{0} & \boldsymbol{0} \\ \boldsymbol{0} & \boldsymbol{Z}_2 & \cdots & \boldsymbol{0} & \boldsymbol{X}_2 \\ \boldsymbol{0} & \boldsymbol{\Delta} & \cdots & \boldsymbol{0} & \boldsymbol{0} \\ \vdots & \vdots & \vdots & \vdots & \vdots \\ \boldsymbol{0} & \boldsymbol{0} & \cdots & \boldsymbol{Z}_M & \boldsymbol{X}_M \\ \boldsymbol{0} & \boldsymbol{0} & \cdots & \boldsymbol{\Delta} & \boldsymbol{0} \end{bmatrix} \quad \text{and} \quad \boldsymbol{y}_e = \begin{bmatrix} \boldsymbol{y}_1 \\ \boldsymbol{0} \\ \boldsymbol{y}_2 \\ \boldsymbol{0} \\ \vdots \\ \boldsymbol{y}_M \\ \boldsymbol{0} \end{bmatrix}. \tag{2.11}$$

Conceptually we could write

$$\left(\widehat{\boldsymbol{b}}_1^T, \ldots, \widehat{\boldsymbol{b}}_M^T, \widehat{\boldsymbol{\beta}}^T\right)^T = (\boldsymbol{X}_e^T \boldsymbol{X}_e)^{-1} \boldsymbol{X}_e^T \boldsymbol{y}_e,$$

but we definitely would *not* want to calculate these values this way. The matrix $\boldsymbol{X}_e$ is sparse and can be very large. If possible we want take advantage of the sparsity and avoid working directly with $\boldsymbol{X}_e$.

Linear regression theory also gives us the conditional maximum likelihood estimate for σ^2

$$\widehat{\sigma}^2(\boldsymbol{\theta}) = \frac{\|\boldsymbol{y}_e - \boldsymbol{X}_e(\widehat{\boldsymbol{b}}_1^T, \ldots, \widehat{\boldsymbol{b}}_M^T, \widehat{\boldsymbol{\beta}}^T)^T\|^2}{N}. \tag{2.12}$$

Notice that the maximum likelihood estimate of σ^2 is the residual sum-of-squares divided by N, not by $N - p$.

Substituting these conditional estimates back into (2.10) provides the profiled likelihood

$$L(\boldsymbol{\theta}) = L(\widehat{\boldsymbol{\beta}}(\boldsymbol{\theta}), \boldsymbol{\theta}, \widehat{\sigma}^2(\boldsymbol{\theta})) = \frac{\exp(-N/2)}{[2\pi\widehat{\sigma}^2(\boldsymbol{\theta})]^{N/2}} \prod_{i=1}^{M} \frac{\text{abs}\,|\boldsymbol{\Delta}|}{\sqrt{\left|\widetilde{\boldsymbol{Z}}_i^T \widetilde{\boldsymbol{Z}}_i\right|}}. \tag{2.13}$$

We do not actually need to calculate the values of $\widehat{\boldsymbol{b}}_1, \ldots, \widehat{\boldsymbol{b}}_M$ or $\widehat{\boldsymbol{\beta}}(\boldsymbol{\theta})$ to evaluate the profiled likelihood. We only need to know the norm of the residual from the augmented least squares problem. The decomposition methods described in §2.2.2 provide us with fast, convenient methods of calculating this.

The pseudo-data representation of the marginal density $p(\boldsymbol{y}_i|\boldsymbol{\beta}, \boldsymbol{\theta}, \sigma^2)$ used in (2.6) is just one way of expressing this density and deriving the likelihood. It is also possible to describe this density as a normal distribution with mean $\boldsymbol{0}$ and a patterned variance–covariance matrix $\boldsymbol{\Sigma}_i$—a representation that is often used to derive the likelihood for the parameters in a linear mixed-effects model. Although we will not use this representation extensively in this chapter, we will use it in Chapter 5, so we present some of this derivation of the likelihood here.

The model (2.1) can be re-expressed as

$$\boldsymbol{y}_i = \boldsymbol{X}_i\boldsymbol{\beta} + \boldsymbol{Z}_i\boldsymbol{b}_i + \boldsymbol{\epsilon}_i = \boldsymbol{X}_i\boldsymbol{\beta} + \boldsymbol{\epsilon}_i^*, \quad i = 1, \ldots, M, \tag{2.14}$$

where $\boldsymbol{\epsilon}_i^* = \boldsymbol{Z}_i\boldsymbol{b}_i + \boldsymbol{\epsilon}_i$. Because the $\boldsymbol{\epsilon}_i^*$ are the sum of two independent multivariate normal random vectors, they are independently distributed as multivariate normal vectors with mean $\boldsymbol{0}$ and variance–covariance matrix $\sigma^2\boldsymbol{\Sigma}_i$, where $\boldsymbol{\Sigma}_i = \boldsymbol{I} + \boldsymbol{Z}_i\boldsymbol{\Psi}\boldsymbol{Z}_i^T/\sigma^2$. It then follows from (2.14) that the $\boldsymbol{y}_i$ are independent multivariate normal random vectors with mean $\boldsymbol{X}_i\boldsymbol{\beta}$ and variance–covariance matrix $\sigma^2\boldsymbol{\Sigma}_i$. That is,

$$p\left(\boldsymbol{y}_i|\boldsymbol{\beta}, \boldsymbol{\theta}, \sigma^2\right) = \left(2\pi\sigma^2\right)^{-\frac{n_i}{2}} \exp\left(\frac{(\boldsymbol{y}_i - \boldsymbol{X}_i\boldsymbol{\beta})^T \boldsymbol{\Sigma}_i^{-1} (\boldsymbol{y}_i - \boldsymbol{X}_i\boldsymbol{\beta})}{-2\sigma^2}\right) |\boldsymbol{\Sigma}_i|^{-\frac{1}{2}}.$$

For a given value of $\boldsymbol{\theta}$, the values of $\boldsymbol{\beta}$ and σ^2 that maximize the likelihood could be written as

$$\widehat{\boldsymbol{\beta}}(\boldsymbol{\theta}) = \left(\sum_{i=1}^M \boldsymbol{X}_i^T\boldsymbol{\Sigma}_i^{-1}\boldsymbol{X}_i\right)^{-1} \sum_{i=1}^M \boldsymbol{X}_i^T\boldsymbol{\Sigma}_i^{-1}\boldsymbol{y}_i,$$

$$\widehat{\sigma}^2(\boldsymbol{\theta}) = \frac{\sum_{i=1}^M \left(\boldsymbol{y}_i - \boldsymbol{X}_i\widehat{\boldsymbol{\beta}}(\boldsymbol{\theta})\right)^T \boldsymbol{\Sigma}_i^{-1} \left(\boldsymbol{y}_i - \boldsymbol{X}_i\widehat{\boldsymbol{\beta}}(\boldsymbol{\theta})\right)}{N}.$$

Computationally these expressions are much more difficult than (2.11) and (2.12). Using these expressions for $\widehat{\boldsymbol{\beta}}(\boldsymbol{\theta})$ and $\widehat{\sigma}^2(\boldsymbol{\theta})$ we could derive the profiled likelihood or log-likelihood.

We present these expressions for completeness only. We prefer to use the expressions from the pseudo-data representation for computation, especially when the pseudo-data representation is combined with orthogonal-triangular decompositions described in the next section.

2.2.2 Orthogonal-Triangular Decompositions

Orthogonal-triangular decompositions of rectangular matrices are a preferred numerical method for solving least squares problems (Chambers, 1977; Kennedy and Gentle, 1980; Thisted, 1988). They are also called *QR decompositions* as the decomposition is often written

$$\boldsymbol{X} = \boldsymbol{Q} \begin{bmatrix} \boldsymbol{R} \\ \boldsymbol{0} \end{bmatrix} = \boldsymbol{Q}_t\boldsymbol{R},$$

where X is an $n \times p$ matrix ($n \geq p$) of rank p, Q is $n \times n$ and orthogonal, R is $p \times p$ and upper triangular, and Q_t (Q-truncated) consists of the first p columns of Q. To say that Q is *orthogonal* means that $Q^T Q = Q Q^T = I$. This implies that $Q_t^T Q_t = I$.

The S function qr is used to create a QR decomposition from a matrix. For example, in §1.4.1 we present a model where the fixed-effects model matrices for each subject are

$$X_i = \begin{bmatrix} 1 & 8 \\ 1 & 10 \\ 1 & 12 \\ 1 & 14 \end{bmatrix}, \quad i = 1, \ldots, 11.$$

We can generate such a matrix in S and create its decomposition by

```
> Xmat <- matrix( c(1, 1, 1, 1, 8, 10, 12, 14), ncol = 2 )
> Xmat
      [,1] [,2]
[1,]    1    8
[2,]    1   10
[3,]    1   12
[4,]    1   14
> Xqr <- qr( Xmat )              # creates a QR structure
> qr.R( Xqr )                    # returns R
      [,1]      [,2]
[1,]    -2 -22.0000
[2,]     0  -4.4721
> qr.Q( Xqr )                    # returns Q-truncated
      [,1]      [,2]
[1,]  -0.5   0.67082
[2,]  -0.5   0.22361
[3,]  -0.5  -0.22361
[4,]  -0.5  -0.67082
> qr.Q( Xqr, complete = TRUE )   # returns the full Q
      [,1]      [,2]      [,3]      [,4]
[1,]  -0.5   0.67082   0.023607   0.54721
[2,]  -0.5   0.22361  -0.439345  -0.71202
[3,]  -0.5  -0.22361   0.807869  -0.21760
[4,]  -0.5  -0.67082  -0.392131   0.38240
```

Although we will write expressions that involve Q, this matrix is not usually evaluated explicitly. Products such as $Q^T y$ or $Q y$ can be calculated directly from information about the decomposition without having to generate this $n \times n$ matrix. See Dongarra, Bunch, Moler and Stewart (1979, Chapter 9) for details. The S functions qr.qty and qr.qy evaluate these products directly.

An important property of orthogonal matrices is that they preserve norms of vectors under multiplication either by Q or by Q^T. That is, the transformation represented by Q is a generalization of a rotation or a reflection in the plane. In particular,

$$\|Q^T y\|^2 = (Q^T y)^T Q^T y = y^T Q Q^T y = y^T y = \|y\|^2.$$

If we apply this to the residual vector in a least squares problem we get

$$
\begin{aligned}
\|y - X\beta\|^2 &= \left\|Q^T (y - X\beta)\right\|^2 \\
&= \|Q^T y - Q^T X\beta\|^2 \\
&= \left\|c - Q^T Q \begin{bmatrix} R \\ 0 \end{bmatrix} \beta\right\|^2 \\
&= \left\|c - \begin{bmatrix} R \\ 0 \end{bmatrix} \beta\right\|^2 \\
&= \|c_1 - R\beta\|^2 + \|c_2\|^2,
\end{aligned}
$$

where $c = \left(c_1^T c_2^T\right)^T = Q^T y$ is the rotated residual vector. The components c_1 and c_2 are of lengths p and $n - p$, respectively.

Because X has rank p, the $p \times p$ matrix R is nonsingular and upper-triangular. The least-squares solution $\widehat{\beta}$ is easily evaluated as the solution to

$$R\widehat{\beta} = c_1$$

and the residual sum-of-squares is $\|c_2\|^2$. Notice that the residual sum-of-squares can be evaluated without having to calculate $\widehat{\beta}$.

2.2.3 Evaluating the Likelihood Through Decompositions

Returning to the linear mixed-effects model, we take an orthogonal-triangular decomposition of the augmented model matrix $\tilde{Z}_i$ from (2.7) as

$$\tilde{Z}_i = Q_{(i)} \begin{bmatrix} R_{11(i)} \\ 0 \end{bmatrix},$$

where $Q_{(i)}$ is $(n_i + q) \times (n_i + q)$ and $R_{11(i)}$ is $q \times q$. Then

$$
\begin{aligned}
\left\|\tilde{y}_i - \tilde{X}_i\beta - \tilde{Z}_i b_i\right\|^2 &= \left\|Q_{(i)}^T \left(\tilde{y}_i - \tilde{X}_i\beta - \tilde{Z}_i b_i\right)\right\|^2 \\
&= \left\|c_{1(i)} - R_{10(i)}\beta - R_{11(i)} b_i\right\|^2 + \left\|c_{0(i)} - R_{00(i)}\beta\right\|^2,
\end{aligned}
$$

where the $q \times p$ matrix $R_{10(i)}$, the $n_i \times p$ matrix $R_{00(i)}$, the q-vector $c_{1(i)}$ and the n_i-vector $c_{0(i)}$ are defined by

$$\begin{bmatrix} R_{10(i)} \\ R_{00(i)} \end{bmatrix} = Q_{(i)}^T \tilde{X}_i \quad \text{and} \quad \begin{bmatrix} c_{1(i)} \\ c_{0(i)} \end{bmatrix} = Q_{(i)}^T \tilde{y}_i.$$

Another way of thinking of these matrices is as components in an orthogonal-triangular decomposition of an augmented matrix

$$\begin{bmatrix} \boldsymbol{Z}_i & \boldsymbol{X}_i & \boldsymbol{y}_i \\ \boldsymbol{\Delta} & \boldsymbol{0} & \boldsymbol{0} \end{bmatrix} = \boldsymbol{Q}_{(i)} \begin{bmatrix} \boldsymbol{R}_{11(i)} & \boldsymbol{R}_{10(i)} & \boldsymbol{c}_{1(i)} \\ \boldsymbol{0} & \boldsymbol{R}_{00(i)} & \boldsymbol{c}_{0(i)} \end{bmatrix},$$

where the reduction to triangular form is halted after the first q columns. (The peculiar numbering scheme for the submatrices and subvectors is designed to allow easy extension to more than one level of random effects.) Returning to the integral in (2.6) we can now remove a constant factor and reduce it to

$$\int \frac{\exp\left[-\left(\|\boldsymbol{y}_i - \boldsymbol{X}_i\boldsymbol{\beta} - \boldsymbol{Z}_i\boldsymbol{b}_i\|^2 + \|\boldsymbol{\Delta}\boldsymbol{b}_i\|^2\right)/2\sigma^2\right]}{\sqrt{2\pi\sigma^2}}\, db_i$$

$$= \exp\left[\frac{\|\boldsymbol{c}_{0(i)} - \boldsymbol{R}_{00(i)}\boldsymbol{\beta}\|^2}{-2\sigma^2}\right] \int \frac{\exp\left[\frac{\|\boldsymbol{c}_{1(i)} - \boldsymbol{R}_{10(i)}\boldsymbol{\beta} - \boldsymbol{R}_{11(i)}\boldsymbol{b}_i\|^2}{-2\sigma^2}\right]}{(2\pi\sigma^2)^{q/2}}\, db_i. \quad (2.15)$$

Because $\boldsymbol{R}_{11(i)}$ is nonsingular, we can perform a change of variable to $\boldsymbol{\phi}_i = (\boldsymbol{c}_{1(i)} - \boldsymbol{R}_{10(i)}\boldsymbol{\beta} - \boldsymbol{R}_{11(i)}\boldsymbol{b}_i)/\sigma$ with differential $d\boldsymbol{\phi}_i = \sigma^{-q}\, \mathrm{abs}\,|\boldsymbol{R}_{11(i)}|\, db_i$ and write the integral as

$$\int \frac{\exp\left(-\|\boldsymbol{c}_{1(i)} - \boldsymbol{R}_{10(i)}\boldsymbol{\beta} - \boldsymbol{R}_{11(i)}\boldsymbol{b}_i\|^2/2\sigma^2\right)}{(2\pi\sigma^2)^{q/2}}\, db_i$$

$$= \frac{1}{\mathrm{abs}\,|\boldsymbol{R}_{11(i)}|} \int \frac{\exp\left(-\|\boldsymbol{\phi}_i\|^2/2\right)}{(2\pi)^{q/2}}\, d\boldsymbol{\phi}_i \quad (2.16)$$

$$= 1/\mathrm{abs}\,|\boldsymbol{R}_{11(i)}|.$$

This is the same result as (2.10) because

$$\sqrt{\left|\tilde{\boldsymbol{Z}}_i^T \tilde{\boldsymbol{Z}}\right|} = \sqrt{\left|\begin{bmatrix} \boldsymbol{R}_{11(i)}^T & \boldsymbol{0}\end{bmatrix} \boldsymbol{Q}_{(i)}^T \boldsymbol{Q}_{(i)} \begin{bmatrix} \boldsymbol{R}_{11(i)} \\ \boldsymbol{0}\end{bmatrix}\right|}$$

$$= \sqrt{\left|\boldsymbol{R}_{11(i)}^T \boldsymbol{R}_{11(i)}\right|}$$

$$= \sqrt{\left|\boldsymbol{R}_{11(i)}^T\right|\left|\boldsymbol{R}_{11(i)}\right|}$$

$$= \sqrt{\left(\left|\boldsymbol{R}_{11(i)}^T\right|\right)^2}$$

$$= \mathrm{abs}\,|\boldsymbol{R}_{11(i)}|.$$

Because $\boldsymbol{R}_{11(i)}$ is triangular, its determinant is simply the product of its diagonal elements.

Substituting (2.16) into (2.15) into (2.6) provides the likelihood as

$$
L\left(\boldsymbol{\beta},\boldsymbol{\theta},\sigma^2|\boldsymbol{y}\right) = \prod_{i=1}^{M} \frac{\exp\left[-\|\boldsymbol{c}_{0(i)} - \boldsymbol{R}_{00(i)}\boldsymbol{\beta}\|^2/2\sigma^2\right]}{(2\pi\sigma^2)^{n_i/2}}\, \text{abs}\left(\frac{|\boldsymbol{\Delta}|}{|\boldsymbol{R}_{11(i)}|}\right)
$$

$$
= \frac{\exp\left(-\sum_{i=1}^{M}\|\boldsymbol{c}_{0(i)} - \boldsymbol{R}_{00(i)}\boldsymbol{\beta}\|^2/2\sigma^2\right)}{(2\pi\sigma^2)^{-N/2}} \prod_{i=1}^{M} \text{abs}\left(\frac{|\boldsymbol{\Delta}|}{|\boldsymbol{R}_{11(i)}|}\right).
$$

The term in the exponent has the form of a residual sum-of-squares for $\boldsymbol{\beta}$ pooled over all the groups. Forming another orthogonal-triangular decomposition

$$
\begin{bmatrix} \boldsymbol{R}_{00(1)} & \boldsymbol{c}_{0(1)} \\ \vdots & \vdots \\ \boldsymbol{R}_{00(M)} & \boldsymbol{c}_{0(M)} \end{bmatrix} = \boldsymbol{Q}_0 \begin{bmatrix} \boldsymbol{R}_{00} & \boldsymbol{c}_0 \\ \boldsymbol{0} & \boldsymbol{c}_{-1} \end{bmatrix} \tag{2.17}
$$

produces the reduced form

$$
L\left(\boldsymbol{\beta},\boldsymbol{\theta},\sigma^2|\boldsymbol{y}\right)
$$
$$
= \left(2\pi\sigma^2\right)^{-N/2} \exp\left(\frac{\|\boldsymbol{c}_{-1}\|^2 + \|\boldsymbol{c}_0 - \boldsymbol{R}_{00}\boldsymbol{\beta}\|^2}{-2\sigma^2}\right) \prod_{i=1}^{M} \text{abs}\left(\frac{|\boldsymbol{\Delta}|}{|\boldsymbol{R}_{11(i)}|}\right). \tag{2.18}
$$

For a given $\boldsymbol{\theta}$, the values of $\boldsymbol{\beta}$ and σ^2 that maximize (2.18) are

$$
\widehat{\boldsymbol{\beta}}(\boldsymbol{\theta}) = \boldsymbol{R}_{00}^{-1}\boldsymbol{c}_0 \quad \text{and} \quad \widehat{\sigma}^2(\boldsymbol{\theta}) = \frac{\|\boldsymbol{c}_{-1}\|^2}{N}, \tag{2.19}
$$

which give the profiled likelihood

$$
L(\boldsymbol{\theta}|\boldsymbol{y}) = L\left(\widehat{\boldsymbol{\beta}}(\boldsymbol{\theta}),\boldsymbol{\theta},\widehat{\sigma}^2(\boldsymbol{\theta})|\boldsymbol{y}\right)
$$
$$
= \left(\frac{N}{2\pi\|\boldsymbol{c}_{-1}\|^2}\right)^{N/2} \exp\left(-\frac{N}{2}\right) \prod_{i=1}^{M} \text{abs}\left(\frac{|\boldsymbol{\Delta}|}{|\boldsymbol{R}_{11(i)}|}\right), \tag{2.20}
$$

or the profiled log-likelihood

$$
\ell(\boldsymbol{\theta}|\boldsymbol{y}) = \log L(\boldsymbol{\theta}|\boldsymbol{y})
$$
$$
= \frac{N}{2}\left[\log N - \log(2\pi) - 1\right] - N\log\|\boldsymbol{c}_{-1}\| + \sum_{i=1}^{M}\log\text{abs}\left(\frac{|\boldsymbol{\Delta}|}{|\boldsymbol{R}_{11(i)}|}\right). \tag{2.21}
$$

The profiled log-likelihood (2.21) is maximized with respect to $\boldsymbol{\theta}$, producing the maximum likelihood estimate $\widehat{\boldsymbol{\theta}}$. The maximum likelihood estimates $\widehat{\boldsymbol{\beta}}$ and $\widehat{\sigma}^2$ are then obtained by setting $\boldsymbol{\theta} = \widehat{\boldsymbol{\theta}}$ in (2.19).

Although technically the random effects b_i are not parameters for the statistical model, they do behave in some ways like parameters and often we want to "estimate" their values. The conditional modes of the random effects, evaluated at the conditional estimate of β, are the *Best Linear Unbiased Predictors* or *BLUPs* of the $b_i, i = 1, \ldots, M$. They can be evaluated, using the matrices from the orthogonal-triangular decompositions, as

$$\widehat{b}_i(\theta) = R_{11(i)}^{-1} \left(c_{1(i)} - R_{10(i)} \widehat{\beta}(\theta) \right). \tag{2.22}$$

In practice, the unknown vector θ is replaced by its maximum likelihood estimate $\widehat{\theta}$, producing estimated BLUPs $\widehat{b}_i(\widehat{\theta})$.

The decomposition (2.17) is equivalent to calculating the QR decomposition of the potentially huge matrix X_e defined in (2.11). If we determined the least-squares solution to (2.11) using an orthogonal-triangular decomposition

$$X_e = Q_e \begin{bmatrix} R_e \\ 0 \end{bmatrix},$$

the triangular part of the decomposition and the leading part of the rotated, augmented response vector would be

$$R_e = \begin{bmatrix} R_{11(1)} & 0 & \cdots & 0 & R_{10(1)} \\ 0 & R_{11(2)} & \cdots & 0 & R_{10(2)} \\ \vdots & \vdots & \vdots & \vdots & \vdots \\ 0 & 0 & \cdots & R_{11(M)} & R_{10(M)} \\ 0 & 0 & 0 & 0 & R_{00} \end{bmatrix} \quad \text{and} \quad c_1 = \begin{bmatrix} c_{1(1)} \\ c_{1(2)} \\ \vdots \\ c_{1(M)} \\ c_0 \end{bmatrix}.$$

Thus, the $\widehat{\beta}(\theta)$ and $\widehat{\sigma}^2(\theta)$ from (2.19) are the same as those from (2.17) and (2.12). The vector c_{-1} is the residual vector in the coordinate system determined by Q_e. Because Q_e is orthogonal, $\|c_{-1}\|^2$ is the residual sum-of-squares for the least squares problem defined by X_e and y_e.

The profiled log-likelihood (2.20) has the same form as (2.13). It consists of three additive components; a constant, a scaled logarithm of the residual sum-of-squares, and a sum of ratios of the logarithms of determinants. In the next section we examine these terms in detail.

2.2.4 Components of the Profiled Log-Likelihood

Returning to the example of the rails data of §1.1, let us consider the different components of the profiled log-likelihood as expressed in (2.21). Recall that the relative precision factor Δ will be a scalar in this case so let us write it as Δ. There are three additive terms in the profiled log-likelihood:

1. The constant $\frac{N}{2}\left[\log N - \log(2\pi) - 1\right]$, which can be neglected for the purposes of optimization.

2. $-N \log \|c_{-1}\|$, a multiple of the logarithm of the norm of a residual vector from the penalized least-squares fit for Δ, X_i, Z_i, and y_i.

3. $\sum_{i=1}^{M} \log \left(\Delta / \operatorname{abs}|R_{11(i)}|\right) = \sum_{i=1}^{M} \log \left(\Delta / \sqrt{|Z_i^T Z_i + \Delta^2|}\right)$. In the general case this is the sum of the logarithms of the ratios of determinants.

In Figure 2.1 we show the two nonconstant terms and the resulting log-likelihood as a function of Δ for the rails example

The shapes of the curves in Figure 2.1 indicate that it would be better to optimize the profiled log-likelihood with respect to $\theta = \log \Delta$ instead of Δ. This transformation will also help to ensure that Δ does not become negative during the course of the iterations of whatever optimization routine we use. In Figure 2.2 we show the components and the log-likelihood as a function of θ. We can see that the log-likelihood is closer to a quadratic with respect to θ than with respect to Δ.

There are patterns in Figure 2.2 that will hold in general for linear mixed-effects models. The log of the norm of the residual is an increasing sigmoidal, or "S-shaped," function with respect to θ. As $\theta \to -\infty$ (or $\Delta \to 0$), this log-norm approaches a horizontal asymptote at a value that corresponds to the log residual norm from an unpenalized regression of the form

$$
y = \begin{bmatrix} y_1 \\ y_2 \\ \vdots \\ y_n \end{bmatrix} = \begin{bmatrix} Z_1 & 0 & \cdots & 0 & X_1 \\ 0 & Z_2 & \cdots & 0 & X_2 \\ \vdots & \vdots & \cdots & \vdots & \vdots \\ 0 & 0 & \cdots & Z_n & X_n \end{bmatrix} \begin{bmatrix} b_1 \\ b_2 \\ \vdots \\ b_n \\ \beta \end{bmatrix} + \epsilon.
$$

At the other extreme, large positive values of θ, and the correspondingly large values of Δ, put such a heavy penalty on the size of the b_i terms in the regression that these are forced to zero. Thus, as $\theta \to \infty$, the penalized residual norm approaches that from a regression of the entire response vector y on the X_i matrices alone.

$$
y = \begin{bmatrix} y_1 \\ y_2 \\ \vdots \\ y_n \end{bmatrix} = \begin{bmatrix} X_1 \\ X_2 \\ \vdots \\ X_n \end{bmatrix} \beta + \epsilon.
$$

In the ratio of determinants term, very large values of Δ will dominate $Z_i^T Z_i$ in the denominator so the ratios approach Δ / Δ and the sum of the

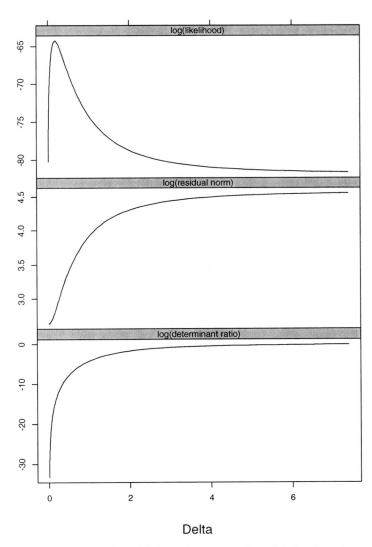

Delta

FIGURE 2.1. The profiled log-likelihood as a function of Δ for the rails example. Two of the components of the log-likelihood, $\log \|c_{-1}\|$, the log of the length of the residual, and $\sum_{i=1}^{M} \log \left(\Delta / \operatorname{abs} |R_{11(i)}|\right)$, the log of the determinant ratios, are shown on the same scale.

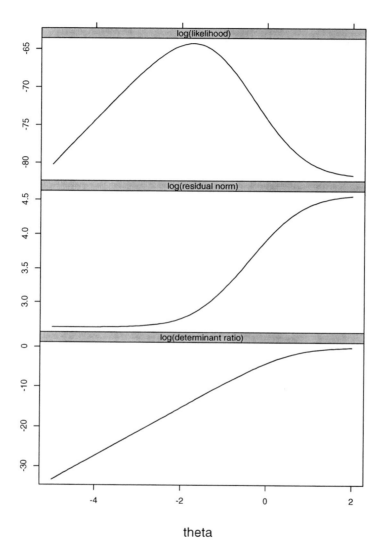

FIGURE 2.2. The profiled log-likelihood as a function of $\theta = \log(\Delta)$ for the rails example. Two of the components of the log-likelihood, $\log \|\boldsymbol{c}_{-1}\|$, the log of the length of the residual, and $\sum_{i=1}^{M} \log \left(\Delta / \operatorname{abs} |\boldsymbol{R}_{11(i)}| \right)$, the log of the determinant ratios, are shown on the same scale.

logarithms approaches zero. Very small values of Δ will have little effect on the denominator so the term has the form

$$\sum_{i=1}^{M} \left(\theta - \log \sqrt{|\boldsymbol{Z}_i^T \boldsymbol{Z}_i|}\right) = M\theta - \sum_{i=1}^{M} \log \sqrt{|\boldsymbol{Z}_i^T \boldsymbol{Z}_i|}.$$

That is, when $\theta \to -\infty$ this term approaches a linear function of θ, as can be seen in Figure 2.2.

Because the log of the determinant ratio approaches a linear function of θ as $\theta \to -\infty$ while the log of the residual norm tends to a finite asymptote, the log-likelihood approaches a linear function of θ. When Δ, and hence θ, becomes large the log-likelihood will usually decrease then approach a constant. This does not always occur, however. For some data sets, the log-likelihood will continue to increase with θ as $\theta \to \infty$. In these cases, the maximum likelihood estimator of σ_b^2 is zero.

Both in the log of the ratio of determinants term and in the log of the norm of the penalized residual term, the effect of Δ is determined by its size relative to the $\boldsymbol{Z}_i$ matrices. Values of Δ that are either much less than or much greater than $\sqrt{\boldsymbol{Z}_i^T \boldsymbol{Z}_i}$ will produce a log-likelihood that is near an asymptote. If there is to be a maximum for finite θ it will have to be near $\theta_0 = \log \Delta_0 = \sum_{i=1}^{M} \log \sqrt{\boldsymbol{Z}_i^T \boldsymbol{Z}_i}/M$. In the case of the rails data $\theta_0 = 0.549$.

2.2.5 Restricted Likelihood Estimation

Maximum likelihood estimates of "variance components," such as σ^2 and σ_b^2 in the rails example, tend to underestimate these parameters. Many analysts prefer the restricted (or residual) maximum likelihood (REML) estimates (Patterson and Thompson, 1971; Harville, 1977) for these quantities.

There are several ways to define the REML estimation criterion. One definition that provides a convenient computational form (Laird and Ware, 1982) is

$$L_R(\boldsymbol{\theta}, \sigma^2 | \boldsymbol{y}) = \int L(\boldsymbol{\beta}, \boldsymbol{\theta}, \sigma^2 | \boldsymbol{y}) \, d\boldsymbol{\beta},$$

which, within a Bayesian framework, corresponds to assuming a locally uniform prior distribution for the fixed effects $\boldsymbol{\beta}$ and integrating them out of the likelihood.

Using (2.18) and the same change-of-variable techniques as in (2.16) gives the log-restricted-likelihood

$$
\ell_R(\boldsymbol{\theta}, \sigma^2 | \boldsymbol{y}) = \log L_R\left(\boldsymbol{\theta}, \sigma^2 \,\middle|\, \boldsymbol{y}\right)
$$

$$
= -\frac{N-p}{2}\log(2\pi\sigma^2) - \frac{\|\boldsymbol{c}_{-1}\|^2}{2\sigma^2} - \log\operatorname{abs}|\boldsymbol{R}_{00}| + \sum_{i=1}^{M}\log\operatorname{abs}\left(\frac{|\boldsymbol{\Delta}|}{|\boldsymbol{R}_{11(i)}|}\right).
$$

This produces the conditional estimate $\widehat{\sigma}_R^2(\theta) = \|\boldsymbol{c}_{-1}\|^2/(N-p)$ for σ^2, from which we obtain the profiled log-restricted-likelihood

$$
\ell_R(\boldsymbol{\theta}|\boldsymbol{y}) = \ell_R(\boldsymbol{\theta}, \widehat{\sigma}_R^2(\theta)|\boldsymbol{y})
$$

$$
= \operatorname{const} - (N-p)\log\|\boldsymbol{c}_{-1}\| - \log\operatorname{abs}|\boldsymbol{R}_{00}| + \sum_{i=1}^{M}\log\operatorname{abs}\left(\frac{|\boldsymbol{\Delta}|}{|\boldsymbol{R}_{11(i)}|}\right).
$$

$$
(2.23)
$$

The components of the profiled log-restricted-likelihood in (2.23) are similar to those in the profiled log-likelihood (2.21) except that the log of the norm of the residual vector has a different multiplier and there is an extra determinant term of $\log\operatorname{abs}|\boldsymbol{R}_{00}| = \log\left|\sum_{i=1}^{M}\boldsymbol{X}_i^T\boldsymbol{\Sigma}_i^{-1}\boldsymbol{X}_i\right|/2$. A plot of the components of the profiled log-restricted-likelihood versus θ for the rails example would be similar in shape to Figure 2.2.

The evaluation of the restricted maximum likelihood estimates is done by optimizing the profiled log-restricted-likelihood (2.23) with respect to $\boldsymbol{\theta}$ only, and using the resulting REML estimate $\widehat{\boldsymbol{\theta}}_R$ to obtain the REML estimate of σ^2, $\widehat{\sigma}_R^2(\widehat{\boldsymbol{\theta}}_R)$. Similarly, the REML estimated BLUPs of the random effects are obtained by replacing $\boldsymbol{\theta}$ with $\widehat{\boldsymbol{\theta}}_R$ in (2.22).

In some ways, it is blurring the definition of the REML criterion to speak of the REML estimate of $\boldsymbol{\beta}$. The REML criterion only depends on $\boldsymbol{\theta}$ and σ. However, it is still useful, and perhaps even sensible, to evaluate the "best guess" at $\boldsymbol{\beta}$ from (2.19) once $\widehat{\boldsymbol{\theta}}_R$ has been determined using the REML criterion.

An important difference between the likelihood function and the restricted likelihood function is that the former is invariant to one-to-one reparameterizations of the fixed effects (e.g., a change in the contrasts representing a categorical variable), while the latter is not. Changing the $\boldsymbol{X}_i$ matrices results in a change in $\log\operatorname{abs}|\boldsymbol{R}_{00}|$ and a corresponding change in $\ell_R(\boldsymbol{\theta}|\boldsymbol{y})$. As a consequence, LME models with different fixed-effects structures fit using REML cannot be compared on the basis of their restricted likelihoods. In particular, likelihood ratio tests are not valid under these circumstances.

2.2.6 Multiple Levels of Random Effects

The likelihood function and the restricted likelihood function for multilevel LME models can be calculated using the same techniques described for the single-level model in §2.2.1, §2.2.3, and §2.2.5. We use the two-level LME model to illustrate the basic steps in the derivation of the multilevel likelihood function.

The likelihood for a model with two levels of random effects is defined as in (2.3), but integrating over both levels of random effects

$$L(\boldsymbol{\beta}, \boldsymbol{\theta}_1, \boldsymbol{\theta}_2, \sigma^2|\boldsymbol{y}) =$$

$$\prod_{i=1}^{M} \int \prod_{j=1}^{M_i} \left[\int p(\boldsymbol{y}_{ij}|\boldsymbol{b}_{ij}, \boldsymbol{b}_i, \boldsymbol{\beta}, \sigma^2)\, p(\boldsymbol{b}_{ij}|\boldsymbol{\theta}_2, \sigma^2)\, d\boldsymbol{b}_{ij} \right] p(\boldsymbol{b}_i|\boldsymbol{\theta}_1, \sigma^2)\, d\boldsymbol{b}_i.$$

$$(2.24)$$

As with a single level of random effects, we can simplify the integrals in (2.24) if we augment the $\boldsymbol{Z}_{ij}$ matrices with $\boldsymbol{\Delta}_2$ and form orthogonal-triangular decompositions of these augmented arrays. This allows us to evaluate the inner integrals. To evaluate the outer integrals we iterate this process.

That is, we first form and decompose the arrays

$$\begin{bmatrix} \boldsymbol{Z}_{ij} & \boldsymbol{Z}_{i,j} & \boldsymbol{X}_{ij} & \boldsymbol{y}_{ij} \\ \boldsymbol{\Delta}_2 & 0 & 0 & 0 \end{bmatrix} = \boldsymbol{Q}_{(ij)} \begin{bmatrix} \boldsymbol{R}_{22(ij)} & \boldsymbol{R}_{21(ij)} & \boldsymbol{R}_{20(ij)} & \boldsymbol{c}_{2(ij)} \\ 0 & \boldsymbol{R}_{11(ij)} & \boldsymbol{R}_{10(ij)} & \boldsymbol{c}_{1(ij)} \end{bmatrix},$$

$$i = 1, \ldots, M, \quad j = 1, \ldots, M_i. \quad (2.25)$$

The matrix $\boldsymbol{R}_{22(ij)}$ will be an upper-triangular matrix of dimension $q_2 \times q_2$. The other arrays in the first row of the decomposition in (2.25) are used only if the conditional estimates $\widehat{\boldsymbol{\beta}}(\boldsymbol{\theta})$ or the conditional modes $\widehat{\boldsymbol{b}}_{ij}(\boldsymbol{\theta})$ and $\widehat{\boldsymbol{b}}_i(\boldsymbol{\theta})$ are required. The arrays in the second row of the decomposition: $\boldsymbol{R}_{11(ij)}$, $\boldsymbol{R}_{10ij}$, and $\boldsymbol{c}_{1(ij)}$ each have n_{ij} rows.

To evaluate the outer integral in (2.24) we again form and decompose an augmented array

$$\begin{bmatrix} \boldsymbol{R}_{11(i1)} & \boldsymbol{R}_{10(i1)} & \boldsymbol{c}_{1(i1)} \\ \vdots & \vdots & \vdots \\ \boldsymbol{R}_{11(iM_i)} & \boldsymbol{R}_{10(iM_i)} & \boldsymbol{c}_{1(iM_i)} \\ \boldsymbol{\Delta}_1 & 0 & 0 \end{bmatrix} = \boldsymbol{Q}_{(i)} \begin{bmatrix} \boldsymbol{R}_{11(i)} & \boldsymbol{R}_{10(i)} & \boldsymbol{c}_{1(i)} \\ 0 & \boldsymbol{R}_{00(i)} & \boldsymbol{c}_{0(i)} \end{bmatrix}$$

$$i = 1, \ldots, M. \quad (2.26)$$

The final decomposition to produce $\boldsymbol{R}_{00}$, $\boldsymbol{c}_0$, and $\boldsymbol{c}_{-1}$ is the same as that in (2.17).

Using the matrices and vectors produced in (2.25), (2.26), and (2.17) and following the same steps as for the single level of nesting we can express the profiled log-likelihood for $\boldsymbol{\theta}_1$ and $\boldsymbol{\theta}_2$ as

$$\ell(\boldsymbol{\theta}_1, \boldsymbol{\theta}_2|\boldsymbol{y}) = \log L(\widehat{\boldsymbol{\beta}}(\boldsymbol{\theta}_1, \boldsymbol{\theta}_2), \boldsymbol{\theta}_1, \boldsymbol{\theta}_2, \widehat{\sigma}^2(\boldsymbol{\theta}_1, \boldsymbol{\theta}_2)|\boldsymbol{y})$$
$$= \text{const} - N \log \|\boldsymbol{c}_{-1}\| + \sum_{i=1}^{M} \log \text{abs} \left(\frac{|\boldsymbol{\Delta}_1|}{|\boldsymbol{R}_{11(i)}|} \right)$$
$$+ \sum_{i=1}^{M} \sum_{j=1}^{M_i} \log \text{abs} \left(\frac{|\boldsymbol{\Delta}_2|}{|\boldsymbol{R}_{22(ij)}|} \right).$$

Similarly, the profiled log-restricted-likelihood is

$$\ell_R(\boldsymbol{\theta}_1, \boldsymbol{\theta}_2|\boldsymbol{y}) = \log L_R(\widehat{\boldsymbol{\beta}}_R(\boldsymbol{\theta}_1, \boldsymbol{\theta}_2), \boldsymbol{\theta}_1, \boldsymbol{\theta}_2, \widehat{\sigma}_R^2(\boldsymbol{\theta}_1, \boldsymbol{\theta}_2)|\boldsymbol{y})$$
$$= \text{const} - (N - p) \log \|\boldsymbol{c}_{-1}\| - \log \text{abs} |\boldsymbol{R}_{00}|$$
$$+ \sum_{i=1}^{M} \log \text{abs} \left(\frac{|\boldsymbol{\Delta}_1|}{|\boldsymbol{R}_{11(i)}|} \right) + \sum_{i=1}^{M} \sum_{j=1}^{M_i} \log \text{abs} \left(\frac{|\boldsymbol{\Delta}_2|}{|\boldsymbol{R}_{22(ij)}|} \right).$$

The calculation methods extend in the obvious way to Q nested levels of random effects.

2.2.7 Parameterizing Relative Precision Factors

In a model with Q nested levels of random effects, there are Q symmetric, positive-definite, variance–covariance matrices $\boldsymbol{\Psi}_k$, $k = 1, \ldots, Q$. For computational purposes we express these in terms of relative precision factors $\boldsymbol{\Delta}_k, k = 1, \ldots, Q$ that satisfy

$$\boldsymbol{\Delta}_k^T \boldsymbol{\Delta}_k = \sigma^2 \boldsymbol{\Psi}_k^{-1}, \quad k = 1, \ldots, Q.$$

To optimize the log-likelihood or log-restricted-likelihood we express the scaled variance–covariance matrices $\boldsymbol{\Psi}_k/\sigma^2$, or equivalently the relative precision factors $\boldsymbol{\Delta}_k$, as a function of unconstrained parameter vectors $\boldsymbol{\theta}_k, k = 1, \ldots, Q$. For example, when Δ is a scalar, as in §2.2.4, we use the unconstrained parameter $\theta = \log \Delta$ when optimizing the log-likelihood.

For the general case where $\boldsymbol{\Psi}_k/\sigma^2$ is a positive-definite, symmetric matrix of size $q \times q$, we parameterize it through its *matrix logarithm*. To define this parameterization, we note that any positive-definite, symmetric matrix $\boldsymbol{A}$ can be expressed as the *matrix exponential* of another symmetric matrix $\boldsymbol{B}$. This means that

$$\boldsymbol{A} = e^{\boldsymbol{B}} = \boldsymbol{I} + \boldsymbol{B} + \frac{\boldsymbol{B}^2}{2!} + \frac{\boldsymbol{B}^3}{3!} + \cdots.$$

If $\boldsymbol{A}$ is the matrix exponential of $\boldsymbol{B}$, then $\boldsymbol{B}$ is the matrix logarithm of $\boldsymbol{A}$.

Suppose $\boldsymbol{A}$ is $q \times q$, symmetric and positive-definite. One way of evaluating its matrix logarithm $\boldsymbol{B}$ is to calculate an eigenvalue-eigenvector decomposition

$$\boldsymbol{A} = \boldsymbol{U}\boldsymbol{\Lambda}\boldsymbol{U}^T,$$

where $\mathbf{\Lambda}$ is $q \times q$ and diagonal while $\mathbf{U}$ is $q \times q$ and orthogonal. If $\mathbf{A}$ is positive-definite, then all the diagonal elements of $\mathbf{\Lambda}$ must be positive. The matrix logarithm of $\mathbf{\Lambda}$ is the diagonal matrix whose diagonal elements are the logarithms of the corresponding elements of $\mathbf{\Lambda}$. We will denote this by $\log \mathbf{\Lambda}$. Finally

$$\mathbf{B} = \log \mathbf{A} = \mathbf{U} \log \mathbf{\Lambda} \mathbf{U}^T.$$

We define $\boldsymbol{\theta}_k$ to be the elements of the upper triangle of the matrix logarithm of $\mathbf{\Psi}_k/\sigma^2$. This gives a nonredundant, unconstrained parameter vector for $\mathbf{\Psi}_k/\sigma^2$.

Other unconstrained parameterizations for $\mathbf{\Psi}_k/\sigma^2$ are used when the matrix is required to have a special structure beyond being symmetric and positive-definite. For example, if $\mathbf{\Psi}_k/\sigma^2$ is to be diagonal and positive-definite then the diagonal elements must all be positive and an unconstrained parameterization uses the logarithms of these diagonal elements.

2.2.8 Optimization Algorithms

Optimization of the profiled log-likelihood or the profiled log-restricted-likelihood of an LME model is usually accomplished through EM iterations or through Newton–Raphson iterations (Laird and Ware, 1982; Lindstrom and Bates, 1988; Longford, 1993).

The EM algorithm (Dempster, Laird and Rubin, 1977) is a popular iterative algorithm for likelihood estimation in models with incomplete data. The EM iterations for the LME model are based on regarding the random effects, such as the $\boldsymbol{b}_i, i = 1, \ldots, M$, as unobserved data. At iteration w we use the current variance–covariance parameter vector, $\boldsymbol{\theta}^{(w)}$, to evaluate the distribution of $\boldsymbol{b}|\boldsymbol{y}$ and derive the expectation of the log-likelihood for a new value of $\boldsymbol{\theta}$ given this conditional distribution. Because we are taking an expectation, this step is called the E step. The M step consists of maximizing this expectation with respect to $\boldsymbol{\theta}$ to produce $\boldsymbol{\theta}^{(w+1)}$. Each iteration of the EM algorithm results in an increase in the log-likelihood function, though a possibly small increase. Efficient implementations of the EM algorithm for LME models are described in Bates and Pinheiro (1998).

The Newton–Raphson algorithm (Thisted, 1988, §4.2.2) is one of the most widely used optimization procedures. It uses a first-order expansion of the score function (the gradient of the log-likelihood function) around the current estimate $\boldsymbol{\theta}^{(w)}$ to produce the next estimate $\boldsymbol{\theta}^{(w+1)}$. Each Newton–Raphson iteration requires the calculation of the score function and its derivative, the Hessian matrix of the log-likelihood. Under general conditions usually satisfied in practice, the Newton–Raphson algorithm converges quadratically. Because the calculation of the Hessian matrix at each iteration may be computationally expensive, simple, quicker to compute approximations are sometimes used, leading to the so-called Quasi–Newton algorithms (Thisted, 1988, §4.3.3.4) .

Any iterative optimization algorithm requires initial values for the parameters. Because we can express both the profiled log-likelihood and the profiled log-restricted-likelihood as a function of the θ parameters, we only need to formulate starting values for θ when performing iterative optimization for LME models. These may be obtained from a previous fit for similar data, or derived from the current data. A general procedure for deriving initial values for θ from the data being fit is described in Bates and Pinheiro (1998) and is implemented in the lme function.

Individual iterations of the EM algorithm are quickly and easily computed. Although the EM iterations generally bring the parameters into the region of the optimum very quickly, progress toward the optimum tends to be slow when near the optimum. Newton–Raphson iterations, on the other hand, are individually more computationally intensive than the EM iterations, and they can be quite unstable when far from the optimum. However, close to the optimum they converge very quickly.

We therefore recommend a hybrid approach of starting with an initial $\theta^{(0)}$, performing a moderate number of EM iterations, then switching to Newton–Raphson iterations. Essentially the EM iterations can be regarded as refining the starting estimates before beginning the more general optimization routine. The lme function implements such a hybrid optimization scheme. It begins by calculating initial estimates of the θ parameters, then uses several EM iterations to get near the optimum, then switches to Newton–Raphson iterations to complete the convergence to the optimum. By default 25 EM iterations are performed before switching to Newton–Raphson iterations.

When fitting an LME model, it is often helpful to monitor the progress of the Newton–Raphson iterations to identify possible convergence problems. This is done by including an optional control argument in the call to lme. The value of control should be a list that can contain any of several flags or settings for the optimization algorithm. One of these flags is msVerbose. When it is set to TRUE or T, diagnostic output on the progress of the Newton–Raphson iterations in the indirect call of the ms function (Bates and Chambers, 1992, §10.2) is produced.

If we set this flag in the first fit for the rails example of §1.1, the diagnostic output is not very interesting because the EM iterations leave the parameter estimates so close to the optimum that convergence of the Newton–Raphson iterations is declared almost immediately.

```
> fm1Rail.lme <- lme( travel ~ 1, data = Rail, random = ~ 1 | Rail,
+          control = list( msVerbose = TRUE ) )
Iteration:  0 ,  1 function calls, F=  61.049
Parameters:
[1] -1.8196
Iteration:  1 ,  2 function calls, F=  61.049
```

```
Parameters:
[1] -1.8196
```

Note that the parameter listed in the iteration output is

$$\widehat{\theta} = \log(\widehat{\Delta}) = \log\left(\widehat{\sigma}/\widehat{\sigma}_b\right) = \log(4.0208/24.805) = -1.8196$$

and this is the only parameter being directly controlled by the optimization algorithm. The function labelled F in the iteration output is the negative of the log-restricted-likelihood but without the constant term $\frac{N-p}{2}\left[\log(N-p) - \log(2\pi) - 1\right]$, which is -0.0396 when $N = 18$ and $p = 1$. Thus, the value of $F = 61.049$ at convergence corresponds to a log-likelihood of $-(61.049 + 0.0396) = -61.089$. Because most optimization algorithms are designed to minimize rather than maximize a function of the parameters, we minimize the negative of the log-likelihood instead of maximizing the log-likelihood.

If we eliminate the EM iterations altogether with another control argument, niterEM, we can observe the progress of the Newton–Raphson iterations for θ.

```
> fm1Rail.lme <- lme( travel ~ 1, data = Rail, random = ~ 1 | Rail,
+        control = list( msVerbose = TRUE, niterEM = 0 ))
Iteration:  0 ,  1  function calls, F=  67.894
Parameters:
[1] -0.43152
Iteration:  1 ,  3  function calls, F=  61.157
Parameters:
[1] -2.0007
Iteration:  2 ,  4  function calls, F=  61.05
Parameters:
[1] -1.8028
Iteration:  3 ,  5  function calls, F=  61.049
Parameters:
[1] -1.8195
```

The algorithm converged to a slightly different value of θ, but with essentially the same value of the log-likelihood.

2.3   Approximate Distributions

Inference on the parameters of a linear mixed-effects model usually relies on approximate distributions for the maximum likelihood estimates and the restricted maximum likelihood estimates derived from asymptotic results.

Pinheiro (1994) has shown that, under certain regularity conditions generally satisfied in practice, the maximum likelihood estimates in the general LME model described in §2.1 are consistent and asymptotically normal. The approximate variance–covariance matrix for the maximum likelihood estimates is given by the inverse of the information matrix (Cox and Hinkley, 1974, §4.8) corresponding to the log-likelihood function $\ell = \ell(\boldsymbol{\beta}, \boldsymbol{\theta}_1, \ldots, \boldsymbol{\theta}_Q, \sigma^2)$. Because

$$\mathrm{E}\left[\partial^2 \ell / \partial \boldsymbol{\beta} \partial \boldsymbol{\theta}_q^T\right] = \mathbf{0}, \; q = 1, \ldots, Q \quad \text{and} \quad \mathrm{E}\left[\partial^2 \ell / \partial \boldsymbol{\beta} \partial \sigma^2\right] = \mathbf{0},$$

the information matrix corresponding to an LME model with Q levels of nesting is block diagonal and, therefore, the maximum likelihood estimates of the fixed effects $\boldsymbol{\beta}$ are asymptotically uncorrelated with the maximum likelihood estimates of $\boldsymbol{\theta}_1, \ldots, \boldsymbol{\theta}_Q$ and σ^2.

The approximate distributions of the maximum likelihood estimates in an LME model with Q levels of nesting are

$$\widehat{\boldsymbol{\beta}} \overset{.}{\sim} \mathcal{N}\left(\boldsymbol{\beta}, \sigma^2 \left[\boldsymbol{R}_{00}^{-1} \boldsymbol{R}_{00}^{-T}\right]\right),$$

$$\begin{bmatrix} \widehat{\boldsymbol{\theta}}_1 \\ \vdots \\ \widehat{\boldsymbol{\theta}}_Q \\ \log \widehat{\sigma} \end{bmatrix} \overset{.}{\sim} \mathcal{N}\left(\begin{bmatrix} \boldsymbol{\theta}_1 \\ \vdots \\ \boldsymbol{\theta}_Q \\ \log \sigma \end{bmatrix}, \; \boldsymbol{\mathcal{I}}^{-1}(\boldsymbol{\theta}_1, \ldots, \boldsymbol{\theta}_Q, \sigma)\right), \quad (2.27)$$

$$\boldsymbol{\mathcal{I}}(\boldsymbol{\theta}_1, \ldots, \boldsymbol{\theta}_Q, \sigma) = -\begin{bmatrix} \frac{\partial^2 \ell}{\partial \boldsymbol{\theta}_1 \partial \boldsymbol{\theta}_1^T} & \frac{\partial^2 \ell}{\partial \boldsymbol{\theta}_2 \partial \boldsymbol{\theta}_1^T} & \cdots & \frac{\partial^2 \ell}{\partial \log \sigma \partial \boldsymbol{\theta}_1^T} \\ \vdots & \vdots & & \vdots \\ \frac{\partial^2 \ell}{\partial \boldsymbol{\theta}_1 \partial \log \sigma} & \frac{\partial^2 \ell}{\partial \boldsymbol{\theta}_2 \partial \log \sigma} & \cdots & \frac{\partial^2 \ell}{\partial^2 \log \sigma} \end{bmatrix},$$

where $\ell = \ell(\boldsymbol{\theta}_1, \ldots, \boldsymbol{\theta}_Q, \sigma^2)$ now denotes the log-likelihood function profiled on the fixed effects, $\boldsymbol{\mathcal{I}}$ denotes the *empirical* information matrix, and $\boldsymbol{R}_{00}$ is defined as in (2.17). We use $\log \sigma$ in place of σ^2 in (2.27) to give an unrestricted parameterization for which the normal approximation tends to be more accurate.

As shown by Pinheiro (1994), the REML estimates in an LME model also are consistent and asymptotically normal, with approximate distributions identical to (2.27) but with ℓ replaced by the log-restricted-likelihood ℓ_R defined in §2.2.5.

In practice, the unknown parameters $\boldsymbol{\theta}_1, \ldots, \boldsymbol{\theta}_Q$ and σ^2 are replaced by their respective ML or REML estimates in the expressions for the approximate variance–covariance matrices in (2.27). The approximate distributions for the maximum likelihood estimates and REML estimates are used to produce hypothesis tests and confidence intervals for the LME model parameters, as described in §2.4.

2.4 Hypothesis Tests and Confidence Intervals

After we have fit a statistical model to the data we usually want to assess the precision of the estimates and the "significance" of various terms in the model or to compare how well one model fits the data relative to another model. This section presents approximate hypothesis tests and confidence intervals for the parameters in an LME model.

2.4.1 Likelihood Ratio Tests

A general method for comparing nested models fit by maximum likelihood is the *likelihood ratio test* (Lehmann, 1986, §1.7). Such a test can also be used with models fit by REML, but only if both models have been fit by REML and if the fixed-effects specification is the same for both models.

One statistical model is said to be *nested* within another model if it represents a special case of the other model. For example, in the analysis of the `Machines` data described in §1.3, we fit one model, `fm1Machine`, with a random effect for `Worker` only, then we fit a second model, `fm2Machine`, with random effects for `Worker` and for `Machine %in% Worker`. The model `fm1Machine` is nested within `fm2Machine` because it represents a special case of `fm2Machine` in which the variance of the `Machine %in% Worker` interaction term is zero.

If L_2 is the likelihood of the more general model (e.g., `fm2Machine`) and L_1 is the likelihood of the restricted model (e.g., `fm1Machine`) we must have $L_2 > L_1$ and, correspondingly, $\log L_2 > \log L_1$. The likelihood ratio test (LRT) statistic

$$2\log(L_2/L_1) = 2[\log(L_2) - \log(L_1)]$$

will be positive. If k_i is the number of parameters to be estimated in model i, then the asymptotic, or "large sample," distribution of the LRT statistic, under the null hypothesis that the restricted model is adequate, is a χ^2 distribution with $k_2 - k_1$ degrees of freedom.

In Chapter 1 we show several examples of likelihood ratio tests performed with the `anova` function. When given two arguments representing fits of nested models, this function displays the LRT statistic in the `L.Ratio` column and gives the p-value from the $\chi^2_{k_2-k_1}$ distribution. The column labelled `df` is the number of parameters in each model. For example, using model fits described in §1.3, we would have

```
> anova( fm1Machine, fm2Machine )
           Model df   AIC    BIC  logLik   Test L.Ratio p-value
fm1Machine     1  5 300.46 310.12 -145.23
fm2Machine     2  6 231.27 242.86 -109.64 1 vs 2  71.191  <.0001
```

The `anova` function also displays the values of the *Akaike Information Criterion* (*AIC*) (Sakamoto et al., 1986) and the *Bayesian Information Criterion* (*BIC*) (Schwarz, 1978). As mentioned in §1.1.1, these are model comparison criteria evaluated as

$$
\begin{aligned}
AIC &= -2\ell\left(\widehat{\boldsymbol{\theta}}|\boldsymbol{y}\right) + 2n_{par}, \\
BIC &= -2\ell\left(\widehat{\boldsymbol{\theta}}|\boldsymbol{y}\right) + n_{par}\log(N)
\end{aligned}
\tag{2.28}
$$

for each model, where n_{par} denotes the number of parameters in the model. Under these definitions, "smaller is better." That is, if we are using AIC to compare two or more models for the same data, we prefer the model with the lowest AIC. Similarly, when using BIC we prefer the model with the lowest BIC. The REML versions of the AIC and the BIC simply replace $\ell(\widehat{\boldsymbol{\theta}}|\boldsymbol{y})$ by $\ell_R(\widehat{\boldsymbol{\theta}}|\boldsymbol{y})$ in (2.28).

We will generally use likelihood-ratio tests to evaluate the significance of terms in the random-effects structure. That is, we fit different nested models in which the random-effects structure changes and apply likelihood-ratio tests. Stram and Lee (1994), using the results of Self and Liang (1987), argued that tests on the random effects structure conducted in this way can be conservative. That is, the p-value calculated from the $\chi^2_{k_2-k_1}$ distribution is greater than it should be. As Stram and Lee (1994) explain, changing from the more general model to the more specific model involves setting the variance of certain components of the random effects to zero, which is on the boundary of the parameter region. The asymptotic results for likelihood ratio tests have to be adjusted for boundary conditions. In the next section we use simulations to demonstrate the effect of these adjustments.

Simulating Likelihood Ratio Test Statistics

One way to check on the distribution of the likelihood ratio test statistic under the null hypothesis is through simulation. The `simulate.lme` function takes two model specifications, the null model and the alternative model. These may be given as lme objects corresponding to each model, or as lists of arguments used to produce such fits. In the latter case, only those characteristics that change between the two models need to be specified in the argument list for the alternative model.

For example, in the analysis of the `OrthoFem` data presented in §1.4.1, the `fm1OrthF` fit to the `OrthoFem` data has the specification

```
> fm1OrthF <- lme( distance ~ age, data = OrthoFem,
+      random = ~ 1 | Subject )
```

while `fm2OrthF` is fit as

```
> fm2OrthF <- update( fm1OrthF, random = ~ age | Subject )
```

Both models correspond to $\boldsymbol{X}_i$ matrices of

$$
\boldsymbol{X}_1 = \boldsymbol{X}_2 = \cdots = \boldsymbol{X}_{11} = \begin{bmatrix} 1 & 8 \\ 1 & 10 \\ 1 & 12 \\ 1 & 14 \end{bmatrix}
$$

with a two-dimensional $\boldsymbol{\beta}$ vector. In fm1OrthF the $\boldsymbol{Z}_i$ matrices are

$$\boldsymbol{Z}_1 = \boldsymbol{Z}_2 = \cdots = \boldsymbol{Z}_{11} = \begin{bmatrix} 1 \\ 1 \\ 1 \\ 1 \\ 1 \end{bmatrix}$$

and the $\boldsymbol{b}_i$ are one-dimensional random vectors with variance $\boldsymbol{\Psi} = \sigma_1^2$. In fm2OrthF the $\boldsymbol{Z}_i$ matrices are

$$\boldsymbol{Z}_1 = \boldsymbol{Z}_2 = \cdots = \boldsymbol{Z}_{11} = \begin{bmatrix} 1 & 8 \\ 1 & 10 \\ 1 & 12 \\ 1 & 14 \end{bmatrix}$$

and the $\boldsymbol{b}_i$ are two-dimensional random vectors with variance–covariance matrix

$$\boldsymbol{\Psi} = \begin{bmatrix} \sigma_1^2 & \sigma_{12} \\ \sigma_{12} & \sigma_2^2 \end{bmatrix}.$$

In the terminology of hypothesis tests, fm1OrthF is the null model and fm2OrthF is the alternative model. In this case, the null model is a special case of the alternative model, with one fewer random effect. The model being fit as fm1OrthF is obtained from the model for fm2OrthF by requiring the last row and column of the 2×2 $\boldsymbol{\Psi}$ to be zero. Although there are three distinct entries in this row and column, these entries are determined by only two parameters because $\boldsymbol{\Psi}$ must be symmetric. Notice that one of these entries, σ_2^2, must be non-negative so setting it to zero corresponds to a boundary condition.

To simulate the likelihood ratio test statistic comparing model fm1OrthF to model fm2OrthF we generate data according to the null model using the parameter values from fm1OrthF. We then fit both the null and the alternative model to each set of simulated data and calculate the likelihood ratio test statistic. This is repeated for nsim cases. By doing this we obtain an empirical distribution of the likelihood ratio test statistic under the null hypothesis. We can then compare the empirical distribution to different χ^2 distributions as in Figure 2.3, which is produced by

```
> orthLRTsim <- simulate.lme( fm1OrthF, fm2OrthF, nsim = 1000 )
> plot( orthLRTsim, df = c(1, 2) )            # produces Figure 2.3
```

Figure 2.3 is a probability–probability plot—similar to a quantile–quantile plot but on the p-value scale, rather than on the scale of the likelihood ratio test (LRT) statistic. The nominal p-values for the simulated LRT statistics, under χ^2 distributions with 1 and 2 degrees of freedom and an equal-weight mixture of those χ^2 distributions (denoted Mix(1,2) in Figure 2.3), for both

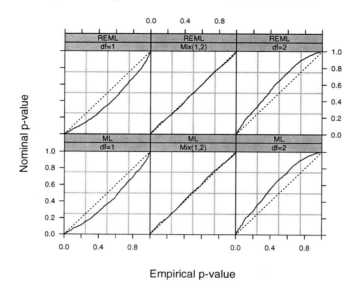

FIGURE 2.3. Plots of the nominal versus empirical p-values for the likelihood ratio test statistic comparing two models for the orthodontic data, female subjects only. The null model, `fm1OrthF`, has a random effect for the intercept only. The alternative model, `fm2OrthF`, has random effects for both the intercept and the slope. The null model was simulated 1000 times, both models were fit to the simulated data, and the likelihood ratio test statistic was calculated for both maximum likelihood and REML estimates. In each panel, the nominal p-values for the LRT statistics under the corresponding distribution are plotted versus the empirical p-values.

ML and REML estimation, are plotted versus the empirical p-values, obtained from the empirical distribution of the simulated LRT statistics.

For both REML and ML estimates, the nominal p-values for the LRT statistics under a χ^2 distribution with 2 degrees of freedom are much greater than the empirical p-values. This is the sense in which the likelihood ratio test using χ_2^2 for the reference distribution will be conservative—the actual p-value is smaller than the p-value that is reported. Stram and Lee (1994) suggest a $0.5\chi_1^2 + 0.5\chi_2^2$ mixture as a reference distribution, which is confirmed in Figure 2.3, for both ML and REML estimation. A χ_1^2 appears to be "anti-conservative" in the sense that the nominal p-values are smaller than the empirical p-values.

The adjustment suggested by Stram and Lee (1994) is not always this successful. According to this adjustment, the null distribution of the likelihood ratio test statistic for comparing `fm1Machine` to `fm2Machine` should have approximately a $0.5\chi_0^2 + 0.5\chi_1^2$ mixture distribution, where χ_0^2 represents a distribution with a point mass at 0. When simulated

```
> machineLRTsim <- simulate.lme(fm1Machine, fm2Machine, nsim= 1000)
```

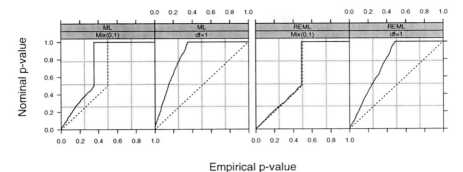

Nominal p-value

Empirical p-value

FIGURE 2.4. Plots of the nominal versus empirical p-values for the likelihood ratio test statistic comparing two models for the Machines data. The null model, fm1Machine, has a random effect for Worker only. The alternative model, fm2Machine, has random effects for the Worker and a random interaction for Machine %in% Worker. Both models were fit to 1000 sets of data simulated from the null model and the likelihood ratio test statistics were calculated.

```
> plot( machineLRTsim, df = c(0, 1),              # produces Figure 2.4
+   layout = c(4,1), between = list(x = c(0, 0.5)) )
```

it produces a distribution for the LRT statistics that closely agrees with the equal-weight mixture in the REML case, but which resembles a $0.65\chi_0^2 + 0.35\chi_1^2$ mixture in the ML case.

It is difficult to come up with general rules for approximating the distribution of the LRT statistic for such nested mixed-effects models. The naive approach of using a χ^2 distribution with the number of degrees of freedom determined by the difference in the number of nonredundant parameters in the models as the reference is easily implemented and tends to be conservative. This is the reference distribution we use to calculate the p-values quoted in the multiargument form of anova. One should be aware that these p-values may be conservative. That is, the reported p-value may be greater than the true p-value for the test and, in some cases, it may be much greater.

2.4.2 Hypothesis Tests for Fixed-Effects Terms

When two nested models differ in the specification of their fixed-effects terms, a likelihood ratio test can be defined for maximum likelihood fits only. As described in §2.2.5 a likelihood ratio test for REML fits is not feasible, because there is a term in the REML criterion that changes with the change in the fixed-effects specification.

Even though a likelihood ratio test for the ML fits of models with different fixed effects can be calculated, we do **not** recommend using such tests. Such

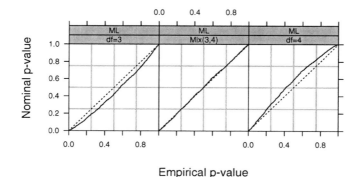

Empirical p-value

FIGURE 2.5. Plots of the nominal versus empirical p-values for the likelihood ratio test statistic comparing two models for the ergoStool data. The alternative model has a fixed effect for Type but the null model does not. The random effects specifications are the same. Both models were fit to 1000 sets of data simulated from the null model and the likelihood ratio test statistics from the maximum likelihood estimates were calculated.

likelihood ratio tests using the standard χ^2 reference distribution tend to be "anticonservative"—sometimes quite badly so.

As an example, consider the ergoStool example analyzed in §1.2.1. Suppose we compare fm1Stool, the model for the ergoStool data with a fixed effect for the Type factor, to a model without a fixed effect for the Type factor.

```
> stoolLRTsim <-
+    simulate.lme( m1 = list(fixed = effort ~ 1, data = ergoStool,
+                            random = ~ 1 | Subject),
+                  m2 = list(fixed = effort ~ Type),
+                  method = "ML", nsim = 1000 )
> plot( stoolLRTsim, df = c(3, 4) )                    # Figure 2.5
```

We can see from Figure 2.5 that, at 3 degrees of freedom, which is the difference in the number of parameters in the two models, the χ^2 distribution gives p-values that are "anticonservative." At 4 degrees of freedom the p-values will be conservative. The nominal p-values for the equal-weight mixture of χ_3^2 and χ_4^2 distributions, represented in the middle panel of Figure 2.5, are in close agreement with the empirical p-values.

In this case the slight anticonservative nature of the reported p-values may not be too alarming. However, as the number of parameters being removed from the fixed effects becomes large, compared to the total number of observations, this inaccuracy in the reported p-values can be substantial. For example, Littell, Milliken, Stroup and Wolfinger (1996, §1.5) provide analyses of data from a partially balanced incomplete block (PBIB) design. The design is similar to the randomized block design in the ergometric experiment described in §1.2 except that not every level of the treatment

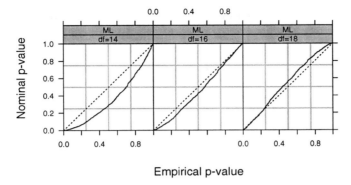

FIGURE 2.6. Plots of the nominal versus empirical p-values for the likelihood ratio test statistic comparing two models for the PBIB data. The alternative model has a fixed effect for Type, but the null model does not. The random effects specifications are the same. Both models were fit to 1000 sets of data simulated from the null model, and the likelihood ratio test statistics from the maximum likelihood estimates were calculated.

appears with every level of the blocking factor. This is the sense in which it is an "incomplete" block design. It is "partially balanced" because every pair of treatments occur together in a block the same number of times. These data are described in greater detail in Appendix A.22 and are given as an object called PBIB that is available with the nlme library.

   The important point with regard to the likelihood ratio tests is that there are 15 levels of the Treatment factor and only 60 observations in total. The blocking factor also has 15 levels. If we simulate the likelihood ratio test and plot the p-values calculated from the χ^2_{14} distribution,

```
>   pbibLRTsim <-
+       simulate.lme( m1 = list( fixed = response ~ 1, data = PBIB,
+                      random = ~ 1 | Block ),
+                m2 = list( fixed = response ~ Treatment ),
+                method = "ML", nsim = 1000 )
> plot( pbibLRTsim, df = c(14,16,18), weights = FALSE ) # Figure 2.6
```

we can see, from Figure 2.6, that the p-values calculated using χ^2_{14} as the reference distribution are seriously "anticonservative."

   Another, perhaps more conventional, approach to performing hypothesis tests involving terms in the fixed-effects specification is to condition on the estimates of the random effects variance–covariance parameters, $\widehat{\boldsymbol{\theta}}$. As described in §2.2.1, for a fixed value of $\boldsymbol{\theta}$, the conditional estimates of the fixed effects, $\widehat{\boldsymbol{\beta}}(\boldsymbol{\theta})$, are determined as standard least-squares estimates. The approximate distribution of the maximum likelihood or the REML estimates of the fixed effects in (2.27) is exact for the conditional estimates $\widehat{\boldsymbol{\beta}}(\boldsymbol{\theta})$.

Conditional tests for the significance of a term in the fixed-effects specification are given by the usual F-tests or t-tests for linear regression models, based on the usual (REML) conditional estimate of the variance

$$\widehat{\sigma}_R^2(\boldsymbol{\theta}) = s^2 = \frac{RSS}{N-p} = \frac{\|\boldsymbol{c}_{-1}\|^2}{N-p}.$$

In practice, the unknown parameter vector $\boldsymbol{\theta}$ is replaced by its maximum likelihood estimate or its REML estimate so the conditional tests hold only approximately.

The conditional t-tests are included in the output of the summary method applied to lme objects. They test the *marginal* significance of each fixed effect coefficient when all other fixed effects are present in the model. For example the t-tests for the fm2Machine model

```
> summary( fm2Machine )
 . . .
Fixed effects: score ~ Machine
             Value Std.Error DF t-value p-value
(Intercept) 59.650    2.1447 36  27.813  <.0001
   Machine1  3.983    1.0885 10   3.660  0.0044
   Machine2  3.311    0.6284 10   5.269  0.0004
 . . .
```

indicate that all the fixed-effects terms are significant.

The conditional F-tests are implemented in the single-argument form of the anova method for fitted models from lme. They test the significance of *terms* in the fixed effects model, which may include several coefficients. By default, the terms are tested *sequentially* in the order they enter the model, but the argument type to anova can be used to specify *marginal* F-tests. For example, to jointly test the significance of all 14 coefficients corresponding to Treatment in the PBIB example we use

```
> fm1PBIB <- lme( response ~ Treatment, data = PBIB, random = ~ 1 )
> anova( fm1PBIB )
            numDF denDF F-value p-value
(Intercept)     1    31  1654.2  <.0001
Treatment      14    31     1.5  0.1576
```

We will compare this result, a p-value of 15.8%, with that from the likelihood ratio test. Because a likelihood ratio test for terms in the fixed-effects specification must be done on ML fits, we first re-fit fm1PBIB using maximum likelihood, then modify the model.

```
> fm2PBIB <- update( fm1PBIB, method = "ML" )
> fm3PBIB <- update( fm2PBIB, response ~ 1 )
> anova( fm2PBIB, fm3PBIB )
        Model df    AIC    BIC  logLik      Test L.Ratio p-value
fm2PBIB     1 17 56.571 92.174 -11.285
fm3PBIB     2  3 52.152 58.435 -23.076 1 vs 2  23.581  0.0514
```

The simulation illustrated in Figure 2.6 shows that the 15.8% p-value from the conditional F-test is much more realistic than the 5.1% p-value from the likelihood ratio test.

For this reason, we prefer the conditional F-tests and t-tests for assessing the significance of terms in the fixed effects.

These conditional tests for fixed-effects terms require denominator degrees of freedom. In the case of the conditional F-tests, the numerator degrees of freedom are also required, being defined by the term itself. The denominator degrees of freedom are determined by the grouping level at which the term is estimated. A term is called *inner* relative to a grouping factor if its value can change within a given level of the grouping factor. A term is *outer* to a grouping factor if its value does not change within levels of the grouping factor. A term is said to be estimated at level i, if it is inner to the $i - 1$st grouping factor and outer to the ith grouping factor. For example, the term Machine in the fm2Machine model is outer to Machine %in% Worker and inner to Worker, so it is estimated at level 2 (Machine %in% Worker). If a term is inner to all Q grouping factors in a model, it is estimated at the level of the within-group errors, which we denote as the $Q + 1$st level.

The intercept, which is the parameter corresponding to the column of 1's in the model matrices X_i, is treated differently from all the other parameters, when it is present. As a parameter it is regarded as being estimated at level 0 because it is outer to all the grouping factors. However, its denominator degrees of freedom are calculated as if it were estimated at level $Q + 1$. This is because the intercept is the one parameter that pools information from all the observations at a level even when the corresponding column in X_i doesn't change with the level.

Letting m_i denote the total number of groups in level i (with the convention that $m_0 = 1$ when the fixed effects model includes an intercept and 0 otherwise, and $m_{Q+1} = N$) and p_i denote the sum of the degrees of freedom corresponding to the terms estimated at level i, the ith level denominator degrees of freedom is defined as

$$denDF_i = m_i - (m_{i-1} + p_i), \quad i = 1, \ldots, Q + 1.$$

This definition coincides with the classical decomposition of degrees of freedom in balanced, multilevel ANOVA designs and gives a reasonable approximation for more general mixed-effects models.

For example, in the fm2Machine model, $Q = 2$, $m_0 = 1$, $m_1 = 6$, $m_2 = 18$, $m_3 = 54$, $p_0 = 1$, $p_1 = 0$, $p_2 = 2$, and $p_3 = 0$, giving $denDF_1 = 5$, $denDF_2 = 10$, and $denDF_3 = 36$.

```
> anova( fm2Machine )
            numDF denDF F-value p-value
(Intercept)     1    36  773.57  <.0001
    Machine     2    10   20.58   3e-04
```

Because `Machine` is estimated at level 2, its denominator degrees of freedom is 10.

Another example is provided by the analysis of the `Oats` data, presented in §1.6, which shows an `anova` of the form

```
> anova( fm2Oats )
               numDF denDF F-value p-value
   (Intercept)     1    51  245.14  <.0001
ordered(nitro)     3    51   41.05  <.0001
       Variety     2    10    1.49  0.2724
```

In this example, $Q = 2$, $m_0 = 1$, $m_1 = 6$, $m_2 = 18$, $m_3 = 72$, $p_0 = 1$, $p_1 = 0$, $p_2 = 2$, and $p_3 = 3$, giving $denDF_1 = 5$, $denDF_2 = 10$, and $denDF_3 = 51$. The `nitro` factor changes within levels of the first-level grouping factor, `Block`, and within levels of the second-level grouping factor, `Variety %in% Block`. Thus, it is inner to each of these grouping factors and is estimated at level 3, with 51 denominator degrees of freedom. By definition, the `Variety` factor cannot change within levels of `Variety %in% Block`, but it changes within levels of `Block`. It is therefore outer to `Variety %in% Block` and inner to `Block`, being estimated at level 2 with 10 denominator degrees of freedom.

2.4.3 Confidence Intervals

Approximate confidence intervals on the variance–covariance components and the fixed effects are obtained using the approximate distributions for the maximum likelihood estimates and the REML estimates described in §2.3 and the conditional t-tests described in §2.4.2.

Letting df_j denote the denominator degrees-of-freedom for the conditional t-test corresponding to the the jth fixed effect based on the $\widehat{\beta}$, an approximate confidence interval of level $1 - \alpha$ for β_j is

$$\widehat{\beta}_j \pm t_{df_j}(1 - \alpha/2)\widehat{\sigma}_R\sqrt{\left[R_{00}^{-1}R_{00}^{-T}\right]_{jj}},$$

where $t_{df_j}(1 - \alpha/2)$ denotes the $1 - \alpha/2$ quantile of the t-distribution with df_j degrees of freedom and $\widehat{\sigma}_R$ denotes the REML estimate of σ. The matrix R_{00} is evaluated at the estimated value of θ.

Confidence intervals on the within-group standard deviation σ are obtained from the approximate distribution in (2.27). Letting $\left[\mathcal{I}^{-1}\right]_{\sigma\sigma}$ represent the last diagonal element of the inverse empirical information matrix defined in (2.27), an approximate confidence interval of level $1 - \alpha$ for σ is

$$\left[\widehat{\sigma}\exp\left(-z(1 - \alpha/2)\sqrt{\left[\mathcal{I}^{-1}\right]_{\sigma\sigma}}\right), \widehat{\sigma}\exp\left(z(1 - \alpha/2)\sqrt{\left[\mathcal{I}^{-1}\right]_{\sigma\sigma}}\right)\right],$$

where $z(1 - \alpha/2)$ denotes the $1 - \alpha/2$ quantile of the standard normal distribution. This confidence interval formulation works for both ML and REML estimates, with the obvious modifications.

Confidence intervals on the variance–covariance components for the random effects are a bit trickier to obtain. In practice, one is interested in getting confidence intervals on the original scale of the elements of $\boldsymbol{\Psi}$ and not in the scale of the unconstrained parameters $\boldsymbol{\theta}$ used in the optimization. For some simple forms of $\boldsymbol{\Psi}$, such as a diagonal structure or a multiple of the identity structure, it is easy to transform the confidence intervals obtained in the unconstrained scale (the logarithm of the standard deviations in the two examples mentioned) back to the original parameter scale (by exponentiating the confidence limits in the case of the diagonal and multiple of the identity structures).

In the case of a general positive-definite $\boldsymbol{\Psi}$, however, usually it is not possible to transform back to the original scale the confidence intervals obtained for the unconstrained parameter used in the optimization. This is true, for example, for the matrix logarithm parameterization described in §2.2.7, when the dimension of $\boldsymbol{\Psi}$ is greater than one.

The approach used in `lme` is to consider a different parameterization for general positive-definite $\boldsymbol{\Psi}$ when calculating confidence intervals. This parameterization, which we call the *natural* parameterization, uses the logarithm of the standard deviations and the generalized logits of the correlations. For a given correlation parameter $-1 < \rho < 1$, its generalized logit is $\log[(1+\rho)/(1-\rho)]$ which can take any value in the real line. We denote by $\boldsymbol{\eta}$ the parameter vector determining the natural parameterization. The elements of $\boldsymbol{\eta}$ are *individually* unconstrained, but not *jointly* so. Therefore, the natural parameterization cannot be used for optimization. However, the elements of $\boldsymbol{\eta}$ can be individually transformed into meaningful parameters in the original scale, so it is a useful parameterization for constructing confidence intervals.

If η_j corresponds to the logarithm of a standard deviation in $\boldsymbol{\Psi}$ and letting $[\boldsymbol{\mathcal{I}}^{-1}]_{jj}$ denote its associated diagonal element in the inverse empirical information matrix, an approximate level $1 - \alpha$ confidence interval for the corresponding standard deviation is

$$\left[\exp\left(\widehat{\eta}_j - z(1 - \alpha/2)\sqrt{[\boldsymbol{\mathcal{I}}^{-1}]_{jj}} \right), \exp\left(\widehat{\eta}_j + z(1 - \alpha/2)\sqrt{[\boldsymbol{\mathcal{I}}^{-1}]_{jj}} \right) \right].$$

An approximate confidence interval for a correlation coefficient represented by η_j in the natural parameter vector is

$$\left[\frac{\exp\left(\widehat{\eta}_j - z(1 - \frac{\alpha}{2})\sqrt{[\boldsymbol{\mathcal{I}}^{-1}]_{jj}} \right) - 1}{\exp\left(\widehat{\eta}_j - z(1 - \frac{\alpha}{2})\sqrt{[\boldsymbol{\mathcal{I}}^{-1}]_{jj}} \right) + 1}, \frac{\exp\left(\widehat{\eta}_j + z(1 - \frac{\alpha}{2})\sqrt{[\boldsymbol{\mathcal{I}}^{-1}]_{jj}} \right) - 1}{\exp\left(\widehat{\eta}_j + z(1 - \frac{\alpha}{2})\sqrt{[\boldsymbol{\mathcal{I}}^{-1}]_{jj}} \right) + 1} \right].$$

2.5   Fitted Values and Predictions

Fitted values, which are the predicted values for the observed responses under the fitted model, are often of interest for model checking. Predicted values for new observations are one of the primary quantities of interest when making inferences from a fitted model.

In a mixed-effects model, fitted values and predictions may be obtained at different levels of nesting, or at the population level. Population level predictions estimate the marginal expected value of the response. For example, letting x_h represent a vector of fixed effects covariates, the marginal expected value of the corresponding response y_h is

$$\text{E}\,[y_h] = x_h^T \beta. \tag{2.29}$$

Predicted values at the kth level of nesting estimate the conditional expectation of the response, given the random effects at levels $\leq k$. For example, letting $z_h(i)$ denote a vector of covariates corresponding to random effects associated with the ith group at the first level of nesting, the *level-1* predictions estimate

$$\text{E}\,[y_h(i)|b_i] = x_h^T \beta + z_h(i)^T b_i. \tag{2.30}$$

Similarly, letting $z_h(i,j)$ denote a covariate vector associated with the jth level-2 group within the ith level-1 group, the *level-2* predicted values estimate

$$\text{E}\,[y_h(i)|b_i, b_{ij}] = x_h^T \beta + z_h(i)^T b_i + z_h(i,j)^T b_{ij}. \tag{2.31}$$

This extends naturally to an arbitrary level of nesting.

The *Best Linear Unbiased Predictors (BLUPs)* of the population expected values and the conditional expectations given the random effects are obtained by replacing, in the expressions defining the expectations, β with its conditional estimate $\widehat{\beta}(\theta)$ and the random effects with their BLUPs. For example, the BLUPs corresponding to the expected values in (2.29), (2.30), and (2.31) are

$$\widehat{y}_h = x_h^T \widehat{\beta}(\theta)$$
$$\widehat{y}_h(i) = x_h^T \widehat{\beta}(\theta) + z_h(i)^T \widehat{b}_i(\theta)$$
$$\widehat{y}_h(i,j) = x_h^T \widehat{\beta}(\theta) + z_h(i)^T \widehat{b}_i(\theta) + z_h(i,j)^T \widehat{b}_{ij}(\theta).$$

In practice, the unknown parameter vector θ is replaced by its maximum likelihood estimate or its REML estimate, producing estimated BLUPs of the expected values.

2.6   Chapter Summary

This chapter presents the theory and computational methods for linear mixed-effects models. We express linear mixed-effects models in the Laird–

Ware formulation. For a single grouping factor that divides the observations into M groups of $n_i, i = 1, \ldots, M$ observations, the model is written

$$y_i = X_i\beta + Z_ib_i + \epsilon_i, \quad i = 1, \ldots, M,$$
$$b_i \sim \mathcal{N}(0, \Psi), \quad \epsilon_i \sim \mathcal{N}(0, \sigma^2 I).$$

The parameters in the model are the p-dimensional fixed effects, β, the $q \times q$ variance–covariance matrix, Ψ, for the random effects, and the variance σ^2 of the "noise" ϵ_i. We estimate these parameters by maximum likelihood (ML) or by restricted maximum likelihood (REML).

Although the random effects $b_i, i = 1, \ldots, M$ are not formally parameters in the model, we will often want to formulate our "best guess" for these values given the data. We use the Best Linear Unbiased Predictors (BLUPs) for this.

For computational purposes the variance–covariance matrix Ψ is re-expressed in terms of the relative precision factor Δ which satisfies

$$\Delta^T\Delta = \Psi/\sigma^2$$

and the matrix Δ is expressed as a function of an unconstrained parameter vector θ.

The *profiled* log-likelihood function with respect to θ can be easily calculated using matrix decompositions. That is, the log-likelihood corresponding to the conditionally best estimates $\widehat{\beta}(\theta)$ and $\widehat{\sigma}^2(\theta)$ can be evaluated as a function of θ alone. This simplifies the problem of optimizing the likelihood to get maximum likelihood estimates because it reduces the dimension of the optimization. The same simplification applies to REML estimation.

We describe approximate distributions for the maximum likelihood estimates and the REML estimates using results from asymptotic theory for linear mixed-effects models.

We compare models that differ in the random effects specification by likelihood ratio tests or by simulation-based parametric bootstrap evaluations.

We assess the significance of terms in the fixed-effects specification by standard linear regression tests conditional on the value of $\widehat{\theta}$. These tests include t-tests for individual coefficients or F-tests for more complicated terms or linear combinations of coefficients. The degrees of freedom for a t-test (or the denominator degrees of freedom for an F-test) depend on whether the factor being considered is inner to the grouping factor (changes within levels of the grouping factor) or outer to the grouping factor (is invariant within levels of the grouping factor).

Approximate confidence intervals for the fixed effects and the variance–covariance parameters are produced from the approximate distributions of the maximum likelihood estimates and REML estimates.

All these results extend to multiple nested levels of random effects. A model with two levels of nested random effects, for example, is written

$$y_{ij} = X_{ij}\beta + Z_{i,j}b_i + Z_{ij}b_{ij} + \epsilon_{ij}, \quad i = 1, \ldots, M, \quad j = 1, \ldots, M_i,$$
$$b_i \sim \mathcal{N}(0, \Psi_1), \quad b_{ij} \sim \mathcal{N}(0, \Psi_2), \quad \epsilon_{ij} \sim \mathcal{N}(0, \sigma^2 I).$$

The two variance–covariance matrices Ψ_1 and Ψ_2 are written in terms of relative precision factors Δ_1 and Δ_2, parameterized by unconstrained parameter vectors θ_1 and θ_2. The profiled log-likelihood or the profiled log-restricted-likelihood, a function of θ_1 and θ_2 only, is maximized to produce the estimates $\widehat{\beta}$, $\widehat{\sigma}^2$, $\widehat{\Psi}_1$, and $\widehat{\Psi}_2$.

Exercises

1. The simulation results presented in Figure 2.4 (p. 87) indicate that the null distribution of the REML likelihood ratio test statistic comparing a null model with a single level of scalar random effects to an alternative model with nested levels of scalar random effects is approximately an equally weighted mixture of a χ^2_0 and a χ^2_1.

   Confirm this result by simulating a LRT statistic on the Oats data, considered in §1.6. The preferred model for those data, fm4Oats, was defined with random = ~Block/Variety. Re-fit this model with random = ~Block. Using this fit as the null model and fm4Oats as the alternative model, obtain a set of simulated LRT statistics with simulate.lme. Plot these simulated LRT statistics setting df = c(0,1) to obtain a plot like Figure 2.4. Are the conclusions from this simulation similar to those from the simulation shown in Figure 2.4?

   Note that simulate.lme must fit both models to nsim simulated sets of data. By default nsim = 1000, which could tie up your computer for a long time. You may wish to set a lower value of nsim if the default number of simulations will take too long.

3
Describing the Structure of Grouped Data

As illustrated by the examples in Chapter 1, we will be modeling data from experiments or studies in which the observations are grouped according to one or more nested classifications. Often this classification is by "Subject" or some similar experimental unit. *Repeated measures* data, *longitudinal* data, and *growth curve* data are examples of this general class of grouped data.

A common and versatile way of organizing data in S is as data.frame objects. These are in the form of tables where each row corresponds to an observation and each column corresponds to one of the variables being observed. We extend the data.frame class to the class of groupedData objects, which are data frames with additional information about the grouping of the observations and, possibly, other special roles of some variables.

In this chapter we describe creating, summarizing and displaying grouped-Data objects with a single level of grouping or with multiple levels of grouping.

3.1 The Display Formula and Its Components

A groupedData object contains the data values themselves, stored as a data frame, and a formula that designates special roles for some of the variables in the data frame. The most important of the special roles is that of a *grouping factor* that divides the observations into the distinct groups of observations. The formula also designates a *response* and, when available, a *primary covariate*. It is given as

```
response ~ primary | grouping
```

where `response` is an expression for the response, `primary` is an expression for the primary covariate, and `grouping` is an expression for the grouping factor. Most often these expressions are simply the name of a variable in the data frame, but they could also be functions of one or more variables. For example, `log(conc)` would be a legitimate expression for the response if `conc` is one of the variables in the data frame.

The `formula` function extracts the formula from a grouped data object. Applied to some of the data sets used in Chapter 1 it produces

```
> formula( Rail )
travel ~ 1 | Rail
> formula( ergoStool )
effort ~ Type | Subject
> formula( Machines )
score ~ Machine | Worker
> formula( Orthodont )
distance ~ age | Subject
> formula( Pixel )
pixel ~ day | Dog/Side
> formula( Oats )
yield ~ nitro | Block
```

Notice that there is not primary covariate in the `Rail` data so we use the constant expression `1` in that position in the formula. In data with multiple, nested grouping factors, such as the `Pixel` data, the grouping factors are separated by "/". Factors that are nested within other factors appear further to the right so an expression like `Dog/Side` indicates that `Side` is nested within `Dog`.

The formula of a grouped data object has the same pattern as the formula used in a call to a trellis graphics function, such as `xyplot`. This is intentional. Because such a formula is available with the data, the `plot` method for objects in the `groupedData` class can produce an informative trellis display from the object alone. It may, in fact, be best to think of the formula stored with the data as a *display formula* for the data because it provides a meaningful default graphical display method for the data.

The `formula` function shown above is an example of an *extractor* function for this class. It returns some property of the object—the display formula in this case—without requiring the user to be aware of how that property is stored. We provide other extractor functions for each of the components of the display formula. The `getGroups` extractor returns the value of the grouping factor. A companion function, `getGroupsFormula`, returns the formula that is evaluated to produce the grouping factor. The extractors for the other components of the display formula are `getResponse` and `getCovariate`,

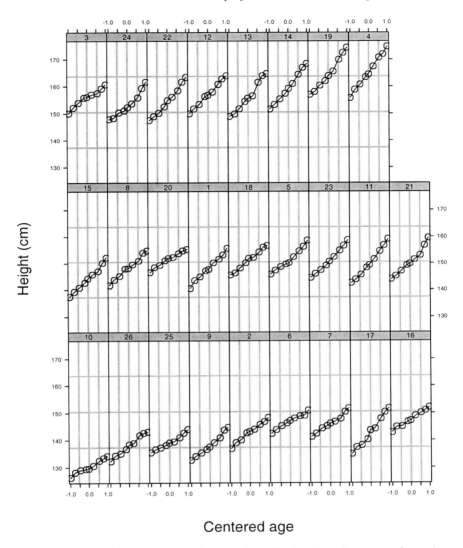

FIGURE 3.1. Heights of 26 boys from Oxford, England, each measured on nine occasions. The ages have been centered and are in an arbitrary unit.

which return numeric vectors or factors, and `getResponseFormula` and `getCovariateFormula`, which return formulas.

It is safer to use these extractor functions instead of checking the display formula for the object and extracting a variable from the object. For example, suppose we wish to check for balance in the `Oxboys` data, shown in Figure 3.1, and consisting of the heights at different ages of several boys from Oxford, England (see Appendix A.19 for more detail). We could use the `table` function on the grouping factor,

```
> table( Oxboys$Subject )
 10 26 25 9 2 6 7 17 16 15 8 20 1 18 5 23 11 21 3 24 22 12 13
  9  9  9 9 9 9 9  9  9  9 9  9 9  9 9  9  9  9 9  9  9  9  9
 14 19 4
  9  9 9
```

to check if each boy's height is recorded the same number of times. To do this, we must know that the grouping factor for the `Oxboys` data is named `Subject`. A form that is easier to remember, and also more general, is

```
> table( getGroups( Oxboys ) )
 10 26 25 9 2 6 7 17 16 15 8 20 1 18 5 23 11 21 3 24 22 12 13
  9  9  9 9 9 9 9  9  9  9 9  9 9  9 9  9  9  9 9  9  9  9  9
 14 19 4
  9  9 9
> unique( table( getGroups( Oxboys ) ) )   # a more concise result
[1] 9
```

Because there are exactly nine observations for each subject, the data are balanced with respect to the number of observations. However, if we also check for balance in the covariate values, we find they are not balanced.

```
> unique( table( getCovariate( Oxboys ), getGroups( Oxboys ) ) )
[1] 1 0
> length( unique( getCovariate( Oxboys ) ) )
[1] 16
```

Further checking reveals that there are 16 unique values of the covariate `age`. The boys are measured at approximately the same ages, but not exactly the same ages. This imbalance could affect some analysis methods for repeated measures data. It does not affect the methods described in this book.

The `isBalanced` function in the nlme library can be used to check a groupedData object for balance with respect to the grouping factor(s) or with respect to the groups and the covariate. It is built from calls to `getGroups` and `table` like those above.

When applied to data with multiple, nested grouping factors, the `get-Groups` extractor takes an optional argument `level`. Levels are counted from the outside inward so, in the `Pixel` data where the grouping is `Dog/Side`, `Dog` is the first level and `Side` is the second level. When the argument `level` specifies a single level the result is returned as a vector

```
> unique( getGroups(Pixel, level = 1) )
 [1] 1  2  3  4  5  6  7  8  9  10
> unique( getGroups(Pixel, level = 2) )
 [1] 1/R  2/R  3/R  4/R  5/R  6/R  7/R  8/R  9/R  10/R 1/L  ...

 1/R < 2/R < 3/R < 4/R < 5/R < ...
```

Notice that the groups at `level = 2`, the "`Side` within `Dog`" factor, are coerced to an ordered factor with distinct levels for each combination of `Side` within `Dog`.

If we extract the groups for multiple levels the result is returned as a data frame with one column for each level. Any inner grouping factors are preserved in their original form in this frame rather than being coerced to an ordered factor with distinct levels as above. For example,

```
> Pixel.groups <- getGroups( Pixel, level = 1:2 )
> class( Pixel.groups )
[1] "data.frame"
> names( Pixel.groups )
[1] "Dog"    "Side"
> unique( Pixel.groups[["Side"]] )
[1] R L
```

In a call to a linear mixed-effects modeling function, `lme` or `lmList`, or to a nonlinear mixed-effects modeling function, `nlsList` or `nlme` (discussed in Chapter 8), the default values for the response, for a covariate, and for the grouping factor are obtained from the formula stored with the data. These are, however, only the default values. They can be overridden with an explicit model formula. For example, during the course of model building we may wish to change our idea of what constitutes the response, say by transforming from a measure of concentration to the logarithm of the concentration. It is not necessary to change the formula stored with the data when doing this. We can use an explicit model formula as an argument to the model fitting function and override the formula stored with the data.

If we do decide to make a permanent change in the formula of a grouped-Data object, we can use the `update` function to do this. For example, we could change the covariate in the PBG data (discussed in §3.2.1 and Appendix A.21) from `dose` to `log(dose)` by updating the object.

```
> formula( PBG )
deltaBP ~ dose | Rabbit
> PBG.log <- update( PBG, formula = deltaBP ~ log(dose) | Rabbit )
> formula(PBG.log)
deltaBP ~ log(dose) | Rabbit
> unique( getCovariate(PBG.log) )
[1] 1.8326 2.5257 3.2189 3.9120 4.6052 5.2983
> unique( getCovariate(PBG) )
[1]   6.25  12.50  25.00  50.00 100.00 200.00
```

3.2 Constructing groupedData Objects

Constructing a groupedData object requires the data to be available as a data frame. There are several ways that data can be imported into S and formed into a data frame. One of the simplest ways is using the `read.table` function on data stored in an external file (Venables and Ripley, 1999, §2.4).

For example, if the Oxford boys' height data are stored in a text file named `oxboys.dat` of the form

```
Subject     age height
      1 -1.0000 140.50
      1 -0.7479 143.40
      1 -0.4630 144.80
      1 -0.1643 147.10
      1 -0.0027 147.70
      1  0.2466 150.20
      1  0.5562 151.70
      1  0.7781 153.30
      1  0.9945 155.80
      2 -1.0000 136.90
         . . .
     26 -0.0027 138.40
     26  0.2466 138.90
     26  0.5562 141.80
     26  0.7781 142.60
     26  1.0055 143.10
```

we can create a data frame with

```
> Oxboys.frm <- read.table( "oxboys.dat", header = TRUE )
> class( Oxboys.frm )        # check the class of the result
[1] "data.frame"
> dim( Oxboys.frm )          # check the dimensions
[1] 234 3
```

The argument `header = TRUE` in the call to `read.table` indicates that the first line of the file is to be used to create the names for the variables in the frame.

The result of `read.table` is of the data.frame class. It has two dimensions: the number of rows (cases) and the number of columns (variables).

A function to create objects of a given class is called a *constructor* for that class. The primary constructor function for a class is often given the same name as the class itself. Thus the default constructor for the groupedData class is the `groupedData` function. Its required arguments are a formula and a data frame. Optional arguments include `labels`, where display labels for the response and the primary covariate can be given, and `units`, where the units of these variables can be given. The default axis labels for data plots are constructed by pasting together components of `labels` and `units`. The reason for separating the units from the rest of the display label is to permit propagation of the units to derived quantities such as the residuals from a fitted model.

For example, reading the Oxboys data from a file, converting it to a groupedData object, and establishing default labels could be accomplished in a single call of

```
> Oxboys <- groupedData( height ~ age | Subject,
+    data = read.table("oxboys.dat", header = TRUE),
+    labels = list(x = "Centered age", y = "Height"),
+    units = list(y = "(cm)") )
> Oxboys                     # display the object
Grouped Data: height ~ age | Subject
    Subject      age height
  1        1 -1.0000 140.50
  2        1 -0.7479 143.40
  3        1 -0.4630 144.80
...
234       26  1.0055 143.10
```

By default the **groupedData** constructor also converts the grouping factor (the factor Subject in the Oxboys data) to an ordered factor. The order is determined by applying a summary function to the response within each group. The default summary function is max, the maximum.

When the grouping factor has been converted to an ordered factor, the panels in the trellis plots are arranged in that order. The order of the panels is from left to right across the rows, starting with the bottom row. In the default ordering described above the maximum value of the response in each panel will increase across the rows starting from the lower left panel. Most of the data plots in this book are ordered in this way.

Conversion of the grouping factor to an ordered factor is done only if the expression for the grouping factor is simply the name of a variable in the data frame. The conversion does not cause the rows of the data frame themselves to be reordered; it merely changes the class of the grouping factor and attaches the ordering to it. One way to examine the ordering is by requesting the unique values of the grouping factor

```
> unique( getGroups( Oxboys ) )
 [1]  1  2  3  4  5  6  7  8  9 10 11 12 13 14 15 16 17 18 19 20
[21] 21 22 23 24 25 26

10 < 26 < 25 < 9 < 2 < 6 < 7 < 17 < 16 < 15 < 8 < 20 < 1 < 18 ...
```

The first value of Subject in the data is 1, but the value of Subject with the smallest maximum height is 10 so this subject's data occupy the lower left panel in Figure 3.1.

The **labels** and **units** arguments are optional. We recommend using them because this makes it easier to create more informative trellis plots.

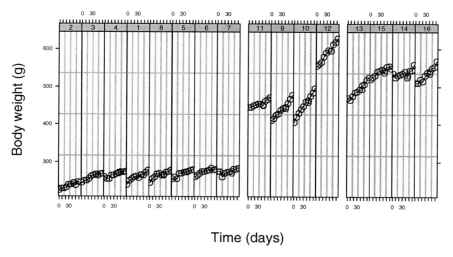

Time (days)

FIGURE 3.2. Weight versus time of rats on three different diets. The first group of eight rats is on the control diet. There were four rats each in the two experimental diets.

3.2.1 Roles of Other Experimental or Blocking Factors

Although the display formula can be used to designate a response, a co-variate, and a grouping factor, these may not be sufficient to describe the structure of the experiment or study completely. Many experiments impose additional structure on the data. For example, the observations in the orthodontic data, described in §1.4.1, are grouped according to the subject on which the measurements were made. In most analyses of this type of data we would want to include the subject's sex as an explanatory factor. Because Sex is a characteristic of the subject, it is invariant within the observations for a single subject. A factor that is invariant within the groups determined by the grouping factor is said to be *outer* to the grouping factor.

When outer factors are present they can, and should, be designated as such when constructing a groupedData object using; for example,

```
outer = ~ Sex
```

Multiple outer factors can be specified by separating their names with "*" in the formula.

Outer factors can be characteristics of the subject (more generally, of the "experimental unit") such as Sex in this example. They can also be experimental factors that are applied to this level of experimental unit. For example, the BodyWeight data, shown in Figure 3.2 and described in Appendix A.3, contain measurements of the weights of 16 rats over time. Eight of the rats were given a control diet, four rats were given one exper-

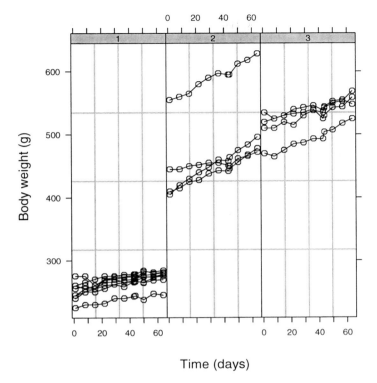

FIGURE 3.3. Weight of rats versus time for three different diets.

imental diet, and four rats were given another experimental diet. The Diet factor is an experimental factor that is outer to the grouping factor Rat.

One benefit of specifying outer factors to the constructor is that the constructor will modify the way in which the groups are ordered. Reordering of the groups is permitted only within the same level of an outer factor (or within the same combination of levels of outer factors, when there is more than one). This ensures that groups at the same levels of all outer factors will be plotted in a set of adjacent panels.

The plot method for the groupedData class allows an optional argument outer that can be either a logical value or a formula. When this argument is used the panels are determined by the factor or combination of factors given in the outer formula. A logical value of TRUE or T can be used instead of a formula, indicating that the outer formula stored with the data should be used to determine the panels in the plot. For example, we can get a stronger visual comparison of the differences between the diets for the BodyWeight data with

```
> plot( BodyWeight, outer = ~ Diet, aspect = 3 )        # Figure 3.3
```

The `aspect` argument, discussed in §3.3.1, is used to enhance the visualization of patterns in the plot. Because this `outer` formula was stored with the `BodyWeight` data we would produce the same plot with

```
> plot( BodyWeight, outer = TRUE, aspect = 3 )
```

In plots grouped by the levels of an `outer` formula, the original grouping factor stored with the data determines which points are associated with each other in the panels. Points in the same group are joined by lines when each panel is a scatter plot of the response versus a continuous covariate. If there is no primary covariate the data will be plotted as a dot plot where points in the same group will be rendered with the same symbol and color.

When there is more than one outer factor in the data the arrangement of the panels depends on the order in which the factors are listed in the `outer` formula for the plot. For example, in Chapter 7 we show several plots of the results of an experiment in which the weights of soybean plants, grown in different experimental plots, were measured several times during the growing season. There were two different varieties of soybeans in the experiment, which was carried out over three consecutive growing seasons. The grouping factor is `Plot` and the outer factors are `Variety` and `Year`. The plot produced by

```
> plot( Soybean, outer = ~ Year * Variety )     # Fig 6.10 (p. 288)
```

arranges panels of the same `Variety` on the same row, making it easy to compare the results for each variety across years. Conversely, the plot produced by

```
> plot( Soybean, outer = ~ Variety * Year )
```

(not shown) arranges panels of the same `Year` on the same row, making it easy to compare varieties within each year.

When specifying outer factors in the constructor or in a plot call we should ensure that they are indeed constant or *invariant* within each level of the grouping factor. The `gsummary` function with the optional argument `invariantsOnly = TRUE` allows us to check this. It returns the values of only those variables that are invariant within each level of the grouping factor. These values are returned as a data frame with one row for each level of the grouping factor. For the `BodyWeight` data we get

```
> gsummary( BodyWeight, invar = TRUE )
  Rat Diet
2   2    1
3   3    1
4   4    1
1   1    1
8   8    1
5   5    1
```

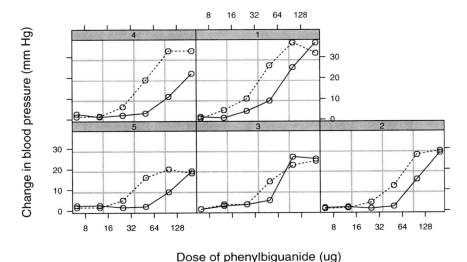

Dose of phenylbiguanide (ug)

FIGURE 3.4. Change in blood pressure versus dose of phenylbiguanide (PBG) for five rabbits. Each rabbit was exposed to increasing doses of PBG, once after treatment with the H_5^3-agonist MDL 72222 and once after treatment with a placebo.

6	6	1
7	7	1
11	11	2
9	9	2
10	10	2
12	12	2
13	13	3
15	15	3
14	14	3
16	16	3

indicating that Diet is an invariant of the grouping structure determined by the Rat factor. Notice that the grouping factor itself, Rat in this case, must be one of the invariants.

An outer factor is a characteristic of the experimental unit or an indicator of a treatment applied to the entire unit. In some experiments one level of a treatment may be applied to the experimental unit, say Subject, for some of the observations, then another level applied for other observations. If such an inner factor is distinct from the primary covariate, we may want to indicate it somehow on a data plot. An example of an inner factor is provided by the phenylbiguanide (PBG) data, shown in Figure 3.4 and described in Appendix A.21. These data are from a *cross-over* trial where each experimental animal was exposed to increasing doses of PBG under two treatments; once with the MD_5-antagonist MDL 72222 and once with a placebo. The change in blood pressure was measured for each dose on each occasion.

In Figure 3.4, produced by

```
> plot( PBG, inner = ~ Treatment, scales = list(x = list(log = 2)))
```

the lines in each panel joint points with the same Treatment. (The scales argument to the plot call will be described in §3.3.) This plot provides a strong indication that the effect of the PBG treatment is to shift the dose–response curve to the right. We will discuss methods for modeling this in Chapter 7.

The PBG data is similar in structure to the Pixel data. In both these data sets there is a continuous response, a continuous covariate (day for Pixel and dose for PBG), a major grouping factor corresponding to the experimental animal (Dog for Pixel and Rabbit for PBG) and a factor that varies within this major grouping factor (Side for Pixel and Treatment for PBG). In the case of the Pixel data we used nested grouping factors to represent this structure. In the case of the PBG data we used a single grouping factor with an inner treatment factor. These two structures are quite similar—in fact, for the purposes of plotting the data, they are essentially equivalent. The choice of one structure or the other is more an indication of how we think the inner factor should be modeled. For the Pixel data we modeled the effect of the Side factor as a random effect because the Side just represents a random selection of lymph node for each Dog. In other experiments there may be important physiological differences between the left side and the right side of the animal so we would model it as a fixed effect. In the PBG data the Treatment factor is a factor with fixed, repeatable levels, and we model it as a fixed effect.

If we decide that an inner factor should be modeled as a random effect we should specify it as part of a nested grouping structure. If it should be a fixed effect we specify it as an inner factor. These choices can be overridden when constructing models.

3.2.2 Constructors for Balanced Data

The ergoStool data is an example of a data set that is balanced both with respect to the grouping factor and with respect to the primary covariate. That is, each Subject has the same number of observations on each Subject and each subject uses each stool type the same number of times.

It can be convenient to represent a balanced, unreplicated set of responses as a matrix. For example,

```
> ergoStool.mat <- asTable( ergoStool )
```

```
> ergoStool.mat
   T1 T2 T3 T4
 8  7 11  8  7
 5  8 11  8  7
 4  7 11 10  9
 9  9 13 10  8
 6  9 11 11 10
 3  7 14 13  9
 7  8 12 12 11
 1 12 15 12 10
 2 10 14 13 12
```

The asTable function, which can only be used with balanced and unreplicated data, produces a table of the responses in the form of a matrix where columns correspond to the unique values of the primary covariate and rows correspond to groups. The dimnames of the matrix are the unique levels of the grouping factor and the covariate.

This table provides a compact representation of balanced data. Often the data from a balanced experiment are provided in the form of a table like this. The balancedGrouped function convert data from a table like this to a groupedData object.

```
> ergoStool.new <- balancedGrouped( effort ~ Type | Subject,
+                                    data = ergoStool.mat )
Warning messages:
  4 missing values generated coercing from character to numeric
         in: as.numeric(dn[[1]])
> ergoStool.new
Grouped Data: effort ~ Type | Subject
    Type Subject effort
 1   T1       8      7
 2   T2       8     11
 3   T3       8      8
 4   T4       8      7
 5   T1       5      8
 ...
36   T4       2     12
```

The formula given as the first argument to balancedGrouped is used to assign names to the response, the primary covariate, and the grouping factor. The data values are extracted from the matrix itself and from its dimnames attribute. The matrix should be arranged so each row contains the data from one group. The covariate values, corresponding to the different columns, can be the levels of a factor or of a continous covariate. If the column names can all be converted to numeric values, the covariate is assumed to be continuous and is coerced to a numeric vector. Otherwise it is left as a factor. The process of checking for numeric values in these names is what generates the warning message in the previous example.

It is a common practice to label the levels of a factor like the Type factor as 1, 2, . . . , which would result in its being coerced to a numeric variable. Unless this is detected and the numeric variable is explicitly converted to a factor, models fit to such data will be nonsensical. It is always a good idea to check that the variables in a groupedData object have the expected classes. We describe how to do this in §3.4.

The optional labels and units arguments can be used with balanced-Grouped just as in the groupedData constructor.

As seen in the example, the balancedGrouped constructor produces an object like any other groupedData object. The matrix of response values is converted to a vector and both the primary covariate and the grouping factor are expanded to have the same length as the response vector. Later manipulations of the object or plots created from the object do not rely on its having been generated from balanced data. This is intentional. Although there is a certain amount of redundancy in storing multiple copies of the same covariate or grouping factor values, it is offset by the flexibility of the data.frame structure in dealing with missing data and with general, unbalanced data.

Many methods for the analysis of longitudinal data or repeated measurements data depend on having balanced data. The analysis techniques described in this book do not.

3.3   Controlling Trellis Graphics Presentations of Grouped Data

Trellis graphics presentations of grouped data allow easy evaluation of the behavior of the response with respect to the primary covariate within each group. They also allow comparisons between groups. Because they are so effective at illustrating both within-group and between-group behavior, they are the default plot method for groupedData objects.

The defaults chosen in the trellis graphics library and in the plot method for groupedData objects will usually provide an informative and visually appealing plot. Sometimes, however, the default plot can be made even more informative by adjusting one or two of the trellis graphics parameters. In this section we describe some of the trellis parameters that are helpful in enhancing plots of grouped data. For a full discussion of the trellis graphics parameters and controls see Becker, Cleveland and Shyu (1996) or the on-line documentation for the trellis library.

3.3.1   Layout of the Trellis Plot

A trellis plot consists of one or more *panels* arranged in a regular array on one or more pages. In the default data plot for a groupedData object with a continuous primary covariate, there is one panel for each level of the grouping factor. The horizonal axis in the panel is the primary covariate, the vertical axis is the response, and the data are represented both as points and by a connecting line. If the primary covariate is a factor, such as in the Machine data, or if there is no primary covariate, such as in the Rail data, the plot is a *dotplot* with one row for each level of the grouping factor. In this case the response is on the horizontal axis.

For numeric covariates the *aspect ratio* of each panel, which is the ratio of the physical size of the vertical axis to that of the horizontal axis, is determined by the 45-degree banking rule described in Cleveland (1994, §3.1). We have found that this rule produces appealing and informative aspect ratios in a wide variety of cases. If you wish to override this choice of aspect ratio, you can give an numerical value as the optional aspect argument in the call to plot. A value greater than 1.0 produces tall, narrow panels, while a value between 0.0 and 1.0 produces short, wide panels.

The arrangement of panels in rows and columns on the page is calculated to make the specified number of panels of the chosen aspect ratio fill as much as possible of the available area. This does not always create a good arrangement for comparing patterns across outer factors. For example, the grouping factor Plant in the CO2 data, described in Appendix A.5, has twelve different levels. The plants themselves come from one of two types and have been subjected to one of two treatments. Because the aspect ratio chosen by the banking rule creates panels that are taller than they are wide, a more-or-less square plot area will be filled with three rows of four panels each (Figure 3.5).

For some combinations of grass type and treatment the panels are spread across more than one row in Figure 3.5. It would be better to keep these combinations on the same row so we can more easily compare treatments and grass types. That is, we would prefer the panels to be arranged in four rows of three or, perhaps, two rows of six. If we have four rows of three, we may wish to indicate visually that the lower two rows represent one type of plant (Québec) and the upper two rows represent the other type (Mississippi). We can do this by specifying a list as the optional between argument. A component named x in this list indicates the sizes of gaps to leave between columns while a component named y specifies gaps between rows. (When forming a between argument for the rows, remember that the trellis convention is to count from the bottom to the top, not from the top to the bottom.) The gaps are given in units of character heights. Generally a gap of 0.5 is sufficient to distinguish groups without using too much space.

An arrangement of the CO2 data in two rows of six panels with a gap between the third and fourth columns, shown in Figure 3.6, is produced by

```
> plot(CO2, layout=c(6,2), between=list(x=c(0,0,0.5,0,0))) # Fig 3.6
```

Assembling the panels on a single page is effective when there is a small or moderate number of groups. If there is a large number of groups, the

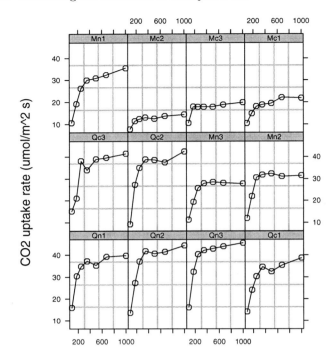

FIGURE 3.5. Carbon dioxide uptake versus ambient CO_2 concentration for *Echinochloa crus-galli* plants, six from Québec and six from Mississippi. Half the plants of each type were chilled overnight before the measurements were taken. The labels of each panel show the origin of the plant (Québec or Mississippi) and the treatment (chilled or nonchilled). This plot shows a default layout of the panels.

single page plot may result in the panels being too small to be informative. In these cases a third component can be added to the `layout` argument causing the plot to be spread over several pages.

   If the number of groups in some level of an outer factor does not fit exactly into the rectangular array, an optional `skip` argument can be used to skip over selected panel locations.

   The use of both of these arguments is illustrated in the code for the figures of the spruce tree growth data (Appendix A.28). There were four groves of trees; two exposed to an ozone-rich atmosphere, and two exposed to a normal atmosphere. In the first and second groves 27 trees were measured, but in the third and fourth groves only 12 and 13 trees were measured, respectively.

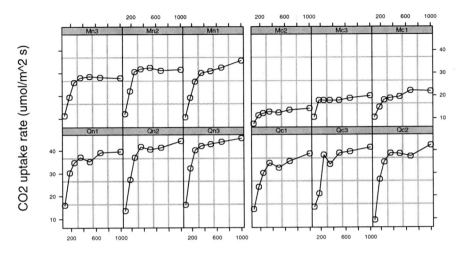

FIGURE 3.6. Carbon dioxide uptake versus ambient CO_2 concentration for *Echinochloa crus-galli* plants. This plot shows an alternative layout of the panels.

The three pages of figures (Figures A.8–A.10, pages 445–447) in an array of four rows of seven columns per page were created with

```
> plot( Spruce, layout = c(7, 4, 3),
+          skip = c(rep(FALSE, 27), TRUE, rep(FALSE, 27), TRUE,
+                   rep(FALSE, 12), rep(TRUE, 2), rep(FALSE,13)) )
```

An alternative arrangement (not shown) of three pages in an array of three rows of nine columns per page can be created with

```
> plot( Spruce, layout = c(9, 3, 3),
+          skip = c(rep(FALSE, 66), TRUE, TRUE, rep(FALSE, 13)) )
```

On the first two pages of this plot the array would be filled. On the third page there would be a gap in the middle of the array to separate the panels corresponding to the two different groves of trees exposed to a normal atmosphere.

3.3.2 Modifying the Vertical and Horizontal Scales

It is often helpful to present quantities such as concentrations on a logarithmic scale. The optional `scales` argument for trellis graphics functions allows specification of logarithmic scales on either the vertical or horizontal axes. The axis annotations can be at powers of 10 or at powers of 2.

For example, the default plot of the DNase assay data shown in Figure 3.7 squeezes most of the data onto the left-hand side of the panel. (These data are described in more detail in Appendix A.7.)

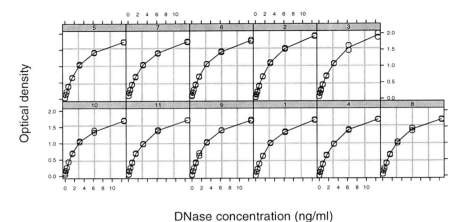

DNase concentration (ng/ml)

FIGURE 3.7. Optical density versus DNase concentration, arithmetic scale

Both the facts that the unique values of the DNase concentration are, for the most part, logarithmically spaced

```
> unique( getCovariate(DNase) )
[1]   0.048828   0.195312   0.390625   0.781250   1.562500   3.125000
[7]   6.250000 12.500000
> log( unique(getCovariate(DNase)), 2 )
[1] -4.35614 -2.35614 -1.35614 -0.35614   0.64386   1.64386   2.64386
[8]   3.64386
```

and the experimenters' desire to fit a logistic response function (described in Chapter 6 and Appendix C.7) to the logarithm of the concentration indicate that we should use a logarithmic scale on the horizontal axis. This change is incorporated in Figure 3.8, produced with

```
> plot( DNase, layout=c(6,2), scales = list(x=list(log=2)) )
```

3.3.3   Modifying the Panel Function

This is an advanced topic. You can consider skipping this section unless you want to modify the way the data are presented within each panel.

The presentation of the data within each panel is controlled by a *panel function*. If no primary covariate is available, or if the primary covariate is a categorical variable, the default panel function for the plot method for groupedData objects is `panel.dotplot`. When the primary covariate is numeric, the default panel function is

```
function(x, y) {
  panel.grid()
  panel.xyplot(x, y)
```

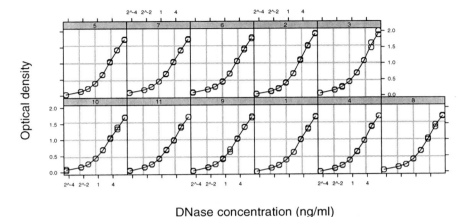

FIGURE 3.8. Optical density versus DNase concentration for eleven runs of an assay. The concentration is shown on a logarithmic scale.

```
y.avg <- tapply(y, x, mean)        # average y for each distinct x
xvals <- as.numeric(names(y.avg))
ord <- order(xvals)
panel.xyplot(xvals[ord], y.avg[ord], type = "l")
}
```

The first two lines of this function draw the background grid and place symbols at the data values. The actual symbol that is drawn is determined by the trellis device that is active when the plot is displayed. It is usually an open circle.

The last four lines of the panel function add a line through the data values. Some care needs to be taken when doing this. In the DNase assay data, for example, there are duplicate observations at each concentration in each run. Rather than "joining the dots," it makes more sense to draw the line through the average response in each set of replicates. In the default panel function `xvals` is defined to be the unique values of x and `y.avg` is calculated as the average of the y values at these distinct x values. Finally, the `xvals` vector is put into increasing order and the line drawn with the points in this order.

This panel function can be overridden with an explicit `panel` argument to the plot call if, for example, you want to omit the background grid. If you do override the default panel function, it would be a good idea to follow the general pattern of this function. In particular, you should draw the grid first (if you choose to do so) then add any points or lines. Also, be careful to handle replicates or unusual ordering of the (x, y) pairs gracefully.

3.3.4   Plots of Multiply-Nested Data

The **plot** method for multiply-nested **grouped**Data objects takes an optional argument **displayLevel** that defines the *display* level to be used. By default, the innermost level of grouping is used as the display level. For the **Pixel** data, this is "**Side** within **Dog**", as shown in Figure 3.9.

```
> plot( Pixel, layout = c(4,5),                    # Figure 3.9
+          between = list(x = c(0,  0.5), y = 0.5) )
```

A more meaningful trellis display of these data, using **Dog** as the display level, is obtained with

```
> plot( Pixel, displayLevel = 1 )                  # Figure 3.10
```

When the display level for a multiply-nested **grouped**Data object is smaller than the maximum grouping level, different actions can be taken with respect to the inner grouping levels. The default action is to preserve, as much as possible, the original structure of the data and use the combination of the inner grouping factors as a single inner factor, as in Figure 3.10. Alternatively, observations within a given value of the display level may be collapsed over some, or all inner levels by giving a **collapse** argument in the **plot** call. Another optional argument, **FUN**, can specify the summary function to be use when collapsing the data. This summary function should take a numeric vector as its argument and return a single numeric value. The default is the **mean** function.

Using **collapse** can help to reduce clutter in a plot. For example, the **Wafer** data, from an experiment in semiconductor manufacturing, give current intensity versus the applied voltage at eight sites on each of ten wafers (see Appendix A.30 for details). Differences between wafers and between sites within wafers are too subtle to be noticeable on a plot at **displayLevel** = 2 that has separate panels for each **Site** within each **Wafer**. Displaying at the **Wafer** level with separate curves for each **Site** within each **Wafer**, as in Figure 3.11, produces a cluttered plot because the eight curves for each wafer nearly overlay each other. It is more informative to examine the mean curve for each wafer and the standard deviation about this mean curve, as in Figures 3.12 and 3.13

```
> plot( Wafer, display = 1, collapse = 1 )          # Fig. 3.12
> plot( Wafer, display = 1, collapse = 1,           # Fig. 3.13
+          FUN = function(x) sqrt(var(x)), layout = c(10,1) )
```

In general there is a trend for the standard deviation about the curve to be greater when the response is greater. This is a common occurrence. A less obvious effect is that the standard deviation is greater in the wafers with overall higher current intensity, even though the differences in the current intensity are small.

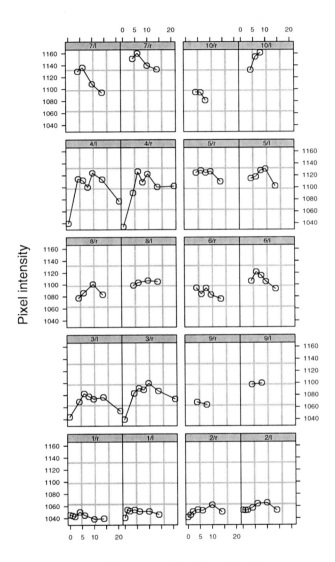

FIGURE 3.9. Mean pixel intensity of lymph nodes in the axillary region versus time by Side within Dog.

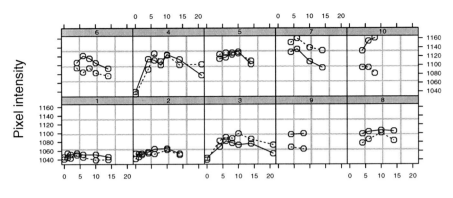

Time post injection (days)

FIGURE 3.10. Mean pixel intensity of lymph nodes in the axillary region versus time by Dog. The two curves on each plot correspond to the left and the right sides of the Dog.

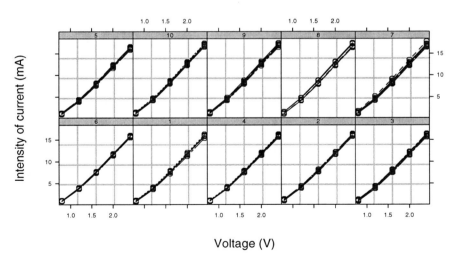

Voltage (V)

FIGURE 3.11. Current versus voltage for the Wafer data. The panels correspond to wafers. With each wafer the current was measured at eight different sites.

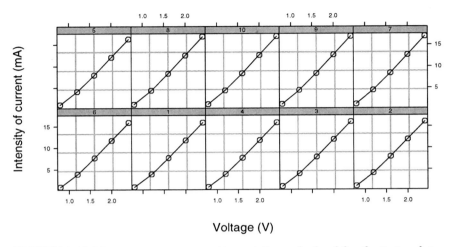

FIGURE 3.12. Mean current versus voltage at the wafer level for the `Wafer` data. Each point on each curve is the average of the current for that voltage at eight sites on the wafer.

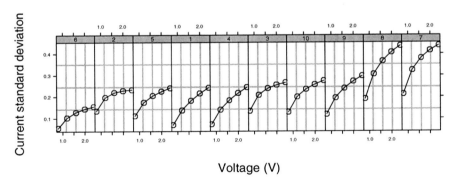

FIGURE 3.13. Standard deviation of current versus voltage at the wafer level for the `Wafer` data. Each point on each curve is the standard deviation of the current at that voltage at eight sites on the wafer.

3.4 Summaries

In addition to creating graphical presentations of the data we may wish to summarize the data numerically, either by group or across groups. In §3.1 we demonstated the use of the `table`, `gsummary`, and `unique` functions to summarize data by group. In this section we expand on the usage of these functions and introduce another groupwise summary function `gapply`.

The `gapply` function is part of the nlme library. It joins several standard S functions in the "apply family." These include `apply`, `lapply`, `tapply`, and `sapply`. They all apply a function to subsections of some data structure and gather the results in some way. Both `lapply` and `sapply` can be used to apply a function to the components of a list. The result of `lapply` is always a list; `sapply` will create a more compact result if possible. Because the groupedData class inherits from the data.frame class and a data frame can be treated as a list whose components are the columns of the frame, we can apply a function to the columns of a groupedData object. For example, we can use `sapply` to check the data.class of the columns of a groupedData object by

```
> sapply( ergoStool, data.class )
    effort      Type    Subject
  "numeric"  "factor"  "ordered"
```

We see that `effort`, the response, is a numeric variable; `Type`, the covariate, is a factor, and `Subject`, the grouping factor, is an ordered factor. We could replace `sapply` with `lapply` and get the same information, but the result would be returned as a list and would not print as compactly.

Checking the data.class of all variables in a data.frame or a groupedData object is an important first step in any data analysis. Because factor levels are often coded as integers, it is a common mistake to leave what should be a factor as a numeric variable. Any linear models using such a "factor" will be meaningless because the factor will be treated as a single numeric variable instead of being expanded into a set of contrasts. Another way of checking on the data.class of variables in a frame is to use the `summary` function shown later in this section.

The `table` and `unique` functions are not solely intended for use with grouped data, but often are useful when working with grouped data. The `gsummary` function, however, is specifically designed for use with grouped-Data objects. In §3.2.1 we used `gsummary` with the optional argument `invariantsOnly=TRUE` to extract only those variables in the groupedData object that are invariant within levels of the grouping factor. These could be experimental factors, like `Diet` in the `BodyWeight` data, or they could simply be additional characteristics of the groups, like `Sex` in the `Orthodont` data. The `Theoph` data are another example of a medical study where the grouping is by `Subject`. They are from a study of the pharmacokinetics of the drug theophylline, which is used to treat asthma. Each subject's weight and dose of theophylline are given in the data. As this is a short-duration study (about 24 hours) and only one dose of theophylline was given, both Wt and Dose are invariants.

```
> gsummary( Theoph, inv = TRUE )
   Subject    Wt Dose
6        6  80.0 4.00
7        7  64.6 4.95
8        8  70.5 4.53
11      11  65.0 4.92
3        3  70.5 4.53
2        2  72.4 4.40
4        4  72.7 4.40
9        9  86.4 3.10
12      12  60.5 5.30
10      10  58.2 5.50
1        1  79.6 4.02
5        5  54.6 5.86
```

Sometimes it is distracting to have the grouping factor itself included as one of the invariants. The optional argument omitGroupingFactor=TRUE suppresses this. The combination of omit = TRUE and inv = TRUE can be used to check if there are any nontrivial invariants. The value returned will be NULL unless there are invariants other than the grouping factor.

```
> gsummary( Theoph, omit = TRUE, inv = TRUE )
     Wt Dose
6  80.0 4.00
7  64.6 4.95
8  70.5 4.53
11 65.0 4.92
3  70.5 4.53
2  72.4 4.40
4  72.7 4.40
9  86.4 3.10
12 60.5 5.30
10 58.2 5.50
1  79.6 4.02
5  54.6 5.86
> is.null(gsummary(Theoph, inv = T, omit = T)) # invariants present
[1] F
> is.null(gsummary(Oxboys, inv = T, omit = T)) # no invariants
[1] T
```

When the invariantsOnly argument is omitted or given the value FALSE, a summary of all the variables in the object is returned. The default summary is a "representative value" for each variable within each group: *numeric* variables are represented by their mean within each group and *non-numeric* variables (e.g. factors) by their modes (the most frequently occuring value) within each group. When multiple modes are present, the first element of the sorted modes is returned.

```
> gsummary( Theoph )
   Subject   Wt Dose     time      conc
 6        6 80.0 4.00 5.888182 3.525455
 7        7 64.6 4.95 5.865455 3.910909
 8        8 70.5 4.53 5.890000 4.271818
11       11 65.0 4.92 5.871818 4.510909
 3        3 70.5 4.53 5.907273 5.086364
 2        2 72.4 4.40 5.869091 4.823636
 4        4 72.7 4.40 5.940000 4.940000
 9        9 86.4 3.10 5.868182 4.893636
12       12 60.5 5.30 5.876364 5.410000
10       10 58.2 5.50 5.915455 5.930909
 1        1 79.6 4.02 5.950000 6.439091
 5        5 54.6 5.86 5.893636 5.782727
```

By giving a numeric summary function, which is a function that calculates a single numerical value from a numeric vector, as the argument FUN, we can produce other summaries. For example, we can check that the Theoph data are sorted according to increasing values of the maximum response with

```
> gsummary( Theoph, FUN = max, omit = TRUE )
     Wt Dose  time   conc
 6 80.0 4.00 23.85   6.44
 7 64.6 4.95 24.22   7.09
 8 70.5 4.53 24.12   7.56
11 65.0 4.92 24.08   8.00
 3 70.5 4.53 24.17   8.20
 2 72.4 4.40 24.30   8.33
 4 72.7 4.40 24.65   8.60
 9 86.4 3.10 24.43   9.03
12 60.5 5.30 24.15   9.75
10 58.2 5.50 23.70  10.21
 1 79.6 4.02 24.37  10.50
 5 54.6 5.86 24.35  11.40
```

This ordering of the subjects does indeed give increasing maximum concentration of theophylline.

The FUN argument to gsummary is applied only to numeric variables in the grouped data object. Any non-numeric variables are represented by their modes within each group. A variable that is invariant within each group is represented by the (single) value that it assumes within each group. In other words, the value returned for each variable is determined according to:

- If the variable is an invariant, its value within each group is returned.

- If the variable is a factor (ordered or unordered) or of mode `character`, the mode for each group is returned.

- If the variable is numeric and not invariant, the summary function `FUN` is applied within each group and those values are returned.

When there are a large number of groups in the data, the result of `gsummary` may itself be so large as to be unwieldy. The `Quinidine` data, described in Appendix A.25, is from a study where thirteen different variables were recorded on 136 subjects.

```
> Quin.sum <- gsummary( Quinidine, omit = TRUE, FUN = mean )
> dim( Quin.sum )
[1] 136  13
```

Upon examining the first few rows

```
> Quin.sum[1:10, ]
        time conc dose interval Age Height  Weight       Race Smoke
109  30.2633   NA   NA       NA  70     67  58.000  Caucasian    no
70    0.7500   NA   NA       NA  68     69  75.000  Caucasian    no
23   52.0263   NA   NA       NA  75     72 108.000  Caucasian   yes
92    8.8571   NA   NA       NA  68     72  65.000  Caucasian   yes
111  18.1638   NA   NA       NA  68     66  56.000      Latin   yes
5    24.3750   NA   NA       NA  62     71  66.000  Caucasian   yes
18  196.8438   NA   NA       NA  87     69  85.375  Caucasian    no
24   31.2500   NA   NA       NA  55     69  89.000      Latin    no
2    12.2000   NA   NA       NA  58     69  85.000      Latin    no
88    4.7900   NA   NA       NA  85     72  77.000  Caucasian    no
      Ethanol     Heart Creatinine    glyco
109      none   No/Mild      >= 50  0.46000
70     former   No/Mild      >= 50  1.15000
23       none   No/Mild      >= 50  0.83875
92     former   No/Mild      >= 50  1.27000
111    former   No/Mild      >= 50  1.23000
5        none    Severe      >= 50  1.39000
18       none   No/Mild       < 50  1.26000
24     former   No/Mild      >= 50  0.57000
2     current  Moderate      >= 50  0.82000
88       none  Moderate      >= 50  0.61000
```

we see some unusual results. The summarized values of `conc`, `dose`, and `interval` are always recorded as missing values (`NA`). A less obvious peculiarity in the data is an apparent inconsistency in the numeric values of `height` and `weight`; for some reason `height` was recorded in inches while `weight` was recorded in kilograms.

Returning to the question of the missing values in `conc`, `dose`, and `interval`, the data from a single subject indicate the reason for this

```
> Quinidine[Quinidine[["Subject"]] == 3, 1:8]
   Subject    time conc dose interval Age Height Weight
17       3    0.00   NA  201       NA  67     69     69
18       3    8.00   NA  201       NA  67     69     69
19       3   16.00   NA  201       NA  67     69     69
20       3   24.00   NA  201       NA  67     69     69
21       3   32.00   NA  201       NA  67     69     69
22       3   41.25  2.4   NA       NA  67     69     69
23       3  104.00   NA  201        8  67     69     69
24       3  113.00  2.3   NA       NA  67     69     69
25       3 3865.00   NA  201        6  67     69     62
26       3 3873.00   NA  201       NA  67     69     62
27       3 3881.00   NA  201       NA  67     69     62
28       3 3889.00   NA  201       NA  67     69     62
29       3 3897.00   NA  201       NA  67     69     62
30       3 3900.00   NA   NA       NA  67     69     62
31       3 3905.00   NA  201       NA  67     69     62
32       3 3909.00  4.7   NA       NA  67     69     62
33       3 4073.00   NA  201        8  67     69     62
```

Each observation is either a record of a dosage or a record of a concentration measurement but never both. Thus, whenever dose is present, conc is missing and whenever conc is present, both dose and interval are missing. Because any subject in the experiment must have at least one dosage record and at least one concentration record, every subject has missing data in the conc, dose, and interval variables.

The default behavior of most summary functions in S is to return the value NA if the input vector contains any missing values. The mean function behaves like this and returns NA for every subject in each of these three variables. The behavior can be overridden in mean (and in several other summary functions) by giving the optional argument na.rm = TRUE. The default value of FUN in gsummary is mean with na.rm = TRUE.

```
> Quin.sum1 <- gsummary( Quinidine, omit = TRUE )
> Quin.sum1[1:10, 1:7]
        time   conc   dose interval     Age Height  Weight
1     92.817 2.2000 268.71   6.0000  60.000     69 106.000
2     12.200 1.2000 166.00       NA  58.000     69  85.000
3   2090.015 3.1333 201.00   7.3333  67.000     69  65.294
4    137.790 3.3667 236.39   7.3333  88.000     66  95.185
5     24.375 0.7000 301.00       NA  62.000     71  66.000
6      3.625 2.6000 166.00   6.0000  76.000     71  93.000
7   1187.320 2.7833 256.22   6.0000  60.097     66  85.484
8     20.019 2.5000 166.00       NA  52.000     71  75.000
9     69.200 3.9500 498.00   6.0000  68.000     70  79.000
10  1717.261 3.1667 201.00   8.0000  73.154     69  79.462
```

Notice that there are still some NA's in the groupwise summary for interval. For these subjects every value of interval was missing.

The function summary can be used with any data frame to summarize the columns according to their class. In particular, we can use it on Quin.sum1 to obtain some summary statistics for each variable with each group (subject) counted only once.

```
> summary( Quin.sum1 )
      time                  conc            dose           interval
Min.    :    0.065   Min.    :0.50   Min.    :  83   Min.    :  5.00
1st Qu.:   19.300   1st Qu.:1.70   1st Qu.:198   1st Qu.:  6.00
Median :   47.200   Median :2.24   Median :201   Median :  6.00
Mean    :  251.000   Mean    :2.36   Mean    :224   Mean    :  6.99
3rd Qu.:  171.000   3rd Qu.:2.92   3rd Qu.:249   3rd Qu.:  8.00
Max.    :5360.000   Max.    :5.77   Max.    :498   Max.    :12.00
                                                    NA's    :29.00

        Age            Height           Weight              Race
Min.    :42.0   Min.    :60.0   Min.    :  41.0   Caucasian:91
1st Qu.:61.0   1st Qu.:67.0   1st Qu.:  67.8   Latin      :35
Median :66.0   Median :70.0   Median :  79.2   Black      :10
Mean    :66.9   Mean    :69.6   Mean    :  79.2
3rd Qu.:73.0   3rd Qu.:72.0   3rd Qu.:  88.2
Max.    :92.0   Max.    :79.0   Max.    :119.0
 Smoke        Ethanol          Heart       Creatinine          glyco
no :94    none    :90   No/Mild :55    < 50 :  36   Min.    :0.390
yes:42    current:16   Moderate:40    >= 50:100   1st Qu.:0.885
          former :30   Severe  :41                 Median :1.170
                                                    Mean    :1.210
                                                    3rd Qu.:1.450
                                                    Max.    :2.990
```

Contrast this with the result of

```
> summary( Quinidine )
  . . .
      time                conc            dose           interval
Min.    :   0   Min.    :  0.40   Min.    :  83   Min.    :   4.00
1st Qu.:  16   1st Qu.:  1.60   1st Qu.:166   1st Qu.:  6.00
Median :  60   Median :  2.30   Median :201   Median :  6.00
Mean    : 373   Mean    :  2.45   Mean    :225   Mean    :  7.11
3rd Qu.: 241   3rd Qu.:  3.00   3rd Qu.:249   3rd Qu.:  8.00
Max.    :8100   Max.    :  9.40   Max.    :603   Max.    : 12.00
                NA's    :1110.00   NA's    :443   NA's    :1222.00
        Age            Height           Weight              Race
Min.    :42.0   Min.    :60.0   Min.    :  41.0   Caucasian:968
1st Qu.:60.0   1st Qu.:67.0   1st Qu.:  69.5   Latin      :384
Median :66.0   Median :69.0   Median :  78.0   Black      :119
Mean    :66.7   Mean    :69.2   Mean    :  79.7
3rd Qu.:74.0   3rd Qu.:72.0   3rd Qu.:  89.0
Max.    :92.0   Max.    :79.0   Max.    :119.0
 Smoke          Ethanol          Heart       Creatinine
no :1024   none    :991   No/Mild :598    < 50 :  418
```

```
yes: 447    current:191   Moderate:375    >= 50:1053
            former :289   Severe  :498
      glyco
Min.   :0.39
1st Qu.:0.93
Median :1.23
Mean   :1.28
3rd Qu.:1.54
Max.   :3.16
```

The first summary tells us that there are 94 nonsmokers in the study and 42 smokers while the second tells us that there are 1024 total observations on the nonsmokers and 447 on the smokers.

Both summaries are useful to us in understanding these data. Because NA's in conc and dose are mutually exclusive, we can see that there are at most 1110 dosage records and at most 443 concentration measurements. Because there are 136 subjects in the study, this means there are on average fewer than three concentration measurements per subject.

We can get an exact count of the number of concentrations by counting the number of nonmissing values in the entire conc variable. One way to do this is

```
> sum( ifelse(is.na(Quinidine[["conc"]]), 0, 1) )
[1] 361
```

This is equivalent to the somewhat more terse expression

```
> sum( !is.na(Quinidine[["conc"]]) )
[1] 361
```

because the logical values TRUE and FALSE are interpreted as 1 and 0 in arithmetic expressions.

A similar expression

```
> sum( !is.na(Quinidine[["dose"]]) )
[1] 1028
```

tells us that there are 1028 dosage events. The 164 rows that are neither dosage records nor concentration measurements are cases where values of other variables changed for the subject.

With only 361 concentration measurements for 136 subjects there are fewer than three concentration measurements per subject. To explore this further, we would like to determine the distribution of the number of concentration measurements per subject. We need to divide the data by subject (or "by group" in our general terminology) and count the number of nonmissing values in the conc variable for each subject. The function gapply is available to perform calculations such as this. It "applies" another function to some or all of the variables in a grouped data object by group.

```
> gapply( Quinidine, "conc", function(x) sum(!is.na(x)) )
 109 70 23 92 111 5 18 24 2 88 91 117 120 13 89 27 53 122 129 132
   1  1  3  1   1 2  3  1 1 1  1   3    2  1  3  1  1   1   2   3
 16 106 15 22 57 77 115 121 123 11 48 126 223 19 38 42 52 56 63 83
  1   1  1  1  3  1   4   1   1  2  2   2   6  1  1  2  1  1  4  1
 104 118 137 17 29 34 46 73 87 103 138 45 44 97 36 37 72 100 8 71
   2   2   1  1  1  1  3  2  2   1   2  3  7  2  2  3  1   3 1  5
 6 14 26 75 20 96 99 134 12 49 67 85 112 127 55 68 124 1 35 47 79
 1  3  1  3  2  3  2   1  1  3  3  1   3   3  6  3   1 2  2  5  3
 95 114 135 105 116 62 65 107 130 66 139 33 80 125 110 128 136 21
  3   2   2   1   3  4  7   4   3  1   3  3  2   1  11   2  11  2
 43 90 102 40 84 98 30 82 93 108 119 32 133 7 9 76 94 58 113 50 39
  1  1   2  2  6  2  1  3  4   1   3  1   2 6 2  6  5  1   2  3  2
 78 25 61 3 64 60 59 10 69 4 81 54 41 74 28 51
 10  2  2 3  4  4  3  6  2 6 11  4  3  3  4  6
```

The result can be a bit confusing when printed in this way. It is a named vector of the counts where the names are the values of the Subject variable. Thus, subjects 124 and 125 both had only a single concentration measurement. To obtain the distribution of the measurements, we apply the table function to this vector

```
> table( gapply(Quinidine, "conc", function(x) sum(!is.na(x))) )
  1  2  3 4 5 6 7 10 11
 46 33 31 9 3 8 2  1  3
```

We see that most of the subjects in the study have very few measurements of the response. A total of 110 out of the 136 subjects have fewer than four response measurements. This is not uncommon in such routine clinical data (data that are collected in the routine course of treatment of patients). A common consequence of having so few observations for most of the subjects is that the information gained from such a study is imprecise compared to that gained from a controlled experiment.

The second argument to gapply is the name or index of the variable in the groupedData object to which we will apply the function. When only a single name or index is given, the value passed to the function to be applied is in the form of a vector. If more than one variable is to be passed, they are passed as a data.frame. If this argument is missing, all the variables in the groupedData object are passed as a data.frame so the effect is to select subsets of the rows corresponding to unique values of the grouping factor from the groupedData object.

To illustrate this, let us determine those subjects for whom there are records that are neither dosage records nor concentration records. For each subject we must determine whether there are any records with both the conc variable and the dose variable missing.

```
> changeRecords <- gapply( Quinidine, FUN = function(frm)
+      any(is.na(frm[["conc"]]) & is.na(frm[["dose"]])) )
```

```
> changeRecords
 109 70 23 92 111 5 18 24 2 88 91 117 120 13 89 27 53 122 129 132
   F  F  F  F   F FF  T FF  F  F   F   F  F  F  F  F   F   F   T
 ...
 78 25 61 3 64 60 59 10 69 4 81 54 41 74 28 51
  F  F  T T  T  F  F  T  FT T  T  T  T  F  T  F
```

As we see, the printed representation as a named vector of length 136 is
not easy to read. It is more informative if we convert the result to a vector
of levels of the Subject factor for which there are records that are neither
dosage records nor concentration records.

```
> sort( as.numeric( names(changeRecords)[changeRecords] ) )
 [1]    3    4    7   10   14   18   28   33   37   40   41   44   45   46   47
[16]   48   50   54   55   61   62   63   64   65   71   75   76   77   79   80
[31]   81   82   84   94   95   96   97   98  110  112  114  118  119  127  132
[46]  133  135  136  139  223
```

Notice that subject 3 is one of those with such a "change" record. We
printed some of the variables from this subject's data on page 123. We can
see there that the record in question is row 30. If we look at this row and
the adjacent rows

```
> Quinidine[29:31,]
Grouped Data: conc ~ time | Subject
   Subject time conc dose interval Age Height Weight     Race
29       3 3897   NA  201       NA  67     69     62 Caucasian
30       3 3900   NA   NA       NA  67     69     62 Caucasian
31       3 3905   NA  201       NA  67     69     62 Caucasian
   Smoke Ethanol   Heart Creatinine glyco
29   yes  former Moderate       < 50  1.71
30   yes  former Moderate       < 50  1.71
31   yes  former Moderate       < 50  1.71
```

we can see that there are no changes in any of the variables except for time
so this row is redundant. We did not cull such redundant rows from the
data set because we wanted to be able directly to compare results based
on these data with analyses done by others.

The data for subject 4 provide a better example.

```
> Quinidine[Quinidine[["Subject"]] == 4, ]
Grouped Data: conc ~ time | Subject
   Subject  time conc dose interval Age Height Weight  Race Smoke
45       4  0.00   NA  332       NA  88     66    103 Black   yes
46       4  7.00   NA  332       NA  88     66    103 Black   yes
47       4 13.00   NA  332       NA  88     66    103 Black   yes
48       4 19.00   NA  332       NA  88     66    103 Black   yes
49       4 21.50  3.1   NA       NA  88     66    103 Black   yes
50       4 85.00   NA  249        6  88     66    103 Black   yes
51       4 91.00  5.8   NA       NA  88     66    103 Black   yes
```

52	4	91.08	NA	249	NA	88	66	103	Black	yes
53	4	97.00	NA	249	NA	88	66	103	Black	yes
54	4	103.00	NA	249	NA	88	66	103	Black	yes
55	4	105.00	NA	NA	NA	88	66	92	Black	yes
56	4	109.00	NA	249	NA	88	66	92	Black	yes
57	4	115.00	NA	249	NA	88	66	92	Black	yes
58	4	145.00	NA	166	NA	88	66	92	Black	yes
59	4	151.00	NA	166	NA	88	66	92	Black	yes
60	4	156.00	3.1	NA	NA	88	66	92	Black	yes
61	4	157.00	NA	166	NA	88	66	92	Black	yes
62	4	163.00	NA	166	NA	88	66	92	Black	yes
63	4	169.00	NA	166	NA	88	66	92	Black	yes
64	4	174.75	NA	201	NA	88	66	92	Black	yes
65	4	177.00	NA	NA	NA	88	66	92	Black	yes
66	4	181.50	3.1	NA	NA	88	66	92	Black	yes
67	4	245.00	NA	201	8	88	66	92	Black	yes
68	4	249.00	NA	NA	NA	88	66	86	Black	yes
69	4	252.50	3.2	NA	NA	88	66	86	Black	yes
70	4	317.00	NA	201	8	88	66	86	Black	yes
71	4	326.00	1.9	NA	NA	88	66	86	Black	yes

	Ethanol	Heart	Creatinine	glyco
45	none	Severe	>= 50	1.48
46	none	Severe	>= 50	1.48
47	none	Severe	>= 50	1.48
48	none	Severe	>= 50	1.48
49	none	Severe	>= 50	1.48
50	none	Severe	>= 50	1.61
51	none	Severe	>= 50	1.61
52	none	Severe	>= 50	1.61
53	none	Severe	>= 50	1.61
54	none	Severe	>= 50	1.61
55	none	Severe	>= 50	1.61
56	none	Severe	>= 50	1.61
57	none	Severe	>= 50	1.61
58	none	Severe	>= 50	1.88
59	none	Severe	>= 50	1.88
60	none	Severe	>= 50	1.88
61	none	Severe	>= 50	1.88
62	none	Severe	>= 50	1.88
63	none	Severe	>= 50	1.88
64	none	Severe	>= 50	1.88
65	none	Severe	>= 50	1.68
66	none	Severe	>= 50	1.68
67	none	Severe	>= 50	1.87
68	none	Severe	>= 50	1.87
69	none	Severe	>= 50	1.87
70	none	Severe	>= 50	1.83
71	none	Severe	>= 50	1.83

Here rows 55, 65, and 68 are in fact "change rows." Both rows 55 and 68 record changes in the subject's weight. Row 65 records a change in the glyco variable (i.e., the serum concentration of alpha-1-acid glycoprotein).

3.5 Chapter Summary

In this chapter we have shown examples of constructing, summarizing, and graphically displaying groupedData objects. These objects include the data, stored as a data frame, and a formula that designates different variables as a response, a primary covariate, and as one or more grouping factors. Other variables can be designated as outer or inner factors relative to the grouping factors. Accessor or extractor functions are available to extract either the formula for these variables or the value of these variables.

Informative and visually appealing trellis graphics displays of the data can be quickly and easily generated from the information that is stored with the data. The regular data summary functions in S can be applied to the data as well as the gsummary and gapply functions that are especially designed for these data.

Informative plots and summaries of the data are very useful for the preliminary phase of the statistical analysis. Many important features of the data are identified at this stage, but usually one is interested in going a step further in the analysis and fitting parametric models, such as the linear mixed-effects models described in the next chapter.

Exercises

1. In Figure 3.6 (p. 113), the twelve panels in the plot of the CO2 data were laid out as two rows of six columns. Create a plot of these data arranged as four rows of three columns. You should insert some space between the second and third rows to separate the panels for the Mississippi plants from those for the Québec plants.

2. Use the outer argument to the plot function to produce a plot of the CO2 data similar to Figure 8.15 (p. 369) where each panel displays the data for all three plants at some combination of Variety and Treatment. Produce two such plots of these data, each laid out as two rows by two columns. One plot should have the rows determined by Variety and the columns by Treatment. This arrangement makes it easy to assess the effect of Treatment within a Variety. The other plot should have rows determined by Treatment and columns by Variety, allowing easy assessment of the effect of Variety within Treatment.

3. The `Dialyzer` data, described in Appendix A.6 and used in some examples in §5.4 and §8.3.3, consists of observations of ultrafiltration rates at different transmembrane pressures on different subjects. The `QB` variable, which indicates the blood flow rate used for the subject, is outer to the grouping factor `Subject`.

   (a) Check if these data are balanced with respect to the number of observations on each `Subject`.

   (b) Check if these data are balanced with respect to the number of observations and the values of the transmembrane pressure. Verify your result using the `isBalanced` function.

   (c) Produce a plot of the ultrafiltration rate versus transmembrane pressure by subject.

   (d) Check with `gsummary` that `QB` is invariant within each `Subject`. Determine which `Subjects` are at which `QB` levels.

   (e) Recreate the plot from part (c) arranging for the panels corresponding to subjects at a `QB` of 200 dl/min to be separated from those at a `QB` of 300 dl/min.

   (f) Use the `outer` argument to `plot` as `outer = TRUE` or `outer = ~QB` to create a plot of ultrafiltration rate versus transmembrane pressure by subject divided into two panels according to blood flow rate. Compare this plot to Figure 5.1 (p. 215). Note the change in the aspect ratio of the panels relative to the plots in parts (c) and (d).

   (g) Use the `aspect` argument in a plot like the last one to produce a plot similar to Figure 5.1.

   (h) Use `gsummary` or `gapply` to determine the maximum observed ultrafiltration rate by subject. Produce dotplots or boxplots of this maximum rate dividing the subjects according to blood flow rate (`QB`). Does the maximum ultrafiltration rate appear to be related to the blood flow rate?

4. In the `DNase` data, described in §3.3.2, there are two measurements of the optical density at each DNase concentration within each run. When such *replicate* observations are available, one can check the assumption of constant variance for the ϵ_{ij} in a linear mixed-effects model by plotting the logarithm of the standard deviation of the replicate measurements versus the logarithm of their average (Box, Hunter and Hunter, 1978, §7.8).

   (a) Determine the average optical density at each set of replicate observations. One way to do this would be to create a copy of the groupedData object `DNase` redefining the display formula to be `density ~1 | Run/conc`. Applying `gsummary` to this object will produce the average optical density for each pair of replicates.

(b) Determine the standard deviation of the optical density at each set of replicate observations. Note that some of these standard deviations should be zero because the replicate observations are equal. Due to numerical round-off some equal replicates may produce a standard deviation that is very small but not exactly zero. Use a boxplot of the logarithm of the standard deviations to check for these. You may wish to replace those small values with zero.

(c) Plot the logarithm of the standard deviation versus the logarithm of the average. Estimate the slope of a straight line fit to these points.

4
Fitting Linear Mixed-Effects Models

As seen in Chapter 1, mixed-effects models provide a flexible and powerful tool for analyzing balanced and unbalanced grouped data. These models have gained popularity over the last decade, in part because of the development of reliable and efficient software for fitting and analyzing them. The linear and nonlinear mixed-effects (nlme) library in S is an example of such software. We describe the lme function from that library in this chapter, as well the methods for displaying and comparing fitted models created by this function.

The first section gives a brief review of the standard linear modeling facilities in S to introduce the general style of the S modeling functions, classes, and methods that are used with the nlme library. The lmList function, used to obtain separate lm fits according to the levels of a grouping variable, is described and illustrated through examples.

The next section describes the linear mixed-effects modeling capabilities in S. The S modeling function lme is described, together with its associated methods. Its use is illustrated through examples, including single- and multilevel grouped data.

After a model has been fit to the data, it is important to examine whether the underlying assumptions appear to be violated. Graphical methods and numerical summaries for assessing the validity of the assumptions in a linear mixed-effects model are described in §4.3.

An "inside-out" model building approach is adopted here, starting with individual fits by group, using plots of the individual coefficients to decide on the random-effects structure, and finally fitting a mixed-effects model to the complete data. We make extensive use of examples to introduce and illustrate the available functions and methods.

In this chapter we will restrict our attention to models in which the within-group errors are independent and have equal variance. Models with more complex within-group covariance structures, such as the heterogeneous $AR(1)$ structure, will be explored in detail in Chapter 5.

4.1   Fitting Linear Models in S with lm and lmList

S offers a variety of functions and methods for fitting and manipulating linear models. The two main modeling functions are lm, for linear regression models, and aov, for analysis of variance models. These two functions have similar syntax and generate similar fitted objects. We concentrate here on lm. A typical call to lm is of the form

```
lm(formula, data)
```

where formula specifies the linear model to be fitted and data gives a data frame in which the variables in formula are to be evaluated. Several other arguments to lm are available and are described in detail in Chambers and Hastie (1992, Chapter 4) and also in Venables and Ripley (1999, Chapter 6).

The formula language used in the formula argument gives lm great flexibility in specifying linear models. The formulas use a version of the syntax defined by Wilkinson and Rogers (1973), which translates a linear model like $y = \beta_0 + \beta_1 x_1 + \beta_2 x_2 + \epsilon$ into the S expression

```
y ~ 1 + x1 + x2
```

The ~ is read "is modeled as." The expression on the left-hand side of the ~ specifies the response to be modeled. The expression on the right-hand side describes the covariates used and the ways they will be combined to form the model. The expression does not include the coefficients (the β's). They are implicit.

The constant term 1 is included by default and does not need to be given explicitly in the model. If a model without an intercept term is desired, a -1 must be included in the formula. The covariates can be factors, ordered factors, continuous variables, matrices, or data frames. Any function of a covariate that returns one of these types of covariate can also be used on the right-hand side of the formula. Function calls are also allowed on the left-hand side of the formula. For example,

```
log(y) ~ exp(x1) + cos(x2)
```

is a valid lm formula. An interaction between two covariates is denoted by the : operator. Nesting of a covariate within a factor is denoted by the

`%in%` operator. More detailed references on the formula language include Chambers and Hastie (1992, Chapter 2) and Venables and Ripley (1999, §6.2).

The `lm` function operates in a style common to most modeling functions in S, in particular `lme` and `nlme`. A call to `lm` returns a fitted object of class `lm` to which several generic functions can be applied. These can display the results of the fit (`print` and `summary`), produce diagnostic plots (`plot`), return predictions (`predict`), extract components (`fitted`, `residuals`, and `coef`), update the original model (`update`), or compare different fitted objects (`anova`).

To illustrate some of these capabilities, we revisit the orthodontic growth curve data of §1.4. Suppose that we initially ignore the grouping structure in the data and fit a single linear regression model of `distance` on `age` to the data from all the subjects. The corresponding call to `lm` is

```
> fm1Orth.lm <- lm( distance ~ age, Orthodont )
```

A brief description of the results is provided by the `print` method which is called implicitly when the fitted object is to be displayed.

```
> fm1Orth.lm                      # equivalent to print( fm1Orth.lm )
Call:
lm(formula = distance~age, data = Orthodont)

Coefficients:
 (Intercept)       age
      16.761  0.66019

Degrees of freedom: 108 total; 106 residual
Residual standard error: 2.5372
```

Diagnostic plots for assessing the quality of the fit are obtained using the method for the `plot` generic function

```
> par( mfrow=c(3,2) )             # arrange 6 separate plots on a page
> plot( fm1Orth.lm )                                  # Figure 4.1
```

There is considerable variability remaining after the fit, as shown in the residual plots. This is not surprising, as we know that the simple linear regression model does not represent the structure of the data well. Apparently there are some observations with unusual influence on the fit, especially observations 39 and 104. Furthermore, the normal probability plot of the residuals suggests that the error distribution has heavier tails than expected from normally distributed variates.

Suppose that we now want to test for possible differences in intercept or in slope between boys and girls. We can use the following model,

$$\text{distance} = \beta_0 + \beta_1 \, \text{Sex} + \beta_2 \, \text{age} + \beta_3 \, \text{age} \times \text{Sex} + \epsilon, \qquad (4.1)$$

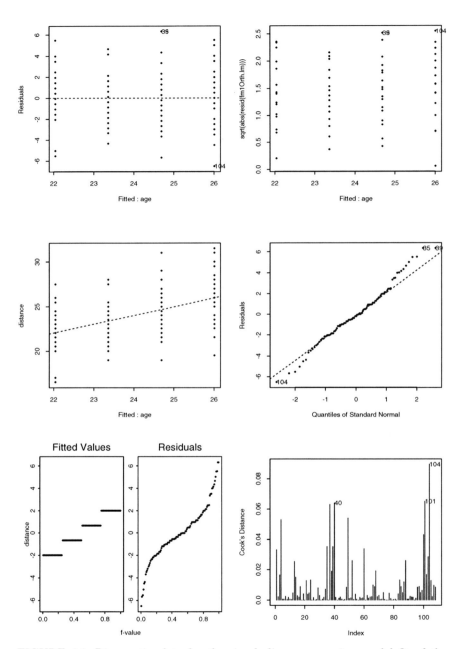

FIGURE 4.1. Diagnostic plots for the simple linear regression model fit of the orthodontic growth curve data.

with Sex representing a binary variable taking values -1 for boys and 1 for girls. The parameters β_1 and β_3 represent, respectively, the intercept and slope gender effects. We can fit this model in S with another call to `lm` or by using `update` on the previous fitted model and redefining the formula.

```
> fm2Orth.lm <- update( fm1Orth.lm, formula = distance ~ Sex*age )
```

The expression `Sex*age` in a linear model formula crosses the `Sex` and `age` factors. This means it generates the main effects for these factors and their interaction. It is equivalent to `Sex + age + Sex:age`.

The `summary` method displays the results in more detail. In particular, it provides information about the marginal significance of the parameter estimates.

```
> summary( fm2Orth.lm )
Call: lm(formula = distance ~Sex + age + Sex:age, data = Orthodont)
Residuals:
   Min    1Q Median   3Q  Max
 -5.62 -1.32 -0.168 1.33 5.25

Coefficients:
              Value Std. Error t value Pr(>|t|)
(Intercept) 16.857    1.109     15.194   0.000
        Sex  0.516    1.109      0.465   0.643
        age  0.632    0.099      6.394   0.000
    Sex:age -0.152    0.099     -1.542   0.126

Residual standard error: 2.26 on 104 degrees of freedom
Multiple R-Squared: 0.423
F-statistic: 25.4 on 3 and 104 d.o.f., the p-value is 2.11e-12

Correlation of Coefficients:
        (Intercept)    Sex    age
    Sex  0.185
    age -0.980      -0.181
Sex:age -0.181      -0.980  0.185
```

The p-values for the `Sex` and `Sex:age` coefficients suggest that there is no gender effect on the orthodontic measurement growth. Because the t-test is only measuring the marginal significance of each term in the model, we should proceed with caution and delete one term at a time from the model. Deleting first the least significant term, `Sex`, we get:

```
> fm3Orth.lm <- update( fm2Orth.lm, formula = . ~ . - Sex )
> summary( fm3Orth.lm )
 . . .
Coefficients:
              Value Std. Error t value Pr(>|t|)
(Intercept) 16.761    1.086     15.432   0.000
```

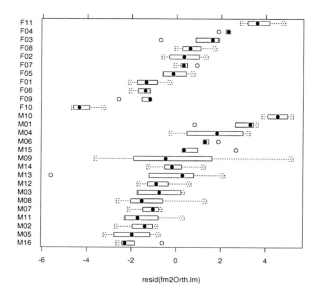

FIGURE 4.2. Residual plots corresponding to the `fm2Orth.lm` fitted object, by subject.

```
       age    0.640    0.097       6.613    0.000
   Sex:age   -0.107    0.020      -5.474    0.000
 . . .
```

By convention, the `.~.` expression represents the `formula` in the object being updated and the `-` operator is used to delete terms from the model.

The `Sex:age` coefficient now becomes very significant, indicating that the growth patterns are different for boys and girls. Because the `lm` fit is not adequate for these data, we will postpone further discussion of these model-building issues until the linear mixed-effects model has been described.

The grouped nature of these data, with repeated measures on each subject at four different years, violates the basic assumption of independence that underlies the statistical methods used in `lm`. Boxplots of the `fm2Orth.lm` residuals by subject show this.

```
> bwplot( getGroups(Orthodont)~resid(fm2Orth.lm) )        # Figure 4.2
```

The most important feature observed in Figure 4.2 is that residuals corresponding to the same subject tend to have the same sign. This indicates the need for a "subject effect" in the model, which is precisely the motivation for linear mixed-effects models.

4.1.1   Separate *lm* Fits per Group: the *lmList* Function

The first step in the model-building process for a linear mixed-effects model, after the functional form of the model has been decided, is choosing which parameters in the model, if any, should have a random-effect component included to account for between-group variation. The `lmList` function and the methods associated with it are useful for this.

A typical call to `lmList` is

```
lmList( formula, data )
```

where the right-hand side of the `formula` consists of two parts separated by the | operator. The first part specifies the linear model to be fitted to each subset of `data`; the second part specifies the grouping factor. Any linear formula allowed in `lm` can also be used as a model formula with `lmList`. The `data` argument gives the data frame in which to find the variables used in `formula`.

Continuing with the analysis of the orthodontic data, we see from a Trellis plot of these data (Figure 1.11, page 31) that a simple linear regression model of `distance` as a function of `age` may be suitable. We fit this by

```
> fm1Orth.lis <- lmList( distance ~ age | Subject, Orthodont )
```

If `data` is a groupedData object (see Chapter 3), the grouping variable can be omitted from `formula`, being extracted from the group formula in `data`.

```
> getGroupsFormula( Orthodont )
~ Subject
```

so an alternative call to `lmList` to obtain the same fitted object is

```
> fm1Orth.lis <- lmList( distance ~ age, Orthodont )
```

Because the `lmList` function is a *generic* function (Chambers and Hastie, 1992, Appendix A) with different methods for arguments of different classes, this same fit can be specified in an even simpler way. If the first argument to `lmList` is a groupedData object, the display formula for this object is used to create a default model formula and to extract the grouping variable expression. Because we are using the same grouping, response, and covariate in our `lmList` fit as in the display formula

```
> formula( Orthodont )
distance ~ age | Subject
```

we can obtain the same fitted model object with the simpler call

```
> fm1Orth.lis <- lmList( Orthodont )
```

Objects returned by `lmList` are of class lmList, for which several display and plot methods are available. Table 4.1 lists some of the most important methods for class lmList. We illustrate the use of some of these methods below.

The `print` method displays minimal information about the fitted object.

TABLE 4.1. Main lmList methods.

augPred	predictions augmented with observed values
coef	coefficients from individual lm fits
fitted	fitted values from individual lm fits
fixef	average of individual lm coefficients
intervals	confidence intervals on coefficients
lme	linear mixed-effects model from lmList fit
logLik	sum of individual lm log-likelihoods
pairs	scatter-plot matrix of coefficients or random effects
plot	diagnostic Trellis plots
predict	predictions for individual lm fits
print	brief information about the lm fits
qqnorm	normal probability plots
ranef	deviations of coefficients from average
resid	residuals from individual lm fits
summary	more detailed information about lm fits
update	update the individual lm fits

```
> fm1Orth.lis
Call:
  Model: distance ~ age | Subject
   Data: Orthodont

Coefficients:
    (Intercept)   age
M16       16.95 0.550
 . . .
F11       18.95 0.675

Degrees of freedom: 108 total; 54 residual
Residual standard error: 1.31
```

The *residual standard error* given in the output is the pooled estimate of
the standard error calculated from the individual lm fits by group. More
detailed output can be obtained using the summary method.

```
> summary( fm1Orth.lis )
Call:
  Model: distance ~ age | Subject
   Data: Orthodont

Coefficients:
   (Intercept)
    Value Std. Error t value   Pr(>|t|)
M16 16.95     3.2882 5.15484 3.6952e-06
```

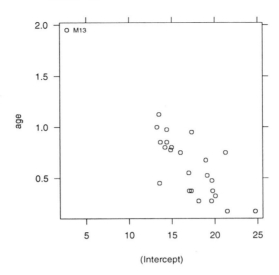

FIGURE 4.3. Pairs plot for `fm1Orth.lis`.

```
M05 13.65     3.2882 4.15124 1.1817e-04
. . .
F11 0.675     0.29293 2.30428 2.5081e-02

Residual standard error: 1.31 on 54 degrees of freedom
```

Diagnostic plots can be obtained using the `plot` method, as described in §4.3.

The main purpose of the preliminary analysis provided by `lmList` is to give an indication of what random-effects structure to use in a linear mixed-effects model. We must decide which random effects to include in a model for the data, and what covariance structure these random effects should have. The `pairs` method provides one view of the random-effects covariance structure.

```
> pairs( fm1Orth.lis, id = 0.01, adj = -0.5 )        # Figure 4.3
```

The `id` argument is used to identify outliers—points outside the estimated probability contour at level `1-id/2` will be marked in the plot. We see that subject `M13` has an unusually low intercept, compensated by a large slope. There appears to be a negative correlation between the intercept and slope estimates. Those with experience analyzing regression models may already have guessed why this pattern occurs. It is because all the data were collected between age 8 and age 14, but the intercept represents the measurement at age 0. This causes a high negative correlation (-0.98) between estimates of the slopes and the intercepts. We can remove this correlation by centering the data. In this case, we fit `distance` as a linear function of

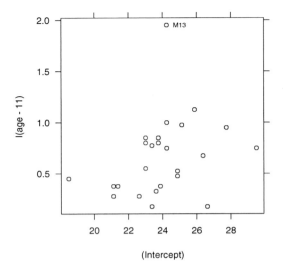

FIGURE 4.4. Pairs plot for `fm2Orth.lis` with ages centered at 11 years.

age - 11. The two quantities being estimated then are the distance at 11 years of age and the slope, or growth rate. We fit this revised model with

```
> fm2Orth.lis <- update( fm1Orth.lis, distance ~ I(age-11) )
```

The corresponding `pairs` plot (Figure 4.4) does not suggest any correlation between the intercept estimates and the slope estimates. It is clear that the orthodontic distance for subject `M13` has grown at an unusually fast rate, but his orthodontic distance at age 11 was about average. Both intercept and slope estimates seem to vary with individual, but to see how significantly they vary among subjects we need to consider the precision of the `lmList` estimates. This can be evaluated with the `intervals` method.

```
> intervals( fm2Orth.lis )

, , (Intercept)
      lower    est.   upper
M16  21.687  23.000  24.313
M05  21.687  23.000  24.313
M02  22.062  23.375  24.688
 . . .
F04  23.562  24.875  26.188
F11  25.062  26.375  27.688

, , I(age - 11)
         lower    est.   upper
M16  -0.037297  0.550  1.1373
M05   0.262703  0.850  1.4373
```

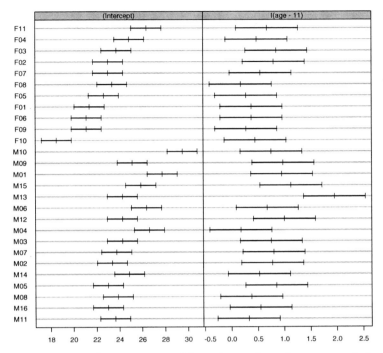

FIGURE 4.5. Ninety-five percent confidence intervals on intercept and slope for each subject in the orthodontic distance growth data.

```
M02  0.187703 0.775 1.3623
. . .
F04 -0.112297 0.475 1.0623
F11  0.087703 0.675 1.2623
```

As often happens, displaying the intervals as a table of numbers is not very informative. We find it much more effective to plot these intervals using

```
> plot( intervals(fm2Orth.lis) )                    # Figure 4.5
```

The individual confidence intervals in Figure 4.5 give a clear indication that a random effect is needed to account for subject-to-subject variability in the intercept. Except for subject M13, all confidence intervals for the slope overlap, so perhaps this parameter can be regarded as a fixed effect in the mixed-effects model. We will explore these questions in §4.2.1, while describing the `lme` function.

To further illustrate the capabilities of `lmList`, we consider data on radio-immunoassays of the protein Insulin-like Growth Factor (IGF-I) presented in Davidian and Giltinan (1995, §3.2.1, p. 65). The data are from quality control radioimmunoassays for ten different lots of a radioactive tracer used

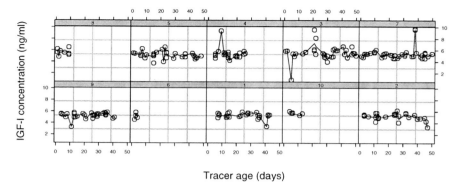

FIGURE 4.6. Estimated concentration of the protein Insulin-like Growth Factor (IGF-I) versus age of radioactive tracer for ten lots of tracer.

in the calibration of IGF-I concentration measurements. They are described in more detail in Appendix A.11.

```
> IGF
Grouped Data: conc ~ age | Lot
    Lot age conc
  1   1   7 4.90
  2   1   7 5.68
 . . .
236  10  11 5.30
237  10  13 5.63
```

This data set, displayed in Figure 4.6, is an example of unbalanced, repeated measures data. We do not consider these data to be longitudinal because different tracer samples are used at each radioimmunoassay. This reduces the potential for serial correlation in the responses.

The primary purpose of the IGF-I experiment was to investigate possible trends in control values with tracer age, which would indicate tracer decay within the usual storage period. We can investigate this by testing if the slope of a simple linear regression model is significantly different from zero. We must account for both the within-lot and the between-lot variability when fitting the model and testing for the significance of the slope. A linear mixed-effects model will do this, but first we investigate the sources of variation in the data by fitting separate regression lines to each lot. As the fixed-effects formula coincides with the display formula in IGF, we can use the simple form of the call to lmList

```
> fm1IGF.lis <- lmList( IGF )
> coef( fm1IGF.lis )
    (Intercept)          age
 9      5.0986   0.0057276
 6      4.6300   0.1700000
```

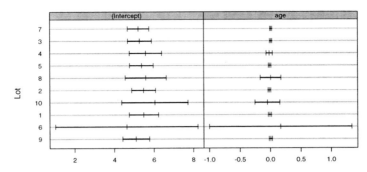

FIGURE 4.7. Ninety-five percent confidence intervals on intercept and slope for each lot in the IGF data.

1	5.4929	-0.0077901
10	6.0516	-0.0473282
2	5.4764	-0.0144271
8	5.5922	0.0060638
5	5.3732	-0.0095140
4	5.5768	-0.0166578
3	5.2788	0.0100830
7	5.2069	0.0093136

A quick look at the individual coefficient estimates indicates that Lot 6 is unusual. It has a low intercept compensated by a high slope. Examination of Figure 4.6 shows that this lot has only four observations and that these are at nearby tracer ages. The coefficients from the individual fit to this lot are unreliable. We will return to this issue in §4.2.1.

The plot of the individual 95% confidence intervals, shown in Figure 4.7, provides some insight about lot-to-lot variation in the parameter estimates.

```
> plot( intervals(fm1IGF.lis) )                    # Figure 4.7
```

Because of the imbalance in the data, these confidence intervals have very different lengths. There is little indication of lot-to-lot variation in either the intercept or the slope estimates, since all confidence intervals overlap. A fixed-effects model seems adequate to represent the IGF data.

```
> fm1IGF.lm <- lm( conc ~ age, data = IGF )
> summary( fm1IGF.lm )
. . .
Coefficients:
            Value Std. Error t value Pr(>|t|)
(Intercept) 5.351    0.104    51.584   0.000
       age -0.001    0.004    -0.170   0.865

Residual standard error: 0.833 on 235 degrees of freedom
. . .
```

There does not appear to be a significant tracer decay within the 50-day period over which the data were collected. A linear mixed-effects model for the IGF data will be considered in the next section.

4.2   Fitting Linear Mixed-Effects Models with `lme`

The general formulation of a linear mixed-effects model, as well as the estimation methods used to fit it, have been described in Chapter 2. In this section, we concentrate on the capabilities available in the nlme library for fitting such models. We initially consider `lme` fits for single-level grouped data with general covariance structures for the random effects. Fitting models with patterned covariance structures for the random effects is described in §4.2.2. In §4.2.3, we describe how to fit multilevel models with `lme`.

4.2.1   Fitting Single-Level Models

We use the `lme` function to fit linear mixed-effects models by maximum likelihood or by restricted maximum likelihood (the default). Several optional arguments can be used with this function, but a typical call is

```
lme( fixed, data, random )
```

The first argument is a two-sided linear formula specifying the fixed effects in the model. The third argument is typically given as a one-sided linear formula, specifying the random effects and the grouping structure in the model. For the orthodontic data, with the ages centered at 11 years, these formulas are:

```
fixed = distance ~ I(age-11), random = ~ I(age-11) | Subject
```

Note that the response is specified only in the `fixed` formula. If the `random` formula is omitted, its default value is taken as the right-hand side of the `fixed` formula. This describes a model in which every fixed effect has an associated random effect. To use this default, `data` must be a groupedData object, so the formula for the grouping structure can be obtained from the display formula.

The argument `data` specifies a data frame in which the variables named in `fixed` and `random` can be evaluated. When `data` inherits from class grouped-Data, the expression defining the grouping structure can be omitted in `random`.

A simple call to `lme` to fit the orthodontic data model is

```
> fm1Orth.lme <- lme( distance ~ I(age-11), data = Orthodont,
+                        random = ~ I(age-11) | Subject )
```

or, because `Orthodont` is a groupedData object and, by default, the random effects have the same form as the fixed effects

TABLE 4.2. Main lme methods.

ACF	empirical autocorrelation function of within-group residuals
anova	likelihood ratio or conditional tests
augPred	predictions augmented with observed values
coef	estimated coefficients for different levels of grouping
fitted	fitted values for different levels of grouping
fixef	fixed-effects estimates
intervals	confidence intervals on model parameters
logLik	log-likelihood at convergence
pairs	scatter-plot matrix of coefficients or random effects
plot	diagnostic Trellis plots
predict	predictions for different levels of grouping
print	brief information about the fit
qqnorm	normal probability plots
ranef	random-effects estimates
resid	residuals for different levels of grouping
summary	more detailed information about the fit
update	update the lme fit
Variogram	semivariogram of within-group residuals

```
> fm1Orth.lme <- lme( distance ~ I(age-11), data = Orthodont )
```

Because the lme function is generic, the model can be described in several different ways. For example, there is an lme method for lmList objects. When an lmList object, such as fm2Orth.lis in §4.1.1, is given as the first argument to lme, it provides default values for fixed, random, and data. We can create the same fitted model with the simple call

```
> fm1Orth.lme <- lme( fm2Orth.lis )
```

One advantage of this method is that initial estimates for the parameters in the profiled (restricted-)likelihood of the mixed-effects model are automatically calculated from the lmList object.

The fitted object is of the lme class, for which several methods are available to display, plot, update, and further explore the estimation results. Table 4.2 lists the most important methods for class lme. We illustrate the use of these methods through the examples in the next sections.

Orthodontic Growth Curve

As for all the classes of objects representing fitted models, the print method for the lme class returns a brief description of the estimation results. It prints the estimates of the standard deviations and the correlations of the

random effects, the within-group standard error, and the fixed effects. For
the fm1Orth.lme object it gives

```
> fm1Orth.lme
Linear mixed-effects model fit by REML
  Data: Orthodont
  Log-restricted-likelihood: -221.32
  Fixed: distance ~ I(age - 11)
  (Intercept) I(age - 11)
      24.023      0.66019

Random effects:
 Formula:  ~ I(age - 11) | Subject
 Structure: General positive-definite
              StdDev   Corr
(Intercept) 2.13433 (Inter
I(age - 11) 0.22643 0.503
   Residual 1.31004

Number of Observations: 108
Number of Groups: 27
```

One of the questions of interest for the orthodontic growth data is whether
boys and girls have different growth patterns. We can assess this by fitting
the model

```
> fm2Orth.lme <- update(fm1Orth.lme,fixed = distance~Sex*I(age-11))
```

Note that lmList cannot be used to test for gender differences in the orthodontic growth data, as it estimates individual coefficients for each subject. In general, we will not be able to use lmList to test for differences due
to factors that are invariant with respect to the groups.

Some more detailed output is supplied by summary.

```
> summary( fm2Orth.lme )
Linear mixed-effects model fit by REML
 Data: Orthodont
     AIC    BIC  logLik
  451.35 472.51 -217.68

Random effects:
 Formula:  ~ I(age - 11) | Subject
 Structure: General positive-definite
              StdDev   Corr
(Intercept) 1.83033 (Inter
I(age - 11) 0.18035 0.206
   Residual 1.31004

Fixed effects: distance ~ Sex + I(age - 11) + Sex:I(age - 11)
```

```
                  Value Std.Error DF t-value p-value
    (Intercept)  23.808   0.38071 79  62.537  <.0001
            Sex  -1.161   0.38071 25  -3.048  0.0054
     I(age - 11)  0.632   0.06737 79   9.381  <.0001
Sex:I(age - 11)  -0.152   0.06737 79  -2.262  0.0264
 Correlation:
                  (Intrc   Sex I(-11)
            Sex 0.185
     I(age - 11) 0.102   0.019
Sex:I(age - 11) 0.019   0.102 0.185


Standardized Within-Group Residuals:
     Min        Q1       Med       Q3       Max
 -3.1681 -0.38594 0.0071041 0.44515 3.8495


Number of Observations: 108
Number of Groups: 27
```

The small *p*-values associated with `Sex` and `Sex:I(age-11)` in the `summary` output indicate that boys and girls have significantly different orthodontic growth patterns.

The `fitted` method is used to extract the fitted values from the `lme` object, using the methodology described in §1.4.2. By default, the *within-group* fitted values, that is, the fitted values corresponding to the individual coefficient estimates, are produced. Population fitted values, based on the fixed-effects estimates alone, are obtained setting the `level` argument to 0 (zero). Both types of fitted values can be simultaneously obtained with

```
> fitted( fm2Orth.lme, level = 0:1 )
    fixed Subject
1 22.616  24.846
2 24.184  26.576
3 25.753  28.307
 . . .
```

Residuals are extracted with the `resid` method, which also takes a `level` argument.

```
> resid( fm2Orth.lme, level = 1 )
    M01      M01      M01      M01      M02      M02      M02      M02
 1.1543 -1.5765 0.69274 0.96198 0.22522 -0.29641 -1.318 0.66034
 . . .
    F10      F10      F10      F10      F11      F11      F11
 -1.2233 0.44296 -0.39073 -0.72443 0.28277 -0.37929 1.4587
    F11
 0.29661
attr(, "label"):
[1] "Residuals (mm)"
```

By default, the raw, or *response* residuals, given by the observed responses minus the fitted values, are calculated. Standardized, or *Pearson* residuals, corresponding to the raw residuals divided by the estimated within-group standard deviation, are obtained using

```
> resid( fm2Orth.lme, level = 1, type = "pearson" )
     M01      M01     M01     M01     M02      M02      M02
 0.88111 -1.2034 0.5288 0.73431 0.17192 -0.22626 -1.0061
 . . .
     F09      F10     F10      F10      F10     F11      F11
-0.76369 -0.93382 0.33813 -0.29826 -0.55298 0.21585 -0.28952
     F11      F11
 1.1135 0.22641
attr(, "label"):
[1] "Standardized residuals"
```

Partial matching of arguments is used throughout the nlme library, so `type = "p"` would suffice in this case.

Predicted values are obtained with the `predict` method. For example, to predict the orthodontic distance for boy M11 and girl F03 at ages 16, 17 and 18, we first define a data frame with the relevant information

```
> newOrth <- data.frame( Subject = rep(c("M11","F03"), c(3, 3)),
+                        Sex = rep(c("Male", "Female"), c(3, 3)),
+                        age = rep(16:18, 2) )
```

and then use

```
> predict( fm2Orth.lme, newdata = newOrth )
    M11     M11     M11     F03     F03    F03
 26.968 27.612 28.256 26.614 27.207 27.8
attr(, "label"):
[1] "Predicted values (mm)"
```

By default, the within-group predictions are produced. To obtain both population and within-group predictions we use

```
> predict( fm2Orth.lme, newdata = newOrth, level = 0:1 )
  Subject predict.fixed predict.Subject
1     M11        28.891          26.968
2     M11        29.675          27.612
3     M11        30.459          28.256
4     F03        25.045          26.614
5     F03        25.525          27.207
6     F03        26.005          27.800
```

The `predict.fixed` column gives the population predictions, while `predict.Subject` gives the within-group predictions. We see that M11 is below the boys' average, while F11 is above the girls' average.

The `fm2Orth.lme` object corresponds to a restricted-maximum likelihood fit, which tends to produce more conservative estimates of the variance components. A maximum likelihood fit is obtained with

```
> fm2Orth.lmeM <- update( fm2Orth.lme, method = "ML" )
> summary( fm2Orth.lmeM )
Linear mixed-effects model fit by maximum likelihood
 Data: Orthodont
      AIC    BIC  logLik
  443.81 465.26 -213.9

Random effects:
 Formula:  ~ I(age - 11) | Subject
 Structure: General positive-definite
                StdDev   Corr
(Intercept) 1.75219 (Inter
I(age - 11) 0.15414 0.234
   Residual 1.31004

Fixed effects: distance ~ Sex + I(age - 11) + Sex:I(age - 11)
                 Value Std.Error DF t-value p-value
   (Intercept) 23.808   0.37332 79  63.775  <.0001
           Sex -1.161   0.37332 25  -3.109  0.0046
   I(age - 11)  0.632   0.06606 79   9.567  <.0001
Sex:I(age - 11) -0.152  0.06606 79  -2.307  0.0237
 . . .
```

As expected, the ML estimates of the random-effects standard deviations are smaller than the corresponding REML estimates. The estimated within-group residual standard deviations are identical, which generally need not occur. In general, the fixed-effects estimates obtained using ML and REML will be similar, though not identical, as in this example. Inferences regarding the fixed effects are essentially the same for the two estimation methods, in this case.

It is instructive, at this point, to compare the individual coefficient estimates obtained with `lmList` to those obtained with `lme`. The function `compareFits` can be used for this. The resulting object has a `plot` method that displays these coefficients side-by-side.

```
> compOrth <-
+         compareFits( coef(fm2Orth.lis), coef(fm1Orth.lme) )
> compOrth
, , (Intercept)
        coef(fm2Orth.lis) coef(fm1Orth.lme)
M16             23.000             23.078
M05             23.000             23.128
M02             23.375             23.455

 . . .
F04             24.875             24.764
F11             26.375             26.156

, , I(age - 11)
M16              0.550              0.59133
```

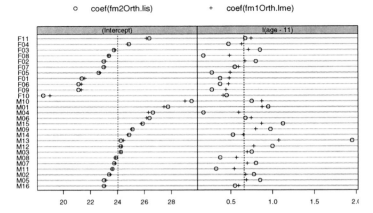

FIGURE 4.8. Individual estimates from an `lmList` fit and from an `lme` fit of the orthodontic distance growth data

```
M05           0.850           0.68579
M02           0.775           0.67469
 . . .
M13           1.950           1.07385
 . . .
F04           0.475           0.63032
F11           0.675           0.74338
> plot( compOrth, mark = fixef(fm1Orth.lme) )          # Figure 4.8
```

The `mark` argument to the `plot` method indicates points in the horizontal axis where dashed vertical lines should be drawn.

The plots in Figure 4.8 indicate that the individual estimates from the `lme` fit tend to be "pulled toward" the fixed-effects estimates, when compared to the `lmList` estimates. This is typical of linear mixed-effects estimation: the individual coefficient estimates from the `lme` fit represent a compromise between the coefficients from the individual fits corresponding to the `lmList` fit and the fixed-effects estimates, associated with the population averages. For this reason, these estimates are often called *shrinkage estimates*, in the sense that they shrink the individual estimates toward the population average.

The shrinkage toward the fixed effects is particularly noticeable for the slope estimate of subject `M13`. As pointed out in §4.2, this subject has an outlying orthodontic growth pattern, which leads to an abnormally high estimated slope in the `lm` fit. The pooling of subjects in the `lme` estimation gives a certain amount of robustness to individual outlying behavior. This feature is better illustrated by the comparison of the predicted values from the two fits, which is obtained with the `comparePred` function.

```
> plot( comparePred(fm2Orth.lis, fm1Orth.lme, length.out = 2),
```

——— fm2Orth.lis   ······· fm1Orth.lme

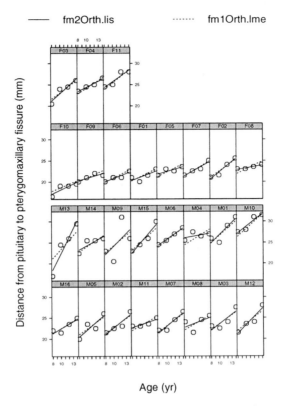

FIGURE 4.9. Individual predicted values from separate `lm` fits and from an `lme` fit of the orthodontic distance growth data

```
+   layout = c(8,4), between = list(y = c(0, 0.5)) )        # Figure 4.9
```

The `length.out` argument specifies the number of predictions for each fitted object. In this case, because the model is a straight line, only two points are needed. The plot of the individual predictions for the `lmList` and `lme` fits, shown in Figure 4.9, clearly indicates the greater sensitivity of the individual `lm` fits to extreme observations.

It is also interesting to compare the `fm2Orth.lme` and the `fmOrth.lmeM` objects, corresponding, respectively, to REML and ML fits of the same model. We compare the estimated random effects for each fit with the `compareFits` function.

```
> plot( compareFits(ranef(fm2Orth.lme), ranef(fm2Orth.lmeM)),
+       mark = c(0, 0) )                                    # Figure 4.10
```

The ML random-effects estimates tend to be closer to zero than the REML estimates, especially the slope random effects. This will usually occur in mixed-effects models, because REML estimation generally produces larger

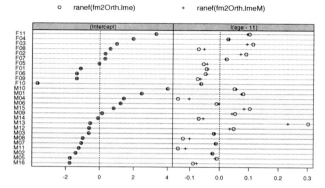

FIGURE 4.10. Individual random-effects estimates from restricted maximum likelihood (REML) and maximum likelihood (ML) fits of the orthodontic distance growth data

estimates for the random-effects variances which, in turn, result in less shrinkage toward zero for the random-effects estimates.

We can also compare the `lme` fit to an `lm` fit without random effects:

```
> fm4Orth.lm <- lm( distance ~ Sex * I(age-11), Orthodont )
> summary( fm4Orth.lm )
. . .
Coefficients:
                  Value Std. Error  t value Pr(>|t|)
    (Intercept)  23.808     0.221   107.731   0.000
            Sex  -1.161     0.221    -5.251   0.000
    I(age - 11)   0.632     0.099     6.394   0.000
Sex:I(age - 11)  -0.152     0.099    -1.542   0.126

Residual standard error: 2.26 on 104 degrees of freedom
. . .
```

The pointwise estimates of the fixed effects are almost identical, but their standard errors are quite different. The `lm` fit has smaller standard errors for the `(Intercept)` and `Sex` fixed effects and larger standard errors for the fixed effects involving `age`. This is because the model used in `lm` ignores the group structure of the data and incorrectly combines the between-group and the within-group variation in the residual standard error. Fixed effects that are associated with invariant factors (factors that do not vary within groups) are actually estimated with less precision than suggested by the `lm` output, because the contribution of the between-group variation to their standard error is larger than that included in the `lm` residual standard error. Conversely, the precision of the fixed effects related to variables that vary within group are less affected by the between-group variation. In the terminology of split-plot experiments, `(Intercept)` and `Sex` are associated with *whole-plot* treatments and should be compared to the whole-plot error, while `I(age-11)` and `Sex:I(age-11)` are related to *subplot* treatments and should be tested against the subplot error.

The `anova` method can be used to compare `lme` and `lm` objects.

```
> anova( fm2Orth.lme, fm4Orth.lm )
            Model df    AIC    BIC  logLik    Test L.Ratio p-value
fm2Orth.lme     1  8 451.35 472.51 -217.68
fm4Orth.lm      2  5 496.33 509.55 -243.17 1 vs 2  50.977  <.0001
```

In this case, as evidenced by the low *p*-value for the likelihood ratio test, the linear mixed-effects model provides a much better description of the data than the linear regression model.

Radioimmunoassays of IGF-I

The linear mixed-effects model corresponding to the simple linear regression of the estimated concentration of IGF-I (y_{ij}) in the jth tracer sample within the ith lot on the age of the tracer sample (x_{ij}) is

$$y_{ij} = (\beta_0 + b_{0i}) + (\beta_1 + b_{1i}) x_{ij} + \epsilon_{ij},$$

$$b_i = \begin{bmatrix} b_{0i} \\ b_{1i} \end{bmatrix} \sim \mathcal{N}\left(0, \Psi\right), \quad \epsilon_{ij} \sim \mathcal{N}\left(0, \sigma^2\right), \tag{4.2}$$

where β_0 and β_1 are, respectively, the fixed effects for the intercept and the slope; the b_i are the random-effects vectors, assumed to be independent for different lots; and the ϵ_{ij} are the independent, identically distributed within-group errors, assumed to be independent of the random effects.

We fit the linear mixed-effects model (4.2) with

```
> fm1IGF.lme <- lme( fm1IGF.lis )
> fm1IGF.lme
Linear mixed-effects model fit by REML
  Data: IGF
  Log-restricted-likelihood: -297.18
  Fixed: conc ~ age
 (Intercept)           age
       5.375 -0.0025337

Random effects:
 Formula: ~ age | Lot
 Structure: General positive-definite
             StdDev    Corr
(Intercept) 0.0823594 (Inter
        age 0.0080862 -1
   Residual 0.8206310

Number of Observations: 237
Number of Groups: 10
```

The fixed-effects estimates are similar to the ones obtained with the single `lm` fit of §4.1.1. The within-group standard errors are also similar in the two fits, which suggests that not much is gained by incorporating random effects into the model. The estimated correlation between the random effects ($\simeq -1$) gives a clear indication that the estimated random-effects covariance matrix is ill-conditioned, suggesting that the model may be overparameterized. The confidence intervals for the standard deviations and correlation coefficient reinforce the indication of overparameterization.

```
> intervals( fm1IGF.lme )
Approximate 95% confidence intervals

 Fixed effects:
                lower        est.      upper
(Intercept)  5.163178   5.3749606 5.5867427
        age -0.012471  -0.0025337 0.0074039

 Random Effects:
  Level: Lot
                           lower        est.     upper
       sd((Intercept))  0.0011710  0.0823594 5.792715
               sd(age)  0.0013177  0.0080862 0.049623
 cor((Intercept),age) -1.0000000 -0.9999640 1.000000

 Within-group standard error:
  lower     est.    upper
 0.7212  0.82063  0.93377
```

The 95% confidence interval for the correlation coefficient covers all possible values for this parameter. There is also evidence of large variability in the estimates of the (Intercept) and age standard deviations. These issues are explored in more detail in §4.3.

The primary question of interest in the IGF-I study is whether the tracer decays with age. We can investigate it with the `summary` method.

```
> summary( fm1IGF.lme )
 . . .
 Fixed effects: conc ~ age
                Value Std.Error  DF t-value p-value
(Intercept)  5.3750   0.10748  226  50.011  <.0001
        age -0.0025   0.00504  226  -0.502  0.6159
 . . .
```

As with the `lm` results of §4.1.1, there is no significant evidence of tracer decay with age, for the 50-day period in which the observations were collected. Note that the standard errors for the estimates are very similar to the ones in the `lm` fit.

The plots of the coefficient estimates corresponding to `fm1IGF.lis` and `fm1IGF.lme` are shown in Figure 4.11. Once again we observe a "shrinkage

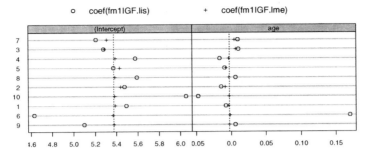

FIGURE 4.11. Individual estimates from separate `lm` fits and from an `lme` fit of the IGF-I radioimmunoassays data.

toward the population mean" pattern for the individual `lme` coefficients. The IGF-I data contain several outlying observations and the dramatic shrinkage in the coefficient estimates observed for some of the lots reflects the greater robustness of the `lme` fit. This is better illustrated by comparing the individual predictions under each fit, as presented in Figure 4.12. The differences in the predicted values for the two fits are particularly dramatic for lots 6 and 10, both of which have observations only over a very limited time range. For lot 6 a single low observation at one of the earliest times causes a dramatic change in the estimate of both the slope and intercept when this lot is fit by itself. When it is combined with the other lots in a mixed-effects model the effect of this single observation is diminished. Also notice that the outlying observations for lots 4, 3, and 7 have very little effect on the parameter estimates because in each of these lots there are several other observations at times both above and below the times of the aberrant observations.

4.2.2 Patterned Variance–Covariance Matrices for the Random Effects: The `pdMat` Classes

The models considered in §4.2.1 do not assume any special form for the random-effects variance–covariance matrix Ψ. In many practical applications, however, we will wish to restrict Ψ to special forms of variance–covariance matrices that are parameterized by fewer parameters. For example, we may be willing to assume that the random effects are independent, in which case Ψ would be diagonal, or that, in addition to being independent, they have the same variance, in which case Ψ would be a multiple of the identity matrix.

The nlme library provides several classes of positive-definite matrices, the pdMat classes, that are used to specify patterned variance–covariance matrices for the random effects. Table 4.3 lists the standard pdMat classes included in the nlme library. The default class of positive-definite matrix

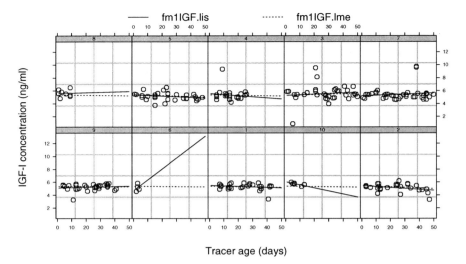

FIGURE 4.12. Individual predicted values from separate `lm` fits and from an `lme` fit of the IGF-I radioimmunoassays data.

TABLE 4.3. Standard `pdMat` classes.

`pdBlocked`	block-diagonal
`pdCompSymm`	compound-symmetry structure
`pdDiag`	diagonal
`pdIdent`	multiple of an identity
`pdSymm`	general positive-definite matrix

for the random effects in the nlme library is `pdSymm`, corresponding to a general symmetric positive-definite matrix.

A function that creates an object of a given class is called a *constructor* for that class. The `pdMat` constructors have the same name as their corresponding classes so, for example, the constructor for the `pdDiag` class is also called `pdDiag`.

Because initial values for Ψ can be derived internally in the `lme` function, the `pdMat` constructors are typically used only to specify a `pdMat` class and a formula for the random-effects model. For example,

```
> pd1 <- pdDiag( ~ age )
> pd1
Uninitialized positive definite matrix structure of class pdDiag
> formula( pd1 )
 ~ age
```

creates an object of class pdDiag, with a formula attribute specifying a random-effects model, but with no initial value assigned to the matrix.

Radioimmunoassays of IGF-I

The pdMat object returned by the constructor is passed through the random argument to the lme function. For example, to specify a diagonal variance–covariance structure for the random effects in the IGF-I example of §4.2.1, we use

```
> fm2IGF.lme <- update( fm1IGF.lme, random = pdDiag(~age) )
> fm2IGF.lme
Linear mixed-effects model fit by REML
  Data: IGF
  Log-restricted-likelihood: -297.4
  Fixed: conc ~ age
  (Intercept)         age
     5.369 -0.0019301

Random effects:
  Formula:  ~ age | Lot
  Structure: Diagonal
          (Intercept)        age Residual
StdDev:   0.00031074 0.0053722   0.8218
  . . .
```

With the exception of the standard deviation for the (Intercept) random effect, all estimates are similar to the ones in fm1IGF.lme. We can compare the two fits with

```
> anova( fm1IGF.lme, fm2IGF.lme )
           Model df    AIC    BIC  logLik   Test L.Ratio p-value
fm1IGF.lme     1  6 606.37 627.12 -297.18
fm2IGF.lme     2  5 604.80 622.10 -297.40 1 vs 2 0.43436  0.5099
```

The large p-value for the likelihood ratio test and the smaller AIC and BIC values for the simpler model fm2IGF.lme indicate that it should be preferred.

Because IGF is a groupedData object, the grouping structure does not need to be given explicitly in random. In cases when both the grouping structure and a pdMat class are to be declared in random, we use a *named list*, with the name specifying the grouping factor.

```
> update( fm1IGF.lme, random = list(Lot = pdDiag(~ age)) )
```

The value argument to the constructor is used to assign a value to the positive-definite matrix. In this case, the random-effects formula needs to be specified through the form argument.

```
> pd2 <- pdDiag( value = diag(2), form = ~ age )
```

```
> pd2
Positive definite matrix structure of class pdDiag representing
      [,1] [,2]
[1,]   1    0
[2,]   0    1
> formula( pd2 )
  ~ age
```

This can be used to provide initial values for the scaled variance–covariance matrix of the random effects, $D = \Psi/\sigma^2$ in the lme call.

```
> lme( conc ~ age, IGF, pdDiag(diag(2), ~age) )
```

Split-Plot Experiment on Varieties of Oats

We now revisit the Oats example of §1.6 and describe alternative ways of analyzing the split-plot data using pdMat classes. The final mixed-effects model resulting from the analysis presented in that section is

$$
\begin{aligned}
y_{ijk} = \beta_0 + \beta_1 N_k + b_i + b_{i,j} + \epsilon_{ijk}, \quad i = 1, \ldots, 6, \\
j = 1, \ldots 3, \quad k = 1, \ldots, 4, \\
\epsilon_{ijk} \sim \mathcal{N}(0, \sigma^2), \quad b_i \sim \mathcal{N}\left(0, \sigma_1^2\right), \quad b_{i,j} \sim \mathcal{N}\left(0, \sigma_2^2\right),
\end{aligned}
\tag{4.3}
$$

where i indexes the *Blocks*, j indexes the *Varieties*, and k indexes the *Nitrogen* concentrations N_k. The intercept is represented by β_0, the Nitrogen slope by β_1 and the yield by y_{ijk}. The b_i denote the Block random effects, the $b_{i,j}$ denote the Variety within Block random effects, and the ϵ_{ijk} denote the within-group errors. This is an example of a two-level mixed-effects model, with the $b_{i,j}$ random effects *nested* within the b_i random effects.

The multilevel model capabilities of lme were used in §1.6 to fit (4.3). We recommend fitting the model this way, as it uses efficient computational algorithms designed specifically for this type of model. Nevertheless, to further illustrate the use of the pdMat classes, we consider equivalent single-level representations of the same model.

By defining

$$
\boldsymbol{y}_i = \begin{bmatrix} y_{i11} \\ y_{i12} \\ \vdots \\ y_{i34} \end{bmatrix}, \quad
\boldsymbol{\epsilon}_i = \begin{bmatrix} \epsilon_{i11} \\ \epsilon_{i12} \\ \vdots \\ \epsilon_{i34} \end{bmatrix}, \quad
\boldsymbol{\beta} = \begin{bmatrix} \beta_0 \\ \beta_1 \end{bmatrix}, \quad
\boldsymbol{b}_i^* = \begin{bmatrix} b_i + b_{i,1} \\ b_i + b_{i,2} \\ b_i + b_{i,3} \end{bmatrix},
$$

$$
\boldsymbol{X}_i = \begin{bmatrix} 1 \\ 1 \\ 1 \end{bmatrix} \otimes \begin{bmatrix} N_1 & 0 & 0 & 0 \\ 0 & N_2 & 0 & 0 \\ 0 & 0 & N_3 & 0 \\ 0 & 0 & 0 & N_4 \end{bmatrix}, \quad
\boldsymbol{Z}_i = \begin{bmatrix} 1 & 0 & 0 \\ 0 & 1 & 0 \\ 0 & 0 & 1 \end{bmatrix} \otimes \begin{bmatrix} 1 \\ 1 \\ 1 \\ 1 \end{bmatrix},
$$

with $\otimes$ denoting the Kronecker product, we can rewrite (4.3) as the single-level model

$$\boldsymbol{y}_i = \boldsymbol{X}_i\boldsymbol{\beta} + \boldsymbol{Z}_i\boldsymbol{b}_i^* + \boldsymbol{\epsilon}_i, \quad \boldsymbol{b}_i^* \sim \mathcal{N}\left(\boldsymbol{0}, \boldsymbol{\Psi}^*\right), \quad \boldsymbol{\epsilon}_i \sim \mathcal{N}\left(\boldsymbol{0}, \sigma^2 \boldsymbol{I}\right), \qquad (4.4)$$

where

$$\boldsymbol{\Psi}^* = \left[\begin{array}{ccc} \sigma_1^2 + \sigma_2^2 & \sigma_1^2 & \sigma_1^2 \\ \sigma_1^2 & \sigma_1^2 + \sigma_2^2 & \sigma_1^2 \\ \sigma_1^2 & \sigma_1^2 & \sigma_1^2 + \sigma_2^2 \end{array}\right].$$

The $\boldsymbol{\Psi}^*$ matrix has a compound symmetry structure, represented in `nlme` by the `pdCompSymm` class. We fit (4.4) with

```
> fm4OatsB <- lme( yield ~ nitro, data = Oats,
+                   random =list(Block = pdCompSymm(~ Variety - 1)))
> summary( fm4OatsB )
Linear mixed-effects model fit by REML
 Data: Oats
      AIC    BIC  logLik
  603.04 614.28 -296.52

Random effects:
 Formula:  ~ Variety - 1 | Block
 Structure: Compound Symmetry
                   StdDev  Corr
VarietyGolden Rain 18.208
VarietyMarvellous  18.208 0.635
   VarietyVictory  18.208 0.635 0.635
         Residual  12.867

Fixed effects: yield ~ nitro
              Value Std.Error DF t-value p-value
(Intercept) 81.872    6.9453 65  11.788  <.0001
      nitro 73.667    6.7815 65  10.863  <.0001
 Correlation:
      (Intr)
nitro -0.293

Standardized Within-Group Residuals:
    Min       Q1      Med      Q3    Max
-1.7438 -0.66475 0.017103 0.54299 1.803
```

Comparing this with the output of `summary(fm4Oats)` in §1.6, we see that, except for the random-effects variance–covariance components, the results are nearly identical.

Verifying the equivalence of the random-effects variance–covariance components requires some extra work. Note that the variance in the compound symmetric matrix $\boldsymbol{\Psi}^*$ is $\sigma_{b^*}^2 = \sigma_1^2 + \sigma_2^2$ and the correlation is

$\rho = \sigma_1^2/\left(\sigma_1^2 + \sigma_2^2\right)$. Therefore, $\sigma_1^2 = \rho\sigma_{b*}^2$ and $\sigma_2^2 = \sigma_{b*}^2 - \sigma_1^2$. We can then derive the REML estimates of σ_1 and σ_2 from fm4OatsB

$$\widehat{\sigma}_1 = \sqrt{0.63471} \times 18.208 = 14.506, \quad \widehat{\sigma}_2 = \sqrt{18.208^2 - 14.506^2} = 11.005,$$

verifying that they are identical to the estimates corresponding to fm4Oats in §1.6. Because the REML estimate of ρ in the summary(fm4OatsB) output was displayed with only three decimal places, we used

```
> corMatrix( fm4OatsB$modelStruct$reStruct$Block )[1,2]
[1] 0.63471
```

to obtain the more accurate $\widehat{\rho}$ used to obtain $\widehat{\sigma}_1$ and $\widehat{\sigma}_2$.

Yet another representation of (4.3) as a single-level model is obtained by defining

$$\boldsymbol{b}_i^{**} = \begin{bmatrix} b_i \\ b_{i,1} \\ b_{i,2} \\ b_{i,3} \end{bmatrix}, \quad \boldsymbol{W}_i = \begin{bmatrix} 1 & 1 & 0 & 0 \\ 1 & 0 & 1 & 0 \\ 1 & 0 & 0 & 1 \end{bmatrix} \otimes \begin{bmatrix} 1 \\ 1 \\ 1 \\ 1 \end{bmatrix},$$

$$\boldsymbol{\Psi}^{**} = \begin{bmatrix} \sigma_1^2 & 0 & 0 & 0 \\ 0 & \sigma_2^2 & 0 & 0 \\ 0 & 0 & \sigma_2^2 & 0 \\ 0 & 0 & 0 & \sigma_2^2 \end{bmatrix}$$

and writing

$$\boldsymbol{y}_i = \boldsymbol{X}_i\boldsymbol{\beta} + \boldsymbol{W}_i\boldsymbol{b}_i^{**} + \boldsymbol{\epsilon}_i, \quad \boldsymbol{b}_i^{**} \sim \mathcal{N}\left(\boldsymbol{0}, \boldsymbol{\Psi}^{**}\right), \quad \boldsymbol{\epsilon}_i \sim \mathcal{N}\left(\boldsymbol{0}, \sigma^2\boldsymbol{I}\right). \quad (4.5)$$

The parameters in (4.3) are equivalent to those in (4.5).

The $\boldsymbol{\Psi}^{**}$ matrix is structured as a block diagonal matrix, with blocks σ_1^2 and $\sigma_2^2\boldsymbol{I}$. The pdBlocked class is used to represent block diagonal matrices in the nlme library. It takes as an argument a list of pdMat objects, specifying the different blocks in the order they appear in the main diagonal of the corresponding block-diagonal matrix. In the case of $\boldsymbol{\Psi}^{**}$, both blocks are expressed as multiples of the identity matrix, represented by the pdIdent class in nlme. We can then fit (4.5) with

```
> fm4OatsC <- lme( yield ~ nitro, data = Oats,
+          random=list(Block=pdBlocked(list(pdIdent(~ 1),
+                        pdIdent(~ Variety-1)))))
> summary( fm4OatsC )
Linear mixed-effects model fit by REML
 Data: Oats
      AIC    BIC  logLik
  603.04 614.28 -296.52

Random effects:
```

```
Composite Structure: Blocked

Block 1: (Intercept)
Formula:  ~ 1 | Block

        (Intercept)
StdDev:     14.505

Block 2: VarietyGolden Rain, VarietyMarvellous, VarietyVictory
Formula:  ~ Variety - 1 | Block
Structure: Multiple of an Identity
        VarietyGolden Rain VarietyMarvellous VarietyVictory
StdDev:                11.005            11.005        11.005
        Residual
StdDev:    12.867


Fixed effects: yield ~ nitro
            Value Std.Error DF t-value p-value
(Intercept) 81.872   6.9451 65  11.788  <.0001
      nitro 73.667   6.7815 65  10.863  <.0001
 Correlation:
      (Intr)
nitro -0.293
 . . .
```

Comparing this to the output of `summary(fm4Oats)` in §1.6 we verify that, as expected, the two fits are nearly identical.

Cell Culture Bioassay with Crossed Random Effects

Data grouped according to *crossed* classification factors induce a variance–covariance structured for the observations which can be flexibly represented by mixed-effects models with crossed random effects. These models can also be fit with the `lme` function. However, unlike in the case of nested random effects, the underlying estimation algorithm is not optimized to take full advantage of the sparse structure of design matrices for crossed random effects.

The crossed random-effects structure is represented in `lme` by a combination of pdBlocked and pdIdent objects. We illustrate its use with an example of a cell culture plate bioassay conducted at Searle, Inc. The data, courtesy of Rich Wolfe and David Lansky, come from a bioassay run on a cell culture plate with two blocks of 30 wells each. The wells in each block are labeled according to six rows and five columns, corresponding to a crossed classification. Within each block, six different samples are randomly assigned to rows and five serial dilutions are randomly assigned to columns. The response variable is the logarithm of the optical density measure on a well. The cells are treated with a compound that they metabolize to produce the stain. Only live cells can make the stain, so the optical density is a

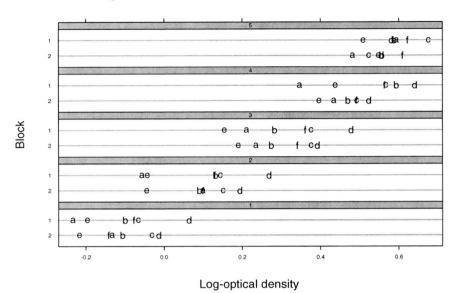

Log-optical density

FIGURE 4.13. Log-optical density measured on the central 60 wells of a cell culture plate. The wells are divided into two blocks with six rows and five columns, with samples being assigned to rows and dilutions to columns. Panels in the plot correspond to the serial dilutions and symbols refer to the samples.

measure of the number of cells that are alive and healthy. These data are described in detail in Appendix A.2 and are included in the nlme library as the groupedData object Assay.

The plot of the log-optical densities, displayed in Figure 4.13, indicates that the response increases with dilution and is generally lower for treatments a and e. There does not appear to be any interactions between sample and dilution.

A full factorial model is used to represent the fixed effects and three random effects are used to account for block, row, and column effects, with the last two random effects nested within block, but crossed with each other. The corresponding mixed-effects model for the log-optical density y_{ijk} in the jth row, kth column of the ith block, for $i = 1, \ldots, 2$, $j = 1, \ldots, 6$, $k = 1, \ldots, 5$, is

$$y_{ijk} = \mu + \alpha_j + \beta_k + \gamma_{jk} + b_i + r_{ij} + c_{ik} + \epsilon_{ijk},$$
$$b_i \sim \mathcal{N}\left(0, \sigma_1^2\right), \quad r_{ij} \sim \mathcal{N}\left(0, \sigma_2^2\right), \quad c_{ik} \sim \mathcal{N}\left(0, \sigma_3^2\right), \quad \epsilon_{ijk} \sim \mathcal{N}\left(0, \sigma^2\right).$$
$$(4.6)$$

The fixed effects in (4.6) are μ, the grand mean, α_j and β_k, the *sample* and *dilution* main effects, and γ_{jk}, the sample-dilution interaction. To ensure identifiability of the fixed effects, it is conventioned that $\alpha_1 = \beta_1 = \gamma_{1k} =$

$\gamma_{j1} = 0$, for $j = 1, \ldots, 6$, $k = 1, \ldots, 5$. The random effects in (4.6) are b_i, the *block* random effect, r_{ij}, the *row within block* random effect, and c_{ik}, the *column within block* random effect. All random effects are assumed independent of each other and independent of the within-groups errors ϵ_{ijk}.

The crossed random-effects structure in (4.6) can alternatively be represented as the random-effects structure corresponding to a single-level model, with *block* as the single grouping variable, and random-effects vector $\boldsymbol{b}_i = (b_i, r_{i1}, \ldots, r_{i6}, c_{i1}, \ldots, c_{i5})^T$, $i = 1, \ldots, 2$, with

$$\text{Var}\,(\boldsymbol{b}_i) = \begin{bmatrix} \sigma_1^2 & 0 & 0 \\ 0 & \sigma_2^2 I & 0 \\ 0 & 0 & \sigma_3^2 I \end{bmatrix}.$$

That is, $\boldsymbol{b}_i$ has a block diagonal variance–covariance matrix, with diagonal blocks given by multiples of the identity matrix. This type of variance–covariance structure is represented in S by a pdBlocked object with pdIdent elements. We fit the linear mixed-effects model (4.6) with `lme` as

```
> ## establishing the desired parameterization for contrasts
> options( contrasts = c("contr.treatment", "contr.poly") )
> fm1Assay <- lme( logDens ~ sample * dilut, Assay,
+     random = pdBlocked(list(pdIdent(~ 1), pdIdent(~ sample - 1),
+                         pdIdent(~ dilut - 1))) )
> fm1Assay
Linear mixed-effects model fit by REML
  Data: Assay
  Log-restricted-likelihood: 38.536
  Fixed: logDens ~ sample * dilut
  (Intercept)  sampleb  samplec  sampled    samplee   samplef   dilut2
     -0.18279 0.080753 0.13398   0.2077  -0.023672 0.073569  0.20443
  dilut3   dilut4   dilut5 samplebdilut2 samplecdilut2
 0.40586 0.57319 0.72064     0.0089389    -0.0084953
 sampleddilut2 sampleedilut2 samplefdilut2 samplebdilut3
    0.0010793    -0.041918     0.019352    -0.025066
 samplecdilut3 sampleddilut3 sampleedilut3 samplefdilut3
    0.018645     0.0039886    -0.027713     0.054316
 samplebdilut4 samplecdilut4 sampleddilut4 sampleedilut4
    0.060789     0.0052598    -0.016486     0.049799
 samplefdilut4 samplebdilut5 samplecdilut5 sampleddilut5
    0.063372    -0.045762     -0.072598     -0.17776
 sampleedilut5 samplefdilut5
    0.013611     0.0040234

Random effects:
 Composite Structure: Blocked

 Block 1: (Intercept)
 Formula:   ~ 1 | Block
```

```
                (Intercept)
StdDev:    0.0098084


  Block 2: samplea, sampleb, samplec, sampled, samplee, samplef
  Formula:   ~ sample - 1 | Block
  Structure: Multiple of an Identity
            samplea  sampleb  samplec  sampled  samplee  samplef
StdDev: 0.025289 0.025289 0.025289 0.025289 0.025289 0.025289


  Block 3: dilut1, dilut2, dilut3, dilut4, dilut5
  Formula:   ~ dilut - 1 | Block
  Structure: Multiple of an Identity
            dilut1     dilut2     dilut3     dilut4     dilut5
StdDev: 0.0091252 0.0091252 0.0091252 0.0091252 0.0091252
            Residual
StdDev: 0.041566


Number of Observations: 60
Number of Groups: 2
```

The REML estimates of the standard deviation components in this example are $\hat{\sigma}_1 = 0.0098$, $\hat{\sigma}_2 = 0.0253$, $\hat{\sigma}_3 = 0.0091$, and $\hat{\sigma} = 0.0416$.

The primary question of interest for this experiment is whether there are significant differences among the fixed effects, which we investigate with

```
> anova( fm1Assay )
               numDF denDF F-value p-value
 (Intercept)       1    29  538.03  <.0001
      sample       5    29   11.21  <.0001
       dilut       4    29  420.80  <.0001
 sample:dilut     20    29    1.61  0.1193
```

As suggested by Figure 4.13, there are significant differences among samples and among dilutions, but no significant interaction between the two factors.

The small estimated standard deviations in the fm1Assay fit suggest that some, or perhaps all, of the random effects can be eliminated from (4.6). However, because our purpose here is just to illustrate the use of lme with crossed random effects, we do not pursue the analysis of the Assay data any further.

New pdMat classes, representing user-defined positive-definite matrix structures, can be added to the set of standard classes in Table 4.3 and used with the lme and nlme functions. For this, one must specify a constructor function, generally with the same name as the class, and, at a minimum, methods for the functions pdConstruct, pdMatrix, and coef. The pdDiag constructor and methods can serve as templates for these.

4.2.3   Fitting Multilevel Models

Linear mixed-effects models with nested grouping factors, generally called multilevel models (Goldstein, 1995) or hierarchical linear models (Bryk and Raudenbush, 1992), can be fitted with the `lme` function, just like single-level models. The only difference is in the specification of the `random` argument, which must provide information about the nested grouping structure and the random-effects model at each level of grouping. We describe the multi-level model capabilities in `lme` through the analyses of two examples from Integrated Circuiting (IC) manufacturing.

Thickness of Oxide Coating on a Semiconductor

Littell et al. (1996) describe data from a passive data collection study in the IC industry in which the thickness of the oxide coating layer was measured on three randomly selected *sites* in each of three *wafers* from each of eight *lots* randomly selected from the population of lots. There are two nested grouping levels in this example: *lot* and *wafer within lot*. The objective of the study was to estimate the variance components associated with the different levels of nesting and the within-group error, to evaluate assignable causes of variability in the oxide deposition process. These data are also described in Appendix A.20 and are included in the `groupedData` object `Oxide` in the `nlme` library.

The plot of the data, shown in Figure 4.14, suggests that the lot-to-lot variability of the oxide layer thickness is greater than the wafer-to-wafer variability within a lot, which, in turn, is greater than the site-to-site variation within a wafer.

A multilevel model to describe the oxide thickness y_{ijk} measured on the kth site of the jth wafer within the ith lot is

$$y_{ijk} = \mu + b_i + b_{i,j} + \epsilon_{ijk}, \quad i = 1, \dots, 8, \quad j, k = 1, 2, 3,$$
$$b_i \sim \mathcal{N}\left(0, \sigma_1^2\right), \quad b_{i,j} \sim \mathcal{N}\left(0, \sigma_2^2\right), \quad \epsilon_{ijk} \sim \mathcal{N}\left(0, \sigma^2\right), \tag{4.7}$$

where the *lot* random effects b_i are assumed to be independent for different i, the *wafer within lot* random effects $b_{i,j}$ are assumed to be independent for different i and j and to be independent of the b_i, and the within-group errors ϵ_{ijk} are assumed to be independent for different i, j, and k and to be independent of the random effects.

The most general form of the argument `random` when `lme` is used to fit a multilevel model is as a *named list* where the names define the grouping factors and the formulas describe the random-effects models at each level. The order of nesting is taken to be the order of the elements in the list, with the outermost level appearing first. In the case of (4.7) we write

```
random = list( Lot = ~ 1, Wafer = ~ 1 )
```

When the random-effects formulas are the same for all levels of grouping, we can replace the named list by a one-sided formula with the common

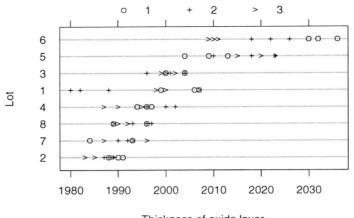

FIGURE 4.14. Thickness of oxide layer measured on different sites of wafers selected from a sample of manufacturing lots. Symbols denote different wafers within the same lot.

random-effects formula and an expression defining the grouping structure separated by a | operator.

```
random = ~ 1 | Lot/Wafer
```

Because Oxide contains this grouping structure in its display formula

```
> formula( Oxide )
Thickness ~ 1 | Lot/Wafer
```

the grouping structure expression can be omitted from random. In fact, because, by default, random is equal to the right-hand side of the fixed formula, ~1 in this case, we can omit random all together from the lme call and fit the model with

```
> fm1Oxide <- lme( Thickness ~ 1, Oxide )
> fm1Oxide
Linear mixed-effects model fit by REML
  Data: Oxide
  Log-restricted-likelihood: -227.01
  Fixed: Thickness ~ 1
  (Intercept)
      2000.2

Random effects:
  Formula:   ~ 1 | Lot
         (Intercept)
StdDev:       11.398
```

```
Formula:   ~ 1 | Wafer %in% Lot
        (Intercept) Residual
StdDev:      5.9888    3.5453
```

```
Number of Observations: 72
Number of Groups:
 Lot Wafer %in% Lot
  8               24
```

The REML estimates of the variance components are the square of the standard deviations in the `fm1Oxide` output: $\widehat{\sigma}_1^2 = 11.398^2 = 129.91$, $\widehat{\sigma}_2^2 = 5.9888^2 = 35.866$, and $\widehat{\sigma}^2 = 3.5453^2 = 12.569$. We can assess the variability in these estimates with the `intervals` method.

```
> intervals( fm1Oxide, which = "var-cov" )
Approximate 95% confidence intervals

  Random Effects:
   Level: Lot
                  lower    est.  upper
 sd((Intercept)) 5.0277 11.398 25.838
   Level: Wafer
                  lower    est.  upper
 sd((Intercept)) 3.4615 5.9888 10.361

  Within-group standard error:
   lower    est.  upper
  2.6719 3.5453 4.7044
```

All intervals are bounded well away from zero, indicating that the two random effects should be kept in (4.7). We can test, for example, if the *wafer within lot* random effect can be eliminated from the model with

```
> fm2Oxide <- update( fm1Oxide, random = ~ 1 | Lot)
> anova( fm1Oxide, fm2Oxide )
         Model df    AIC    BIC  logLik   Test L.Ratio p-value
fm1Oxide     1  4 462.02 471.07 -227.01
fm2Oxide     2  3 497.13 503.92 -245.57 1 vs 2   37.11  <.0001
```

The very high value of the likelihood ratio test statistic confirms that the significance of that term in the model.

As with single-level fits, estimated BLUPs of the individual coefficients are obtained using `coef`, but, because of the multiple grouping levels, a `level` argument is used to specify the desired grouping level. For example, to get the estimated average oxide layer thicknesses by lot, we use

```
> coef( fm1Oxide, level = 1 )
   (Intercept)
1      1996.7
2      1988.9
```

```
3         2001.0
4         1995.7
5         2013.6
6         2019.6
7         1992.0
8         1993.8
```

while the estimated average thicknesses per wafer are obtained with

```
> coef( fm1Oxide, level = 2 )          # default when level is omitted
     (Intercept)
1/1        2003.2
1/2        1984.7
1/3        2001.1
 . . .
8/2        1995.2
8/3        1990.7
```

The `level` argument is used similarly with the methods `fitted`, `predict`, `ranef`, and `resid`, with the difference that multiple levels can be simultaneously specified. For example, to get the estimated random effects at both grouping levels we use

```
> ranef( fm1Oxide, level = 1:2 )
Level: Lot
   (Intercept)
1    -3.46347
2   -11.22164
 . . .
8    -6.38538

Level: Wafer %in% Lot
     (Intercept)
1/1    6.545993
1/2  -11.958939
 . . .
8/3   -3.074863
```

These methods are further illustrated in §4.3, when we describe tools for assessing the adequacy of fitted models.

Manufacturing of Analog MOS Circuits

The `Wafer` data, introduced in §3.3.4 and shown in Figure 4.15, provide another example of multilevel data in IC manufacturing and are used here to illustrate the capabilities in `lme` when covariates are used in a multilevel model. As described in Appendix A.30, these data come from an experiment conducted at the Microelectronics Division of Lucent Technologies to study different sources of variability in the manufacturing of analog MOS circuits. The intensity of current (in mA) at 0.8, 1.2, 1.6, 2.0, and 2.4

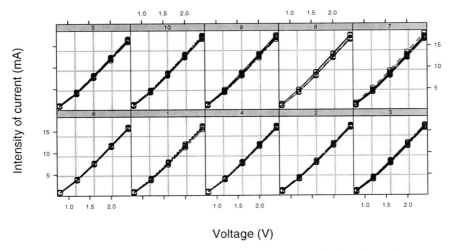

FIGURE 4.15. Current versus voltage curves for each site, by wafer.

V was measured on 80μm$\times0.6\mu$m n-channel devices. Measurements were made on eight sites in each of ten wafers selected from the same lot. Two levels of nesting are present in these data: *wafer* and *site within wafer*. The main objective of the experiment was to construct an empirical model for simulating the behavior of similar circuits.

From Figure 4.15, it appears that current can be modeled as a quadratic function of voltage. We initially consider a *full* mixed-effects model, with all terms having random effects at both the *wafer* and the *site within wafer* levels. The corresponding multilevel model for the intensities of current y_{ijk} at the kth level of voltage v_k in the jth site within the ith wafer is expressed, for $i = 1, \ldots, 10$, $j = 1, \ldots, 8$, and $k = 1, \ldots, 5$ as

$$y_{ijk} = (\beta_0 + b_{0i} + b_{0i,j}) + (\beta_1 + b_{1i} + b_{1i,j})\, v_k + (\beta_2 + b_{2i} + b_{2i,j})\, v_k^2 + \epsilon_{ijk},$$

$$\boldsymbol{b}_i = \begin{bmatrix} b_{0i} \\ b_{1i} \\ b_{2i} \end{bmatrix} \sim \mathcal{N}\left(0, \boldsymbol{\Psi}_1\right), \quad \boldsymbol{b}_{i,j} = \begin{bmatrix} b_{0i,j} \\ b_{1i,j} \\ b_{2i,j} \end{bmatrix} \sim \mathcal{N}\left(\mathbf{0}, \boldsymbol{\Psi}_2\right),$$

$$\epsilon_{ijk} \sim \mathcal{N}\left(0, \sigma^2\right).$$

$$\text{(4.8)}$$

The parameters β_0, β_1, and β_2 are the fixed effects in the quadratic model, $\boldsymbol{b}_i$ is the *wafer-level* random-effects vector, $\boldsymbol{b}_{i,j}$ is the *site within wafer-level* random-effects vector, and ϵ_{ijk} is the within-group error. As usual, the $\boldsymbol{b}_i$ are assumed to be independent for different i, the $\boldsymbol{b}_{i,j}$ are assumed to be independent for different i, j and independent of the $\boldsymbol{b}_i$, and the ϵ_{ijk} are assumed to be independent for different i, j, k and independent of the random effects.

The large number of parameters in (4.8)—twelve variance–covariance components for the random effects—makes the optimization of the profiled log-restricted-likelihood quite difficult and unstable. To make the optimization more stable during this initial model-building phase, we simplify (4.8) by assuming that Ψ_1 and Ψ_2 are diagonal matrices.

```
> fm1Wafer <- lme( current ~ voltage + voltage^2, data = Wafer,
+               random = list(Wafer = pdDiag(~voltage + voltage^2),
+                             Site = pdDiag(~voltage + voltage^2)))
> summary( fm1Wafer )
Linear mixed-effects model fit by REML
 Data: Wafer
       AIC      BIC  logLik
  -281.51 -241.67 150.75

Random effects:
 Formula: ~ voltage + voltage^2 | Wafer
 Structure: Diagonal
        (Intercept) voltage I(voltage^2)
StdDev:  0.00047025 0.18717      0.025002

 Formula: ~ voltage + voltage^2 | Site %in% Wafer
 Structure: Diagonal
        (Intercept) voltage I(voltage^2) Residual
StdDev:  0.00038085 0.13579    1.5202e-05  0.11539

Fixed effects: current ~ voltage + voltage^2
               Value Std.Error  DF t-value p-value
 (Intercept) -4.4612  0.051282 318 -86.992  <.0001
     voltage  5.9034  0.092700 318  63.683  <.0001
I(voltage^2)  1.1704  0.022955 318  50.987  <.0001
 Correlation:
             (Intr) voltag
     voltage -0.735
I(voltage^2)  0.884 -0.698

Standardized Within-Group Residuals:
     Min       Q1      Med      Q3     Max
 -1.8967 -0.53534 0.024858 0.79853 1.7777

Number of Observations: 400
Number of Groups:
 Wafer Site %in% Wafer
    10              80
```

Because Wafer is a groupedData object and the random-effects model is identical for both levels of grouping, we could have used

```
    random = pdDiag( ~ voltage + voltage^2 )
```

in the `lme` call to produce `fm1Wafer`. In this case, the object specified in `random` is repeated for all levels of grouping.

The very small estimated standard deviations for the `(Intercept)` random effect at both levels of grouping and for the `voltage^2` random effect at the `Site %in% Wafer` level suggest that these terms could be eliminated from (4.8). Before pursuing this any further, we should assess the adequacy of the fitted model. This is considered in detail in §4.3 and reveals that important terms are omitted from the fixed-effects model in (4.8). We therefore postpone this discussion until §4.3 and proceed with the analysis of `fm1Wafer` in this section to further illustrate the use of lme methods with multilevel objects.

As with single-level objects, the `fitted` method is used to extract the fitted values, with the `level` argument being used to specify the desired level(s) of grouping. For example, to get the population level fitted values, we use

```
> fitted( fm1Wafer, level = 0 )
       1      1      1      1      1      1      1      1      1
  1.0106 4.3083 7.9805 12.027 16.448 1.0106 4.3083 7.9805 12.027
 . . .
      10     10     10     10
  4.3083 7.9805 12.027 16.448
attr(, "label"):
[1] "Fitted values (mA)"
```

Similarly, residuals are extracted using the `resid` method. For example, the `Wafer` and `Site %in% Wafer` residuals are obtained with

```
> resid( fm1Wafer, level = 1:2 )
         Wafer        Site
  1  0.0615008   0.0680629
  2 -0.1898559  -0.1800129
 . . .
399  0.0051645   0.1187074
400 -0.2076543  -0.0714028
```

The `predict` method is used to obtain predictions for new observations. For example, to obtain the predicted currents at 1.0 V, 1.5 V, 3.0 V and 3.5 V for Wafer 1, we first construct a data frame with the relevant information

```
> newWafer <-
+     data.frame( Wafer = rep(1, 4), voltage = c(1, 1.5, 3, 3.5) )
```

and then use

```
> predict( fm1Wafer, newWafer, level = 0:1 )
  Wafer predict.fixed predict.Wafer
1     1        2.6126        2.4014
```

2	1	7.0273	6.7207
3	1	23.7826	23.2314
4	1	30.5381	29.9192

Note that, because no predictions were desired at the Site %in% Wafer level, Site did not need to be specified in newWafer. If we are interested in getting predictions for a specific site, say 3, within Wafer 1, we can use

```
> newWafer2 <- data.frame( Wafer = rep(1, 4), Site = rep(3, 4),
+                          voltage = c(1, 1.5, 3, 3.5) )
> predict( fm1Wafer, newWafer2, level = 0:2 )
  Wafer Site predict.fixed predict.Wafer predict.Site
1     1  1/3        2.6126        2.4014        2.4319
2     1  1/3        7.0273        6.7207        6.7666
3     1  1/3       23.7826       23.2314       23.3231
4     1  1/3       30.5381       29.9192       30.0261
```

These methods will be used extensively in the next section, when we describe methods for assessing the adequacy of the fitted models.

4.3 Examining a Fitted Model

Before making inferences about a fitted mixed-effects model, we should check whether the underlying distributional assumptions appear valid for the data. There are two basic distributional assumptions for the mixed-effects models considered in this chapter:

> *Assumption 1* - the within-group errors are independent and identically normally distributed, with mean zero and variance σ^2, and they are independent of the random effects.

> *Assumption 2* - the random effects are normally distributed, with mean zero and covariance matrix Ψ (not depending on the group) and are independent for different groups;

The nlme library provides several methods for assessing the validity of these assumptions. The most useful of these methods are based on plots of the residuals, the fitted values, and the estimated random effects. The validity of the distributional assumptions may also be formally assessed using hypothesis tests. However, only rarely do the conclusions of a hypothesis test about some assumption in the model contradict the information displayed in a diagnostic plot.

4.3.1 Assessing Assumptions on the Within-Group Error

Because the assumptions above break down into several smaller assumptions, several diagnostic plots are needed to properly assess their validity. We start by considering Assumption 1 on the within-group error term.

Dependencies among the within-group errors are usually modeled with correlation structures, which will be discussed in detail in §5.3, where methods for assessing the assumption of independence among the within-group errors will also be described. In this section, we concentrate on methods for assessing the assumption that the within-group errors are normally distributed, are centered at zero, and have constant variance.

The primary quantities used to assess the adequacy of Assumption 1 are the *within-group residuals*, defined as the difference between the observed response and the within-group fitted value. Conditional on the random-effects variance–covariance components, the within-group residuals are the BLUPs of the within-group errors. In practice, the within-group residuals are only estimated BLUPs, as the random-effects variance–covariance components need to be replaced with their estimates. Nevertheless, they generally provide good surrogates for the within-group errors and can be used to qualitatively assess the validity of Assumption 1. Other quantities used for assessing Assumption 1 graphically include the within-group fitted values, the observed values, and any covariates of interest.

The `plot` method for the lme class is the primary tool for obtaining diagnostic plots for Assumption 1. It takes several optional arguments, but a typical call is

```
plot( object, formula )
```

where `object` is an lme object and `formula` is a formula object describing the components to be used in the plot. The general expression for `formula` is y˜x|g where y and x define, respectively, the vertical and horizontal axes for the plot and g is an optional factor (or a set of factors separated by * operators) defining the panels of a Trellis display of y˜x. Any variables, or functions of variables, which can be evaluated in the data frame used to fit `object`, are allowed for y, x, and g, provided the resulting plot is a valid `Trellis` plot. The symbol "." is reserved to represent the fitted `object` itself in the `formula` definition. For example, `resid(.)` can be used in `formula` to represent `resid(object)`. We illustrate the capabilities of the `plot` method through the analysis of the examples described in earlier sections of this chapter. Details about the arguments to the `lmList` and lme plot methods are given in the help files in Appendix B.

Orthodontic Growth Curve

The first residual plot we consider is the boxplot of residuals by group. For the fm2Orth.lme object of §4.2.1 we use

```
> plot( fm2Orth.lme, Subject~resid(.), abline = 0 )     # Figure 4.16
```

The argument abline = 0 indicates that a vertical line at zero should be added to the plot. This plot is useful for verifying that the errors are centered at zero (i.e., $E\left[\epsilon\right] = \mathbf{0}$), have constant variance across groups

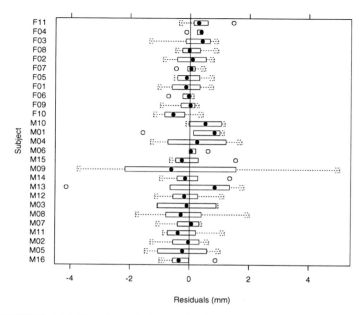

FIGURE 4.16. Boxplots of the residuals for `fm2Orth.lme` by subject.

$(\mathrm{Var}\,(\epsilon_{ij}) = \sigma^2)$, and are independent of the group levels. Figure 4.16 indicates that the residuals are centered at zero, but that the variability changes with group. Because there are only four observations per subject, we cannot rely too much on the individual boxplots for inference about the within-group variances. We observe an outlying observation for subject M13 and large residuals for subject M09. A pattern suggested by the individual boxplots is that there is more variability among boys (the lower 16 boxplots) than among girls (the upper 11 boxplots). We can get a better feeling for this pattern by examining the plot of the standardized residuals versus fitted values by gender, shown in Figure 4.17

```
> plot( fm2Orth.lme, resid(., type = "p") ~ fitted(.) | Sex,
+        id = 0.05, adj = -0.3 )                        # Figure 4.17
```

The `type = "p"` argument to the `resid` method specifies that the standardized residuals should be used. The `id` argument specifies a critical value for identifying observations in the plot (standardized residuals greater than the `1-id/2` standard normal quantile in absolute value are identified in plot). By default, the group labels are used to identify the observations. The argument `adj` controls the position of the identifying labels. It is clear from Figure 4.17 that the variability in the orthodontic distance measurements is greater among boys than among girls. Within each gender the variability seems to be constant. The outlying observations for subjects M09 and M13 are evident.

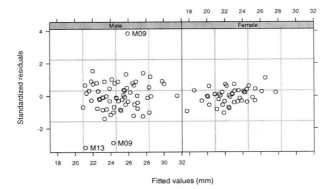

FIGURE 4.17. Scatter plots of standardized residuals versus fitted values for fm2Orth.lme by gender.

A more general model to represent the orthodontic growth data allows different variances by gender for the within-group error. The lme function allow the modeling of heteroscesdasticity of the within-error group via a weights argument. This topic will be covered in detail in §5.2, but, for now, it suffices to know that the varIdent variance function structure allows different variances for each level of a factor and can be used to fit the heteroscedastic model for the orthodontic growth data as

```
> fm3Orth.lme <-
+    update( fm2Orth.lme, weights = varIdent(form = ~ 1 | Sex) )
> fm3Orth.lme
Linear mixed-effects model fit by REML
  Data: Orthodont
  Log-restricted-likelihood: -207.15
  Fixed: distance ~ Sex + I(age - 11) + Sex:I(age - 11)
  (Intercept)    Sex I(age - 11) Sex:I(age - 11)
       23.808 -1.1605     0.63196        -0.15241

Random effects:
 Formula:  ~ I(age - 11) | Subject
 Structure: General positive-definite
            StdDev   Corr
(Intercept) 1.85498 (Inter
I(age - 11) 0.15652 0.394
   Residual 1.62959

Variance function:
 Structure: Different standard deviations per stratum
 Formula:  ~ 1 | Sex
 Parameter estimates:
 Male  Female
    1 0.40885
```

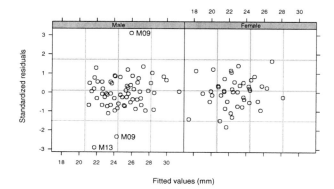

FIGURE 4.18. Scatter plots of standardized residuals versus fitted values for the heteroscedastic fit of `fm3Orth.lme` by gender.

```
Number of Observations: 108
Number of Groups: 27
```

The parameters for `varIdent` give the ratio of the stratum standard errors to the within-group standard error. To allow identifiability of the parameters, the within-group standard error is equal to the first stratum standard error. For the orthodontic data, the standard error for the girls is about 41% of that for the boys. The remaining estimates are very similar to the ones in the homoscedastic fit `fm2Orth.lme`. We can assess the adequacy of the heteroscedastic fit by re-examining plots of the standardized residuals versus the fitted values by gender, shown in Figure 4.18. The standardized residuals in each gender now have about the same variability. We can still identify the outlying observations, corresponding to subjects M09 and M13. Overall, the standardized residuals are small, suggesting that the linear mixed-effects model was successful in explaining the orthodontic growth curves This is better seen by looking at a plot of the observed responses versus the within-group fitted values.

```
> plot( fm3Orth.lme, distance ~ fitted(.),
+        id = 0.05, adj = -0.3 )                    # Figure 4.19
```

The `fm3Orth.lme` fitted values are in close agreement with the observed orthodontic distances, except for the three extreme observations on subjects M09 and M13.

The need for an heteroscedastic model for the orthodontic growth data can be formally tested with the `anova` method.

```
> anova( fm2Orth.lme, fm3Orth.lme )
            Model df    AIC    BIC  logLik    Test L.Ratio p-value
fm2Orth.lme     1  8 451.35 472.51 -217.68
fm3Orth.lme     2  9 432.30 456.09 -207.15 1 vs 2  21.059  <.0001
```

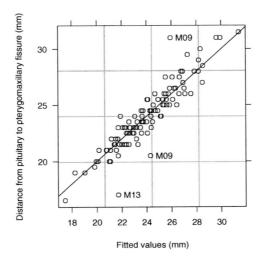

FIGURE 4.19. Observed versus fitted values plot for `fm3Orth.lme`.

The very small p-value of the likelihood ratio statistic confirms that the heteroscedastic model explains the data significantly better than the homoscedastic model.

The assumption of normality for the within-group errors can be assessed with the normal probability plot of the residuals, produced by the `qqnorm` method. A typical call to `qqnorm` is of the form

```
qqnorm( object, formula )
```

where `object` is an lme object and `formula` is a one-sided formula object of the form `~x|g`. As in the `plot` method, the symbol `"."` represents the fitted lme object. The x term in formula can be either the residuals (`resid(.)`), or the random effects (`ranef(.)`) associated with the fit. In this section, we consider only the case where x defines a vector of residuals. The random-effects case is considered in §4.3.2. The g term in `formula` defines an optional factor (or set of factors joined by `*`) determining the panels for a Trellis display.

For example, to obtain the normal plots of the residuals corresponding to `fm3Orth.lme` by gender, we use

```
> qqnorm( fm3Orth.lme, ~resid(.) | Sex )              # Figure 4.20
```

Once again, we observe the three outlying points, but for the rest of the observations the normality assumption seems plausible.

Radioimmunoassays of IGF-I

We initially consider the plot of the standardized residuals versus fitted values by `Lot` for the `fm2IGF.lme` object, obtained with

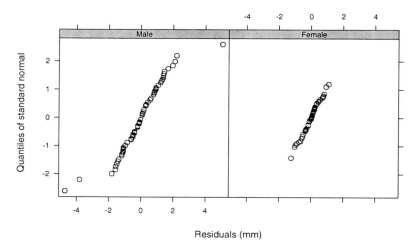

FIGURE 4.20. Normal plot of residuals for the `fm3Orth.lme` lme fit.

```
> plot( fm2IGF.lme, resid(., type = "p") ~ fitted(.) | Lot,
+        layout = c(5,2) )                              # Figure 4.21
```

The residuals are centered around zero and seem to have similar variability across lots. There are some outliers in the data, most noticeably for Lots 3 and 7.

We assess the normality of the within-group errors with a `qqnorm` plot of the residuals.

```
> qqnorm( fm2IGF.lme, ~ resid(.),
+          id = 0.05, adj = -0.75 )                     # Figure 4.22
```

The normal plot in Figure 4.22 suggests that the distribution of the within-group errors has heavier tails than expected under normality, but is also symmetric around zero. Perhaps a mixture of normal distributions or a t-distribution with a moderate number of degrees of freedom would model the distribution of the within-group error more adequately. However, as the heavier tails seem to be distributed symmetrically, the estimates of the fixed effects should not change substantially under either a mixture model or a t-model. The heavier tails tend to inflate the estimate of the within-group standard error under the Gaussian model, leading to more conservative tests for the fixed effects, but, because the p-value for the hypothesis that the decay of tracer activity with age is zero is quite high (0.673), the main conclusion should remain unchanged under either a mixture or a t-model.

Thickness of Oxide Coating on a Semiconductor

The plot of the within-group standardized residuals (`level = 2` in this case) versus the within-group fitted values is the default display produced by the `plot` method. Therefore,

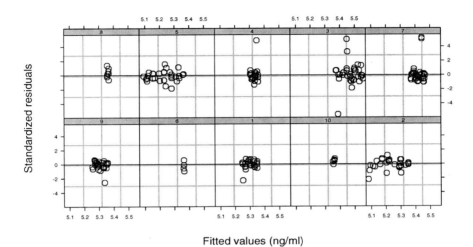

FIGURE 4.21. Scatter plots of standardized residuals versus fitted values for the `fm2IGF.lme` fit, by lot.

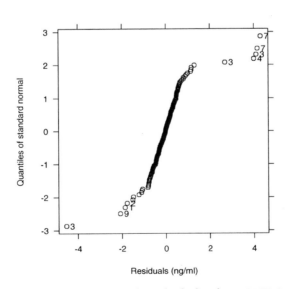

FIGURE 4.22. Normal plot of residuals for the `fm2IGF.lme` fit.

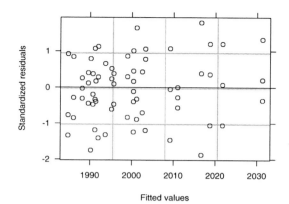

FIGURE 4.23. Scatter plot of the standardized within-group residuals versus the within-group fitted values for the fm1Oxide fit.

```
> plot( fm1Oxide )                              # Figure 4.23
```

results in the plot shown in Figure 4.23, which does not indicate any departures from the within-group errors assumptions: the residuals are symmetrically distributed around zero, with approximately constant variance.

By default, the **qqnorm** method produces a normal plot of the within-group standardized residuals. Hence,

```
> qqnorm( fm1Oxide )                            # Figure 4.24
```

gives the normal plot in Figure 4.24, which indicates that the assumption of normality for the within-group errors is plausible.

Manufacturing of Analog MOS Circuits

The plot of the within-group residuals versus voltage by wafer, shown in Figure 4.25, shows a clear periodic pattern for the residuals.

```
> plot( fm1Wafer, resid(.) ~ voltage | Wafer )     #  Figure 4.25
```

We can enhance the visualization of this pattern by adding a **loess** smoother (Cleveland, Grosse and Shyu, 1992) to each panel, using

```
> plot( fm1Wafer, resid(.) ~ voltage | Wafer,
+        panel = function(x, y, ...) {
+                panel.grid()
+                panel.xyplot(x, y)
+                panel.loess(x, y, lty = 2)
+                panel.abline(0, 0)
+                } )                            # Figure 4.26
```

The **panel** argument to the **plot** method overwrites the default panel function, allowing customized displays.

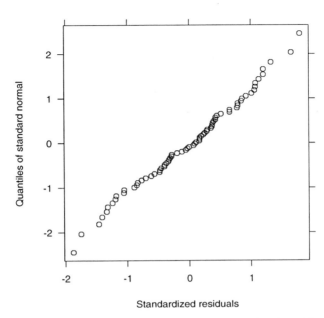

FIGURE 4.24. Normal plot of within-group standardized residuals for the fm1Oxide fit.

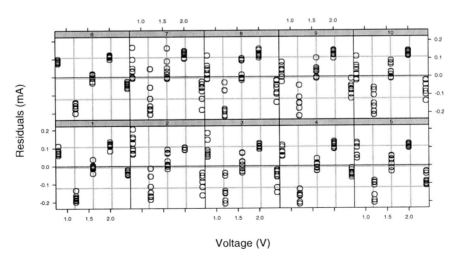

FIGURE 4.25. Scatter plots of within-group residuals versus voltage by wafer for the fm1Wafer fit.

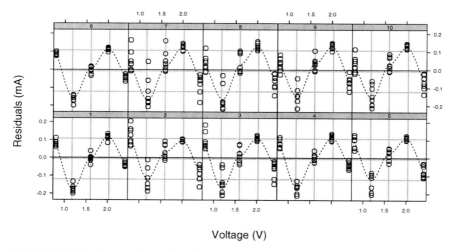

FIGURE 4.26. Scatter plots of within-group residuals versus voltage by wafer for the `fm1Wafer` fit. A `loess` smoother has been added to each panel to enhance the visualization of the residual pattern.

The same periodic pattern appears in all panels of Figure 4.26, with a period T of approximately 1.5 V. Noting that the residuals are centered around zero, this periodic pattern can be represented by the cosine wave

$$\beta_3 \cos{(\omega v)} + \beta_4 \sin{(\omega v)}, \tag{4.9}$$

where β_3 and β_4 determine the *amplitude* ($= \sqrt{\beta_3^2 + \beta_4^2}$) and ω is the *frequency* of the cosine wave. We can incorporate this pattern into model (4.8) by rewriting the fixed-effects model for the expected value of y_{ijk} as

$$\mathrm{E}\,[y_{ijk}] = \beta_0 + \beta_1 v_k + \beta_2 v_k^2 + \beta_3 \cos{(\omega v_k)} + \beta_4 \sin{(\omega v_k)}. \tag{4.10}$$

Because ω is unknown, (4.10) formally gives an example of a *nonlinear* mixed-effects model, discussed later in Chapter 8. To proceed with the model building analysis, we assume, for now, that ω is a purely fixed effect (i.e., does not vary with wafer and/or site) and estimate it by fitting (4.9) to the within-group residuals of `fm1Wafer`, using the `nls` function in S (Bates and Chambers, 1992). By replacing ω in (4.10) with its nonlinear least-squares estimate $\widehat{\omega}$, we are left with a linear mixed-effects model, which can then be fitted with `lme`.

Initial values for β_3, β_4, and ω are needed in the iterative algorithm used in `nls`. The frequency ω and the period T are related by the formula $\omega = 2\pi/T$, giving the initial estimate $\omega_0 = 2\pi/1.5 = 4.19 \ V^{-1}$ for ω. For a fixed $\omega = \omega_0$, (4.9) is a linear function of β_3 and β_4 and we can use `lm` to obtain initial estimates for these parameters.

```
> attach( Wafer )                    # making variables in Wafer available
> coef(lm(resid(fm1Wafer) ~ cos(4.19*voltage)+sin(4.19*voltage)-1))
 cos(4.19 * voltage) sin(4.19 * voltage)
            -0.051872              0.1304
> nls( resid(fm1Wafer) ~ b3*cos(w*voltage) + b4*sin(w*voltage),
+       start = list(b3 = -0.0519, b4 = 0.1304, w = 4.19) )
Residual sum of squares : 0.72932
parameters:
      b3        b4        w
 -0.11173 0.077682 4.5679
formula: resid(fm1Wafer) ~ b3*cos(w*voltage) + b4*sin(w*voltage)
400 observations
> detach( )
```

The resulting estimate for the frequency is $\widehat{w} = 4.5679 \ V^{-1}$.

We refit model (4.8), with the fixed-effects model replaced by (4.10) and ω held constant at $\widehat{\omega}$, with

```
> fm2Wafer <- update( fm1Wafer,
+       fixed = . ~ . + cos(4.5679*voltage) + sin(4.5679*voltage),
+       random = list(Wafer=pdDiag(~voltage+voltage^2),
+               Site=pdDiag(~voltage+voltage^2)) )
> summary( fm2Wafer )
Linear mixed-effects model fit by REML
 Data: Wafer
       AIC      BIC logLik
  -1232.6 -1184.9 628.31

Random effects:
 Formula:  ~ voltage + voltage^2 | Wafer
 Structure: Diagonal
          (Intercept) voltage I(voltage^2)
StdDev:       0.12888 0.34865      0.049074

 Formula:  ~ voltage + voltage^2 | Site %in% Wafer
 Structure: Diagonal
          (Intercept) voltage I(voltage^2) Residual
StdDev:      0.039675 0.23437      0.047541 0.011325

Fixed effects: current ~ voltage + I(voltage^2) +
    cos(4.5679 * voltage) + sin(4.5679 * voltage)
                        Value Std.Error  DF t-value p-value
          (Intercept) -4.2554   0.04223 316 -100.76  <.0001
              voltage  5.6224   0.11416 316   49.25  <.0001
          I(voltage^2)  1.2585   0.01696 316   74.21  <.0001
 cos(4.5679 * voltage) -0.0956   0.00112 316  -85.05  <.0001
 sin(4.5679 * voltage)  0.1043   0.00150 316   69.42  <.0001
 . . .
```

The . ˜ . in the fixed formula is an abbreviated form for the fixed-effects formula in the original lme object, fm1Wafer. This convention is also available in other S modeling functions, such as lm and aov. The random argument was included in the call to update to prevent the estimated random-effects parameters from fm1Wafer to be used as initial estimates (these give bad initial estimates in this case and may lead to convergence problems).

The very high t-values for the sine and cosine terms in the summary output indicate a significant increase in the quality of the fit when these terms are included in the fixed-effects model. The estimated standard deviations for the random effects are quite different from the ones in fm1Wafer and now suggest that there is significant *wafer-to-wafer* and *site-to-site* variation in all random effects in the model. The estimated within-group standard deviation for fm2Wafer is about ten times smaller than that of fm1Wafer, giving further evidence of the greater adequacy of (4.10). We assess the variability in the estimates with

```
> intervals( fm2Wafer )
Approximate 95% confidence intervals

  Fixed effects:
                          lower       est.      upper
         (Intercept) -4.338485  -4.255388  -4.172292
             voltage  5.397744   5.622357   5.846969
        I(voltage^2)  1.225147   1.258512   1.291878
  cos(4.5679 * voltage) -0.097768 -0.095557 -0.093347
  sin(4.5679 * voltage)  0.101388  0.104345  0.107303

  Random Effects:
   Level: Wafer
                        lower      est.     upper
    sd((Intercept))  0.065853  0.128884  0.25225
        sd(voltage)  0.174282  0.348651  0.69747
    sd(I(voltage^2)) 0.023345  0.049074  0.10316
   Level: Site
                        lower      est.     upper
    sd((Intercept))  0.017178  0.039675  0.091635
        sd(voltage)  0.175311  0.234373  0.313332
    sd(I(voltage^2)) 0.035007  0.047541  0.064564

  Within-group standard error:
     lower      est.      upper
  0.0085375  0.011325  0.015023
```

The plots of the within-group residuals versus voltage, by wafer, shown in Figure 4.27, indicate that there is still some periodicity left in some of the wafers, suggesting that random effects may be needed to accommodate the variation in ω. We postpone this discussion until Chapter 8, when nonlinear mixed-effects models are described. Note that the absolute values of the

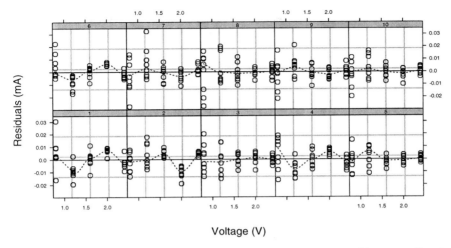

FIGURE 4.27. Scatter plots of within-group residuals versus within-group fitted values by wafer for the `fm2Wafer` fit. A `loess` smoother has been added to each panel to enhance the visualization of the residual pattern.

within-groups residuals in Figure 4.27 are about an order of magnitude smaller than the ones in Figure 4.26.

The normal plot of the within-group residuals for `fm2Wafer`, obtained with

```
> qqnorm( fm2Wafer )                              # Figure 4.28
```

and shown in Figure 4.28, does not indicate any violations from the assumption of normality for the within-group errors.

4.3.2   Assessing Assumptions on the Random Effects

In this section, we describe diagnostic methods for assessing Assumption 2, on the distribution of the random effects. The `ranef` method is used to extract the estimated BLUPs of the random effects from lme objects. These are the primary quantities for assessing the distributional assumptions about the random effects.

Two types of diagnostic plots will be used to investigate departures from Assumption 2:

- `qqnorm`—normal plot of estimated random effects for checking marginal normality and identifying outliers;

- `pairs`—scatter plot matrix of the estimated random effects for identifying outliers and checking the assumption of homogeneity of the random effects covariance matrix;

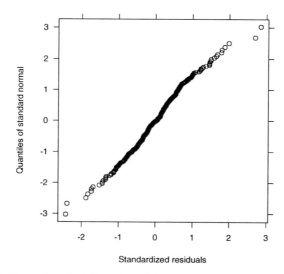

FIGURE 4.28. Normal probability plot of the within-group standardized residuals for the `fm2Wafer` fit.

We illustrate the use of these display methods by continuing the analysis of the examples in §4.3.1.

Orthodontic Growth Curve

We first consider the homoscedastic fitted object `fm2Orth.lme` and investigate the marginal normality of the corresponding random effects using the normal plots in Figure 4.29.

```
> qqnorm( fm2Orth.lme, ~ranef(.),
+          id = 0.10, cex = 0.7 )                    # Figure 4.29
```

As in the residual plots of §4.3.1, the `id` argument specifies a critical value for identifying points in the plot (standardized random-effects estimates greater than the `1-id/2` standard normal quantile in absolute value are identified). By default, the group labels are used to identify the points. The assumption of normality seems reasonable for both random effects, though there is some asymmetry in the distribution of the `(Intercept)` random effects. A few outliers appear to be present in both random effects: `F10`, `F11`, and `M10` for the intercept and `M13` for the slope.

To investigate the homogeneity of the random-effects distribution for boys and girls, we use the `pairs` method to obtain scatter plots of the random-effects estimates by gender, as shown in Figure 4.30.

```
> pairs( fm2Orth.lme, ~ranef(.) | Sex,
+          id = ~ Subject == "M13", adj = -0.3 )      # Figure 4.30
```

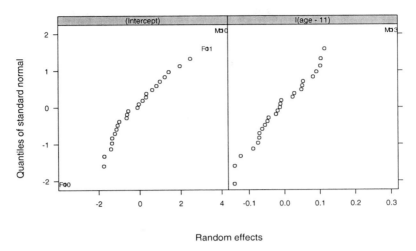

FIGURE 4.29. Normal plot of estimated random effects for the homoscedastic `fm2Orth.lme` fit.

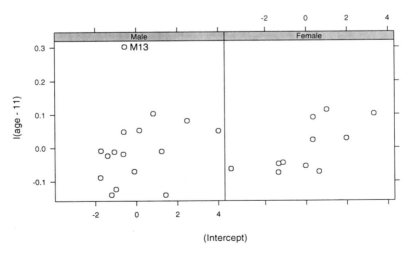

FIGURE 4.30. Scatter plot of estimated random effects for the homoscedastic `fm2Orth.lme` fit.

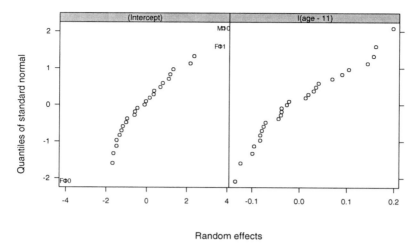

FIGURE 4.31. Normal plot of estimated random effects for the heteroscedastic fm3Orth.lme lme fit.

The id argument is given as a formula object, which is used to identify subject M13 in the plot. Alternatively, it can be given as a numeric value such that points outside the normal distribution contour of level 1-id/2 are identified in the plot. Except for the pair corresponding to subject M13, the estimated random effects in the two groups seem to have similar distributions.

As seen in §4.3.1, the within-group variance appears to be larger among boys than among girls for the Orthodont data and the heteroscedastic fit fm3Orth.lme gives a better representation of the data. We consider now the diagnostic plots for the random effects in this model. The normal probability plots of the estimated random effects (Figure 4.31) are similar to those in Figure 4.29 with one important exception: the estimated random effect for the slope of subject M13 is no longer identified in the plot as an outlier. In mixed-effects estimation, there is a trade-off between the within-group variability and the between-group variability, when accounting for the overall variability in the data. The use of a common within-group variance in fm2Orth.lme leads to an increase in the estimated between-group variability, which in turn allows the random-effects estimates to be pulled away by outliers. The heteroscedastic model in fm3Orth.lme accommodates the impact of the boys' outlying observations in the within-group variances estimation and reduces the estimated between-group variability, thus increasing the degree of *shrinkage* in the random-effects estimates. In this case, the use of different within-group variances by gender adds a certain degree of robustness to the lme fit.

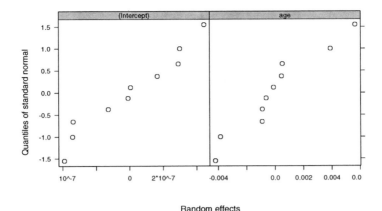

FIGURE 4.32. Normal plot of estimated random effects for the homoscedastic fm2IGF.lme fit.

The pairs plots by gender for Orthodont.lme3, not included here, do not suggest any departures from the assumption of homogeneity of the random-effects distribution.

Radioimmunoassays of IGF-I

The normal plots of the estimated random effects for the fm2IGF.lme fit, shown in Figure 4.32, do not indicate any departures from normality or any outlying subjects. They do, however, suggest that there is very little variability in the (Intercept) random effect, as the estimated random effects are on the order of 10^{-7}.

The relative variability of each random effect with respect to the corresponding fixed-effect estimates can be calculated as

```
> fm2IGF.lme
 . . .
  Fixed: conc ~ age
  (Intercept)           age
       5.369  -0.0019301

Random effects:
 Formula: ~ age | Lot
  Structure: Diagonal
        (Intercept)         age Residual
 StdDev:  0.00031074 0.0053722    0.8218
 . . .
> c( 0.00031074, 0.0053722 )/abs( fixef(fm2IGF.lme) )
  (Intercept)        age
  5.7876e-05 2.7833
```

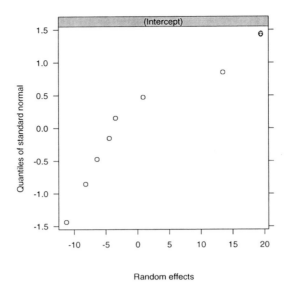

FIGURE 4.33. Normal plot of estimated `Lot` random effects for the `fm1Oxide` fit.

The relative variability of the `(Intercept)` random effect is only 0.006%, while the relative variability of the `age` random effect is about 278.3%, suggesting that the former could be dropped from the model.

```
> fm3IGF.lme <- update( fm2IGF.lme, random = ~ age - 1 )
> anova( fm2IGF.lme, fm3IGF.lme )
          Model df   AIC    BIC logLik   Test   L.Ratio p-value
fm2IGF.lme     1  5 604.8 622.10 -297.4
fm3IGF.lme     2  4 602.8 616.64 -297.4 1 vs 2 1.0881e-05  0.9974
```

The high p-value for the likelihood ratio test indicates that the random effect for the intercept does not contribute significantly to the fit of the `IGF` data.

The normal plot of the estimated random effects for the reduced model `fm3IGF.lme`, not included here, does not suggest any violations of the normality assumption and does not show any outlying values.

Thickness of Oxide Coating on a Semiconductor

Normal probability plots of the estimated random effects must be examined at each level of grouping when assessing the adequacy of a multilevel model fit. The normal plot of the estimated `Lot` random effects for the `fm1Oxide` fit, shown in Figure 4.33, is obtained with

```
> qqnorm( fm1Oxide, ~ranef(., level = 1), id=0.10 )    # Figure 4.33
```

The `level` argument to the `ranef` method is required in this case, as, by default, `ranef` returns a list with the estimated random effects at each level

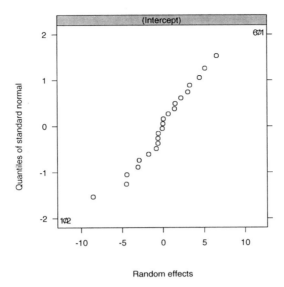

FIGURE 4.34. Normal plot of estimated `Wafer %in% Lot` random effects for the `fm1Oxide` fit.

of grouping, which will cause an error in `qqnorm`. Because there are only eight random effects at the `Lot` level, it is difficult to identify any patterns in Figure 4.33. Lot 6 is indicated as a potential outlier, which is consistent with the plot of the data in Figure 4.14, where this lot is shown to have the thickest oxide layers.

The normal plot of the `Wafer %in% Lot` random effects, shown in Figure 4.34, and obtained with

```
> qqnorm( fm1Oxide, ~ranef(., level = 2), id=0.10 )    # Figure 4.34
```

does not indicate any departures from normality. There is some mild evidence that Wafers 1/2 and 6/1 may be outliers.

Manufacturing of Analog MOS Circuits

The pairs plot of the estimated `Wafer` random effects for the `fm2Wafer` fit, shown in Figure 4.35, suggests that the random effects at that level are correlated.

The `fm2Wafer` fit uses a diagonal structure for the variance–covariance matrix of the `Wafer` random effects, which is equivalent to assuming that these random effects are independent. We test the independence assumption by fitting the model with a general positive-definite structure for the variance–covariance matrix of the `Wafer` random effects and comparing it to the `fm2Wafer` fit using the `anova` method.

```
> fm3Wafer <- update( fm2Wafer,
```

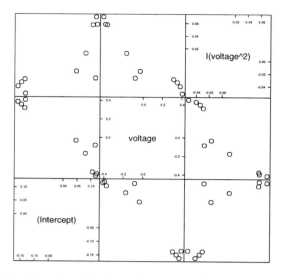

FIGURE 4.35. Scatter plot of estimated `Wafer` random effects for the `fm2Wafer` fit.

```
+                     random = list(Wafer = ~voltage+voltage^2,
+                                   Site = pdDiag(~voltage+voltage^2)))

> fm3Wafer
 . . .
Random effects:
 Formula:  ~ voltage + voltage^2 | Wafer
 Structure: General positive-definite
              StdDev    Corr
 (Intercept) 0.131622 (Intr) voltag
     voltage 0.359244 -0.967
I(voltage^2) 0.051323  0.822 -0.940

 Formula:  ~ voltage + voltage^2 | Site %in% Wafer
 Structure: Diagonal
         (Intercept) voltage I(voltage^2) Residual
StdDev:    0.033511 0.21831     0.045125 0.011832
 . . .
> anova( fm2Wafer, fm3Wafer )
         Model df     AIC      BIC logLik   Test L.Ratio p-value
fm2Wafer     1 12 -1232.6 -1184.9 628.31
fm3Wafer     2 15 -1267.0 -1207.3 648.50 1 vs 2  40.378  <.0001
```

There is a very significant increase in the log-restricted-likelihood, as evidenced by the large value for the likelihood ratio test, indicating that the more general model represented by `fm3Wafer` gives a better fit.

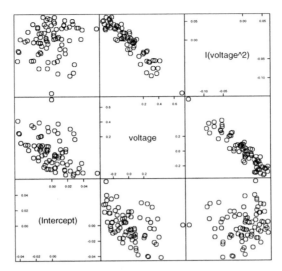

FIGURE 4.36. Scatter plot of estimated random effects at the `Site %in% Wafer` level for the `fm3Wafer` fit.

The pairs plot for the estimated `Site %in% Wafer` random effects corresponding to `fm3Wafer` in Figure 4.36 indicate that there is a strong negative correlation between the `voltage` and `voltage^2` random effects, but no substantial correlation between either of these random effects and the `(Intercept)` random effects. A block-diagonal matrix can be used to represent such covariance structure, with the `(Intercept)` random effect corresponding to one block and the `voltage` and `voltage^2` random effects corresponding to another block.

```
> fm4Wafer <- update( fm3Wafer,
+     random = list(Wafer = ~ voltage + voltage^2,
+             Site = pdBlocked(list(~1, ~voltage+voltage^2 - 1))))
> fm4Wafer
 . . .
Random effects:
 Formula:  ~ voltage + voltage^2 | Wafer
 Structure: General positive-definite
             StdDev    Corr
 (Intercept) 0.131807 (Intr) voltag
     voltage 0.354746 -0.967
I(voltage^2) 0.049957  0.814 -0.935

 Composite Structure: Blocked

 Block 1: (Intercept)
 Formula:  ~ 1 | Site %in% Wafer
```

```
          (Intercept)
StdDev:    0.066562


  Block 2: voltage, I(voltage^2)
  Formula:  ~ voltage + voltage^2 - 1 | Site %in% Wafer
  Structure: General positive-definite
                  StdDev    Corr
       voltage 0.2674061 voltag
  I(voltage^2) 0.0556441 -0.973
     Residual 0.0091086
  . . .
> anova( fm3Wafer, fm4Wafer )
         Model df     AIC      BIC logLik    Test L.Ratio p-value
fm3Wafer     1 15 -1267.0 -1207.3 648.50
fm4Wafer     2 16 -1461.6 -1398.0 746.82 1 vs 2  196.65  <.0001
```

The small p-value for the likelihood ratio test indicates that the `fmWafer4` model fits the data significantly better than the model represented by `fm3Wafer`. This will be the final model consider in this chapter for the `Wafer` data. Later in §8 we revisit this example, fitting a nonlinear mixed-effects model to the data.

The normal plot of the estimated `Site %in% Wafer` random effects corresponding to `fm4Wafer`, shown in Figure 4.37, does not suggest any significant departure from the assumption of normality for these random effects. There is some moderate evidence that Sites 1/6, 7/3 and 8/1 may be outliers.

```
> qqnorm( fm4Wafer, ~ranef(., level = 2), id = 0.05,
+         cex = 0.7, layout = c(3, 1) )          # Figure 4.37
```

4.4 Chapter Summary

This chapter describes the capabilities available in the nlme library for fitting and analyzing linear mixed-effects models with uncorrelated, homoscedastic within-group errors. The lme function, for fitting linear mixed-effects models, is described in detail and its various capabilities and associated methods are illustrated through the analyses of several real data examples, covering single-level models, multilevel nested models, and models with crossed random effects.

The model-building approach developed in this chapter follows an "inside-out" strategy, using individual lm fits, obtained with the lmList function, to construct more sophisticated linear mixed-effects models. A rich, integrated suite of diagnostic plots to assess model assumptions is described and illustrated through examples.

The class of mixed-effects models which can be fit with lme is greatly extended by the availability of patterned random-effects variance–covariance

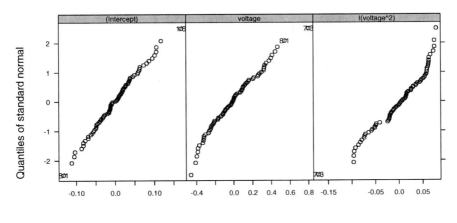

FIGURE 4.37. Normal plot of estimated `Site %in% Wafer` random effects for the `fm4Wafer` fit.

structures. These are implemented in S through pdMat classes, which can be extended with user defined classes.

The linear mixed-effects model considered in this chapter is extended in two different ways later in the book. In Chapter 5, the assumption of uncorrelated, homoscedastic within-group errors is relaxed, and variance functions and correlation structures are introduced the model heteroscedasticity and within-group dependence. The assumption of linearity for $E\left[y_i|b_i\right]$ is relaxed in Chapter 8, when nonlinear mixed-effects models are described.

Exercises

1. In §1.3.3 (p. 27) we fit the model

   ```
   > fm3Machine <- update(fm1Machine, random = ~Machine-1|Worker)
   ```

   The estimated correlations of the components of the random-effects vectors b_i in fm3Machine, 0.803, 0.623, and 0.771, are similar in magnitude. The estimated standard deviations of these components, 4.08, 8.63, and 4.39 are comparable to each other although not as close as the estimated correlations. Together, these suggest a compound symmetry structure for Ψ as described in §4.2.2.

   (a) Fit a model like that of fm3Machine, but with a compound symmetry structure. This is similar to the model fm40atsB in §4.2.2. Examine numerical summaries and residual plots for this model.

    (b) Compare this fitted model to `fm3Machine` with `anova`. Is the greater number of parameters in `fm3Machine` producing a significantly better fit?

    (c) Compare the `pdCompSymm` model to `fm2Machine` from §1.3.3 with `anova`. Note that the likelihoods and the numbers of parameters are identical.

    (d) The `pdCompSymm` model is equivalent to `fm2Machine`. Use `VarCorr` to extract the estimates of the variance components from each model. What is the relationship between the estimated variance components of the two models?

2. There are some unusual residuals for subjects `M09` and `M13` in several of the fits of the Orthodont data. See, for example, Figure 4.17 (p. 177). An examination of the original data, say in Figure 4.9 (p. 153), shows some suspicious observations on these subjects. The observation at age 8 for subject `M13` is unusually low. The four observations for subject `M09` decrease, then increase dramatically, then decrease again. The two observations at ages 10 and 12 appear to be incorrect.

    (a) Repeat the stages of fitting a linear mixed-effects models to the Orthodont data with the suspicious observations for subjects `M13` and `M09` removed. Do you arrive at different conclusions regarding the models?

    (b) There are other subjects like `M09` in the Orthodont data for whom the observed `distance` decreases with increasing age. Because it is unlikely that this measurement on the same child would get smaller over time, these are probably misrecorded data or unusually large measurement errors.

        i. Find the number of pairs of measurements that represent decreases within a subject. You can begin with `gapply(Ortho-dont, "distance", diff)` which returns a list of successive differences of the `distance` by `Subject`.

        ii. What should be done with these aberrant data points?

        iii. If you handle the suspect data by deleting some observations, repeat as much of this chapter's analysis of Orthodont as is feasible with the modified data. (Note that if you decide to delete some of the suspect data, the resulting data will be unbalanced. Is it still possible to fit mixed-effects models to such unbalanced data?)

3. When analyzing the Assay data (Figure 4.13, p. 164), we noted that the interaction term in the fixed effects for the fitted model `fm1Assay` is not significant.

(a) Refit the model as, say, `fm2Assay` with the interaction term removed from the fixed effects. Are the estimates of the other parameters changed noticeably? Can you compare this fitted model to `fm1Assay` with `anova` and assess the significance of the interaction term?

(b) Use `VarCorr` to extract the estimates of the variance components from `fm2Assay`. Notice that the estimated variance component for the columns on the plate, indexed by `dilut` within `Block`, is very small. Refit the model as, say, `fm3Assay` with this variance component eliminated. You can express the `random` argument for this reduced model in at least three ways: (i) using `pdBlocked` and `pdIdent` as in `fm1Assay`, (ii) using `pdCompSymm` as in `fm4OatsB`, or (iii) using nested random effects for `Block` and for `sample` within `Block`. Experiment with these different representations. Demonstrate that that the fitted models for these three representations are indeed equivalent. For which of the three representations does `lme` converge most easily? Compare one of these fitted models to `fm2Assay` using `anova`. Is the variance component for the columns significant?

(c) Extract the estimates of the variance components from the fitted model `fm3Assay` and examine the confidence intervals on the standard deviations of the random effects. Do the variance components appear to be significantly greater than zero? Fit a model, say `fm4Assay`, with a single variance component for the `Block`. Compare it to `fm3Assay` with `anova`.

(d) Can the random effects be eliminated entirely? Fit a model, say `fm5Assay`, using just the fixed-effects terms from `fm4Assay`. Because this model does not have any random-effects terms, you will need to use `lm` or `gls` to fit it.

(e) Compare models `fm2Assay`, `fm3Assay`, `fm4Assay`, and `fm5Assay` with `anova`. Which model provides the simplest adequate representation for these data?

(f) Notice that the `dilut` factor represents serial dilutions that will be equally spaced on the logarithm scale. Does converting `dilut` to an ordered factor, so the contrasts used for this factor will be polynomial contrasts, suggest further simplifications of the model?

5

Extending the Basic Linear Mixed-Effects Model

The linear mixed-effects model formulation used in Chapters 2 and 4 allows considerable flexibility in the specification of the random-effects structure, but restricts the within-group errors to be independent, identically distributed random variables with mean zero and constant variance. As illustrated in Chapters 2 and 4, this *basic* linear mixed-effects model provides an adequate model for many different types of grouped data observed in practice. However, there are many applications involving grouped data for which the within-group errors are *heteroscedastic* (i.e., have unequal variances) or are *correlated* or are both heteroscedastic and correlated.

In this chapter, we extend the basic linear mixed-effects model to allow heteroscedastic, correlated within-group errors. We describe how the lme function can be used to fit the extended linear mixed-effects model, illustrating its various capabilities through examples. We also introduce a new modeling function, gls—for *generalized least squares*, to fit models with heteroscedastic and correlated within-group errors, but with no random effects.

In §5.1, we show how the estimation and computational methods of Chapter 2 can be adapted to the extended model formulation and describe how the variance–covariance structure of the within-group errors can be decomposed into two independent components: a *variance* structure and a *correlation* structure. Variance function models to represent the variance structure component of the within-group errors are described in §5.2 and their use with lme is illustrated through examples. Classes of correlation models to represent the correlation structure of the within-group errors are presented and have their use illustrated in §5.3. In §5.4, the `gls` function is described and illustrated through examples.

5.1   General Formulation of the Extended Model

As described in §2.1.1, the basic single-level linear mixed-effects model (2.1) assumes that the within-group errors ϵ_i are independent $\mathcal{N}\left(0, \sigma^2 I\right)$ random vectors. The extended single-level linear mixed-effects model relaxes this assumption by allowing heteroscedastic and correlated within-group errors, being expressed as

$$\boldsymbol{y}_i = \boldsymbol{X}_i \boldsymbol{\beta} + \boldsymbol{Z}_i \boldsymbol{b}_i + \boldsymbol{\epsilon}_i, \quad i = 1, \ldots, M,$$
$$\boldsymbol{b}_i \sim \mathcal{N}(\boldsymbol{0}, \boldsymbol{\Psi}), \quad \boldsymbol{\epsilon}_i \sim \mathcal{N}\left(\boldsymbol{0}, \sigma^2 \boldsymbol{\Lambda}_i\right), \quad i = 1, \ldots, M, \tag{5.1}$$

where the $\boldsymbol{\Lambda}_i$ are positive-definite matrices parametrized by a fixed, generally small, set of parameters $\boldsymbol{\lambda}$. As in the basic linear mixed-effects model of §2.1, the within-group errors ϵ_i are assumed to be independent for different i and independent of the random effects $\boldsymbol{b}_i$. The σ^2 is factored out of the $\boldsymbol{\Lambda}_i$ for computational reasons (it can then be eliminated from the profiled likelihood function).

Similarly, the extended two-level linear mixed-effects model generalizes the basic two-level model (2.2) described in §2.1.2 by letting

$$\epsilon_{ij} \sim \mathcal{N}\left(\boldsymbol{0}, \sigma^2 \boldsymbol{\Lambda}_{ij}\right), \quad i = 1, \ldots, M, \quad j = 1, \ldots, M_i,$$

where the $\boldsymbol{\Lambda}_{ij}$ are positive-definite matrices parametrized by a fixed $\boldsymbol{\lambda}$ vector. This readily generalizes to a multilevel model with Q levels of random effects. For simplicity, we concentrate for the remainder of this section on the extended single-level model (5.1), but the results we obtain are easily generalizable to multilevel models with an arbitrary number of levels of random effects.

5.1.1   Estimation and Computational Methods

Because $\boldsymbol{\Lambda}_i$ is positive-definite, it admits an invertible square-root $\boldsymbol{\Lambda}_i^{1/2}$ (Thisted, 1988, §3), with inverse $\boldsymbol{\Lambda}_i^{-1/2}$, such that

$$\boldsymbol{\Lambda}_i = \left(\boldsymbol{\Lambda}_i^{1/2}\right)^T \boldsymbol{\Lambda}_i^{1/2} \quad \text{and} \quad \boldsymbol{\Lambda}_i^{-1} = \boldsymbol{\Lambda}_i^{-1/2} \left(\boldsymbol{\Lambda}_i^{-1/2}\right)^T.$$

Letting

$$\boldsymbol{y}_i^* = \left(\boldsymbol{\Lambda}_i^{-1/2}\right)^T \boldsymbol{y}_i, \qquad \boldsymbol{\epsilon}_i^* = \left(\boldsymbol{\Lambda}_i^{-1/2}\right)^T \boldsymbol{\epsilon}_i,$$
$$\boldsymbol{X}_i^* = \left(\boldsymbol{\Lambda}_i^{-1/2}\right)^T \boldsymbol{X}_i, \qquad \boldsymbol{Z}_i^* = \left(\boldsymbol{\Lambda}_i^{-1/2}\right)^T \boldsymbol{Z}_i, \tag{5.2}$$

and noting that

$$\epsilon_i^* \sim \mathcal{N}\left[\left(\boldsymbol{\Lambda}_i^{-1/2}\right)^T \mathbf{0}, \sigma^2 \left(\boldsymbol{\Lambda}_i^{-1/2}\right)^T \boldsymbol{\Lambda}_i \boldsymbol{\Lambda}_i^{-1/2}\right] = \mathcal{N}\left(\mathbf{0}, \sigma^2 \boldsymbol{I}\right),$$

we can rewrite (5.1) as

$$\boldsymbol{y}_i^* = \boldsymbol{X}_i^* \boldsymbol{\beta} + \boldsymbol{Z}_i^* \boldsymbol{b}_i + \epsilon_i^*, \quad i = 1, \ldots, M,$$
$$\boldsymbol{b}_i \sim \mathcal{N}(\mathbf{0}, \boldsymbol{\Psi}), \quad \epsilon_i^* \sim \mathcal{N}\left(\mathbf{0}, \sigma^2 \boldsymbol{I}\right), \quad i = 1, \ldots, M. \tag{5.3}$$

That is, $\boldsymbol{y}_i^*$ is described by a basic linear mixed-effects model.

Because the differential of the linear transformation $\boldsymbol{y}_i^* = (\boldsymbol{\Lambda}_i^{-1/2})^T \boldsymbol{y}_i$ is simply $d\boldsymbol{y}_i^* = \left|\boldsymbol{\Lambda}_i^{-1/2}\right| d\boldsymbol{y}_i$, the likelihood function L corresponding to the extended linear mixed-effects model (5.1) is expressed as

$$L\left(\boldsymbol{\beta}, \boldsymbol{\theta}, \sigma^2, \boldsymbol{\lambda} | \boldsymbol{y}\right) = \prod_{i=1}^M p\left(\boldsymbol{y}_i | \boldsymbol{\beta}, \boldsymbol{\theta}, \sigma^2, \boldsymbol{\lambda}\right)$$

$$= \prod_{i=1}^M p\left(\boldsymbol{y}_i^* | \boldsymbol{\beta}, \boldsymbol{\theta}, \sigma^2, \boldsymbol{\lambda}\right) \left|\boldsymbol{\Lambda}_i^{-1/2}\right| = L\left(\boldsymbol{\beta}, \boldsymbol{\theta}, \sigma^2, \boldsymbol{\lambda} | \boldsymbol{y}^*\right) \prod_{i=1}^M \left|\boldsymbol{\Lambda}_i^{-1/2}\right|, \tag{5.4}$$

where $p(\cdot)$ denotes a probability density function.

The likelihood function $L\left(\boldsymbol{\beta}, \boldsymbol{\theta}, \sigma^2, \boldsymbol{\lambda} | \boldsymbol{y}^*\right)$ corresponds to a basic linear mixed-effects model and, therefore, all results presented in §2.2 also apply to it. The compact representation of the profiled log-likelihood based on orthogonal-triangular decompositions, described in §2.2.3, can also be used with the likelihood $L\left(\boldsymbol{\beta}, \boldsymbol{\theta}, \sigma^2, \boldsymbol{\lambda} | \boldsymbol{y}^*\right)$, leading to numerically efficient algorithms for maximum likelihood estimation.

The restricted likelihood corresponding to the extended model (5.1) is defined, as in §2.2.5, by integrating out the fixed effects from the likelihood.

$$L_R\left(\boldsymbol{\theta}, \sigma^2, \boldsymbol{\lambda} | \boldsymbol{y}\right) = \int L\left(\boldsymbol{\beta}, \boldsymbol{\theta}, \sigma^2, \boldsymbol{\lambda} | \boldsymbol{y}\right) d\boldsymbol{\beta} = L_R\left(\boldsymbol{\theta}, \sigma^2, \boldsymbol{\lambda} | \boldsymbol{y}^*\right) \prod_{i=1}^M \left|\boldsymbol{\Lambda}_i^{-1/2}\right|.$$

The function $L_R\left(\boldsymbol{\beta}, \boldsymbol{\theta}, \sigma^2, \boldsymbol{\lambda} | \boldsymbol{y}^*\right)$ corresponds to a restricted likelihood function of a basic linear mixed-effects model. Hence, the results in §2.2.5 can be used to obtain a numerically efficient representation of the profiled log-restricted-likelihood.

5.1.2   An Extended Linear Model with No Random Effects

The variance–covariance matrix of the response vector $\boldsymbol{y}_i$ in the extended linear mixed-effects model (5.1),

$$\mathrm{Var}\left(\boldsymbol{y}_i\right) = \boldsymbol{\Sigma}_i = \sigma^2 \left(\boldsymbol{Z}_i \boldsymbol{D} \boldsymbol{Z}_i^T + \boldsymbol{\Lambda}_i\right)$$

has two components that can be used to model heteroscedasticity and correlation: a *random-effects* component, given by $\boldsymbol{Z}_i \boldsymbol{D} \boldsymbol{Z}_i^T$, and a *within-group* component, given by $\boldsymbol{\Lambda}_i$. In practice, these two components may "compete" with each other in the model specification, in the sense that similar $\boldsymbol{\Sigma}_i$ matrices may result from a more complex random-effects component being added to a simpler within-group component (say $\boldsymbol{\Lambda}_i = \boldsymbol{I}$), or a simpler random-effects component (say $\boldsymbol{Z}_i \boldsymbol{D} \boldsymbol{Z}_i^T = \sigma_b^2 \boldsymbol{1} \boldsymbol{1}^T$) being added to a more complex within-group component. There will generally be a trade-off between the complexity of the two components of $\boldsymbol{\Sigma}_i$ and some care must be exercised to prevent nonidentifiability, or near nonidentifiability, of the parameters in the model.

In some applications, one may wish to avoid incorporating random effects in the model to account for dependence among observations, choosing to use the within-group component $\boldsymbol{\Lambda}_i$ to directly model variance–covariance structure of the response. This results in the simplified version of the extended linear mixed-effects model (5.1)

$$\boldsymbol{y}_i = \boldsymbol{X}_i \boldsymbol{\beta} + \boldsymbol{\epsilon}_i, \quad \boldsymbol{\epsilon}_i \sim \mathcal{N}\left(\boldsymbol{0}, \sigma^2 \boldsymbol{\Lambda}_i\right), \quad i = 1, \ldots, M. \tag{5.5}$$

Estimation under this model has been studied extensively in the linear regression literature (Draper and Smith, 1998; Thisted, 1988), usually assuming that the $\boldsymbol{\Lambda}_i$ are known, being referred to as the *generalized least squares* problem.

Using the same transformations as in (5.2), we can re-express (5.5) as a "classic" linear regression model.

$$\boldsymbol{y}_i^* = \boldsymbol{X}_i^* \boldsymbol{\beta} + \boldsymbol{\epsilon}_i^*, \quad \boldsymbol{\epsilon}_i^* \sim \mathcal{N}\left(\boldsymbol{0}, \sigma^2 \boldsymbol{I}\right), \quad i = 1, \ldots, M. \tag{5.6}$$

Hence, for fixed $\boldsymbol{\lambda}$, the maximum likelihood estimators of $\boldsymbol{\beta}$ and σ^2 are obtained by solving an *ordinary least-squares* problem. Letting $\boldsymbol{X}^*$ denote the matrix obtained by stacking up the $\boldsymbol{X}_i^*$ matrices, the conditional MLEs of $\boldsymbol{\beta}$ and σ^2 are

$$\widehat{\boldsymbol{\beta}}(\boldsymbol{\lambda}) = \left[(\boldsymbol{X}^*)^T \boldsymbol{X}^*\right]^{-1} (\boldsymbol{X}^*)^T \boldsymbol{y}^*,$$
$$\widehat{\sigma^2}(\boldsymbol{\lambda}) = \frac{\left\|\boldsymbol{y}^* - \boldsymbol{X}^* \widehat{\boldsymbol{\beta}}(\boldsymbol{\lambda})\right\|^2}{N}. \tag{5.7}$$

The profiled log-likelihood corresponding to (5.5), which is a function of $\boldsymbol{\lambda}$ only, is obtained by replacing $\boldsymbol{\beta}$ and σ^2 in the full log-likelihood by their conditional MLEs (5.7), giving

$$\ell\left(\boldsymbol{\lambda}|\boldsymbol{y}\right) = \text{const} - N \log \left\|\boldsymbol{y}^* - \boldsymbol{X}^* \widehat{\boldsymbol{\beta}}(\boldsymbol{\lambda})\right\| - \frac{1}{2} \sum_{i=1}^{M} \log |\boldsymbol{\Lambda}_i|. \tag{5.8}$$

Maximum likelihood estimates are obtained by first optimizing the profiled log-likelihood (5.8) over $\boldsymbol{\lambda}$ and then replacing $\boldsymbol{\lambda}$ in (5.7) with its MLE $\widehat{\boldsymbol{\lambda}}$, to get the MLEs for $\boldsymbol{\beta}$ and σ^2. Orthogonal-triangular decomposition methods similar to the ones described in §2.2.3 can be used to obtain more compact and numerically efficient representations of the profiled log-likelihood (5.8) and the conditional MLEs (5.7), being left to the reader as an exercise.

The restricted likelihood for the extended linear model (5.5) is defined as in the extended linear mixed-effects model (5.1), by integrating the parameter vector $\boldsymbol{\beta}$ out of the full likelihood. Using the results in Harville (1977), we write the profiled log-restricted-likelihood corresponding to (5.5) as

$$
\ell_R(\boldsymbol{\lambda}|\boldsymbol{y}) = \text{const} - (N - p)\log\left\|\boldsymbol{y}^* - \boldsymbol{X}^*\widehat{\boldsymbol{\beta}}(\boldsymbol{\lambda})\right\|
$$

$$
- \frac{1}{2}\log\left|(\boldsymbol{X}^*)^T\boldsymbol{X}^*\right| - \frac{1}{2}\sum_{i=1}^{M}\log|\boldsymbol{\Lambda}_i|,
$$

with p denoting the dimension of $\boldsymbol{\beta}$. The REML estimator of σ^2 is defined as in (5.7), but with the N in the denominator replaced by $N - p$.

The modeling function gls in the nlme library is used to fit the extended linear model (5.5), using either maximum likelihood or restricted maximum likelihood. We describe the use of this function in §5.4, after presenting the capabilities available in the nlme library for modeling heteroscedasticity and correlation.

5.1.3 Decomposing the Within-Group Variance–Covariance Structure

The $\boldsymbol{\Lambda}_i$ matrices can always be decomposed into a product of simpler matrices

$$
\boldsymbol{\Lambda}_i = \boldsymbol{V}_i\boldsymbol{C}_i\boldsymbol{V}_i, \tag{5.9}
$$

where $\boldsymbol{V}_i$ is diagonal and $\boldsymbol{C}_i$ is a correlation matrix, that is, a positive-definite matrix with all diagonal elements equal to one. The matrix $\boldsymbol{V}_i$ in (5.9) is not uniquely defined, as we can multiply any number of its rows by -1 and still get the same decomposition. To ensure uniqueness, we require that all the diagonal elements of $\boldsymbol{V}_i$ be positive.

It is easy to verify that

$$
\text{Var}(\epsilon_{ij}) = \sigma^2[\boldsymbol{V}_i]_{jj}^2, \qquad \text{cor}(\epsilon_{ij}, \epsilon_{jk}) = [\boldsymbol{C}_i]_{jk},
$$

so that $\boldsymbol{V}_i$ describes the variance and $\boldsymbol{C}_i$ describes the correlation of the within-group errors ϵ_i. This decomposition of $\boldsymbol{\Lambda}_i$ into a *variance structure* component and a *correlation structure* component is convenient both theoretically and computationally. It allows us to model and develop code for the two structures separately and to combine them into a flexible family of models for the within-group variance–covariance. In §5.2 we describe variance function structures to represent the variance component $\boldsymbol{V}_i$ and in §5.3 we present correlation structures for the correlation component $\boldsymbol{C}_i$.

5.2   Variance Functions for Modeling Heteroscedasticity

Variance functions are used to model the variance structure of the within-group errors using covariates. They have been studied in detail in the context of mixed-effects models by Davidian and Giltinan (1995) and in the context of the extended linear model (5.5) by Carroll and Ruppert (1988).

Following Davidian and Giltinan (1995, Ch. 4), we define the general variance function model for the within-group errors in the extended single-level linear mixed-effects model (5.1) as

$$\text{Var}\left(\epsilon_{ij}|\boldsymbol{b}_i\right) = \sigma^2 g^2 \left(\mu_{ij}, \boldsymbol{v}_{ij}, \boldsymbol{\delta}\right), \quad i = 1, \ldots, M, \quad j = 1, \ldots, n_i, \quad (5.10)$$

where $\mu_{ij} = \text{E}\left[y_{ij}|\boldsymbol{b}_i\right]$, $\boldsymbol{v}_{ij}$ is a vector of *variance covariates*, $\boldsymbol{\delta}$ is a vector of *variance parameters* and $g(\cdot)$ is the variance function, assumed continuous in $\boldsymbol{\delta}$. For example, if the within-group variability is believed to increase with some power of the absolute value of a covariate v_{ij}, we can write the variance model as

$$\text{Var}\left(\epsilon_{ij}|\boldsymbol{b}_i\right) = \sigma^2 \left|v_{ij}\right|^{2\delta}.$$

The variance function in this case is $g(x, y) = |x|^y$ and the covariate v_{ij} can be the expected value μ_{ij}.

The single-level variance function model (5.10) can be generalized to multilevel models. For example, the variance function model for a two-level model is

$$\text{Var}\left(\epsilon_{ijk}|\boldsymbol{b}_{i,j}, \boldsymbol{b}_{ij}\right) = \sigma^2 g^2 \left(\mu_{ijk}, \boldsymbol{v}_{ijk}, \boldsymbol{\delta}\right),$$
$$i = 1, \ldots, M, \quad j = 1, \ldots, M_i, \quad k = 1, \ldots, n_{ij},$$

where $\mu_{ijk} = \text{E}\left[y_{ijk}|\boldsymbol{b}_{i,j}, \boldsymbol{b}_{ij}\right]$. We concentrate, for the remainder of this section, in the single-level model (5.1), but all results presented here easily generalize to multilevel models.

The variance function formulation (5.10) is very flexible and intuitive, because it allows the within-group variance to depend on the fixed effects, $\boldsymbol{\beta}$, and the random effects, $\boldsymbol{b}_i$, through the expected values, μ_{ij}. However, as discussed in Davidian and Giltinan (1995, Ch. 4), it poses some theoretical and computational difficulties, as the within-group errors and the random effects can no longer assumed to be independent. Under the assumption that $\text{E}\left[\boldsymbol{\epsilon}_i|\boldsymbol{b}_i\right] = \boldsymbol{0}$, it is easy to verify that $\text{Var}\left(\epsilon_{ij}\right) = \text{E}\left[\text{Var}\left(\epsilon_{ij}|\boldsymbol{b}_i\right)\right]$, so that the dependence on the unobserved random effects can be avoided by integrating them out of the variance model. Because the variance function g is generally nonlinear in $\boldsymbol{b}_i$, integrating the random effects out of the variance model (5.10) does not lead to a computationally feasible optimization procedure. Instead, we proceed as in Davidian and Giltinan (1995, Ch. 6), and use an approximate variance model in which the expected values μ_{ij} are replaced by their BLUPs $\widehat{\mu}_{ij} = x_{ij}^T \beta + z_{ij}^T \widehat{b}_i$, where x_{ij} and z_{ij} denote, respectively, the jth rows of X_i and Z_i,

$$\mathrm{Var}\,(\epsilon_{ij}) \simeq \sigma^2 g^2 \left(\widehat{\mu}_{ij}, v_{ij}, \delta\right), \quad i = 1, \ldots, M, \quad j = 1, \ldots, n_i. \quad (5.11)$$

Under this approximation, the within-group errors are assumed independent of the random effects, as in (5.1), and the results in §5.1.1 can still be used. Note that, if the conditional variance model (5.10) does not depend on μ_{ij}, (5.11) gives the exact marginal variance and no approximation is required.

When the conditional variance model (5.10) depends on μ_{ij}, the optimization algorithm follows an "iteratively reweighted" scheme: for given $\beta^{(t)}$, $\theta^{(t)}$, $\lambda^{(t)}$, the corresponding BLUPs $\widehat{\mu}_{ij}^{(t)}$ are obtained and held fixed while the objective function is optimized to produce new estimates $\beta^{(t+1)}$, $\theta^{(t+1)}$, $\lambda^{(t+1)}$ which, in turn, give updated BLUPs $\widehat{\mu}_{ij}^{(t+1)}$, with the process iterating until convergence. The resulting estimates approximate the (restricted) maximum likelihood estimates. When the variance model does not involve μ_{ij}, the (restricted) likelihood can be directly optimized, producing the exact (restricted) maximum likelihood estimates.

Variance functions for the extended linear model (5.5) are similarly defined, but, because no random effects are present, the model for the marginal variance does not involve any approximations, being expressed as

$$\mathrm{Var}\,(\epsilon_{ij}) = \sigma^2 g^2 \left(\mu_{ij}, v_{ij}, \delta\right), \quad i = 1, \ldots, M, \quad j = 1, \ldots, n_i, \quad (5.12)$$

where $\mu_{ij} = \mathrm{E}\,[y_{ij}] = x_{ij}^T \beta$ and v_{ij} and δ are defined as in (5.10). This coincides with the variance function model proposed in Carroll and Ruppert (1988, §3). When the variance model (5.12) depends on μ_{ij}, the optimization algorithm parallels the linear mixed-effects model iteratively reweighted scheme described before: for given $\beta^{(t)}$ and $\lambda^{(t)}$ estimated $\mu_{ij}^{(t)}$ are obtained and held fixed while the objective function is optimized to produce new estimates $\beta^{(t+1)}$ and $\lambda^{(t+1)}$, which, in turn, give updated $\mu_{ij}^{(t+1)}$, with the process iterating until convergence. This estimation procedure is known in the literature (Carroll and Ruppert, 1988, §3.2) as *pseudo-likelihood* estimation (when the objective function corresponds to a log-likelihood), or *pseudo-restricted-likelihood* estimation (when the objective function corresponds to a log-restricted-likelihood). If the variance model (5.12) does not involve μ_{ij}, the (restricted) likelihood can be directly optimized and the exact (restricted) maximum likelihood estimates obtained.

TABLE 5.1. Standard varFunc classes.

varFixed	fixed variance
varIdent	different variances per stratum
varPower	power of covariate
varExp	exponential of covariate
varConstPower	constant plus power of covariate
varComb	combination of variance functions

5.2.1   Variance Functions in nlme: The varFunc Classes

The nlme library provides a set of classes of variance functions, the var-Func classes, that are used to specify within-group variance models in either the extended linear mixed-effects model (5.1), or the extended linear model (5.5). Table 5.1 lists the standard varFunc classes included in the nlme library. The varFunc constructors have the same name as their corresponding classes.

The two main arguments for most of the varFunc constructors are value and form. The first specifies the values of the variance parameters δ and the second is a one-sided formula specifying the variance covariate v and, optionally, a stratification variable for the variance parameters—different parameters are used for each level of the stratification variable. For example, to specify age as a variance covariate and to have different variance parameters for each level of Sex, we would use

    form = ~ age | Sex

The fitted object may be referenced in form by the symbol ., so, for example,

    form = ~ fitted(.)

specifies the fitted values as the variance covariate.

Several methods are available for each varFunc class, including initial-ize, which initializes the variance covariates and stratification variables, and varWeights, which extracts the *variance weights*, defined as the inverse of the variance function values. We now describe and illustrate each of the standard varFunc classes.

varFixed

This class represents a variance function with no parameters and a single variance covariate, being used when the within-group variance is known up to a proportionality constant. The varFixed constructor takes a single argument, value, which is a one-sided formula that evaluates to the desired variance model. No stratification variables or expressions involving the symbol . are allowed in value. For example, suppose it is known that the within-group variance increases linearly with `age`

$$\text{Var}\left(\epsilon_{ij}\right) = \sigma^2 \text{age}_{ij},$$

corresponding to a variance function

$$g\left(\text{age}_{ij}\right) = \sqrt{\text{age}_{ij}}$$

and being represented as

```
> vf1Fixed <- varFixed( ~ age )
```

The variance covariate is calculated in the `initialize` method, which, besides the varFixed object, also requires a `data` argument naming a data frame in which to evaluate the variance covariate formula. Initialization of a varFunc object is generally done inside the modeling function (e.g., `lme`, `gls`) which uses it.

```
> vf1Fixed <- initialize( vf1Fixed, data = Orthodont )
> varWeights( vf1Fixed )
   [1] 0.35355 0.31623 0.28868 0.26726 0.35355 0.31623 0.28868
   . . .
```

The variance weights in this case are given by $1/\sqrt{\text{age}_{ij}}$.

varIdent

This class represents a variance model with different variances for each level of a stratification variable s, taking values in the set $\{1, 2, \ldots, S\}$,

$$\text{Var}\left(\epsilon_{ij}\right) = \sigma^2 \delta_{s_{ij}}^2, \tag{5.13}$$

corresponding to the variance function

$$g\left(s_{ij}, \boldsymbol{\delta}\right) = \delta_{s_{ij}}.$$

The variance model (5.13) uses $S+1$ parameters to represent S variances and, therefore, is not identifiable. To achieve identifiability, we need to impose some restriction on the variance parameters $\boldsymbol{\delta}$. We use $\delta_1 = 1$, so that δ_l, $l = 2, \ldots, S$ represent the ratio between the standard deviations of the lth stratum and the first stratum. By definition, $\delta_l > 0$, $l = 2, \ldots, S$.

The main arguments to the `varIdent` constructor are `value`, a named numeric vector or a named list specifying the values and the strata of the $\boldsymbol{\delta}$ parameters that are allowed to vary in the optimization, and `form`, a one-sided formula of the form `~1|s` specifying the stratification variable `s`. Some, or all, of the variance parameters may be set to fixed values (which do not vary during the optimization) using the argument `fixed`. This may be given either as named numeric vector, or a named list, specifying the values and the strata of the $\boldsymbol{\delta}$ that are to remain fixed during the optimization.

For example, to specify and initialize a variance function with different variances per gender for the `Orthodont` data described in §1.4.1, with the initial value of the δ parameter corresponding to `Female` set to 0.5, we use

```
> vf1Ident <- varIdent( c(Female = 0.5), ~ 1 | Sex )
> vf1Ident <- initialize( vf1Ident, Orthodont )
> varWeights( vf1Ident )
 Male Male Male Male Male Male Male Male Male Male Male Male
    1    1    1    1    1    1    1    1    1    1    1    1
 . . .
 Female Female
      2      2
```

If the ratio between the Female and Male standard deviations, given by δ, is to be kept fixed at 0.5 and not allowed to vary during the optimization, we can use instead

```
> vf2Ident <- varIdent( form =  ~ 1 | Sex, fixed = c(Female = 0.5))
> vf2Ident <- initialize( vf2Ident, Orthodont )
> varWeights( vf2Ident )
 Male Male Male Male Male Male Male Male Male Male Male Male
    1    1    1    1    1    1    1    1    1    1    1    1
 . . .
 Female Female
      2      2
```

It is possible to specify several stratification variables simultaneously in form, separated by the * operator. The levels of the different stratification variables are pasted together and a different δ is used for each combination of levels. For example, to specify a variance function with different variances for each age and Sex combination we can use

```
> vf3Ident <- varIdent( form = ~ 1 | Sex * age )
> vf3Ident <- initialize( vf3Ident, Orthodont )
> varWeights( vf3Ident )
 Male*8 Male*10 Male*12 Male*14 Male*8 Male*10 Male*12 Male*14
      1       1       1       1      1       1       1       1
 . . .
 Female*12 Female*14
         1         1
```

By default, all variance parameters are initialized to 1, corresponding to equal variance weights of 1 being assigned to all strata.

varPower

The variance model represented by this class is

$$\text{Var}\left(\epsilon_{ij}\right) = \sigma^2 \left|v_{ij}\right|^{2\delta}, \tag{5.14}$$

corresponding to the variance function

$$g\left(v_{ij}, \delta\right) = \left|v_{ij}\right|^{\delta},$$

which is a power of the absolute value of the variance covariate. The parameter δ is unrestricted (i.e., may take any value in the real line) so (5.14) can model cases where the variance increases or decreases with the absolute value of the variance covariate. Note that, when $v_{ij} = 0$ and $\delta > 0$, the variance function is 0 and the variance weight is undefined. Therefore, this class of variance functions should not be used with variance covariates that may assume the value 0.

The main arguments to the `varPower` constructor are `value` and `form`, which specify, respectively, an initial value for δ, when this is allowed to vary in the optimization, and a one-sided formula with the variance covariate. By default, `value = 0`, corresponding to equal variance weights of 1, and `form = ~fitted(.)`, corresponding to a variance covariate given by the fitted values. For example, to specify a variance model with the δ parameter initially set to 1, and allowed to vary in the optimization, and the fitted values as the variance covariate, we use

```
> vf1Power <- varPower( 1 )
> formula( vf1Power )
~ fitted(.)
```

The `fixed` argument can be used to set δ to a fixed value, which does not change in the optimization. For example,

```
> vf2Power <- varPower( fixed = 0.5 )
```

specifies a model in which the variance increases linearly with the fitted values.

An optional stratification variable, or several stratification variables separated by `*`, may be included in the `form` argument, with a different δ being used for each stratum. This corresponds to the following generalization of (5.14)

$$\mathrm{Var}\,(\epsilon_{ij}) = \sigma^2 \, |v_{ij}|^{2\delta_{s_{ij}}}\,, \quad g\,(v_{ij}, s_{ij}, \boldsymbol{\delta}) = |v_{ij}|^{\delta_{s_{ij}}}\,.$$

When a stratification variable is included in `form`, the arguments `value` and `fixed` must be either named vectors, or named lists. For example, to specify a model for the `Orthodont` data in which the variance increases linearly with the fitted values for `Males` and stays constant for `Females`, with both parameters held fixed during the optimization, we use

```
> vf3Power <- varPower( form = ~ fitted(.) | Sex,
+    fixed = list(Male = 0.5, Female = 0) )
```

varExp

The variance model represented by this class is

$$\mathrm{Var}\,(\epsilon_{ij}) = \sigma^2 \exp\,(2\delta v_{ij})\,, \tag{5.15}$$

corresponding to the variance function

$$g(v_{ij}, \delta) = \exp(\delta v_{ij}),$$

which is an exponential function of the variance covariate. The parameter δ is unrestricted, so (5.15) can model cases where the variance increases or decreases with the variance covariate. There are no restrictions on the variance covariate, which, in particular, may take the value 0.

The arguments value, form, and fixed to the varExp constructor are defined and declared as in varPower, assuming also the same default values. An optional stratification variable, or stratification variables separated by *, may be specified in form, corresponding to the following generalization of (5.15)

$$\text{Var}(\epsilon_{ij}) = \sigma^2 \exp(2\delta_{s_{ij}} v_{ij}), \quad g(v_{ij}, s_{ij}, \boldsymbol{\delta}) = \exp(\delta_{s_{ij}} v_{ij}).$$

As with varPower, when a stratification variable is included in form, the arguments value and fixed must be either named vectors, or named lists. For example, a variance model in which the Male variance increases exponentially with age, but the Female variance is held at a constant value is expressed as

```
> vf1Exp <- varExp( form = ~ age | Sex, fixed = c(Female = 0) )
```

varConstPower

The variance model represented by this class is

$$\text{Var}(\epsilon_{ij}) = \sigma^2 \left(\delta_1 + |v_{ij}|^{\delta_2}\right)^2, \qquad (5.16)$$

corresponding to the variance function

$$g(v_{ij}, \boldsymbol{\delta}) = \delta_1 + |v_{ij}|^{\delta_2},$$

which is a constant plus a power of the absolute value of the variance covariate. δ_1 is restricted to be positive and δ_2 is unrestricted. If $\delta_2 > 0$, which will generally be the case in practice, the varConstPower variance function is approximately constant and equal to δ_1, when the variance covariate is close to 0, and increases with the absolute value of the variance covariate as it gets away from 0. This generally gives a more realistic model than the varPower model, in cases when the variance covariate takes values close or equal to 0.

Initial values for parameters that are allowed to vary during the optimization are specified through the arguments const and power. The former is used for δ_1 and the latter for δ_2. By default, const = 1 and power = 0, corresponding to constant variance weights equal to 1. The form argument is defined as in varPower and varExp, assuming the same default value. The argument `fixed` is given as a list with components `const` and `power` and may be used to set either, or both, of the variance parameters to a fixed value. For example, to specify a variance function with δ_1 fixed at the value 1, with δ_2 allowed to vary but initialized to 0.5, and the fitted values as the variance covariate, we use

```
> vf1ConstPower <- varConstPower( power = 0.5,
+          fixed = list(const = 1) )
```

An optional stratification variable, or several stratification variables separated by `*`, may be included in the `form` argument, with different δ_1 and δ_2 being used for each stratum. This corresponds to the following generalization of (5.16)

$$\text{Var}\,(\epsilon_{ij}) = \sigma^2 \left(\delta_{1,s_{ij}} + |v_{ij}|^{\delta_{2,s_{ij}}} \right)^2, \quad g\,(v_{ij}, s_{ij}, \boldsymbol{\delta}) = \delta_{2,s_{ij}} + |v_{ij}|^{\delta_{2,s_{ij}}}.$$

When a stratification variable is included in `form`, the arguments `const`, `power` and the components of `fixed` must be either named vectors, or named lists.

varComb

This class allows the combination of two, or more, variance models, by multiplying together the corresponding variance functions. For example, a variance model in which the variance is proportional to an exponential function of a variance covariate, but in which the proportionality constant varies according to the levels of an stratification variable s can be expressed as a product of a `varIdent` variance function and a `varExp` variance function.

$$\text{Var}\,(\epsilon_{ij}) = \sigma^2 \delta_{1,s_{ij}}^2 \exp\,(2\delta_2 v_{ij}) = \sigma^2 g_1^2\,(s_{ij}, \boldsymbol{\delta}_1)\, g_2^2\,(v_{ij}, \delta_2)\,,$$
$$g_1\,(s_{ij}, \boldsymbol{\delta}_1) = \delta_{1,s_{ij}}, \qquad g_2\,(v_{ij}, \delta_2) = \exp\,(\delta_2 v_{ij})\,. \tag{5.17}$$

The `varComb` constructor can take any number of `varFunc` objects as arguments. For example, to represent (5.17), with `Sex` as the stratification variable and `age` as the variance covariate, we use

```
> vf1Comb <- varComb( varIdent(c(Female = 0.5), ~ 1 | Sex),
+                      varExp(1, ~ age) )
> vf1Comb <- initialize( vf1Comb, Orthodont )
> varWeights( vf1Comb )
       1     2         3         4     5     6         7         8     9
   0.125  0.1  0.083333  0.071429  0.125  0.1  0.083333  0.071429  0.125
  . . .
     98        99      100  101 102       103       104  105 106       107
   0.2  0.16667  0.14286  0.25  0.2  0.16667  0.14286  0.25  0.2  0.16667
        108
   0.14286
```

The varComb variance weights correspond to the product of the individual variance weights of each of its varFunc objects. In the case of vf1Comb, these are given by $1/\sqrt{\mathrm{age}_{ij}}$ for Male (observations 1–64) and $2/\sqrt{\mathrm{age}_{ij}}$ for Female (observations 65–108).

New varFunc classes, representing user-defined variance functions, can be added to the set of standard classes in Table 5.1 and used with the modeling functions in the nlme library. For this, one must specify a constructor function, generally with the same name as the class, and, at a minimum, methods for the functions coef, coef<-, and initialize. The varPower constructor and methods can serve as templates for these.

5.2.2   Using Variance Functions with lme

Variance functions are specified in the lme function using the weights argument. By default, weights = NULL, corresponding to a homoscedastic variance model for the within-group errors. Variance models can be specified in weights either as a one-sided formula, in which case it is passed as the single argument to the varFixed constructor, or as a varFunc object, created using the standard constructors described in §5.2.1, or a user-defined constructor. In this section, we describe the use of variance models in lme through the analysis of two examples of grouped data with heteroscedastic within-group errors.

High-Flux Hemodialyzer Ultrafiltration Rates

Vonesh and Carter (1992) describe and analyze data measured on high-flux hemodialyzers to assess their *in vivo* ultrafiltration characteristics. The ultrafiltration rates (in ml/hr) of 20 high-flux dialyzers were measured at 7 ascending transmembrane pressures (in dmHg). The *in vitro* evaluation of the dialyzers used bovine blood at flow rates of either 200 dl/min or 300 dl/min. These data are also described in Appendix A.6 and are included in the nlme library as the groupedData object Dialyzer.

The plots of the ultrafiltration rates versus transmembrane pressure by bovine blood flow rate, displayed in Figure 5.1, reveal that the ultrafiltration rate increases with transmembrane pressure and that higher ultrafiltration rates are attained with the 300 dl/min blood flow dialyzers. These plots also indicate that the variability in the ultrafiltration rates increases with transmembrane pressure.

Vonesh and Carter (1992) use a nonlinear model to represent the relationship between ultrafiltration rate and transmembrane pressure. An alternative analysis is presented in Littell et al. (1996), who compare several linear mixed-effects models and extended linear models to represent the ultrafiltration rate y_{ij} at the jth transmembrane pressure x_{ij} for the ith subject. The best linear mixed-effects model indicated by their analysis

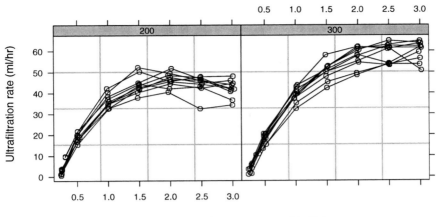

FIGURE 5.1. Hemodialyzer ultrafiltration rates (in ml/hr) measured at 7 different transmembrane pressures (in dmHg) on 20 high-flux dialyzers. *In vitro* evaluation of dialyzers based on bovine blood flow rates of 200 dl/min and 300 dl/min.

is

$$y_{ij} = (\beta_0 + \gamma_0 Q_i + b_{0i}) + (\beta_1 + \gamma_1 Q_i + b_{1i}) x_{ij}$$
$$+ (\beta_2 + \gamma_2 Q_i + b_{2i}) x_{ij}^2 + (\beta_3 + \gamma_3 Q_i) x_{ij}^3 + (\beta_4 + \gamma_4 Q_i) x_{ij}^4 + \epsilon_{ij},$$
$$(5.18)$$

$$b_i = \begin{bmatrix} b_{0i} \\ b_{1i} \\ b_{2i} \end{bmatrix} \sim \mathcal{N}(\mathbf{0}, \mathbf{\Psi}), \quad \epsilon_{ij} \sim \mathcal{N}(0, \sigma^2),$$

where Q_i is a binary variable taking values -1 for 200 dl/min hemodialyzers and 1 for 300 dl/min hemodialyzers; $\beta_0, \beta_1, \beta_2, \beta_3$, and β_4 are, respectively, the intercept, linear, quadratic, cubic, and quartic fixed effects averaged over the levels of Q; γ_i is the blood flow effect associated with the fixed effect β_i; b_i is the vector of random effects, assumed independent for different i; and ϵ_{ij} is the within-group error, assumed independent for different i, j and independent of the random effects.

We fit the homoscedastic linear mixed-effects model (5.18) with

```
> fm1Dial.lme <-
+   lme(rate ~(pressure + pressure^2 + pressure^3 + pressure^4)*QB,
+        Dialyzer, ~ pressure + pressure^2)
> fm1Dial.lme
Linear mixed-effects model fit by REML
  Data: Dialyzer
  Log-restricted-likelihood: -326.39
  Fixed: rate ~(pressure + pressure^2 + pressure^3 + pressure^4)*QB
```

```
(Intercept) pressure I(pressure^2) I(pressure^3) I(pressure^4)
  -16.598    88.673      -42.732         9.2165        -0.77561
      QB pressure:QB I(pressure^2):QB I(pressure^3):QB
-0.63174     0.31044             1.5741            0.050928
I(pressure^4):QB
    -0.085967
```

```
Random effects:
 Formula:  ~ pressure + pressure^2 | Subject
 Structure: General positive-definite
              StdDev   Corr
   (Intercept) 1.4989 (Intr) pressr
      pressure 4.9074 -0.507
I(pressure^2) 1.4739  0.311 -0.944
      Residual 1.8214
```

```
Number of Observations: 140
Number of Groups: 20
```

The primary tool for investigating within-group heteroscedasticity is the plots of residuals against the fitted values and other candidate variance covariates. In the hemodialyzer example, the transmembrane pressure is a natural candidate for the variance covariate. The corresponding residuals plot, obtained with

```
> plot( fm1Dial.lme, resid(.) ~ pressure, abline = 0 ) # Figure 5.2
```

and displayed in Figure 5.2, confirms that the within-group variability increases with transmembrane pressure.

Because of its flexibility, the varPower variance function is a common choice for modeling monotonic heteroscedasticity, when the variance covariate is bounded away from zero (transmembrane pressure varies in the data between 0.235 dmHg and 3.030 dmHg). The corresponding model only differs from (5.18) in that the within-group errors are allowed to be heteroscedastic with variance model

$$\text{Var}\,(\epsilon_{ij}) = \sigma^2 x_{ij}^{2\delta} \tag{5.19}$$

and we fit it with

```
> fm2Dial.lme <- update( fm1Dial.lme,
+                      weights = varPower(form = ~ pressure) )
> fm2Dial.lme
Linear mixed-effects model fit by REML
  Data: Dialyzer
  Log-restricted-likelihood: -309.51
  Fixed: rate ~(pressure + pressure^2 + pressure^3 + pressure^4)*QB
  (Intercept) pressure I(pressure^2) I(pressure^3) I(pressure^4)
     -17.68    93.711      -49.186        12.245         -1.2426
```

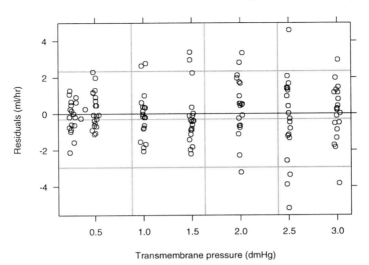

FIGURE 5.2. Plot of residuals versus transmembrane pressure for the homoscedastic fitted object `fm1Dial.lme`.

```
      QB pressure:QB  I(pressure^2):QB  I(pressure^3):QB
 -0.92072       1.3528          0.48071          0.49118
 I(pressure^4):QB
       -0.14624

Random effects:
 Formula:  ~ pressure + pressure^2 | Subject
 Structure: General positive-definite
             StdDev   Corr
   (Intercept) 1.8569 (Intr) pressr
      pressure 5.3282 -0.522
I(pressure^2) 1.6483  0.362 -0.954
      Residual 1.2627

Variance function:
 Structure: Power of variance covariate
 Formula:  ~ pressure
 Parameter estimates:
   power
 0.74923
Number of Observations: 140
Number of Groups: 20
```

The **anova** method can be used to test the significance of the heteroscedastic model (5.19).

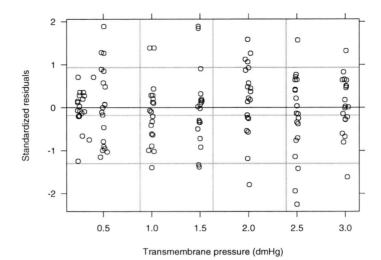

FIGURE 5.3. Plot of standardized residuals versus transmembrane pressure for the varPower fitted object fm2Dial.lme.

```
> anova( fm1Dial.lme, fm2Dial.lme )
            Model df   AIC    BIC  logLik   Test L.Ratio p-value
fm1Dial.lme     1 17 686.78 735.53 -326.39
fm2Dial.lme     2 18 655.01 706.63 -309.51 1 vs 2   33.77  <.0001
```

As expected, there is a highly significant increase in the log-likelihood associated with the inclusion of the varPower variance function. The plot of the standardized residuals, defined as $r_{ij} = (y_{ij} - \widehat{y}_{ij}) / (\widehat{\sigma} x_{ij}^{\widehat{\delta}})$ in this case, versus the variance covariate is used to graphically assess the adequacy of the variance model.

```
> plot( fm2Dial.lme, resid(., type = "p") ~ pressure,
+        abline = 0 )                              # Figure 5.3
```

The resulting plot, displayed in Figure 5.3, reveals a reasonably homogeneous pattern of variability for the standardized residuals, indicating that the varPower model successfully describes the within-group variance.

We assess the variability of the variance parameter estimate $\widehat{\delta}$ with the intervals method.

```
> intervals( fm2Dial.lme )
. . .
 Variance function:
        lower    est.   upper
power  0.4079 0.74923 1.0906
. . .
```

The resulting confidence interval is based on a normal approximation for the distribution of the (restricted) maximum likelihood estimators, with

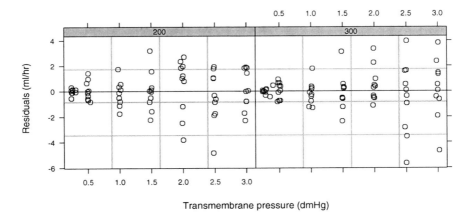

FIGURE 5.4. Plot of residuals versus transmembrane pressure by bovine blood flow level for the varPower fitted object fm2Dial.lme.

an approximate variance–covariance matrix given by the inverse of the observed information matrix evaluated at the converged values.

An interesting question about the hemodialyzer data is whether the within-group variability depends on the bovine blood flow level. We can investigate this graphically with the plots of the raw residuals versus transmembrane pressure by blood flow level, obtained with

```
> plot(fm2Dial.lme, resid(.) ~ pressure|QB, abline = 0)    # Fig. 5.4
```

and displayed in Figure 5.4. Note that the pattern of increasing variability is still present in the raw residuals, as we did not transform the data, but, instead, incorporated the within-group heteroscedasticity in the model. The heteroscedastic patterns seem the same for both blood flow levels. We can test this formally using

```
> fm3Dial.lme <- update(fm2Dial.lme,
+                    weights=varPower(form = ~ pressure | QB))
> fm3Dial.lme
. . .
Variance function:
 Structure: Power of variance covariate, different strata
 Formula:  ~ pressure | QB
 Parameter estimates:
     200      300
 0.64775 0.83777
. . .
> anova( fm2Dial.lme, fm3Dial.lme )
           Model df    AIC    BIC  logLik   Test L.Ratio p-value
fm2Dial.lme     1 18 655.01 706.63 -309.51
fm3Dial.lme     2 19 656.30 710.78 -309.15 1 vs 2 0.71091  0.3991
```

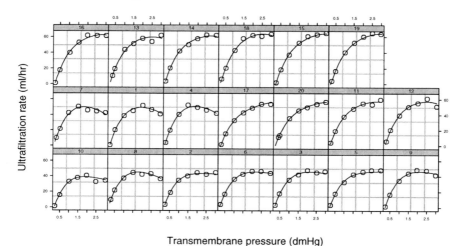

Transmembrane pressure (dmHg)

FIGURE 5.5. Plot of predicted ultrafiltration rates versus transmembrane pressure by subject corresponding to the fitted object fm2Dial.lme.

As expected, there is no evidence of a significant effect of blood flow in the within-group variability.

Because the transmembrane pressure takes values relatively close to zero, we may wish to investigate if a varConstPower variance model

$$\text{Var}\left(\epsilon_{ij}\right) = \sigma^2 \left(\delta_1 + x_{ij}^{\delta_2}\right)^2$$

would give a more appropriate representation of the within-group variance.

```
> fm4Dial.lme <- update(fm2Dial.lme,
+                         weights = varConstPower(form = ~ pressure))
> anova( fm2Dial.lme, fm4Dial.lme )
            Model df    AIC     BIC  logLik    Test L.Ratio p-value
fm2Dial.lme     1 18 655.01 706.63 -309.51
fm4Dial.lme     2 19 656.85 711.34 -309.43 1 vs 2 0.15915  0.6899
```

The nearly identical log-likelihood values indicate that there is no need for the extra parameter associated with varConstPower and that the simpler varPower model should be maintained.

A final assessment of the heteroscedastic version of the linear mixed-effects model (5.18) with within-group variance model (5.19) is provided by the plot of the augmented predictions by subject, obtained with

```
> plot( augPred(fm2Dial.lme), grid = T )          # Figure 5.5
```

and displayed in Figure 5.5. The predicted values closely match the observed ultrafiltration rates, attesting the adequacy of the model.

One of the questions of interest for the hemodialyzer data is whether the ultrafiltration characteristics differ with the evaluation blood flow rates, which is suggested by the plots in Figure 5.1. The anova method can be used to test the significance of the terms associated with the evaluation blood flow rates, in the order they were entered in the model (a sequential type of test).

```
> anova( fm2Dial.lme )
. . .
                  numDF denDF F-value p-value
    (Intercept)       1   112   552.9  <.0001
       pressure       1   112  2328.6  <.0001
  I(pressure^2)       1   112  1174.6  <.0001
  I(pressure^3)       1   112   359.9  <.0001
  I(pressure^4)       1   112    12.5  0.0006
             QB       1    18     4.8  0.0414
    pressure:QB       1   112    80.1  <.0001
I(pressure^2):QB      1   112     1.4  0.2477
I(pressure^3):QB      1   112     2.2  0.1370
I(pressure^4):QB      1   112     0.2  0.6840
. . .
```

The large p-values associated with terms of degree greater than or equal to 2 involving the variable QB suggest that they are not needed in the model. We can verify their *joint* significance with

```
> anova( fm2Dial.lme, Terms = 8:10 )
F-test for: I(pressure^2):QB, I(pressure^3):QB, I(pressure^4):QB
    numDF denDF F-value p-value
1       3   112  1.2536  0.2939
```

The large p-value for the F-test confirms that these terms could be eliminated from the model.

Body Weight Growth in Rats

As a second example to illustrate the use of variance functions with lme, we revisit the BodyWeight data introduced in §3.2.1 and described in Hand and Crowder (1996, Table A.1), on the body weights of rats measured over 64 days. The body weights of the rats (in grams) are measured on day 1 and every seven days thereafter, until day 64, with an extra measurement on day 44. There are three groups of rats, each on a different diet. These data are also described in Appendix A.3 and are included in the nlme library as the groupedData object BodyWeight.

The plots of the body weights versus time by diet, shown in Figure 5.6, indicate strong differences among the three diet groups. There is also evidence of a rat in diet group 2 with an unusually high initial body weight.

The body weights appear to grow linearly with time, possibly with different intercepts and slopes for each diet, and with intercept and slope

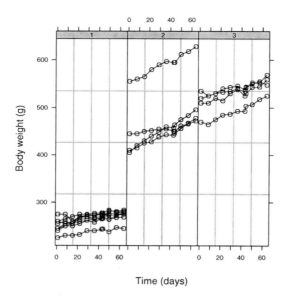

Time (days)

FIGURE 5.6. Body weights of rats measured over a period of 64 days. The rats are divided into three groups on different diets.

random effects to account for rat-to-rat variation. We express this as the linear mixed-effects model

$$y_{ij} = (\beta_0 + \gamma_{02}D_{2i} + \gamma_{03}D_{3i} + b_{0i}) + (\beta_1 + \gamma_{12}D_{2i} + \gamma_{13}D_{3i} + b_{1i}) t_{ij} + \epsilon_{ij},$$

$$\boldsymbol{b}_i = \begin{bmatrix} b_{0i} \\ b_{1i} \end{bmatrix} \sim \mathcal{N}\left(\boldsymbol{0}, \boldsymbol{\Psi}\right), \quad \epsilon_{ij} \sim \mathcal{N}\left(0, \sigma^2\right), \qquad (5.20)$$

where D_{2i} is a binary variable taking the value 1 if the ith rat receives Diet 2; D_{3i} is a binary indicator variable for Diet 3; β_0 and β_1 are, respectively, the average intercept and the average slope for rats under Diet 1; γ_{0k} is the average difference in intercept between rats under Diet k and rats under Diet 1; γ_{1k} is the average difference in slope between rats under Diet k and rats under Diet 1; $\boldsymbol{b}_i$ is the vector of random effects, assumed independent for different i; and ϵ_{ij} is the within-group error, assumed independent for different i, j and independent of the random effects.

To fit (5.20), we first need to reset the contrasts option to use the contr.treatment parameterization for factors, as described in §1.2.1,

```
> options( contrasts = c("contr.treatment", "contr.poly") )
```

and then use

```
> fm1BW.lme <- lme( weight ~ Time * Diet, BodyWeight,
+                   random = ~ Time )
```

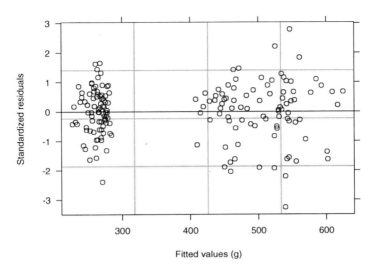

FIGURE 5.7. Plot of standardized residuals versus fitted values for the homo-scedastic fitted object fm1BW.lme.

```
> fm1BW.lme
Linear mixed-effects model fit by REML
  Data: BodyWeight
  Log-restricted-likelihood: -575.86
  Fixed: weight ~ Time * Diet
  (Intercept)    Time  Diet2  Diet3 TimeDiet2 TimeDiet3
       251.65 0.35964 200.67 252.07   0.60584   0.29834

Random effects:
 Formula:   ~ Time | Rat
 Structure: General positive-definite
            StdDev    Corr
(Intercept) 36.93907 (Inter
       Time  0.24841 -0.149
   Residual  4.44361

Number of Observations: 176
Number of Groups: 16
```

The plot of the standardized residuals versus the fitted values, displayed in Figure 5.7, gives clear indication of within-group heteroscedasticity. Because the fitted values are bounded away from zero, we can use the varPower variance function to model the heteroscedasticity.

```
> fm2BW.lme <- update( fm1BW.lme, weights = varPower() )
> fm2BW.lme
Linear mixed-effects model fit by REML
  Data: BodyWeight
```

```
Log-restricted-likelihood: -570.96
Fixed: weight ~ Time * Diet
(Intercept)    Time  Diet2  Diet3 TimeDiet2 TimeDiet3
      251.6 0.36109 200.78 252.17   0.60182   0.29523

Random effects:
 Formula:  ~ Time | Rat
 Structure: General positive-definite
               StdDev   Corr
(Intercept) 36.89887 (Inter
       Time  0.24373 -0.147
   Residual  0.17536

Variance function:
 Structure: Power of variance covariate
 Formula:  ~ fitted(.)
 Parameter estimates:
   power
 0.54266
Number of Observations: 176
Number of Groups: 16
```

Note that the `form` argument did not need to be specified in the call to varPower, because its default value, ~`fitted(.)`, corresponds to the desired variance covariate.

The plot of the standardized residuals versus fitted values for the heteroscedastic fit corresponding to `fm2BW.lme`, displayed in Figure 5.8, indicates that the varPower variance function adequately represents the within-group heteroscedasticity.

We can test the significance of the variance parameter in the varPower model using the **anova** method, which, as expected, strongly rejects the assumption of homoscedasticity (i.e., $\delta = 0$).

```
> anova( fm1BW.lme, fm2BW.lme )
          Model df    AIC    BIC  logLik   Test L.Ratio p-value
fm1BW.lme     1 10 1171.7 1203.1 -575.86
fm2BW.lme     2 11 1163.9 1198.4 -570.96 1 vs 2  9.7984  0.0017
```

The primary question of interest for the `BodyWeight` data is whether the growth rates differ significantly among diets. Because of the parametrization used in (5.20), the summary method only provides tests for differences between Diets 1 and 2 (γ_{12} in (5.20)) and between Diets 1 and 3 (γ_{13} in (5.20)).

```
> summary( fm2BW.lme )
 . . .
Fixed effects: weight ~ Time * Diet
             Value Std.Error  DF t-value p-value
(Intercept) 251.60    13.068 157  19.254  <.0001
```

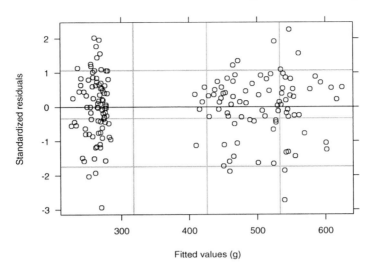

FIGURE 5.8. Plot of standardized residuals versus fitted values for the varPower
fitted object `fm2BW.lme`.

Time	0.36	0.088	157	4.084	0.0001
Diet2	200.78	22.657	13	8.862	<.0001
Diet3	252.17	22.662	13	11.127	<.0001
TimeDiet2	0.60	0.155	157	3.871	0.0002
TimeDiet3	0.30	0.156	157	1.893	0.0602

. . .

There appears to be a significant increase in growth rate associated with
Diet 2 (`TimeDiet2`) and a borderline significant increase in growth rate for
Diet 3 (`TimeDiet3`). We can test the difference in growth rates between Diets
2 and 3 using the `anova` method.

```
> anova( fm2BW.lme, L = c(TimeDiet2 = 1, TimeDiet3 = -1) )
F-test for linear combination(s)
 TimeDiet2 TimeDiet3
         1        -1
  numDF denDF F-value p-value
1     1   157  2.8608  0.0927
```

The argument `L` is used to specify contrasts of coefficients to be tested as
equal to zero. The names of the elements in `L` must correspond to coefficients
in the model. There does not seem to be a significant difference in growth
rate between Diets 2 and 3.

5.3 Correlation Structures for Modeling Dependence

Correlation structures are used to model dependence among observations. In the context of mixed-effects models and extended linear models, they are used to model dependence among the within-group errors. Historically, correlation structures have been developed for two main classes of data: *time-series* data and *spatial* data. The former is generally associated with observations indexed by an integer-valued *time* variable, while the latter refers primarily to observations indexed by a two-dimensional *spatial location* vector, taking values in the real plane.

To establish a general framework for correlation structures, we assume that the within-group errors ϵ_{ij} are associated with *position* vectors $\boldsymbol{p}_{ij}$. For time series data, the $\boldsymbol{p}_{ij}$ are typically integer scalars, while for spatial data they are generally two-dimensional coordinate vectors. The correlation structures considered in this book are assumed to be *isotropic* (Cressie, 1993, §2.3.1); that is, the correlation between two within-group errors $\epsilon_{ij}, \epsilon_{ij'}$ is assumed to depend on the corresponding position vectors $\boldsymbol{p}_{ij}, \boldsymbol{p}_{ij'}$ only through some distance between them, say $d\left(\boldsymbol{p}_{ij}, \boldsymbol{p}_{ij'}\right)$, and not on the particular values they assume. The general within-group correlation structure for single-level grouping is expressed, for $i = 1, \ldots, M$ and $j, j' = 1, \ldots, n_i$, as

$$\text{cor}\left(\epsilon_{ij}, \epsilon_{ij'}\right) = h\left[d(\boldsymbol{p}_{ij}, \boldsymbol{p}_{ij'}), \boldsymbol{\rho}\right], \qquad (5.21)$$

where $\boldsymbol{\rho}$ is a vector of *correlation parameters* and $h(\cdot)$ is a correlation function taking values between -1 and 1, assumed continuous in $\boldsymbol{\rho}$, and such that $h\left(0, \boldsymbol{\rho}\right) = 1$, that is, if two observations have identical position vectors, they are the same observation and therefore have correlation 1.

The single-level correlation model (5.21) can be easily generalized to multilevel grouping. For example, the correlation model for two nested levels of grouping is

$$\text{cor}\left(\epsilon_{ijk}, \epsilon_{ijk'}\right) = h\left[d(\boldsymbol{p}_{ijk}, \boldsymbol{p}_{ijk'}), \boldsymbol{\rho}\right],$$

$$i = 1, \ldots, M, \quad j = 1, \ldots, M_i, \quad k, k' = 1, \ldots, n_{ij}.$$

Note that the correlation model applies to within-group errors within the same innermost level of grouping. We concentrate, for the remainder of this section, in the single-level correlation model (5.21), but all correlation structures presented here can be easily extended to multilevel models.

5.3.1 Serial Correlation Structures

Serial correlation structures are used to model dependence in time-series data, that is, data observed sequentially over time and indexed by a one-dimensional position vector. Serial correlation structures for linear models without random effects have been extensively studied by Box, Jenkins and Reinsel (1994). In the context of linear mixed-effects models, they are described in detail in Jones (1993).

We simplify the isotropy assumption and assume that the serial correlation model depends on the one-dimensional positions $p_{ij}, p_{ij'}$ only through their absolute difference. The general serial correlation model is then defined as

$$\text{cor}\left(\epsilon_{ij}, \epsilon_{ij'}\right) = h\left(\left|p_{ij} - p_{ij'}\right|, \boldsymbol{\rho}\right).$$

In the context of time-series data, the correlation function $h\left(\cdot\right)$ is referred to as the *autocorrelation* function. The *empirical autocorrelation function* (Box et al., 1994, §3), a nonparametric estimate of the autocorrelation function, provides a useful tool for investigating serial correlation in time-series data. Let $r_{ij} = \left(y_{ij} - \widehat{y}_{ij}\right)/\widehat{\sigma}_{ij}$, denote the standardized residuals from a fitted mixed-effects model, with $\sigma_{ij}^2 = \text{Var}\left(\epsilon_{ij}\right)$. The empirical autocorrelation at lag l is defined as

$$\widehat{\rho}\left(l\right) = \frac{\sum_{i=1}^{M}\sum_{j=1}^{n_i-l} r_{ij}r_{i(j+l)}/N(l)}{\sum_{i=1}^{M}\sum_{j=1}^{n_i} r_{ij}^2/N(0)}, \tag{5.22}$$

where $N(l)$ is the number of residual pairs used in the summation defining the numerator of $\widehat{\rho}(l)$.

Serial correlation structures typically require that the data be observed at integer time points and do not easily generalize to continuous position vectors. We describe below some of the most common serial correlation structures used in practice, all of which are implemented in the nlme library.

Compound Symmetry

This is the simplest serial correlation structure, which assumes equal correlation among all within-group errors pertaining to the same group. The corresponding correlation model is

$$\text{cor}\left(\epsilon_{ij}, \epsilon_{ij'}\right) = \rho, \quad \forall j \neq j', \quad h(k, \rho) = \rho, \quad k = 1, 2, \dots, \tag{5.23}$$

where the single correlation parameter ρ is generally referred to as the *intraclass* correlation coefficient.

The variance–covariance matrix for the ith response vector in a single-level linear mixed-effects model with independent and identically distributed within-group errors, with variance σ^2, and a single intercept random effect, with variance σ_b^2, is $\sigma^2\boldsymbol{I} + \sigma_b^2\boldsymbol{11}^T$, corresponding to the correlation matrix $\sigma^2/\left(\sigma_b^2 + \sigma^2\right)\boldsymbol{I} + \sigma_b^2/\left(\sigma_b^2 + \sigma^2\right)\boldsymbol{11}^T$. This is equivalent to a compound symmetry structure with intraclass correlation $\rho = \sigma_b^2/\left(\sigma_b^2 + \sigma^2\right)$, indicating that the correlation structure defined by this linear mixed-effects model is a particular case of (5.23). The two correlation models are not equivalent, however, because the intraclass correlation in the linear mixed-effects model can only take values between 0 and 1, while (5.23) allows ρ to take negative values (to have a positive-definite compound symmetry correlation structure, it is only required that $\rho > -1/\left[\max_{i \leq M} (n_i) - 1\right]$).

The compound symmetry correlation model tends to be too simplistic for practical applications involving time-series data, as, in general, it is more realistic to assume a model in which the correlation between two observations decreases, in absolute value, with their distance. It is a useful model for applications involving short time series per group, or when all observations within a group are collected at the same time, as in split-plot experiments.

General

This structure represents the other extreme in complexity to the compound symmetry structure. Each correlation in the data is represented by a different parameter, corresponding to the correlation function

$$h(k, \boldsymbol{\rho}) = \rho_k, \quad k = 1, 2, \ldots . \tag{5.24}$$

Because the number of parameters in (5.24) increases quadratically with the maximum number of observations within a group, this correlation structure will often lead to overparameterized models. When there are relatively few observations per group, the general correlation structure is useful as an exploratory tool to determine a more parsimonious correlation model.

Autoregressive–Moving Average

This family of correlation structures, described in detail in Box et al. (1994), includes different classes of linear stationary models: *autoregressive* models, *moving average* models, and mixture of autoregressive–moving average models. These are also called *Box and Jenkins* models.

Autoregressive–moving average correlation models assume that the data are observed at integer time points and, for simplicity, we use the notation ϵ_t to refer to an observation taken at time t. The distance, or *lag*, between two observations ϵ_t and ϵ_s is given by $|t - s|$. So lag-1 refers to observations one time unit apart and so on.

Autoregressive models express the current observation as a linear function of previous observations plus a homoscedastic noise term, a_t, centered at 0 ($\mathrm{E}\,[a_t] = 0$) and assumed independent of the previous observations.

$$\epsilon_t = \phi_1 \epsilon_{t-1} + \cdots + \phi_p \epsilon_{t-p} + a_t. \tag{5.25}$$

The number of past observations included in the linear model (5.25), p, is called the *order* of the autoregressive model, which is denoted by $AR(p)$. There are p correlation parameters in an $AR(p)$ model, given by $\boldsymbol{\phi} = (\phi_1, \ldots, \phi_p)$.

The $AR(1)$ model is the simplest (and one of the most useful) autoregressive model. Its correlation function decreases in absolute value exponentially with lag.

$$h(k, \phi) = \phi^k, \quad k = 0, 1, \ldots . \tag{5.26}$$

The single correlation parameter, ϕ, represents the lag-1 correlation and takes values between -1 and 1. The $AR(1)$ model is one of the few serial correlation structures that can be generalized to continuous time measurements. We define the *continuous time $AR(1)$* correlation function (Jones, 1993, 3.3), denoted $CAR(1)$, as

$$h(s, \phi) = \phi^s, \quad s \geq 0, \quad \phi \geq 0. \tag{5.27}$$

Note that the single correlation parameter ϕ in (5.27) must be non-negative.

For autoregressive models of order greater than 1, the correlation function does not admit a simple representation as in (5.26), being defined recursively through the difference equation (Box et al., 1994, §3.2.2)

$$h(k, \phi) = \phi_1 h(|k - 1|, \phi) + \cdots + \phi_p h(|k - p|, \phi), \quad k = 1, 2, \ldots .$$

Moving average correlation models assume that the current observation is a linear function of independent and identically distributed noise terms.

$$\epsilon_t = \theta_1 a_{t-1} + \cdots + \theta_q a_{t-q} + a_t. \tag{5.28}$$

The number of noise terms included in the linear model (5.28), q, is called the order of the moving average model, which is denoted by $MA(q)$. There are q correlation parameters in an $MA(q)$ model, given by $\boldsymbol{\theta} = (\theta_1, \ldots, \theta_q)$.

The correlation function for an $MA(q)$ model is

$$h(k, \boldsymbol{\theta}) = \begin{cases} \frac{\theta_k + \theta_1 \theta_{k-1} + \cdots + \theta_{k-q} \theta_q}{1 + \theta_1^2 + \cdots + \theta_q^2}, & k = 1, \ldots, q, \\ 0, & k = q+1, q+2, \ldots . \end{cases}$$

Observations more than q time units apart are uncorrelated, as they do not share any common noise terms a_t.

Strategies for estimating the order of autoregressive and moving average models in time series applications are discussed in Box et al. (1994, §3).

Mixed autoregressive–moving average models, called $ARMA$ models, are obtained by combining together an autoregressive model and a moving average model.

$$\epsilon_t = \sum_{i=1}^{p} \phi_i \epsilon_{t-i} + \sum_{j=1}^{q} \theta_j a_{t-j} + a_t.$$

There are $p + q$ correlation parameters $\boldsymbol{\rho}$ in an $ARMA(p, q)$ model, corresponding to the combination of the p autoregressive parameters $\boldsymbol{\phi} =$

$(\phi_1, \ldots, \phi_p)$ and the q moving average parameters $\boldsymbol{\theta} = (\theta_1, \ldots, \theta_q)$. By convention, $ARMA(p, 0) = AR(p)$ and $ARMA(0, q) = MA(q)$, so that both autoregressive and moving average models are particular examples of the general $ARMA$ model.

The correlation function for an $ARMA(p, q)$ model behaves like the correlation function of an $AR(p)$ model for lags greater than q and like an $AR(p)$ correlation function plus a term related to the moving average part of the model, between lags 1 and q. It is obtained using the recursive relations

$$h(k, \boldsymbol{\rho}) = \begin{cases} \phi_1 h(|k - 1|, \boldsymbol{\rho}) + \cdots + \phi_p h(|k - p|, \boldsymbol{\rho}) + \\ \quad \theta_1 \psi(k - 1, \boldsymbol{\rho}) + \cdots + \theta_q \psi(k - q, \boldsymbol{\rho}), & k = 1, \ldots, q, \\ \phi_1 h(|k - 1|, \boldsymbol{\rho}) + \cdots + \phi_p h(|k - p|, \boldsymbol{\rho}), & k = q + 1, q + 2, \ldots, \end{cases}$$

where $\psi(k, \boldsymbol{\phi}, \boldsymbol{\theta}) = \mathrm{E}\left[\epsilon_{t-k} a_t\right] / \mathrm{Var}\left(\epsilon_t\right)$. Note that $\psi(k, \boldsymbol{\phi}, \boldsymbol{\theta}) = 0$, $k = 1, 2, \ldots$, as, in this case, ϵ_{t-k} and a_t are independent and $\mathrm{E}\left[a_t\right] = 0$.

5.3.2 Spatial Correlation Structures

Spatial correlation structures were originally proposed to model dependence in data indexed by continuous two-dimensional position vectors, such as *geostatistical data, lattice data,* and *point patterns.* Because the isotropic spatial correlation structures we consider are continuous functions of some distance between position vectors, they are easily generalized to any finite number of position dimensions. In particular, they can be used with time series data. The basic reference for spatial correlation structures used with linear models with no random effects is Cressie (1993). Spatial correlation structures in the context of mixed-effects models are described at length in Diggle et al. (1994).

For simplicity of notation, we denote by $\epsilon_{\boldsymbol{x}}$ the observation taken at position $\boldsymbol{x} = (x_1, \ldots, x_r)^T$. Any distance metric may be used with isotropic spatial correlation structures, the most common being the *Euclidean,* or L_2, distance, defined as $d_E\left(\epsilon_{\boldsymbol{x}}, \epsilon_{\boldsymbol{y}}\right) = \sqrt{\sum_{i=1}^r (x_i - y_i)^2}$. Other popular choices are the *Manhattan,* or L_1, distance $d_{\mathrm{Man}}\left(\epsilon_{\boldsymbol{x}}, \epsilon_{\boldsymbol{y}}\right) = \sum_{i=1}^r |x_i - y_i|$ and the *maximum* distance $d_{\mathrm{Max}}\left(\epsilon_{\boldsymbol{x}}, \epsilon_{\boldsymbol{y}}\right) = \max_{i=1, \ldots, r} |x_i - y_i|$.

Semivariogram

Spatial correlation structures are generally represented by their *semivariogram,* instead of their correlation function (Cressie, 1993, §2.3.1). The semivariogram of an isotropic spatial correlation structure with a distance function $d(\cdot)$ is defined as

$$\gamma\left[d\left(\epsilon_{\boldsymbol{x}}, \epsilon_{\boldsymbol{y}}\right), \boldsymbol{\lambda}\right] = \tfrac{1}{2}\mathrm{Var}\left(\epsilon_{\boldsymbol{x}} - \epsilon_{\boldsymbol{y}}\right) = \tfrac{1}{2}\mathrm{E}\left[\epsilon_{\boldsymbol{x}} - \epsilon_{\boldsymbol{y}}\right]^2, \tag{5.29}$$

with the last equality following from $\mathrm{E}\left[\epsilon_{\boldsymbol{x}}\right] = \mathrm{E}\left[\epsilon_{\boldsymbol{y}}\right] = 0$. The within-group errors can be standardized to have unit variance, without changing their correlation structure. So, without loss of generality, we assume that $\text{Var}(\epsilon_x) = 1$, $\forall x$. In this case, $\gamma(\cdot)$ will depend only on the correlation parameters $\boldsymbol{\rho}$ and it is easy to verify that

$$\gamma(s, \boldsymbol{\rho}) = 1 - h(s, \boldsymbol{\rho}).$$

It follows from $h(0, \boldsymbol{\rho}) = 1$ that $\gamma(0, \boldsymbol{\rho}) = 0$. To account for abrupt changes at very small distances, it is desirable, in some applications, to allow a discontinuity in $\gamma(\cdot)$ at 0, so that $\gamma(s, \boldsymbol{\rho}) \to c_0$, when $s \downarrow 0$, with $0 < c_0 < 1$. This is called a *nugget effect* in the spatial statistics literature (Cressie, 1993, §2.3.1). In terms of the correlation function, the nugget effect translates into $h(s, \boldsymbol{\rho}) \to 1 - c_0$ as $s \downarrow 0$. It is easy to obtain a correlation function that incorporates a nugget effect from a correlation function that is continuous in s.

$$h_{\text{nugg}}(s, c_0, \boldsymbol{\rho}) = \begin{cases} (1 - c_0)\, h_{\text{cont}}(s, \boldsymbol{\rho}), & s > 0, \\ 0, & s = 0. \end{cases}$$

The standardized residuals $r_{ij} = (y_{ij} - \widehat{y}_{ij})/\widehat{\sigma}_{ij}$, with $\sigma_{ij}^2 = \text{Var}(\epsilon_{ij})$, are the primary quantities used for estimating the semivariogram. The *classical* estimator of the semivariogram (Matheron, 1962) is

$$\widehat{\gamma}(s) = \frac{1}{2N(s)} \sum_{i=1}^{M} \sum_{d(\boldsymbol{p}_{ij}, \boldsymbol{p}_{ij'})=s} (r_{ij} - r_{ij'})^2, \tag{5.30}$$

where $N(s)$ denotes the number of residual pairs at a distance s of each other. Because $\widehat{\gamma}(s)$ uses the squared differences between residual pairs, it can be quite sensitive to outliers. Furthermore, because each residual r_{ij} appears in $n_i - 1$ squared differences in (5.30), a single outlier can affect the estimation of the semivariogram at several distances. A robust estimator of the semivariogram, proposed by Cressie and Hawkins (1980), uses the square-root differences to reduce the influence of outliers.

$$\bar{\gamma}(s) = \left(\frac{1}{2N(s)} \sum_{i=1}^{M} \sum_{d(\boldsymbol{p}_{ij}, \boldsymbol{p}_{ij'})=s} |r_{ij} - r_{ij'}|^{1/2} \right)^4 / (0.457 + 0.494/N(s)). \tag{5.31}$$

Some Isotropic Variogram Models

Cressie (1993, §2.3.1) describes an extensive collection of isotropic variogram models and give conditions for their validity. The single-parameter models in Table 5.2 are a subset of the collection in Cressie (1993), with the *Linear* variogram model modified so that it is bounded in s. $I(s < \rho)$ denotes a binary variable taking value 1 when $s < \rho$ and 0 otherwise. Most

TABLE 5.2. Some isotropic variogram models for spatial correlation structures.

Exponential	$\gamma(s, \rho) = 1 - \exp\left(-s/\rho\right)$
Gaussian	$\gamma(s, \rho) = 1 - \exp\left[-(s/\rho)^2\right]$
Linear	$\gamma(s, \rho) = 1 - (1 - s/\rho) I(s < \rho)$
Rational quadratic	$\gamma(s, \rho) = (s/\rho)^2 / \left[1 + (s/\rho)^2\right]$
Spherical	$\gamma(s, \rho) = 1 - \left[1 - 1.5(s/\rho) + 0.5(s/\rho)^3\right] I(s < \rho)$

of the models in Table 5.2 are also described in Littell et al. (1996, §9.3.1).

The correlation parameter ρ is generally referred to as the *range* in the spatial statistics literature (Littell et al., 1996, §9.3).

For one-dimensional position vectors, the exponential spatial correlation structure is equivalent to the $CAR(1)$ structure (5.27). This is easily verified by defining $\phi = \exp(-1/\rho)$ and noting the correlation function associated with the exponential structure is expressed as $h(s, \phi) = \phi^s$. The exponential correlation model can be regarded as a multivariate generalization of the the $CAR(1)$ model.

Correlation functions for the structures in Table 5.2 may be obtained using the relation $h(s, \rho) = 1 - \gamma(s, \rho)$. A nugget effect c_0 may be added to any of the variogram models, using

$$\gamma_{\text{nugg}}(s, c_o, \rho) = \begin{cases} c_0 + (1 - c_0)\gamma(s, \rho), & s > 0, \\ 0, & s = 0. \end{cases}$$

Figure 5.9 displays plots of the semivariograms models in Table 5.2, corresponding to a range of $\rho = 1$ and a nugget effect of $c_0 = 0.1$ The semivariograms increase monotonically with distance and vary between 0 and 1, corresponding to non-negative correlation functions that decrease monotonically with distance. All of the spatial correlation models in Table 5.2 are implemented in the nlme library as corStruct classes, described in the next section.

5.3.3   *Correlation Structures in nlme: The* corStruct *Classes*

The nlme library provides a set of classes of correlation structures, the corStruct classes, which are used to specify within-group correlation models in either the extended linear mixed-effects model (5.1), or the extended linear model (5.5). Table 5.3 lists the standard corStruct classes in the nlme library. The corStruct constructors have the same name as their corresponding classes.

The two main arguments to most of the corStruct constructors are value and form. The first specifies the values of the correlation parameters and

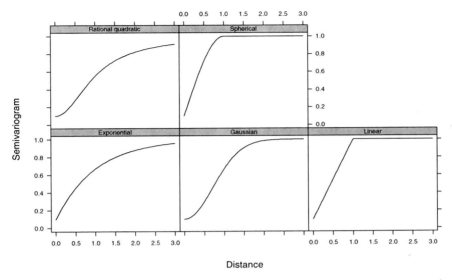

FIGURE 5.9. Plots of semivariogram versus distance for the isotropic spatial correlation models in Table 5.2 with range = 1 and nugget effect = 0.1.

the second is a one-sided formula specifying the position vector and, optionally, a grouping variable for the data—observations in different groups are assumed independent. For example, to specify age as a position variable and to have Subject defining the grouping of the data, we use

```
form = ~ age | Subject
```

A two-dimensional position vector with coordinates x and y is specified with

```
form = ~ x + y
```

The argument fixed, available in all corStruct constructors, may be used to specify fixed correlation structures, which coefficients are not allowed to change during the numerical optimization in the modeling functions. If fixed = TRUE, the coefficients in the structure are fixed. Default is fixed = FALSE.

Several methods are available for each corStruct class, including initialize, which initializes position vectors and grouping variables, and corMatrix, which extracts the within-group correlation matrices. We now describe and illustrate the standard corStruct classes.

corCompSymm

This class implements the compound symmetry correlation structure (5.23. The argument value is used to initialize the intraclass correlation coefficient, assuming a default value of 0. Because the compound symmetry

TABLE 5.3. Standard corStruct classes.

corCompSymm	compound symmetry
corSymm	general
corAR1	autoregressive of order 1
corCAR1	continuous-time $AR(1)$
corARMA	autoregressive-moving average
corExp	exponential
corGaus	Gaussian
corLin	linear
corRatio	rational quadratic
corSpher	spherical

correlation model does not depend on the position of the observation, but just on the group to which it belongs, the form argument is used only to specify the grouping structure. For example,

```
> cs1CompSymm <- corCompSymm( value = 0.3, form = ~ 1 | Subject )
```

specifies a compound symmetry structure with intraclass correlation of 0.3 and grouping defined by Subject. The variable used on the left hand side of the | operator in form is ignored, so

```
> cs2CompSymm <- corCompSymm( value = 0.3, form = ~ age | Subject )
```

gives a corCompSymm object identical to cs1CompSymm. By default, form = ~1, implying that all observations belong to the same group.

The initialize method is used to initialize the grouping factor and the correlation matrices per group. It takes an argument data, naming a data frame in which to evaluate the variables in form.

```
> cs1CompSymm <- initialize( cs1CompSymm, data = Orthodont )
> corMatrix( cs1CompSymm )
$MO1:
     [,1] [,2] [,3] [,4]
[1,]  1.0  0.3  0.3  0.3
[2,]  0.3  1.0  0.3  0.3
[3,]  0.3  0.3  1.0  0.3
[4,]  0.3  0.3  0.3  1.0
  .  .  .
```

Typically, initialize is only called from within the modeling function using the corStruct object.

corSymm

This class implements the general correlation structure (5.24). The argument value is used to initialize the correlation parameters, being given as a numeric vector with the lower diagonal elements of the the largest correlation matrix represented by the corSymm object stacked columnwise. For example, to represent the correlation matrix

$$\begin{bmatrix} 1.0 & 0.2 & 0.1 & -0.1 \\ 0.2 & 1.0 & 0.0 & 0.2 \\ 0.1 & 0.0 & 1.0 & 0.0 \\ -0.1 & 0.2 & 0.0 & 1.0 \end{bmatrix} \tag{5.32}$$

we use

```
value = c( 0.2, 0.1, -0.1, 0, 0.2, 0 )
```

The correlations specified in value must define a positive-definite correlation matrix. By default, value = numeric(0), which leads to initial values of 0 being assigned to all correlations in the initialize method.

The argument form specifies a one-sided formula with the position variable and, optionally, a grouping variable. The position variable defines the indices of the correlation parameters for each observation and must evaluate to an integer vector, with nonrepeated values per group, such that its unique values, when sorted, form a sequence of consecutive integers. By default, the position variable in form is 1, in which case the order of the observations within the group is used to index the correlation parameters.

For example, to specify a general correlation correlation structure with initial correlation matrix as in (5.32), observation order within the group as the position variable, and grouping variable Subject, we use

```
> cs1Symm <- corSymm( value = c(0.2, 0.1, -0.1, 0, 0.2, 0),
+                     form = ~ 1 | Subject )
> cs1Symm <- initialize( cs1Symm, data = Orthodont )
> corMatrix( cs1Symm )
$MO1:
     [,1] [,2] [,3] [,4]
[1,]  1.0  0.2  0.1 -0.1
[2,]  0.2  1.0  0.0  0.2
[3,]  0.1  0.0  1.0  0.0
[4,] -0.1  0.2  0.0  1.0
```

corAR1

This class implements an autoregressive correlation structure of order 1, for integer position vectors. The argument value initializes the single correlation parameter ϕ, which takes values between -1 and 1, and, by default, is set to 0. The argument form is a one-sided formula specifying the position variable and, optionally, a grouping variable. The position variable must evaluate to an integer vector, with nonrepeated values per group, but its values are not required to be consecutive, so that missing time points are naturally accommodated. By default, form = ~1, implying that the order of the observations within the group be used as the position variable.

For example, to specify an $AR(1)$ correlation structure with $\phi = 0.8$, position variable given by observation order within-group, and grouping variable Subject, we use

```
> cs1AR1 <- corAR1( 0.8, form = ~ 1 | Subject )
> cs1AR1 <- initialize( cs1AR1, data = Orthodont )
> corMatrix( cs1AR1 )
$MO1:
       [,1] [,2] [,3]   [,4]
[1,] 1.000 0.80 0.64 0.512
[2,] 0.800 1.00 0.80 0.640
[3,] 0.640 0.80 1.00 0.800
[4,] 0.512 0.64 0.80 1.000
```

As described in §5.3.1, the $AR(1)$ model is equivalent to an $ARMA(1,0)$ model, so that the corARMA class can also be used to represent an corAR1 object. However, the corAR1 methods are designed to take advantage of the particular structure of the $AR(1)$ model, and are substantially more efficient than the corresponding corARMA methods.

corCAR1

This class implements the continuous time $AR(1)$ structure (5.27). Its arguments are defined as in corAR1, but the position variable can be any continuous variable with non-repeated values per group and the correlation parameter ϕ can only take positive values.

corARMA

This corStruct class is used to specify autoregressive–moving average models for the within-group errors. The argument value specifies the values of the autoregressive and moving average parameters. If both autocorrelation and moving average parameters are present, the former should precede the latter in value. By default, all correlation parameters are set to 0, corresponding to uncorrelated within-group errors. The form argument is a one-sided formula defining the position variable and, optionally, a grouping variable. The position variable must evaluate to an integer vector, with non-repeated elements per group. Its values do not need to be consecutive, so that missing time points are allowed. By default, form = ~1, implying that the order of the observations within the group be used as the position variable.

Two additional arguments, p and q, are used, respectively, to specify the order of the autoregressive model and the order of the moving average model. Using p > 0 and q = 0 specifies an $AR(p)$ model, while p = 0 and q > 0 specifies an $MA(q)$ model. By default, p = 0 and q = 0.

For example, to specify an $MA(1)$ model with parameter $\theta = 0.4$, position variable given by the observation order within the group, and groups defined by Subject, we use

```
> cs1ARMA <- corARMA( 0.4, form = ~ 1 | Subject, q = 1 )
> cs1ARMA <- initialize( cs1ARMA, data = Orthodont )
> corMatrix( cs1ARMA )
$M01:
         [,1]    [,2]    [,3]    [,4]
[1,] 1.00000 0.34483 0.00000 0.00000
[2,] 0.34483 1.00000 0.34483 0.00000
[3,] 0.00000 0.34483 1.00000 0.34483
[4,] 0.00000 0.00000 0.34483 1.00000
```

An $ARMA(1,1)$ model with parameters $\phi = 0.8$ and $\theta = 0.4$ is specified
with

```
> cs2ARMA <- corARMA( c(0.8, 0.4), form = ~ 1 | Subject, p=1, q=1 )
> cs2ARMA <- initialize( cs2ARMA, data = Orthodont )
> corMatrix( cs2ARMA )
$M01:
        [,1]  [,2]  [,3]    [,4]
[1,] 1.0000 0.880 0.704 0.5632
[2,] 0.8800 1.000 0.880 0.7040
[3,] 0.7040 0.880 1.000 0.8800
[4,] 0.5632 0.704 0.880 1.0000
```

Spatial corStruct Classes

The corStruct classes representing spatial correlation structures are corExp,
corGaus, corLin, corRatio, and corSpher. As the corresponding constructors
all have the same syntax, we will show examples for corExp only.

The argument value is used to specify values for the range ρ and the
nugget effect c_0, in this order. The range only takes positive values and
the nugget effect can only vary between 0 and 1. By default, value =
numeric(0), in which case the range is initialized to 90% of the minimum
between-pairs distance and the nugget effect is initialized to 0.1.

The argument form is a one-sided formula specifying a position vector
and, optionally, a grouping variable. The coordinates of the position vector must be numeric variables, but are otherwise unrestricted. By default,
form = ~1, translating into a one-dimensional position vector given by the
observation order within the group.

The argument nugget determines whether a nugget effect should be included in the correlation model. If TRUE, a nugget effect is included. Its
default value is FALSE, corresponding to no nugget effect in the model. The
argument metric is a character string specifying a metric to be used for calculating the between-pairs distances. Possible values include "euclidean",
"maximum", and "manhattan", corresponding to the metrics described in
§5.3.2.

To illustrate the use of the spatial corStruct classes, we consider an arti-
ficial data frame spatDat, with columns x and y.

```
> spatDat
     x    y
1 0.00 0.00
2 0.25 0.25
3 0.50 0.50
4 0.75 0.75
5 1.00 1.00
```

An exponential spatial correlation structure based on the Euclidean distance between x and y, with range equal to 1, and no nugget effect is constructed and initialized with

```
> cs1Exp <- corExp( 1, form = ~ x + y )
> cs1Exp <- initialize( cs1Exp, spatDat )
> corMatrix( cs1Exp )
         [,1]    [,2]    [,3]    [,4]    [,5]
[1,] 1.00000 0.70219 0.49307 0.34623 0.24312
[2,] 0.70219 1.00000 0.70219 0.49307 0.34623
[3,] 0.49307 0.70219 1.00000 0.70219 0.49307
[4,] 0.34623 0.49307 0.70219 1.00000 0.70219
[5,] 0.24312 0.34623 0.49307 0.70219 1.00000
```

To calculate the distances in the Manhattan (L_1) metric, we use

```
> cs2Exp <- corExp( 1, form = ~ x + y, metric = "man" )
> cs2Exp <- initialize( cs2Exp, spatDat )
> corMatrix( cs2Exp )
         [,1]    [,2]    [,3]    [,4]    [,5]
[1,] 1.00000 0.60653 0.36788 0.22313 0.13534
[2,] 0.60653 1.00000 0.60653 0.36788 0.22313
[3,] 0.36788 0.60653 1.00000 0.60653 0.36788
[4,] 0.22313 0.36788 0.60653 1.00000 0.60653
[5,] 0.13534 0.22313 0.36788 0.60653 1.00000
```

Note that because partial matches are used on the value of the metric argument, we only gave the first three characters of "manhattan" in the call.

A nugget effect of 0.2 is added to the correlation structure using

```
> cs3Exp <- corExp( c(1, 0.2), form = ~ x + y, nugget = T )
> cs3Exp <- initialize( cs3Exp, spatDat )
> corMatrix( cs3Exp )
         [,1]    [,2]    [,3]    [,4]    [,5]
[1,] 1.00000 0.56175 0.39445 0.27698 0.19449
[2,] 0.56175 1.00000 0.56175 0.39445 0.27698
[3,] 0.39445 0.56175 1.00000 0.56175 0.39445
[4,] 0.27698 0.39445 0.56175 1.00000 0.56175
[5,] 0.19449 0.27698 0.39445 0.56175 1.00000
```

New corStruct classes, representing user-defined correlation structures, can be added to the set of standard classes in Table 5.3 and used with the modeling functions in the nlme library. For this, one must specify a constructor function, generally with the same name as the class, and, at a minimum, methods for the functions coef, corMatrix, and initialize. The corAR1 constructor and methods can serve as templates for these.

5.3.4   Using Correlation Structures with lme

Correlation structures are specified in lme through the correlation argument. By default, correlation = NULL, corresponding to uncorrelated within-group errors. Correlation structures are specified as corStruct objects, created using the standard constructors described in §5.3.3, or a user-defined corStruct constructor. In this section, we describe the use of correlation models in lme through the analysis of two examples of grouped data with correlated within-group errors.

When assessing the adequacy of a correlation model, it is often useful to consider diagnostic plots of the *normalized* residuals, defined as $r_i = \hat{\sigma}^{-1} (\widehat{\Lambda}_i^{-1/2})^T (y_i - \widehat{y}_i)$, where $\hat{\sigma}^2 \widehat{\Lambda}_i$ denotes the estimated variance–covariance matrix for the i within-group errors. If the within-group variance–covariance model is correct, the normalized residuals should be approximately distributed as independent $\mathcal{N}(0, I)$ random vectors.

Counts of Ovarian Follicles

Pierson and Ginther (1987) report on a study of the number of ovarian follicles larger than 10 mm in diameter detected in eleven different mares at several times in their estrus cycles. The data were recorded daily from three days before ovulation until three days after the next ovulation. The measurement times for each mare are scaled so that the ovulations for each mare occur at times 0 and 1. These data are also described in Appendix A.18 and are included in the nlme library as the groupedData object Ovary.

The plots of the number of follicles versus time per mare, shown in Figure 5.10, suggest a periodic behavior for the number of follicles over time.

Preliminary analyses indicate that the following linear mixed-effects model provides a reasonable representation for the number of follicles y_{ij} for the ith mare at time t_{ij}.

$$y_{ij} = (\beta_0 + b_{0i}) + (\beta_1 + b_{1i}) \sin(2\pi t_{ij}) + \beta_2 \cos(2\pi t_{ij}) + \epsilon_{ij},$$

$$b_i = \begin{bmatrix} b_{0i} \\ b_{1i} \end{bmatrix} \sim \mathcal{N}\left(0, \operatorname{diag}\left(\sigma_0^2, \sigma_1^2\right)\right), \quad \epsilon_{ij} \sim \mathcal{N}\left(0, \sigma^2\right), \tag{5.33}$$

where β_0, β_1, and β_2 are the fixed effects, b_i is the random effects vector, assumed independent for different mares, and ϵ_{ij} is the within-group error, assumed independent for different i, j and independent of the random effects. The random effects b_{0i} and b_{1i} are assumed to be independent with variances σ_0^2 and σ_1^2, respectively.

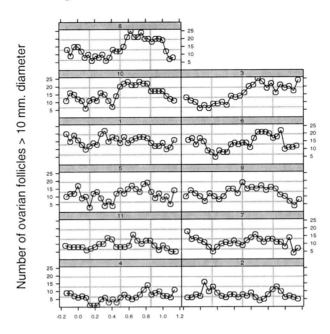

Time in estrus cycle

FIGURE 5.10. Number of ovarian follicles greater than 10 mm in diameter detected in mares at various times in their estrus cycles. The times have been scaled so the ovulations occur at times 0 and 1.

We fit the linear mixed-effects model (5.33) with

```
> fm1Ovar.lme <- lme( follicles ~ sin(2*pi*Time) + cos(2*pi*Time),
+                     data = Ovary, random = pdDiag(~sin(2*pi*Time)) )
> fm1Ovar.lme
Linear mixed-effects model fit by REML
  Data: Ovary
  Log-restricted-likelihood: -813.04
  Fixed: follicles ~ sin(2 * pi * Time) + cos(2 * pi * Time)
  (Intercept) sin(2 * pi * Time) cos(2 * pi * Time)
      12.182           -3.2985            -0.86237

Random effects:
 Formula: ~ sin(2 * pi * Time) | Mare
 Structure: Diagonal
        (Intercept) sin(2 * pi * Time) Residual
StdDev:      3.0521             2.0793   3.1129

Number of Observations: 308
Number of Groups: 11
```

The observations in the Ovary data were collected at equally spaced calendar times. When the calendar time was converted to the ovulation cycle scale, the intervals between observations remained very similar, but no longer identical. Therefore, when considered in the scale of the within-group observation order, the Ovary data provides an example of *time-series* data. We use it here to illustrate the modeling of serial correlation structures in lme.

The ACF method for the lme class obtains the empirical autocorrelation function (5.22) from the residuals of an lme object.

```
> ACF( fm1Ovar.lme )
     lag        ACF
1     0   1.000000
2     1   0.379480
3     2   0.179722
4     3   0.035693
5     4   0.059779
6     5   0.002097
7     6   0.064327
8     7   0.071635
9     8   0.048578
10    9   0.027782
11   10  -0.034276
12   11  -0.077204
13   12  -0.161132
14   13  -0.196030
15   14  -0.289337
```

Empirical autocorrelations at larger lags tend to be less reliable, because they are estimated with fewer residual pairs. We control the number of lags for which to calculate the empirical autocorrelations in ACF with the argument maxLag. A plot of the empirical autocorrelation function, displayed in Figure 5.11, is obtained with

```
> plot(ACF(fm1Ovar.lme,  maxLag = 10), alpha = 0.01)   # Figure 5.11
```

The argument alpha specifies the significance level for approximate two-sided critical bounds for the autocorrelations (Box et al., 1994), given by $\pm z(1 - \alpha/2)/\sqrt{N(l)}$, with $z(p)$ denoting the standard normal quantile of order p and $N(l)$ defined as in (5.22).

The empirical autocorrelations in Figure 5.11 are significantly different from 0 at the first two lags, decrease approximately exponentially for the first four lags, and stabilize at nonsignificant levels for larger lags. This suggests that an $AR(1)$ model may be suitable for the within-group correlation and we fit it with

```
> fm2Ovar.lme <- update( fm1Ovar.lme, correlation = corAR1() )
Linear mixed-effects model fit by REML
 Data: Ovary
```

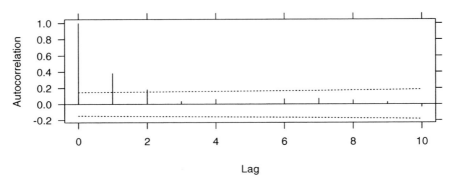

FIGURE 5.11. Empirical autocorrelation function corresponding to the standardized residuals of the fm1Ovar.lme object.

```
    Log-restricted-likelihood: -774.72
    Fixed: follicles ~ sin(2 * pi * Time) + cos(2 * pi * Time)
    (Intercept) sin(2 * pi * Time) cos(2 * pi * Time)
        12.188              -2.9853            -0.87776

Random effects:
 Formula:  ~ sin(2 * pi * Time) | Mare
 Structure: Diagonal
         (Intercept) sin(2 * pi * Time) Residual
StdDev:      2.8585              1.258   3.5071

Correlation Structure: AR(1)
 Formula: ~ 1 | Mare
 Parameter estimate(s):
    Phi
 0.5722
Number of Observations: 308
Number of Groups: 11
```

Note that no arguments need to be passed to corAR1 in this case, as its default formula ~1 specifies the position variable as the within-group order of the observations, which is what is desired for the fm2Ovar.lme model.

Because the fm1Ovar.lme model is nested within the fm2Ovar.lme model (corresponding to $\phi = 0$) we can compare them using a likelihood ratio test.

```
> anova( fm1Ovar.lme, fm2Ovar.lme )
            Model df    AIC    BIC  logLik   Test L.Ratio p-value
fm1Ovar.lme     1  6 1638.1 1660.4 -813.04
fm2Ovar.lme     2  7 1563.4 1589.5 -774.72 1 vs 2  76.634  <.0001
```

The very significant p-value for the likelihood ratio test indicates that the $AR(1)$ provides a substantially better fit of the data than the independent errors model (5.33), suggesting that within-group serial correlation is present in Ovary. We can assess the precision of the correlation parameter estimate in fm2Ovar.lme with the intervals method.

```
> intervals( fm2Ovar.lme )
. . .
 Correlation structure:
        lower    est.    upper
Phi 0.36607  0.5722  0.72481
. . .
```

Consistently with the likelihood ratio test results, the confidence interval on ϕ indicates that it is significantly different from 0.

The autocorrelation pattern in Figure 5.11 is also consistent with that of an $MA(2)$ model, in which only the first two lags have nonzero correlations. We fit this model with

```
> fm3Ovar.lme <- update(fm1Ovar.lme, correlation = corARMA(q = 2))
> fm3Ovar.lme
. . .
Correlation Structure: ARMA(0,2)
 Formula: ~ 1 | Mare
 Parameter estimate(s):
  Theta1   Theta2
 0.47524  0.25701
. . .
```

The $AR(1)$ and $MA(2)$ models are not nested and, therefore, cannot be compared through a likelihood ratio test. They can, however, be compared via their information criterion statistics.

```
> anova( fm2Ovar.lme, fm3Ovar.lme, test = F )
            Model df    AIC     BIC   logLik
fm2Ovar.lme     1  7 1563.4  1589.5 -774.72
fm3Ovar.lme     2  8 1571.2  1601.0 -777.62
```

Even though it has one fewer parameter than the $MA(2)$ model, the $AR(1)$ model is associated with a larger log-restricted-likelihood, which translates into smaller AIC and BIC, making it the preferred model of the two.

Because the fixed- and random-effects models in (5.33) use a continuous time scale, we investigate if a continuous time $AR(1)$ model would provide a better representation of the within-group correlation, using the corCAR1 class.

```
> fm4Ovar.lme <- update( fm1Ovar.lme,
+                     correlation = corCAR1(form = ~Time) )
> anova( fm2Ovar.lme, fm4Ovar.lme, test = F )
            Model df    AIC     BIC   logLik
fm2Ovar.lme     1  7 1563.4  1589.5 -774.72
fm4Ovar.lme     2  7 1565.5  1591.6 -775.77
```

No grouping variable needs to be specified in the call to corCAR1, as the innermost grouping variable in the lme object is used by default. As indicated by the AIC and the BIC in the anova output, the $AR(1)$ model provides a better representation of the within-group correlation than its continuous-time version.

An "intermediate" model between the $AR(1)$ and the $MA(2)$ models is the $ARMA(1,1)$ model, which has an exponentially decaying autocorrelation function for lags ≥ 2, but allows more flexibility in the first autocorrelation. We fit it with

```
> fm50var.lme <- update(fm10var.lme, corr = corARMA(p = 1, q = 1))
> fm50var.lme
 . . .
Correlation Structure: ARMA(1,1)
 Formula: ~ 1 | Mare
 Parameter estimate(s):
    Phi1    Theta1
 0.78716 -0.27957
 . . .
```

The $AR(1)$ model is nested within the $ARMA(1,1)$ model (corresponding to $\theta_1 = 0$) and we can use anova to compare the two fits through a likelihood ratio test.

```
> anova( fm20var.lme, fm50var.lme )
            Model df    AIC    BIC  logLik   Test L.Ratio p-value
fm20var.lme     1  7 1563.4 1589.5 -774.72
fm50var.lme     2  8 1559.9 1589.7 -771.95 1 vs 2  5.5537  0.0184
```

The low p-value for the likelihood ratio test indicates that the $ARMA(1,1)$ model provides a better fit of the data.

We can assess the adequacy of the $ARMA(1,1)$ model using the empirical autocorrelation function of the normalized residuals.

```
> plot( ACF(fm50var.lme,  maxLag = 10, resType = "n"),
+             alpha = 0.01 )                              # Figure 5.12
```

No significant autocorrelations are observed in Figure 5.12, indicating that the normalized residuals behave like uncorrelated noise, as expected under the appropriate correlation model.

Body Weight Growth in Rats

We revisit the BodyWeight example of §5.2.2 to illustrate the use of corStruct classes in lme in combination with variance functions. As described in §5.2.2, the observations in the BodyWeight data are not equally spaced in time, as an extra observation is taken at 44 days. We use the spatial correlation corStruct classes to use fit continuous-time within-group correlation models, which naturally accommodate the imbalance in the data.

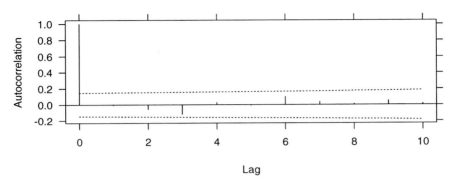

FIGURE 5.12. Empirical autocorrelation function corresponding to the normalized residuals of the `fm50var.lme` object.

The `Variogram` method for the lme class estimates the sample semivariogram from the residuals of the lme object. The arguments `resType` and `robust` control, respectively, what type of residuals should be used (`"pearson"` or `"response"`) and whether the robust algorithm (5.31) or the classical algorithm (5.30) should be used to estimate the semivariogram. The defaults are `resType = "pearson"` and `robust = FALSE`, so that classical estimates of the semivariogram are obtained from the standardized residuals. The argument `form` is a one-sided formula specifying the position vector to be used for the semivariogram calculations.

```
> Variogram( fm2BW.lme, form = ~ Time )
      variog dist n.pairs
1  0.34508    1      16
2  0.99328    6      16
3  0.76201    7     144
4  0.68496    8      16
5  0.68190   13      16
6  0.95118   14     128
7  0.89959   15      16
8  1.69458   20      16
9  1.12512   21     112
10 1.08820   22      16
11 0.89693   28      96
12 0.93230   29      16
13 0.85144   35      80
14 0.75448   36      16
15 1.08220   42      64
16 1.56652   43      16
17 0.64378   49      48
18 0.67350   56      32
19 0.58663   63      16
```

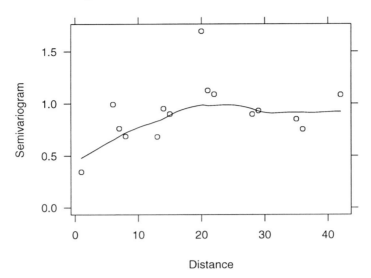

FIGURE 5.13. Sample semivariogram estimates corresponding to the standard-ized residuals of the `fm2BW.lme` object. A loess smoother is added to the plot to enhance the visualization of patterns in the semivariogram.

The columns in the data frame returned by `Variogram` represent, respec-tively, the sample semivariogram, the distance, and the number of residual pairs used in the estimation. Because of the imbalance in the time measure-ments, the number of residual pairs used at each distance varies consider-ably, making some semivariogram estimates more reliable than others. In general, the number of residual pairs used in the semivariogram estimation decreases with distance, making the values at large distances unreliable. We can control the maximum distance for which semivariogram estimates should be calculated using the argument `maxDist`.

A graphical representation of the sample semivariogram is obtained with the `plot` method for class Variogram.

```
> plot( Variogram(fm2BW.lme, form = ~ Time,
+                 maxDist = 42) )                        # Figure 5.13
```

The resulting plot, shown in Figure 5.13, includes a *loess* smoother (Cleveland et al., 1992) to enhance the visualization of semivariogram patterns. The semivariogram seems to increase with distance up to 20 days and then sta-bilizes around 1. We initially use an exponential spatial correlation model for the within-group errors, fitting it with

```
> fm3BW.lme <- update( fm2BW.lme, corr = corExp(form = ~ Time) )
> fm3BW.lme

  . . .

Correlation Structure: Exponential spatial correlation
```

```
Formula: ~ Time | Rat
Parameter estimate(s):
 range
 4.8862
Variance function:
 Structure: Power of variance covariate
 Formula:  ~ fitted(.)
 Parameter estimates:
   power
 0.59436
 . . .
```

Note that the model corresponding to `fm3BW.lme` includes both a variance function and a correlation structure. We assess the variability in the spatial correlation parameter estimate with the `intervals` method.

```
> intervals( fm3BW.lme )
 . . .
Correlation structure:
      lower   est.  upper
range 1.852 4.8862 12.891
 . . .
```

The confidence intervals is bounded away from zero, suggesting that the spatial correlation model produced a significantly better fit. We can also test this using the `anova` method.

```
> anova( fm2BW.lme, fm3BW.lme )
          Model df    AIC    BIC  logLik   Test L.Ratio p-value
fm2BW.lme     1 11 1163.9 1198.4 -570.96
fm3BW.lme     2 12 1145.1 1182.8 -560.57 1 vs 2  20.781  <.0001
```

The likelihood ratio test also indicates that the corExp model fits the data significantly better than the independent errors model corresponding to `fm2BW.lme`.

The semivariogram plot in Figure 5.13 gives some indication that a nugget effect may be present in the data. We can test it with

```
> fm4BW.lme <- update( fm3BW.lme,
+                     corr = corExp(form =  ~ Time, nugget = T) )
> anova( fm3BW.lme, fm4BW.lme )
          Model df    AIC    BIC  logLik   Test    L.Ratio
fm3BW.lme     1 12 1145.1 1182.8 -560.57
fm4BW.lme     2 13 1147.1 1187.9 -560.57 1 vs 2 0.00043111
          p-value
fm3BW.lme
fm4BW.lme   0.9834
```

The nearly identical log-likelihood values indicate that a nugget effect is, in fact, not needed.

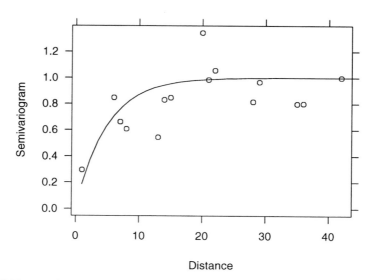

FIGURE 5.14. Sample semivariogram estimates corresponding to the standardized residuals of the `fm2BW.lme` object. The fitted semivariogram corresponding
to `fm3BW.lme` is added to plot.

When an lme object includes a spatial corStruct object, we can further
assess the adequacy of the correlation model with the `plot` method. In this
case, instead of a loess smoother, the fitted semivariogram corresponding
to the corStruct object is displayed in the plot, along with the sample
variogram estimates.

```
> plot( Variogram(fm3BW.lme, form = ~ Time,
+                  maxDist = 42) )                    # Figure 5.14
```

The fitted semivariogram agrees reasonably well with the sample variogram
estimates.

We can also assess the adequacy of the exponential spatial correlation
model by investigating the sample semivariogram for the normalized residuals.

```
> plot( Variogram(fm3BW.lme, form = ~ Time, maxDist = 42,
+                  resType = "n", robust = T) )       # Figure 5.15
```

The robust semivariogram estimator is used to reduce the influence of an
outlying value at distance 1 on the loess smoother. The sample semivariogram estimates in Figure 5.15 appear to vary randomly around the $y = 1$
line, suggesting that the normalized residuals are approximately uncorrelated and, hence, the corExp model is adequate.

We may compare the corExp model to other spatial correlation models,
using the `update` method and the `anova` method. As the models are not

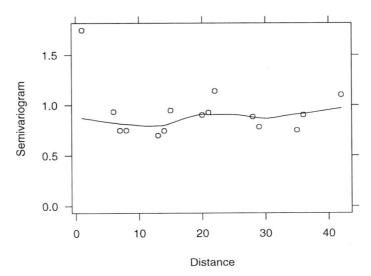

FIGURE 5.15. Sample semivariogram estimates corresponding to the normalized residuals of the `fm3BW.lme` object. A loess smoother is added to the plot to enhance the visualization of patterns in the semivariogram.

nested, they have to be compared based on the information criteria AIC and BIC.

```
> fm5BW.lme <- update( fm3BW.lme, corr = corRatio(form = ~ Time) )
> fm6BW.lme <- update( fm3BW.lme, corr = corSpher(form = ~ Time) )
> fm7BW.lme <- update( fm3BW.lme, corr = corLin(form = ~ Time) )
> fm8BW.lme <- update( fm3BW.lme, corr = corGaus(form = ~ Time) )
> anova( fm3BW.lme, fm5BW.lme, fm6BW.lme, fm7BW.lme, fm8BW.lme )
          Model df    AIC    BIC  logLik
fm3BW.lme     1 12 1145.1 1182.8 -560.57
fm5BW.lme     2 12 1148.8 1186.4 -562.38
fm6BW.lme     3 12 1150.8 1188.4 -563.39
fm7BW.lme     4 12 1150.8 1188.4 -563.39
fm8BW.lme     5 12 1150.8 1188.4 -563.39
```

The corExp fit has the smallest AIC and BIC and seems the most adequate within-group correlation model for the `BodyWeight` data, among the spatial correlation models considered.

5.4   Fitting Extended Linear Models with `gls`

The general formulation of the extended linear model, as well as the estimation methods used to fit it, have been described in §5.1.2. In this section,

TABLE 5.4. Main gls methods.

ACF	empirical autocorrelation function of residuals
anova	likelihood ratio or conditional tests
augPred	predictions augmented with observed values
coef	estimated coefficients for expected response model
fitted	fitted values
intervals	confidence intervals on model parameters
logLik	log-likelihood at convergence
plot	diagnostic Trellis plots
predict	predicted values
print	brief information about the fit
qqnorm	normal probability plots
resid	residuals
summary	more detailed information about the fit
update	update the gls fit
Variogram	semivariogram of residuals

we concentrate on the capabilities available in the nlme library for fitting such models.

The gls function is used to fit the extended linear model (5.5), using either maximum likelihood, or restricted maximum likelihood. It can be viewed as an lme function without the argument random. Several arguments are available in gls, but typical calls are of the form

```
gls( model, data, correlation )    # correlated errors
gls( model, data, weights )        # heteroscedastic errors
gls( model, data, correlation, weights )  # both
```

The first argument, model, is a two-sided linear formula specifying the model for the expected value of the response. Correlation and weights are used, as in lme, to define, respectively, the correlation model and the variance function model for the error term. Data specifies a data frame in which the variables named in model, correlation, and weights can be evaluated.

The fitted object returned by gls inherits from class gls, for which several methods are available to display, plot, update, and further explore the estimation results. Table 5.4 lists the most important methods for class gls. The use of the gls function and its associated methods is described and illustrated through the examples in the next sections.

Orthodontic Growth Curve

The Orthodont data were analyzed in §4.2 and §4.3 using a linear mixed-effects model. We describe here an alternative analysis based on the extended linear model (5.5).

Because of the small number of observations per subject, it is feasible for these data to fit a linear model with a general variance–covariance structure for the errors. The corresponding extended linear model for the ith subject orthodontic distance at age j, $i = 1, \dots, 27$ and $j = 1, \dots, 4$, is written as

$$\text{distance}_{ij} = \beta_0 + \beta_1 \, \text{Sex} + \beta_2 \left(\text{age}_j - 11\right) + \beta_3 \left(\text{age}_j - 11\right) \text{Sex} + \epsilon_{ij},$$

$$\epsilon_i = \begin{bmatrix} \epsilon_{i1} \\ \epsilon_{i2} \\ \epsilon_{i3} \\ \epsilon_{i4} \end{bmatrix} \sim \mathcal{N}\left(0, \Lambda_i\right), \qquad \Lambda_i = \begin{bmatrix} \sigma_1^2 & \sigma_{12} & \sigma_{13} & \sigma_{14} \\ \sigma_{12} & \sigma_2^2 & \sigma_{23} & \sigma_{24} \\ \sigma_{13} & \sigma_{23} & \sigma_3^2 & \sigma_{34} \\ \sigma_{14} & \sigma_{24} & \sigma_{34} & \sigma_4^2 \end{bmatrix}, \quad (5.34)$$

with Sex representing a binary variable taking values -1 for boys and 1 for girls. The parameters β_1 and β_3 represent, respectively, the intercept and slope gender effects. The variance–covariance matrix Λ_i is assumed the same for all subjects.

We fit (5.34) with `gls` using a combination of the corSymm correlation class and the varIdent variance function class.

```
> fm1Orth.gls <- gls( distance ~ Sex * I(age - 11), Orthodont,
+                     correlation = corSymm(form = ~ 1 | Subject),
+                     weights = varIdent(form = ~ 1 | age) )
```

In this case, because `Orthodont` is a groupedData object with grouping variable `Subject`, the argument `form` could be omitted in the call to corSymm.

The `print` method gives some basic information about the fit.

```
> fm1Orth.gls
Generalized least squares fit by REML
  Model: distance ~ Sex * I(age - 11)
  Data: Orthodont
  Log-restricted-likelihood: -213.66

Coefficients:
  (Intercept)    Sex I(age - 11) Sex:I(age - 11)
       23.801 -1.136      0.6516         -0.17524

Correlation Structure: General
 Formula:  ~ 1 | age

 Parameter estimate(s):
 Correlation:
       1     2     3
 2 0.568
 3 0.659 0.581
 4 0.522 0.725 0.740
Variance function:
 Structure: Different standard deviations per stratum
 Formula:  ~ 1 | age
```

```
Parameter estimates:
8      10     12      14
1 0.8792 1.0747 0.95872
Degrees of freedom: 108 total; 104 residual
Residual standard error: 2.329
```

The correlation estimates are similar, suggesting that a compound symmetry structure may be a suitable correlation model. We explore this further with the `intervals` method.

```
> intervals( fm1Orth.gls )
Approximate 95% confidence intervals

Coefficients:
                   lower       est.      upper
   (Intercept)  23.06672   23.80139   24.536055
           Sex  -1.87071   -1.13605   -0.401380
    I(age - 11)  0.52392    0.65160    0.779289
Sex:I(age - 11) -0.30292   -0.17524   -0.047552

Correlation structure:
            lower      est.    upper
cor(1,2) 0.098855 0.56841 0.83094
cor(1,3) 0.242122 0.65878 0.87030
cor(1,4) 0.021146 0.52222 0.81361
cor(2,3) 0.114219 0.58063 0.83731
cor(2,4) 0.343127 0.72510 0.90128
cor(3,4) 0.382248 0.73967 0.90457

Variance function:
     lower     est. upper
10 0.55728 0.87920 1.3871
12 0.71758 1.07468 1.6095
14 0.61253 0.95872 1.5005

Residual standard error:
 lower est. upper
1.5985 2.329 3.3933
```

All confidence intervals for the correlation parameters overlap, corroborating the compound symmetry assumption. We can test it formally by updating the fitted object and using the `anova` method.

```
> fm2Orth.gls <-
+    update(fm1Orth.gls, corr = corCompSymm(form = ~ 1 | Subject))
> anova( fm1Orth.gls, fm2Orth.gls )
            Model df    AIC    BIC  logLik   Test L.Ratio p-value
fm1Orth.gls     1 14 455.32 492.34 -213.66
fm2Orth.gls     2  9 452.74 476.54 -217.37 1 vs 2  7.4256  0.1909
```

The large p-value for the likelihood ratio statistics reported in the `anova` output confirms the compound symmetry model.

The confidence intervals for the variance function parameters corresponding to `fm2Orth.gls`,

```
> intervals( fm2Orth.gls )
 . . .
 Variance function:
      lower    est.   upper
10  0.56377 0.86241 1.3193
12  0.68132 1.03402 1.5693
14  0.60395 0.92045 1.4028
 . . .
```

all include the value 1, suggesting that the variability does not change with age. The high p-value for the associated likelihood ratio test confirms this assumption.

```
> fm3Orth.gls <- update( fm2Orth.gls, weights = NULL )
> anova( fm2Orth.gls, fm3Orth.gls )
            Model df    AIC    BIC  logLik   Test L.Ratio p-value
fm2Orth.gls     1  9 452.74 476.54 -217.37
fm3Orth.gls     2  6 448.53 464.40 -218.26 1 vs 2  1.7849  0.6182
```

As with other modeling functions, the `plot` method is used to assess the assumptions in the model. Its syntax is identical to the `plot` method for class `lme`. For example, to examine if the error variance is the same for boys and girls we may examine the plot of the normalized residuals versus age by gender, obtained with

```
> plot( fm3Orth.gls, resid(., type = "n") ~ age | Sex )  # Fig. 5.16
```

and displayed in Figure 5.16. It is clear that there is more variability among boys than girls, which we can represent in the model with the `varIdent` variance function class.

```
> fm4Orth.gls <- update( fm3Orth.gls,
+                        weights = varIdent(form = ~ 1 | Sex) )
> anova( fm3Orth.gls, fm4Orth.gls )
            Model df    AIC    BIC  logLik   Test L.Ratio p-value
fm3Orth.gls     1  6 448.53 464.40 -218.26
fm4Orth.gls     2  7 438.96 457.47 -212.48 1 vs 2  11.569   7e-04
```

As expected, the likelihood ratio test gives strong evidence in favor of the heteroscedastic model.

The `qqnorm` method is used to assess the assumption of normality for the errors. Its syntax is identical to the corresponding `lme` method. The normal plots of the normalized residuals by gender, obtained with

```
> qqnorm( fm4Orth.gls, ~resid(., type = "n") )        # Figure 5.17
```

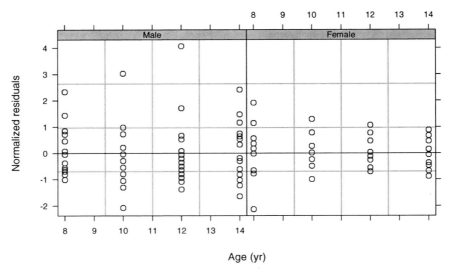

FIGURE 5.16. Scatter plots of normalized residuals versus age by gender for the fm3Orth.gls fitted object.

and displayed in Figure 5.17, do not indicate serious departures from normality and confirm that the variance function model included in fm4Orth.gls was successful in accommodating the error heteroscedasticity.

It is interesting, at this point, to compare the gls fit corresponding to fm4Orth.gls to the lme fit corresponding to fm3Orth.lme, obtained in §4.3.1. Because the corresponding models are not nested, a likelihood ratio test is nonsensical. However, the information criteria can be compared, as the fixed effects models are identical for the two fits. The anova method can be used to compare gls and lme objects.

```
> anova( fm3Orth.lme, fm4Orth.gls, test = F )
            Model df    AIC    BIC  logLik
fm3Orth.lme     1  9 432.30 456.09 -207.15
fm4Orth.gls     2  7 438.96 457.47 -212.48
```

The lme fit has the smallest AIC and BIC and, therefore, seems to give a better representation of the Orthodont data.

The choice between an lme model and a gls model should take into account more than just information criteria and likelihood tests. A mixed-effects model has a hierarchical structure which, in many applications, provides a more intuitive way of accounting for within-group dependency than the direct modeling of the marginal variance–covariance structure of the response in the gls approach. Furthermore, the mixed-effects estimation gives, as a byproduct, estimates for the random effects, which may be of interest in themselves. The gls model focuses on marginal inference and is more appealing when a hierarchical structure for the data is not believed to be

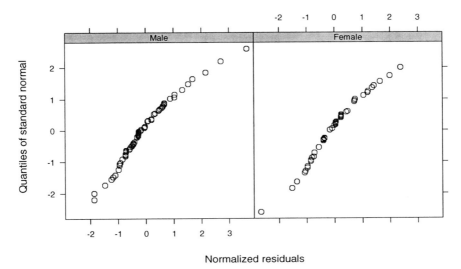

FIGURE 5.17. Normal plots plots of normalized residuals by gender for the `fm4Orth.gls` fitted object.

present, or is not relevant in the analysis, and one is more interested in parameters associated with the error variance–covariance structure, as in time-series analysis and spatial statistics.

High-Flux Hemodialyzer Ultrafiltration Rates

The hemodialyzer data were analyzed in §5.2.2 to illustrate the use of variance functions with linear mixed-effects models. An alternative analysis, using the `gls` function, is presented here.

We initially consider a linear model with the same expected response as the linear mixed-effects model (5.18) and independent, homoscedastic errors.

$$y_{ij} = (\beta_0 + \gamma_0 Q_i) + (\beta_1 + \gamma_1 Q_i)\, x_{ij} + (\beta_2 + \gamma_2 Q_i+)\, x_{ij}^2 +$$
$$(\beta_3 + \gamma_3 Q_i)\, x_{ij}^3 + (\beta_4 + \gamma_4 Q_i)\, x_{ij}^4 + \epsilon_{ij}, \quad \epsilon_{ij} \sim \mathcal{N}\left(0, \sigma^2\right), \tag{5.35}$$

where Q_i is a binary variable taking values -1 for 200 dl/min hemodialyzers and 1 for 300 dl/min hemodialyzers; β_0, β_1, β_2, β_3, and β_4 are, respectively, the intercept, linear, quadratic, cubic, and quartic coefficients averaged over the levels of Q; γ_i is the blood flow effect associated with the coefficient β_i; and ϵ_{ij} is the error term.

We may fit the linear model (5.35) using the `gls` function

```
> fm1Dial.gls <-
+   gls(rate ~(pressure + pressure^2 + pressure^3 + pressure^4)*QB,
+       Dialyzer)
```

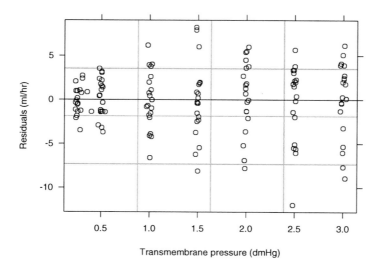

FIGURE 5.18. Plot of residuals versus transmembrane pressure for the homo-scedastic fitted object `fm1Dial.gls`.

Because no variance functions and correlation structures are used, the gls fit is equivalent to an lm fit, in this case.

The plot of the residuals versus transmembrane pressure, obtained with

```
> plot( fm1Dial.gls, resid(.) ~ pressure,
+          abline = 0 )                              # Figure 5.18
```

and shown in Figure 5.18, displays the same pattern of heteroscedasticity observed for the within-group residuals of the lme object `fm1Dial.lme`, presented in Figure 5.2.

As in the lme analysis of §5.2.2, we choose the flexible variance function class varPower to model the heteroscedasticity in the response, and test its significance using the anova method.

```
> fm2Dial.gls <- update( fm1Dial.gls,
+                          weights = varPower(form = ~ pressure) )
> anova( fm1Dial.gls, fm2Dial.gls)
            Model df    AIC     BIC  logLik   Test L.Ratio p-value
fm1Dial.gls     1 11 768.10 799.65 -373.05
fm2Dial.gls     2 12 738.22 772.63 -357.11 1 vs 2  31.882  <.0001
```

As expected, the likelihood ratio test strongly rejects the assumption of homoscedasticity. The plot of the standard residuals corresponding to `fm2Dial.gls` versus pressure, shown in Figure 5.19, indicates that the power variance function adequately represents the heteroscedasticity in the data.

Because the hemodialyzer ultrafiltration rates are measured sequentially over time on the same subjects, the within-subject observations are likely

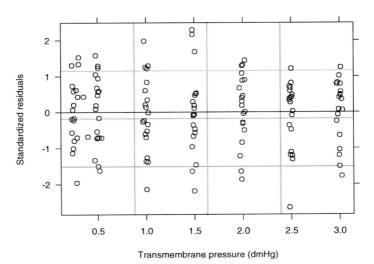

FIGURE 5.19. Plot of standardized residuals versus transmembrane pressure for the heteroscedastic fitted object `fm2Dial.gls`.

to be correlated. In §5.2.2, random effects were used to account for the within-group correlation. We may, alternatively, use a correlation structure to directly model the association among the within-subject errors. Because the measurements in this example are equally spaced in time, the empirical autocorrelation function can be used to investigate within-subject correlation. The `ACF` method for the `gls` class has a syntax similar to the corresponding `lme` method, but includes a `form` argument which allows the specification of a time covariate, to define the lags between observations, and a grouping variable, to specify a partition for the residuals.

```
> ACF( fm2Dial.gls, form = ~ 1 | Subject )
    lag      ACF
1     0 1.000000
2     1 0.770851
3     2 0.632301
4     3 0.408306
5     4 0.200737
6     5 0.073116
7     6 0.077802
```

The empirical ACF values indicate that the within-group observations are correlated, and that the correlation decreases with lag. As usual, it is more informative to look at a plot of the empirical ACF, displayed in Figure 5.20 and obtained with

```
> plot( ACF( fm2Dial.gls, form = ~ 1 | Subject),
+       alpha = 0.01 )                              # Figure 5.20
```

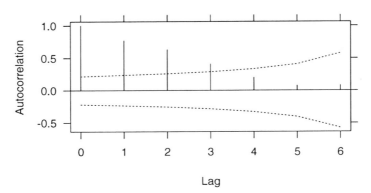

FIGURE 5.20. Empirical autocorrelation function corresponding to the standard-ized residuals of the `fm2Dial.gls` object.

The ACF pattern observed in Figure 5.20 suggests that an $AR(1)$ model may be appropriate to describe it. The corAR1 class is used to represent it.

```
> fm3Dial.gls <- update( fm2Dial.gls,
+                   corr = corAR1(0.771, form = ~ 1 | Subject) )
> fm3Dial.gls
Generalized least squares fit by REML
  Model: rate ~(pressure + pressure^2 + pressure^3 + pressure^4)*QB
  Data: Dialyzer
  Log-restricted-likelihood: -308.34

Coefficients:
  (Intercept) pressure I(pressure^2) I(pressure^3) I(pressure^4)
      -16.818   92.334       -49.265         11.4       -1.0196
       QB pressure:QB I(pressure^2):QB I(pressure^3):QB
  -1.5942    1.7054          2.1268          0.47972
  I(pressure^4):QB
         -0.22064

Correlation Structure: AR(1)
 Formula: ~ 1 | Subject
 Parameter estimate(s):
      Phi
 0.75261
Variance function:
 Structure: Power of variance covariate
 Formula: ~ pressure
 Parameter estimates:
   power
 0.51824
Degrees of freedom: 140 total; 130 residual
Residual standard error: 3.0463
```

The initial value for the $AR(1)$ parameter is given by the lag-1 empirical autocorrelation. The `form` argument is used to specify the grouping variable `Subject`.

The `intervals` method is used to assess the variability in the estimates.

```
> intervals( fm3Dial.gls )
Approximate 95% confidence intervals

 Coefficients:
                      lower       est.        upper
      (Intercept)  -18.8968  -16.81845  -14.740092
         pressure   81.9144   92.33423  102.754073
    I(pressure^2)  -63.1040  -49.26515  -35.426279
    I(pressure^3)    4.5648   11.39967   18.234526
    I(pressure^4)   -2.1248   -1.01964    0.085557
               QB   -4.7565   -1.59419    1.568141
      pressure:QB  -13.6410    1.70544   17.051847
 I(pressure^2):QB  -17.9484    2.12678   22.201939
 I(pressure^3):QB   -9.3503    0.47972   10.309700
 I(pressure^4):QB   -1.8021   -0.22064    1.360829

 Correlation structure:
        lower    est.    upper
 Phi 0.56443 0.75261 0.86643

 Variance function:
          lower    est.    upper
 power 0.32359 0.51824 0.71289

 Residual standard error:
    lower    est.   upper
   2.3051  3.0463  4.0259
```

The confidence interval on the correlation parameter ϕ is bounded away from zero, indicating that the $AR(1)$ model provides a significantly better fit. We can confirm it with the likelihood ratio test.

```
> anova( fm2Dial.gls, fm3Dial.gls )
            Model df    AIC    BIC  logLik    Test L.Ratio p-value
fm2Dial.gls     1 12 738.22 772.63 -357.11
fm3Dial.gls     2 13 642.67 679.95 -308.34 1 vs 2  97.546  <.0001
```

No significant correlations are observed in the plot of the empirical autocorrelation function for the normalized residuals of `fm3Dial.gls`, displayed in Figure 5.21, indicating that the $AR(1)$ adequately represents the within-subject dependence.

The gls model corresponding to `fm3Dial.gls` may be compared to the best lme model for the `Dialyzer` data in §5.2.2, corresponding to the `fm2Dial.lme` object. As the models are not nested, only the information criterion statistics can be compared.

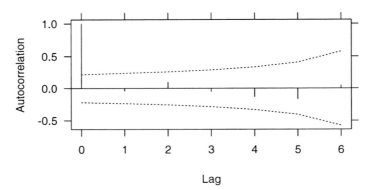

FIGURE 5.21. Empirical autocorrelation function for the normalized residuals of the `fm3Dial.gls` object.

```
> anova( fm3Dial.gls, fm2Dial.lme, test = F )
            Model df    AIC    BIC  logLik
fm3Dial.gls     1 13 642.67 679.95 -308.34
fm2Dial.lme     2 18 655.01 706.63 -309.51
```

The two log-likelihoods are very similar, suggesting that the models give equivalent representations of the data. Because the gls model has five fewer parameters than the lme model, its information criterion statistics take smaller values, suggesting it is a better model. However, as pointed out in the previous example, the choice between a gls and an lme should take other factors in consideration, besides the information criteria.

Wheat Yield Trials

Stroup and Baenziger (1994) describe an agronomic experiment to compare the yield of 56 different varieties of wheat planted in four blocks arranged according to a randomized complete complete block design. All 56 varieties of wheat were used in each block. The latitude and longitude of each experimental unit in the trial were also recorded. These data are described in greater detail in Appendix A.31, being included in the nlme library as the groupedData object `Wheat2`.

The plot of the wheat yields for each variety by block, shown in Figure 5.22, suggests that a *block effect* is present in the data. As pointed out by Littell et al. (1996, §9.6.2), the large number of plots within each block makes the assumption of within-block homogeneity unrealistic. A better representation of the dependence among the experimental units may be obtained via spatial correlation structures that use the information on their latitude and longitude. The corresponding extended linear model for the ith wheat variety yield in the jth block, y_{ij}, for $i = 1, \ldots, 56$, $j = 1, \ldots, 4$,

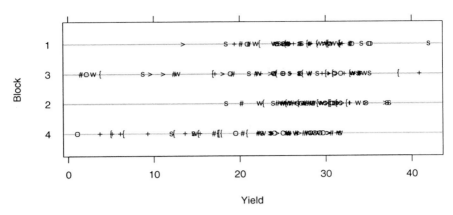

FIGURE 5.22. Yields of 56 different varieties of wheat for each block of a randomized complete block design.

is given by

$$y_{ij} = \tau_i + \epsilon_{ij}, \quad \epsilon = \mathcal{N}\left(\mathbf{0}, \sigma^2 \mathbf{\Lambda}\right), \tag{5.36}$$

where τ_i denotes the average yield for variety i and ϵ_{ij} denotes the error term, assumed to be normally distributed with mean 0 and with variance–covariance matrix $\sigma^2 \mathbf{\Lambda}$.

To explore the structure of $\mathbf{\Lambda}$, we initially fit the linear model (5.36) with the errors assumed independent and homoscedastic, i.e., $\mathbf{\Lambda} = \mathbf{I}$.

```
> fm1Wheat2 <- gls( yield ~ variety - 1, Wheat2 )
```

As described in §5.3.2, the sample semivariogram of the standardized residuals is the primary tool for investigating spatial correlation in the errors. The variogram method for class gls is used to obtain the sample semivariogram for the residuals of a gls object. Its syntax is identical to that of the Variogram method for lme objects.

```
> Variogram( fm1Wheat2, form = ~ latitude + longitude )
      variog     dist n.pairs
 1  0.36308   4.3000    1212
 2  0.40696   5.6080    1273
 3  0.45366   8.3863    1256
 4  0.51639   9.3231    1245
 5  0.57271  10.5190    1254
 6  0.58427  12.7472    1285
 7  0.63854  13.3929    1175
 8  0.65123  14.7635    1288
 9  0.73590  16.1818    1290
10  0.73797  17.3666    1187
11  0.75081  18.4567    1298
```

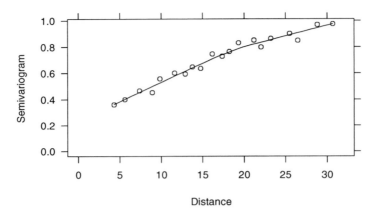

FIGURE 5.23. Sample semivariogram estimates corresponding to the standardized residuals of the `fm1Wheat2` object. A loess smoother is added to the plot to enhance the visualization of patterns in the semivariogram.

```
12 0.88098 20.2428    1226
13 0.81019 21.6335    1281
14 0.86199 22.6736    1181
15 0.86987 24.6221    1272
16 0.85818 26.2427    1223
17 0.97145 28.4542    1263
18 0.98778 30.7877    1228
19 1.09617 34.5879    1263
20 1.34146 39.3641    1234
```

The sample semivariogram increases with distance, indicating, as expected, that the observations are spatially correlated. The graphical representation of the sample semivariogram produced by the `plot` method, displayed in Figure 5.23, allows easier interpretation of the spatial correlation pattern.

```
> plot( Variogram(fm1Wheat2, form = ~ latitude + longitude,
+        maxDist = 32), xlim = c(0,32) )              # Figure 5.23
```

The distance is "censored" at 32, using the `maxDist` argument, to avoid the less reliable estimates associated with distant plots. A nugget effect of about 0.2 seems to be present and the semivariogram appears to approach 1 around distance 28.

Littell et al. (1996, §9.6.2) use the spherical correlation structure to model the spatial correlation in these data. We can fit it using the corSpher class. Initial values for the range and the nugget effect are obtained from Figure 5.23, noting that, in corSpher, the range is the distance at which the semivariogram first equals 1.

```
> fm2Wheat2 <- update( fm1Wheat2, corr = corSpher(c(28, 0.2),
+                    form = ~ latitude + longitude, nugget = T) )
```

```
> fm2Wheat2
Generalized least squares fit by REML
  Model: yield ~ variety - 1
  Data: Wheat2
  Log-restricted-likelihood: -533.93

Coefficients:
 varietyARAPAHOE varietyBRULE varietyBUCKSKIN varietyCENTURA
          26.659        25.85          34.848         25.095
. . .

 varietySIOUXLAND varietyTAM107 varietyTAM200 varietyVONA
           25.656         22.77        18.764       24.782

Correlation Structure: Spherical spatial correlation
 Formula: ~ latitude + longitude

 Parameter estimate(s):
  range  nugget
 27.457 0.20931
Degrees of freedom: 224 total; 168 residual
Residual standard error: 7.4106
```

An alternative model suggested by the slow increase of the semivariogram with distance in Figure 5.23 is the rational quadratic model of §5.3.2. An initial value for the range in this model can also be obtained from the semivariogram plot, noting that, when distance = range the semivariogram in the rational quadratic model is equal to $(1 + \text{nugget})/2$. For a nugget effect of 0.2, this gives an initial estimate for the range of about 12.5 (the approximate distance in Figure 5.23 at which the semivariogram is 0.6). The corRatio class is used to fit the rational quadratic model.

```
> fm3Wheat2 <- update( fm1Wheat2,
+                corr = corRatio(c(12.5, 0.2),
+                    form = ~ latitude + longitude, nugget = T) )
> fm3Wheat2
Generalized least squares fit by REML
  Model: yield ~ variety - 1
  Data: Wheat2
  Log-restricted-likelihood: -532.64

Coefficients:
 varietyARAPAHOE varietyBRULE varietyBUCKSKIN varietyCENTURA
          26.546       26.284          35.037         24.867
. . .

 varietySIOUXLAND varietyTAM107 varietyTAM200 varietyVONA
            25.74        22.476        18.693       25.046

Correlation Structure: Rational quadratic spatial correlation
 Formula: ~ latitude + longitude
```

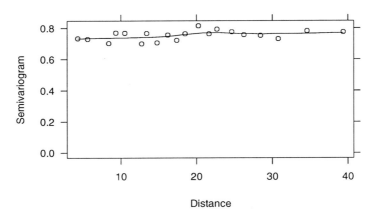

FIGURE 5.24. Sample semivariogram estimates corresponding to the normalized residuals of the `fm3Wheat2` object. A loess smoother is added to the plot to enhance the visualization of patterns in the semivariogram.

```
Parameter estimate(s):
 range nugget
13.461 0.1936
Degrees of freedom: 224 total; 168 residual
Residual standard error: 8.8463
> anova( fm2Wheat2, fm3Wheat2 )
          Model df    AIC    BIC  logLik
fm2Wheat2     1 59 1185.9 1370.2 -533.93
fm3Wheat2     2 59 1183.3 1367.6 -532.64
```

The smaller AIC and BIC values for the rational quadratic model indicate that it gives a better representation of the correlation in the data than the spherical model. We can test the significance of the spatial correlation parameters comparing the `fm3Wheat2` fit to the fit with independent errors corresponding to `fm1Wheat2`.

```
> anova( fm1Wheat2, fm3Wheat2 )
          Model df    AIC    BIC  logLik  Test L.Ratio p-value
fm1Wheat2     1 57 1354.7 1532.8 -620.37
fm3Wheat2     2 59 1183.3 1367.6 -532.64 1 vs 2  175.46  <.0001
```

The large value of the likelihood ratio test statistics gives strong evidence against the assumption of independence.

We can verify the adequacy of the corRatio model by examining the plot of the sample semivariogram for the normalized residuals of `fm3Wheat2`.

```
> plot( Variogram(fm3Wheat2, resType = "n") )        # Figure 5.24
```

No patterns are observed in the plot of the sample semivariogram, suggesting that the rational quadratic model is adequate.

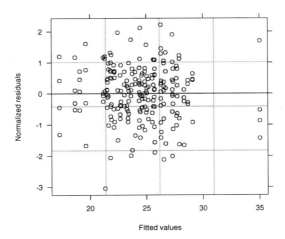

FIGURE 5.25. Scatter plot of normalized residuals versus fitted values for the
fitted object `fm3Wheat2`.

The normalized residuals are also useful for investigating heteroscedasticity and departures from normality. For example, the plot of the normalized
residuals versus the fitted values, displayed in Figure 5.25 and obtained
with

```
> plot( fm3Wheat2, resid(., type = "n") ~ fitted(.),
+        abline = 0 )                              # Figure 5.25
```

does not indicate any heteroscedastic patterns. The normal plot of the
normalized residuals in Figure 5.26 is obtained with

```
> qqnorm( fm3Wheat2, ~ resid(., type = "n") )      # Figure 5.26
```

No significant departures from the assumption of normality are observed
in this plot.

The `anova` method can be used to assess differences between the wheat
varieties. For example, to test if there are any differences between varieties,
we first have to reparametrize (5.36) in terms of an intercept plus contrasts
between the varieties and then use `anova`.

```
> fm4Wheat2 <- update( fm3Wheat2, model = yield ~ variety )
> anova( fm4Wheat2 )
Denom. DF: 168
            numDF F-value p-value
(Intercept)     1  30.405  <.0001
    variety    55   1.851  0.0015
```

The small p-value of the F-test for `variety` indicates that there are significant differences between varieties. We can test specific contrasts using the

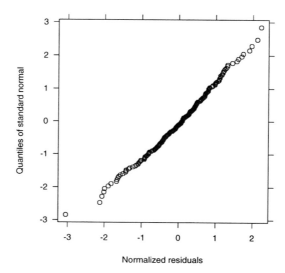

FIGURE 5.26. Normal plots plots of normalized residuals for the fitted object `fm3Wheat2`.

argument `L` to `anova`, with the original *cell means* parametrization. For example, to test the difference between the first and the third wheat varieties we use

```
> anova( fm3Wheat2, L = c(-1, 0, 1) )
Denom. DF: 168
  F-test for linear combination(s)
  varietyARAPAHOE varietyBUCKSKIN
               -1               1
   numDF F-value p-value
1      1  7.6966  0.0062
```

The small p-value for the F-test, combined with the coefficient estimates displayed previously, indicates that the BUCKSKIN variety has significant higher yields than the ARAPAHOE variety. Similar analyses can be obtained for other linear contrasts of the model coefficients.

5.5 Chapter Summary

In this chapter the linear mixed-effects model of Chapters 2 and 4 is extended to include heteroscedastic, correlated within-group errors. We show how the estimation and computational methods of Chapter 2 can be extended to this more general linear mixed-effects model. We introduce several classes of *variance functions* to characterize heteroscedasticity and several classes of *correlation structures* to represent serial and spatial correlation, and describe how variance functions and correlations structures can be combined to flexibly model the within-group variance–covariance structure.

We illustrate, through several examples, how the lme function is used to fit the extended linear mixed-effects model and describe a suite of S classes and methods to implement variance functions (varFunc) and correlation structures (corStruct). Any of these classes, or others defined by users, can be used with lme to fit extended linear mixed-effects models.

An extended linear model with heteroscedastic, correlated errors is introduced and a new modeling function to fit it, gls, is described. This extended linear model can be thought of as an extended linear mixed-effects model with no random effects, and any of the varFunc and corStruct classes available with lme can also be used with gls. Several examples are used the illustrate the use of gls and its associated methods.

Exercises

1. The within-group heteroscedasticity observed for the fm1BW.lme fit of the BodyWeight data in §5.2.2 was modeled using the power variance function (varPower) with the fitted values as the variance covariate. An alternative approach, which is explored in this exercise, is to allow different variances for each Diet.

    (a) Plot the residuals of fm1BW.lme versus Diet (use plot(fm1BW.lme, resid(.) ~ as.integer(Diet), abline = 0)). Note that the variability for Diets 1 and 2 are similar, but the residuals for Diet 3 have larger variability.

    (b) Update the fm1BW.lme fit allowing different variances per Diet (use weights = varIdent(form = ~1|Diet)). To get a fit that can be compared to fm1BW.lme, remember to set the contrasts parameterization to "contr.helmert". Obtain confidence intervals on the variance function coefficients using intervals. Do they agree with the conclusions from the plot of the residuals versus Diet? Explain.

    (c) Compare the fit with the varIdent variance function to fm1BW.lme using anova. Compare it also to fm2BW.lme, the fit with the varPower variance function. Which variance function model is preferable? Why?

    (d) Use the gls function described in §5.4 to fit a model with a varPower variance function in the fitted values and a corCAR1 correlation structure in Time, but with no random effects. Compare the resulting fit to the fm3BW.lme fit of §5.3.4 using anova. Are the two models nested? Why?

2. A multilevel LME analysis of the Oats data was presented in §1.6 and an equivalent single-level analysis using a block-diagonal Ψ matrix (pdBlocked class) was discussed in §4.2.2. This exercise illustrates yet another possible approach for analyzing split-plot experiments such as the Oats data.

   (a) Fit an LME model to the Oats data using the same fixed-effects model as in the fm4Oats fit of §1.6, a single random intercept at the Block level, and a general correlation structure (corSymm) at the Variety within Block level (use random = ~1|Block and corr = corSymm(form = ~1|Block/Variety)).

   (b) Obtain the confidence intervals on the within-group correlations associated with the corSymm structure. Note that all intervals overlap and the estimates are of similar magnitude. This suggests a compound symmetry correlation structure. Update the previous fit using corr = corCompSymm(form=~1|Block/Variety). Assess the validity of the corCompSymm model using anova.

   (c) Compare the fit obtained in the previous item to the fm4Oats fit. Note that the log-likelihoods are nearly identical and so are the estimated fixed effects and the estimated standard error for Blocks $(\hat{\sigma}_1)$. The two models are actually identical: verify that the estimated within-group variance for the lme fit with the corCompSymm structure is equal to the sum of the estimated Variety within Block variance $(\hat{\sigma}_2^2)$ and the estimated within-group variance $(\hat{\sigma}^2)$ in the fm4Oats fit and that the estimated correlation for the corCompSymm structure is equal to $\hat{\sigma}_2^2/(\hat{\sigma}_2^2+\hat{\sigma}^2)$.

   (d) Use gls to fit a model with corCompSymm correlation structure at the Variety within Block level, but with no random effects and use it to assess the significance of the "Block effect" in the Oats experiment.

3. Obtain a gls fit of the Ovary data using the same fixed-effects model and correlation structure as in the fm5Ovar.lme fit of §5.3.4 but with no random effects (use corr = corARMA(form=~1|Mare, p=1, q=1) to define the grouping structure). Compare the fit with fm5Ovar.lme using anova. Is there significant evidence for keeping the random effects in the model? Can you give an explanation for what you found?

4. As mentioned in §5.2, new varFunc classes representing user-defined variance functions can be incorporated into the nlme library and used with any of its modeling function. This exercise illustrates how a new varFunc class can be created.

We use a variance function based on a linear combination of covariates, which we denote varReg. Letting v denote a vector of variance covariates and δ the variance parameters, the varReg variance model and variance function are defined as

$$\text{Var}\left(\epsilon\right) = \sigma^2 \exp\left(2v^T\delta\right), \qquad g\left(v,\delta\right) = \exp\left(v^T\delta\right),$$

where, as in §5.2, ϵ represents the random variable whose variance is being modeled. Note that the variance parameters δ are unconstrained and, for identifiability, the linear model in the variance function can not have an intercept (it would be confounded with σ). The varReg class extends the varExp class defined in §5.2 by allowing more than one variance covariate to be used. The varReg class is frequently used in the analysis of dispersion effects in robust designs (Wolfinger and Tobias, 1998).

(a) Write a constructor for the varReg class. It should contain at least two arguments: value with initial values for the variance parameters δ (set the default to numeric(0) to indicate non-initialized structures) and form with a linear formula defining the variance covariates. To simplify things, assume that the fitted object can not be used to define a variance covariate for this class, that no stratification of parameters will be available and that no parameters can be made fixed in the estimation. You can use the varExp constructor as a template.

(b) Next you need to write an initialize method. This takes two required arguments object, the varReg object, and data, a data frame in which to evaluate the variance covariates. For consistency with the generic function, include a ... at the end of the argument list. The initialize method should obtain the model matrix corresponding to formula(object), evaluated on data, and save it as an attribute for later calculations. As mentioned before, the model matrix should not have an (Intercept) column; check if one is present and remove it if necessary. You should make sure that the parameters are initialized (if no starting values are given in the constructor, initialize them to 0, corresponding to an homoscedastic variance model). Additionally, the "logLik" and "weights" attributes of the returned object need to be initialized. Note that the weights are simply exp(-modMat %*% coefs), where modMat represents the model matrix for the variance covariates and coefs the initialized coefficients. You can use the initialize.varExp method as a template.

(c) The coef method for the varReg class is very simple: for consistency with other varFunc coef methods, it takes three arguments: object (the varReg object), unconstrained, and allCoef. Because

δ is unconstrained and we are not allowing any parameters to be fixed, the method will always return `as.vector(object)`.

(d) The `coef<-` method takes two arguments: `object`, the `varReg` object, and `value`, the new values for the coefficients. This method is also used to update the value of the `"logLik"` and `"weights"` attributes, after the coefficients in `object` have been updated. You can use `coef<-.varExp` as a template.

(e) Write a `summary.varReg` method to provide a description of the `varReg` variance function when it is the `varReg` object is printed (usually as part of the output of some modeling function). Use `summary.varExp` as an example.

(f) Test your class by fitting an LME model to the `BodyWeight` data with a different variances for each `Diet` (you can use `weights = varReg(form = ~Diet)`, because the `initialize` method will remove the `(Intercept)` column of the model matrix).

Part II

Nonlinear Mixed-Effects Models

6

Nonlinear Mixed-Effects Models: Basic Concepts and Motivating Examples

This chapter gives an overview of the nonlinear mixed-effects (NLME) model, introducing its main concepts and ideas through the analysis of real-data examples. The emphasis is on presenting the motivation for using NLME models when analyzing grouped data, while introducing some of the key features in the nlme library for fitting and analyzing such models. This chapter serves as an *appetizer* for the material covered in the last two chapters of the book: the theoretical foundations and computational methods for NLME models described in Chapter 7 and the nonlinear modeling facilities available in the nlme library, described in detail in Chapter 8.

6.1 LME Models vs. NLME Models

The first and possibly most important question about NLME models is *why would one want to use them?* This question, of course, also applies to nonlinear regression models in general as does the answer: *interpretability, parsimony*, and *validity beyond the observed range of the data*.

When choosing a regression model to describe how a response variable varies with covariates, one always has the option of using models, such as polynomial models, that are linear in the parameters. By increasing the order of a polynomial model, one can get increasingly accurate approximations to the true, usually nonlinear, regression function, *within the observed range of the data*. These *empirical* models are based only on the observed relationship between the response and the covariates and do not include any theoretical considerations about the underlying mechanism producing the data.

Nonlinear models, on the other hand, are often *mechanistic*, i.e., based on a model for the mechanism producing the response. As a consequence, the model parameters in a nonlinear model generally have a natural physical interpretation. Even when derived empirically, nonlinear models usually incorporate known, theoretical characteristics of the data, such as asymptotes and monotonicity, and in these cases, can be considered as *semi-mechanistic* models. A nonlinear model generally uses fewer parameters than a competitor linear model, such as a polynomial, giving a more *parsimonious* description of the data. Nonlinear models also provide more reliable predictions for the response variable outside the observed range of the data than, say, polynomial models would.

To illustrate these differences between linear and nonlinear models, let us consider a simple example in which the expected height h_t of a tree at time t follows a three-parameter *logistic* growth model.

$$h_t = \phi_1 / \{1 + \exp\left[-(t - \phi_2)/\phi_3\right]\}. \tag{6.1}$$

As described in Appendix C.7, the parameters in (6.1) have a physical interpretation: ϕ_1 is the asymptotic height; ϕ_2 is the time at which the tree reaches half of its asymptotic height; and ϕ_3 is the time elapsed between the tree reaching half and $1/(1 + e^{-1}) \simeq 3/4$ of its asymptotic height. The logistic model (6.1) is linear in one parameter, ϕ_1, but nonlinear in ϕ_2 and ϕ_3.

To make the example more concrete, suppose that $\phi_1 = 3$, $\phi_2 = 1$, and $\phi_3 = 1.2$ and that we initially want to model the tree growth for $0.4 \le t \le 1.6$. The logistic curve, shown as a solid line in Figure 6.1, is approximated very well in the interval $[0.4, 1.6]$ by the fifth-degree polynomial

$$h_t \simeq -2.2911 + 16.591t - 44.411t^2 + 56.822t^3 - 31.514t^4 + 6.3028t^5$$

obtained as a least-squares fit to equally spaced t values in the interval $[0.4, 1.6]$. The polynomial fit, shown as a dashed line in Figure 6.1, is virtually indistinguishable from the logistic curve within this interval.

Unlike the coefficients in the logistic model, the coefficients in the polynomial approximation do not have any physical interpretation. Also, the linear polynomial model uses twice as many parameters as the logistic model to give comparable fitted values. Finally, the polynomial approximation is unreliable outside the interval $[0.4, 1.6]$. Figure 6.2, displaying the two curves over the extended interval $[0, 2]$, shows the dramatic differences between the curves outside the original range. We would expect growth curves to follow a pattern more like the logistic model than like the polynomial model.

Nonlinear mixed-effects models extend linear mixed-effects models by allowing the regression function to depend nonlinearly on fixed and random effects. Because of its greater flexibility, an NLME model is generally more

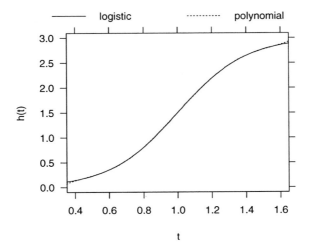

FIGURE 6.1. Logistic curve with parameters $\phi_1 = 3$, $\phi_2 = 1$, and $\phi_3 = 1.2$ and its fifth-order polynomial approximation over the interval $[0.4, 1.6]$.

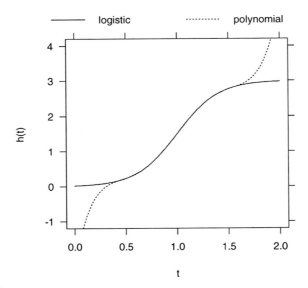

FIGURE 6.2. Logistic curve with parameters $\phi_1 = 3$, $\phi_2 = 1$, and $\phi_3 = 1.2$ and its fifth-order polynomial approximation over the interval $[0.4, 1.6]$, plotted over the interval $[0, 2]$.

interpretable and parsimonious than a competitor empirical LME model based, say, on a polynomial function. Also, the predictions obtained from an NLME model extend more reliably outside the observed range of the data.

The greater flexibility of NLME models does not come without cost, however. Because the random effects are allowed to enter the model non-linearly, the marginal likelihood function, obtained by integrating the joint density of the response and the random effects with respect to the random effects, does not have a closed-form expression, as in the LME model. As a consequence, an approximate likelihood function needs to be used for the estimation of parameters, leading to more computationally intensive estimation algorithms and to less reliable inference results. These issues are described and discussed in detail in Chapter 7.

An important practical difference between NLME and LME models is that the former require starting estimates for the fixed-effects coefficients. Determining reasonable starting estimates for the parameters in a nonlinear model is somewhat of an art, although some general recommendations are available (Bates and Watts, 1988, §3.2). In many applications, the same nonlinear model is to be used several times with similar datasets. In these cases, it is worthwhile to program the steps used to obtain starting estimates into a function, which can then be used to produce starting estimates from many datasets. Such *self-starting* nonlinear models are described and illustrated in §8.1.2.

There are far more similarities than differences between LME and NLME models. Both models are used with grouped data and serve the same purpose: to describe a response variable as a function of covariates, taking into account the correlation among observations in the same group. Random effects are used to represent within-group dependence in both LME and NLME models, and the assumptions about the random effects and the within-group errors are identical in the two models.

The same "inside-out" model building strategy used for LME models in Part I is used here with NLME models: whenever feasible, we begin by getting separate fits of a model for each group, then examining these individual fits to see which coefficients appear to be common to all groups and which coefficients seem to vary among groups. We then proceed to fit an overall mixed-effects model to the data, using random-effects terms if they seem warranted. Diagnostics plots are then used to assess the model's assumptions and to decide on refinements of the initial model.

The S methods for displaying, plotting, comparing, and updating nlme fitted objects will look very familiar to the readers of Part I. Because of the similarities between LME and NLME models, most of the lme methods described in §4.2.1 can be used, without changes, with nlme objects. The nonlinear mixed-effects models modeling functions and methods included in the nlme library are described and illustrated in detail in Chapter 8. The examples in the next sections serve to illustrate some of the basic concepts

of NLME models and to introduce the capabilities provided by the nlme
library for analyzing them.

6.2 Indomethicin Kinetics

The data for our first example come from a laboratory study on the pharmacokinetics of the drug indomethicin (Kwan, Breault, Umbenhauer, McMahon and Duggan, 1976). Six human volunteers received bolus intravenous
injections of the same dose of indomethicin and had their plasma concentrations of the drug (in mcg/ml) measured 11 times between 15 minutes
and 8 hours postinjection. These data are described in more detail in Appendix A.12 and are also analyzed in Davidian and Giltinan (1995, §2.1).
The indomethicin data are included in the nlme library as the groupedData
object Indometh.

   The plot of the Indometh data, obtained with

```
> plot( Indometh )                              # Figure 6.3
```

and displayed in Figure 6.3, reveals a familiar pattern in plots of grouped
data: the concentration curves have a similar shape, but differ among individuals. As indicated in Figure 6.3, these data are balanced, as the serum
concentrations were measured at the same time points for all six individuals.

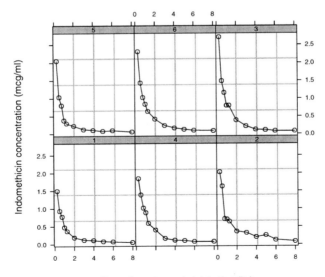

FIGURE 6.3. Concentration of indomethicin over time for six subjects following
intravenous injection.

A common method of modeling pharmacokinetic data is to represent the human body as a system of compartments in which the drug is transferred according to first-order kinetics (Gibaldi and Perrier, 1982). In such a *compartment model* the concentration of the drug over time in the different compartments is determined by a linear system of differential equations, whose solution can be expressed as a linear combination of exponential terms. The mechanistic model for the indomethicin concentration is from a two-compartment model. It expresses the expected concentration $E(y_t)$ at time t as a linear combination of two exponentials

$$E(y_t) = \phi_1 \exp(-\phi_2 t) + \phi_3 \exp(-\phi_4 t), \quad \phi_2 > 0, \ \phi_4 > 0. \tag{6.2}$$

The biexponential model (6.2) is not *identifiable*, in the sense of having a unique vector of parameters associated with a given set of predictions, because the parameters in the two exponential terms may be exchanged without changing the predictions. Identifiability is ensured by requiring that $\phi_2 > \phi_4$ so that the first exponential term determines the initial elimination phase of the drug (the sharp decreases in the individual curves of Figure 6.3). The terminal elimination phase is primarily determined by the second exponential term.

Model (6.2) is linear in multipliers ϕ_1 and ϕ_3, but nonlinear in the rate constants ϕ_2 and ϕ_4. Because the rate constants must be positive to be physically meaningful, we reparameterize (6.2) in terms of the log-rate constants $\phi_2' = \log \phi_2$ and $\phi_4' = \log \phi_4$.

When introducing LME models in §1.1 we demonstrated that the LME model can be considered as a compromise between a single linear model that ignores the grouping in the data and a linear model that provides separate fits for each of the groups. These linear models were fit using `lm` and `lmList`, respectively. To illustrate why NLME models are useful for datasets like the indomethicin data, we will repeat that development using the nonlinear model-fitting functions `nls` and `nlsList`.

If we ignore the grouping of the concentration measurements according to individual and fit a single nonlinear model to all the data we express the indomethicin concentration y_{ij} in individual i at time t_j is

$$y_{ij} = \phi_1 \exp\left[-\exp(\phi_2') t_j\right] + \phi_3 \exp\left[-\exp(\phi_4') t_j\right] + \epsilon_{ij}, \tag{6.3}$$

where the error terms ϵ_{ij} are assumed to be independently distributed as $\mathcal{N}(0, \sigma^2)$. Nonlinear regression models with independent, identically distributed Gaussian errors, like (6.3), are fit by *nonlinear least squares* (Bates and Watts, 1988; Seber and Wild, 1989), as implemented in the S function `nls` (Bates and Chambers, 1992).

Starting values for the parameters in the biexponential model are usually obtained through the method of *peeling*, as described in Appendix C.4.1. However, we need not be concerned with obtaining initial estimates for the biexponential model parameters, as the self-starting model function

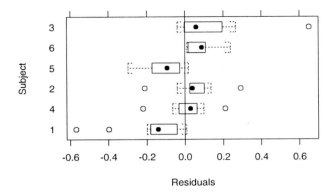

FIGURE 6.4. Boxplots of residuals by subject for fm1Indom.nls, a nonlinear least squares fit of a two-compartment model to the indomethicin data ignoring the subject dependence.

SSbiexp, described in Appendix C.4, produces them automatically from the data. A call to nls to fit the biexponential model (6.3) using a self-starting function is almost as simple as a call to lm to fit a linear model.

```
> fm1Indom.nls <- nls( conc ~ SSbiexp(time, A1, lrc1, A2, lrc2),
+    data = Indometh )
> summary(fm1Indom.nls)

Formula: conc ~ SSbiexp(time, A1, lrc1, A2, lrc2)

Parameters:
         Value Std. Error  t value
   A1  2.77342    0.25323  10.9522
 lrc1  0.88627    0.22222   3.9882
   A2  0.60663    0.26708   2.2713
 lrc2 -1.09209    0.40894  -2.6705

Residual standard error: 0.174489 on 62 degrees of freedom
. . .
> plot( fm1Indom.nls, Subject ~ resid(.), abline = 0 )  # Figure 6.4
```

The correspondence between the parameters in fm1Indom.nls and in (6.3) is $\phi_1 = $ A1, $\phi_2' = $ lrc1, $\phi_3 = $ A2, $\phi_4' = $ lrc2.

The boxplots of the residuals by individual in Figure 6.4 are similar to those from the lm fit to the Rails data (Figure 1.2). In Figure 6.4 the residuals tend to be mostly negative for some subjects and mostly positive for others although the pattern is not as pronounced as that in Figure 1.2. Because a single concentration curve is used for all subjects, the individual differences noticed in Figure 6.3 are incorporated in the residuals, thus inflating the residual standard error. Probably the most important drawback of using an nls model with grouped data is that it prevents us from understanding the true structure of the data and from considering different sources of variability that are of interest in themselves. For example, in the indomethicin study, an important consideration in determining an adequate therapeutic regime for the drug is knowing how the concentration profiles vary among individuals.

To fit a separate biexponential model to each subject, thus allowing the individual effects to be incorporated in the parameter estimates, we express the model as

$$ y_{ij} = \phi_{1i} \exp\left[-\exp\left(\phi'_{2i}\right) t_j\right] + \phi_{3i} \exp\left[-\exp\left(\phi'_{4i}\right) t_j\right] + \epsilon_{ij}, \qquad (6.4) $$

where, as before, the ϵ_{ij} are independent $\mathcal{N}(0, \sigma^2)$ errors and use the nlsList function

```
> fm1Indom.lis <- nlsList(conc ~ SSbiexp(time, A1, lrc1, A2, lrc2),
+    data = Indometh )
> fm1Indom.lis
Call:
  Model: conc ~ SSbiexp(time, A1, lrc1, A2, lrc2) | Subject
   Data: Indometh

Coefficients:
       A1      lrc1      A2       lrc2
1  2.0293  0.57938  0.19154  -1.78783
4  2.1979  0.24249  0.25481  -1.60153
2  2.8277  0.80143  0.49903  -1.63508
5  3.5663  1.04095  0.29170  -1.50594
6  3.0022  1.08811  0.96840  -0.87324
3  5.4677  1.74968  1.67564  -0.41226

Degrees of freedom: 66 total; 42 residual
Residual standard error: 0.07555
```

We can see that there is considerable variability in the individual parameter estimates and that the residual standard error is less than one-half that from the nls fit. The boxplots of the residuals by subject, shown in Figure 6.5, indicate that the individual effects have been accounted for in the fitted nlsList model.

The nlsList model is at the other extreme of the flexibility spectrum compared to the nls model: it uses 24 coefficients to represent the individual concentration profiles and does not take into account the obvious similarities among the individual curves, indicated in Figure 6.3. The nlsList model is useful when one is interested in modeling the behavior of a particular, fixed set of individuals, but it is not adequate when the observed individuals are to be treated as a sample from a population of similar individuals, which constitutes the majority of applications involving grouped data. In this case, the interest is in estimating the average behavior of an

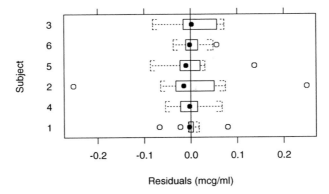

FIGURE 6.5. Boxplots of residuals by subject for `fm1Indom.lis`, a set of nonlinear regression fits of a two-compartment model to the indomethicin data where each subject's data is fit separately.

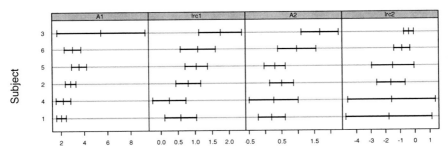

FIGURE 6.6. Ninety-five percent confidence intervals on the biexponential model parameters for each individual in the indomethicin data.

individual in the population and the variability among and within individuals, which is precisely what mixed-effects models are designed to do.

The plot of the individual confidence intervals for the coefficients in the `nlsList` model, shown in Figure 6.6, gives a better idea about their variability among subjects.

```
> plot( intervals(fm1Indom.lis) )                # Figure 6.6
```

The terminal phase log-rate constants, ϕ_{4i}, do not seem to vary substantially among individuals, but the remaining parameters do.

Recall that in `lmList` fits to balanced data, the lengths of the confidence intervals on a parameter were the same for all the groups (see Figure 1.12, p. 33, or Figure 1.13, p. 34). This does not occur in an `nlsList` fit because the approximate standard errors used to produce the confidence intervals in a nonlinear least squares fit depend on the parameter estimates (Seber and Wild, 1989, §5.1).

To introduce the concepts of fixed and random effects in a nonlinear mixed-effects model, it is useful to re-express the model (6.4) as

$$
\begin{aligned}
y_{ij} = \left[\bar{\phi}_1 + \left(\phi_{1i} - \bar{\phi}_1 \right) \right] \exp \left\{ - \exp \left[\bar{\phi}'_2 + \left(\phi'_{2i} - \bar{\phi}'_2 \right) \right] t_j \right\} \\
+ \left[\bar{\phi}_3 + \left(\phi_{3i} - \bar{\phi}_3 \right) \right] \exp \left\{ - \exp \left[\bar{\phi}'_4 + \left(\phi'_{4i} - \bar{\phi}'_4 \right) \right] t_j \right\} + \epsilon_{ij},
\end{aligned}
\tag{6.5}
$$

where $\bar{\phi}$ denotes the average of the individual parameters. The `nlsList` model treats the deviations of the individual coefficients from their mean as parameters to be estimated. Mixed-effects models, on the other hand, represent these deviations from the mean value of the coefficients (the *fixed effects*) as *random effects*, treating the individuals as a sample from a population. The nonlinear mixed-effects version of (6.5) is

$$
\begin{aligned}
y_{ij} = \left(\beta_1 + b_{1i} \right) \exp \left[- \exp \left(\beta_2 + b_{2i} \right) t_j \right] \\
+ \left(\beta_3 + b_{3i} \right) \exp \left[- \exp \left(\beta_4 + b_{4i} \right) t_j \right] + \epsilon_{ij}.
\end{aligned}
\tag{6.6}
$$

The fixed effects β_1, β_2, β_3, and β_4 represent the mean values of the parameters in the population of individuals. The individual deviations are represented by the random effects b_{1i}, b_{2i}, b_{3i}, and b_{4i}, which are assumed to be distributed normally with mean $\mathbf{0}$ and variance–covariance matrix $\mathbf{\Psi}$. Random effects corresponding to different individuals are assumed to be independent. The within-group errors ϵ_{ij} are assumed to be independently distributed as $\mathcal{N}(0, \sigma^2)$ and to be independent of the random effects.

The nonlinear mixed-effects model (6.6) gives a compromise between the rigid `nls` model (6.3) and the overparameterized `nlsList` model (6.4). It accommodates individual variations through the random effects, but ties the different individuals together through the fixed effects and the variance–covariance matrix $\mathbf{\Psi}$.

A crucial step in the model-building of mixed-effects models is deciding which of the coefficients in the model need random effects to account for their between-subject variation and which can be treated as purely fixed effects. Plots of individual confidence intervals obtained from an `nlsList` fit, like the one shown in Figure 6.6, are often useful for that purpose. In the case of the indomethicin data, the individual confidence intervals suggest that the b_{4i} random effect for the terminal elimination phase lograte constant in (6.6) is not needed.

An alternative model-building strategy is to start with a model with random effects for all parameters and then examine the fitted object to decide which, if any, of the random effects can be eliminated from the model. One problem with this approach is that, when a general positive-definite structure is assumed for the random effects variance–covariance matrix $\mathbf{\Psi}$, the number of parameters to estimate increases with the square of the number of random effects. In cases where the number of random effects is large relative to the number of individuals, as in the indomethicin data, it is generally recommended to use a diagonal $\mathbf{\Psi}$ initially, to prevent convergence problems with an overparameterized model. We apply this approach to the indomethicin data.

```
> fm1Indom.nlme <- nlme( fm1Indom.lis,
+   random = pdDiag(A1 + lrc1 + A2 + lrc2 ~ 1) )
> fm1Indom.nlme
Nonlinear mixed-effects model fit by maximum likelihood
  Model: conc ~ SSbiexp(time, A1, lrc1, A2, lrc2)
  Data: Indometh
  Log-likelihood: 54.592
  Fixed: list(A1 ~ 1, lrc1 ~ 1, A2 ~ 1, lrc2 ~ 1)
     A1   lrc1     A2   lrc2
 2.8276 0.7733 0.46104 -1.345

Random effects:
  Formula: list(A1 ~ 1, lrc1 ~ 1, A2 ~ 1, lrc2 ~ 1)
  Level: Subject
  Structure: Diagonal
             A1    lrc1      A2       lrc2 Residual
StdDev: 0.57136 0.15811 0.11154 7.2051e-11 0.081496

Number of Observations: 66
Number of Groups: 6
```

The nlme function extracts the information about the model to fit, the parameters to estimate, and the starting estimates for the fixed effects from the fm1Indom.lis object.

The near-zero estimate for the standard deviation of the lrc2 random effect suggests that this term could be dropped from the model. The remaining estimated standard deviations suggest that the other random effects should be kept in the model. We can test if the lrc2 random effect can be removed from the model by updating the fit and using anova.

```
> fm2Indom.nlme <- update( fm1Indom.nlme,
+   random = pdDiag(A1 + lrc1 + A2 ~ 1) )
> anova( fm1Indom.nlme, fm2Indom.nlme )
              Model df     AIC     BIC logLik    Test    L.Ratio
fm1Indom.nlme     1  9 -91.185 -71.478 54.592
fm2Indom.nlme     2  8 -93.185 -75.668 54.592 1 vs 2 6.2637e-06
              p-value
fm1Indom.nlme
fm2Indom.nlme   0.998
```

The two fits give nearly identical log-likelihoods, confirming that lrc2 can be treated as a purely fixed effect.

To further explore the variance–covariance structure of the random effects that are left in fm2Indom.nlme, we update the fit using a general positive-definite Ψ matrix.

```
> fm3Indom.nlme <- update( fm2Indom.nlme, random = A1+lrc1+A2 ~ 1 )
```

```
> fm3Indom.nlme
 . . .
Random effects:
 Formula: list(A1 ~ 1, lrc1 ~ 1, A2 ~ 1)
 Level: Subject
 Structure: General positive-definite
           StdDev   Corr
      A1 0.690406 A1    lrc1
    lrc1 0.179030 0.932
      A2 0.153669 0.471 0.118
Residual 0.078072
 . . .
```

The large correlation between the A1 and lrc1 random effects and the small correlation between these random effects and the A2 random effect suggest that a block-diagonal Ψ could be used to represent the variance–covariance structure of the random effects.

```
> fm4Indom.nlme <- update( fm3Indom.nlme,
+    random = pdBlocked(list(A1 + lrc1 ~ 1, A2 ~ 1)) )
> anova( fm3Indom.nlme, fm4Indom.nlme )
              Model df    AIC     BIC logLik    Test L.Ratio
fm3Indom.nlme     1 11 -94.945 -70.859 58.473
fm4Indom.nlme     2  9 -97.064 -77.357 57.532 1 vs 2  1.8809
              p-value
fm3Indom.nlme
fm4Indom.nlme   0.3904
```

The large p-value for the likelihood ratio test and the smaller values for the AIC and BIC corroborate the block-diagonal variance–covariance structure. Allowing the A1 and lrc1 random effects to be correlated causes a significant improvement in the log-likelihood.

```
> anova( fm2Indom.nlme, fm4Indom.nlme )
              Model df    AIC     BIC logLik    Test L.Ratio
fm2Indom.nlme     1  8 -93.185 -75.668 54.592
fm4Indom.nlme     2  9 -97.064 -77.357 57.532 1 vs 2  5.8795
              p-value
fm2Indom.nlme
fm4Indom.nlme   0.0153
```

The plot of the standardized residuals versus the fitted values corresponding to fm4Indom.nlme, presented in Figure 6.7, does not indicate any departures from the NLME model assumptions, except for two possible outlying observations for Individual 2.

```
> plot( fm4Indom.nlme, id = 0.05, adj = -1 )          # Figure 6.7
```

No significant departures from the assumption of normality for the within-group errors is observed in the normal probability plot of the standardized residuals of fm4Indom.nlme, shown in Figure 6.8.

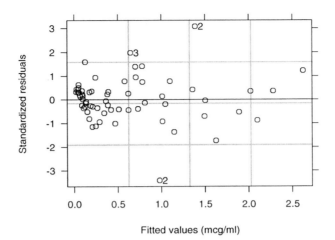

FIGURE 6.7. Scatter plot of standardized residuals versus fitted values for fm4Indom.nlme.

```
> qqnorm( fm4Indom.nlme )                              # Figure 6.8
```

A final assessment of the adequacy of the fm4Indom.nlme model is given by the plot of the augmented predictions in Figure 6.9. For comparison, and to show how individual effects are accounted for in the NLME model, both the population predictions (corresponding to random effects equal to zero) and the within-group predictions (obtained using the estimated random effects) are displayed.

```
> plot( augPred(fm4Indom.nlme, level = 0:1) )          # Figure 6.9
```

Note that the within-group predictions are in close agreement with the observed concentrations, illustrating that the NLME model can accommodate individual effects.

We conclude that fm4Indom.nlme provides a good representation of the concentration profiles in the indomethicin data. Its summary

```
> summary( fm4Indom.nlme )
Nonlinear mixed-effects model fit by maximum likelihood
  Model: conc ~ SSbiexp(time, A1, lrc1, A2, lrc2)
 Data: Indometh
      AIC      BIC logLik
  -97.064 -77.357 57.532

Random effects:
 Composite Structure: Blocked

 Block 1: A1, lrc1
 Formula: list(A1 ~ 1, lrc1 ~ 1)
```

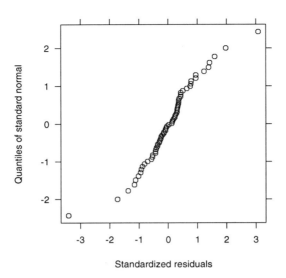

FIGURE 6.8. Normal plot of standardized residuals for the `fm4Indom.nlme` nlme fit.

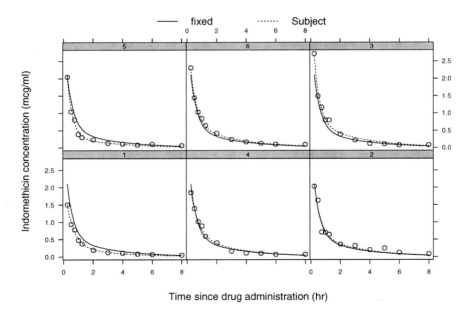

FIGURE 6.9. Population predictions (`fixed`), within-group predictions (`Subject`), and observed concentrations of indomethicin (circles) versus time since injection for `fm4Indom.nlme`.

```
Level: Subject
Structure: General positive-definite
        StdDev   Corr
  A1 0.69496 A1
lrc1 0.17067 0.905


Block 2: A2
Formula: A2 ~ 1 | Subject
            A2 Residual
StdDev: 0.18344 0.078226


Fixed effects: list(A1 ~ 1, lrc1 ~ 1, A2 ~ 1, lrc2 ~ 1)
        Value Std.Error DF t-value p-value
  A1   2.8045   0.31493 57  8.9049  <.0001
lrc1   0.8502   0.11478 57  7.4067  <.0001
  A2   0.5887   0.13321 57  4.4195  <.0001
lrc2  -1.1029   0.16954 57 -6.5054  <.0001
 . . .
```

shows that the fixed-effects estimates are similar to the parameter estimates in the nls fit fm1Indom.nls. The approximate standard errors for the fixed effects are substantially different from and, except for A1, considerably smaller than those from the nls fit. The estimated within-group standard error is slightly larger than the residual standard error in the nlsList fit fm1Indom.lis.

6.3 Growth of Soybean Plants

The example discussed in this section illustrates an important area of application of NLME models: growth curve data. It also introduces the concept of using covariates to explain between-group variability in NLME models.

The soybean data, displayed in Figure 6.10, are described in Davidian and Giltinan (1995, §1.1.3, p. 7) as "Data from an experiment to compare growth patterns of two genotypes of soybeans: Plant Introduction #416937 (P), an experimental strain, and Forrest (F), a commercial variety." The average leaf weight (in grams) of six plants chosen at random from each plot was measured at approximately weekly intervals, between two and eleven weeks after planting. The experiment was carried out over three different planting years: 1988, 1989, and 1990. Eight plots were planted with each genotype in each planting year, giving a total of forty-eight plots in the study. These data are available in the nlme library as the groupedData object Soybean and are also described in Appendix A.27.

```
> Soybean[1:3, ]
Grouped Data: weight ~ Time | Plot
```

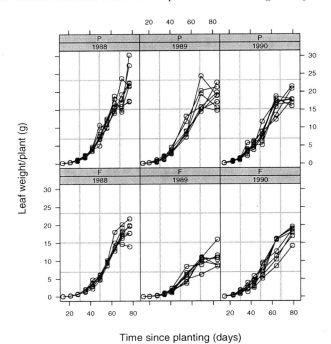

FIGURE 6.10. Average leaf weight per plant of two genotypes of soybean versus time since planting, over three different planting years. Within each year data were obtained on eight plots of each variety of soybean.

```
    Plot Variety Year Time weight
1 1988F1       F 1988   14  0.106
2 1988F1       F 1988   21  0.261
3 1988F1       F 1988   28  0.666
> plot( Soybean, outer = ~ Year * Variety )              # Figure 6.10
```

The average leaf weight per plant in each plot is measured the same number of times, but at different times, making the data unbalanced.

There is considerable variation in the growth curves among plots, but the same overall S-shaped pattern is observed for all plots. This nonlinear growth pattern is well described by the three parameter logistic model (6.1), introduced in §6.1. The self-starting function SSlogis, described in Appendix C.7, can be used to automatically generate starting estimates for the parameters in an nlsList fit.

```
> fm1Soy.lis <- nlsList( weight ~ SSlogis(Time, Asym, xmid, scal),
+    data = Soybean )
Error in nls(y ~ 1/(1 + exp((xmid - x)/scal)..: singular gradient
       matrix
Dumped
```

```
> fm1Soy.lis
Call:
  Model: weight ~ SSlogis(Time, Asym, xmid, scal) | Plot
  Data: Soybean

Coefficients:
          Asym     xmid     scal
1988F4  15.1513  52.834   5.1766
1988F2  19.7455  56.575   8.4067
. . .
1989P8      NA       NA       NA
. . .
1990P5  19.5438  51.148   7.2920
1990P2  25.7873  62.360  11.6570

Degrees of freedom: 404 total; 263 residual
Residual standard error: 1.0438
```

The error message from `nls` indicates that convergence was not attained for one of the plots, `1989P8`. The `nlsList` function is able to recover from such nonconvergence problems and carry on with subsequent `nls` fits. Missing values (`NA`) are assigned to the coefficients of the nonconverging fits. The coefficients in `fm1Soy.lis` are related to the logistic model parameters as follows: $\phi_1 =$ `Asym`, $\phi_2 =$ `xmid`, and $\phi_3 =$ `scal`

Analysis of the individual confidence intervals for `fm1Soy.lis` suggests that random effects are needed for all of the parameters in the logistic model. The corresponding nonlinear mixed-effects model for the average leaf weight per plant y_{ij} in plot i at t_{ij} days after planting is

$$y_{ij} = \frac{\phi_{1i}}{1 + \exp\left[-(t_{ij} - \phi_{2i})/\phi_{3i}\right]} + \epsilon_{ij},$$

$$\phi_i = \begin{bmatrix} \phi_{1i} \\ \phi_{2i} \\ \phi_{3i} \end{bmatrix} = \begin{bmatrix} \beta_1 \\ \beta_2 \\ \beta_3 \end{bmatrix} + \begin{bmatrix} b_{1i} \\ b_{2i} \\ b_{3i} \end{bmatrix} = \boldsymbol{\beta} + \boldsymbol{b}_i, \tag{6.7}$$

$$\boldsymbol{b}_i \sim \mathcal{N}(\boldsymbol{0}, \boldsymbol{\Psi}), \quad \epsilon_{ij} \sim \mathcal{N}(0, \sigma^2).$$

The fixed effects $\boldsymbol{\beta}$ represent the mean values of the parameters ϕ_i in the population and the random effects $\boldsymbol{b}_i$ represent the deviations of the ϕ_i from their mean values. The random effects are assumed to be independent for different plots and the within-group errors ϵ_{ij} are assumed to be independent for different i, j and to be independent of the random effects.

Because the number of plots in the soybean data, 48, is large compared to the number of random effects in (6.7), we use a general positive-definite $\boldsymbol{\Psi}$ for the initial NLME model. Because we can extract information about the model, the parameters to estimate, and the starting values for the fixed effects from the `fm1Soy.lis` object we can fit model (6.7) with the simple call

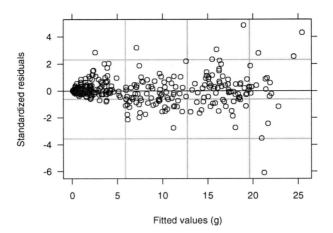

Fitted values (g)

FIGURE 6.11. Scatter plot of standardized residuals versus fitted values for fm1Soy.nlme.

```
> fm1Soy.nlme <- nlme( fm1Soy.lis )
> fm1Soy.nlme
Nonlinear mixed-effects model fit by maximum likelihood
  Model: weight ~ SSlogis(Time, Asym, xmid, scal)
  Data: Soybean
  Log-likelihood: -739.84
  Fixed: list(Asym ~ 1, xmid ~ 1, scal ~ 1)
   Asym   xmid   scal
 19.253 55.02 8.4033

Random effects:
 Formula: list(Asym ~ 1, xmid ~ 1, scal ~ 1)
 Level: Plot
 Structure: General positive-definite
         StdDev  Corr
    Asym 5.2011 Asym  xmid
    xmid 4.1974 0.721
    scal 1.4047 0.711 0.958
Residual 1.1235

Number of Observations: 412
Number of Groups: 48
```

The plot of the standardized residuals versus the fitted values in Figure 6.11 shows a pattern of increasing variability for the within-group errors. We model the within-group heteroscedasticity using the *power* variance function, described in §5.2.1 and represented in S by the varPower class.

```
> fm2Soy.nlme <- update( fm1Soy.nlme, weights = varPower() )
```

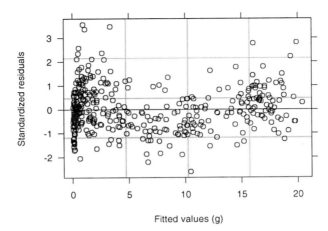

FIGURE 6.12. Standardized residuals versus fitted values in the `nlme` fit of the soybean data, with heteroscedastic error.

```
> anova( fm1Soy.nlme, fm2Soy.nlme )
            Model df     AIC     BIC  logLik    Test L.Ratio p-value
fm1Soy.nlme     1 10  1499.7  1539.9 -739.84
fm2Soy.nlme     2 11   737.3   781.6 -357.66 1 vs 2  764.35  <.0001
```

The heteroscedastic model provides a much better representation of the data. We can assess the adequacy of the *power* variance function by again plotting the standardized residuals against the fitted values as in Figure 6.12. The power variance function seems to model adequately the within-plot heteroscedasticity.

The primary question of interest for the soybean data is the possible relationship between the growth pattern of the soybean plants and the experimental factors `Variety` and `Year`. Plots of estimates of the random effects are useful for exploring the relationship between the individual growth patterns and the experimental factors.

```
> plot(ranef(fm2Soy.nlme, augFrame = T),
+       form = ~ Year * Variety, layout = c(3,1))        # Figure 6.13
```

In Figure 6.13 all three parameters seem to vary with year and variety. It appears that the asymptote (`Asym`) and the scale (`scal`) are larger for the P variety than for the F variety and that this difference is more pronounced in 1989. The time at which half of the asymptotic leaf weight is attained (`xmid`) appears to be smaller for the P variety than for the F variety.

The `fixed` argument to `nlme` allows linear modeling of parameters with respect to covariates. For example, we can model the dependence of all three parameters on `Year` with

```
> soyFix <- fixef( fm2Soy.nlme )
```

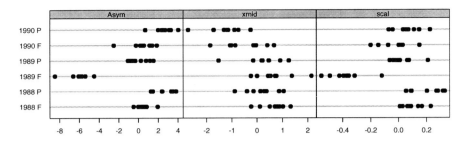

Random effects

FIGURE 6.13. Estimates of the random effects by `Year` and `Variety` in the `nlme` fit of the soybean data.

```
> options( contrasts = c("contr.treatment", "contr.poly") )
> fm3Soy.nlme <- update( fm2Soy.nlme,
+     fixed = Asym + xmid + scal ~ Year,
+     start = c(soyFix[1], 0, 0, soyFix[2], 0, 0, soyFix[3], 0, 0) )
> fm3Soy.nlme
 . . .
  Log-likelihood: -325.02
  Fixed: Asym + xmid + scal ~ Year
 Asym.(Intercept) Asym.Year1989 Asym.Year1990 xmid.(Intercept)
           20.222        -6.3775        -3.4995           54.118
  xmid.Year1989 xmid.Year1990 scal.(Intercept) scal.Year1989
         -2.4696        -4.8764           8.0515       -0.93374
  scal.Year1990
        -0.66884

Random effects:
 Formula: list(Asym ~ 1, xmid ~ 1, scal ~ 1)
 Level: Plot
 Structure: General positive-definite
                     StdDev    Corr
Asym.(Intercept) 2.3686896 Asy(I) xmd(I)
xmid.(Intercept) 0.5863454 -0.997
scal.(Intercept) 0.0043059 -0.590  0.652
        Residual 0.2147634

Variance function:
 Structure: Power of variance covariate
 Formula:  ~ fitted(.)
 Parameter estimates:
   power
 0.95187
 . . .
```

The default parameterization for contrasts is changed to contr.treatment
to allow easier interpretation of the fixed effects included in the model:
they represent differences from the year 1988. Note that, because the fixed-
effects model has changed, new starting values for the fixed effects must be
provided.

We can assess the significance of Year for the fixed effects model using
anova with a single argument.

```
> anova( fm3Soy.nlme )
                  numDF denDF F-value p-value
Asym.(Intercept)      1   356   402.4  <.0001
      Asym.Year       2   356   105.0  <.0001
xmid.(Intercept)      1   356  9641.2  <.0001
      xmid.Year       2   356    10.2  <.0001
scal.(Intercept)      1   356  8378.5  <.0001
      scal.Year       2   356    11.7  <.0001
```

As suggested by Figure 6.13, Year has a very significant effect on the growth
pattern of the soybean plants.

The estimated standard deviation for the scal random effect in the
fm3Soy.nlme fit is only 0.004, corresponding to an estimated coefficient of
variation with respect to the scal.(Intercept) fixed effect of only 0.05%.
This suggests that scal can be treated as a purely fixed effect. When we do
refit the model dropping the scal random effect, we get a p-value of 0.99
in the likelihood ratio test. It often happens that creating a better-fitting
model for the fixed effects, by including their dependence on covariates, reduces the need for random-effects terms. In these cases, the between-group
parameter variation is mostly being explained by the covariates included
in the model.

Proceeding sequentially in the model-building process by examining plots
of the estimated random effects against the experimental factors, testing
for the inclusion of covariates and for the elimination of random effects, we
end up with the following model, in which the only random effect is that
for Asym.

```
> summary( fm4Soy.nlme )
 . . .
     AIC    BIC  logLik
  616.32 680.66 -292.16

Random effects:
 Formula: Asym ~ 1 | Plot
         Asym.(Intercept) Residual
StdDev:           1.0349  0.21804

Variance function:
 Structure: Power of variance covariate
 Formula:  ~ fitted(.)
```

```
Parameter estimates:
 power
0.9426
Fixed effects: list(Asym ~ Year * Variety, xmid ~ Year + Variety,
                    scal ~ Year)
                    Value Std.Error  DF t-value p-value
      Asym.(Intercept) 19.434   0.9537 352  20.379  <.0001
         Asym.Year1989 -8.842   1.0719 352  -8.249  <.0001
         Asym.Year1990 -3.707   1.1768 352  -3.150  0.0018
         Asym.Variety   1.623   1.0380 352   1.564  0.1188
   Asym.Year1989Variety 5.571   1.1704 352   4.760  <.0001
   Asym.Year1990Variety 0.147   1.1753 352   0.125  0.9004
      xmid.(Intercept) 54.815   0.7548 352  72.622  <.0001
         xmid.Year1989 -2.238   0.9718 352  -2.303  0.0218
         xmid.Year1990 -4.970   0.9743 352  -5.101  <.0001
         xmid.Variety  -1.297   0.4144 352  -3.131  0.0019
      scal.(Intercept)  8.064   0.1472 352  54.762  <.0001
         scal.Year1989 -0.895   0.2013 352  -4.447  <.0001
         scal.Year1990 -0.673   0.2122 352  -3.172  0.0016
 . . .
```

The residual plots for `fm4Soy.nlme` do not indicate any violations in the NLME model assumptions. An overall assessment of the adequacy of the model is provided by the plot of the augmented predictions in Figure 6.14, which indicates that the `fm4Soy.nlme` model describes the individual growth patterns of the soybean plots well.

6.4   Clinical Study of Phenobarbital Kinetics

The data for the last example in this chapter come from a clinical pharmacokinetic study of the drug phenobarbital, used for preventing seizures (Grasela and Donn, 1985). The study followed 59 preterm infants during the first 16 days after birth. Each infant received one or more intravenous injections of phenobarbital. As part of routine clinical monitoring, blood samples were drawn from the infants at irregular time intervals to determine the serum concentration of phenobarbital. The number of concentration measurements per individual varies between 1 and 6, with a total of 155 measurements for all 59 individuals. This is typical of clinical pharmacokinetic studies: there are relatively few observations on each of a large number of individuals. In addition to the concentration and dosing information, each infant's birth weight (in kilograms) and 5-minute Apgar score (which gives an overall indication of health of the newborn) are provided. These data have been analyzed in Grasela and Donn (1985), Boeckmann, Sheiner and Beal (1994), Davidian and Giltinan (1995, §6.6), and Littell et al. (1996, §12.5).

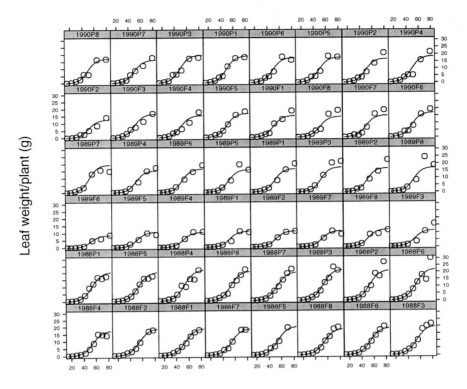

Time since planting (days)

FIGURE 6.14. Average leaf weights of soybean plants versus time since planting and their within-group predictions from the model fm4Soy.nlme.

The phenobarbital data, displayed in Figure 6.15, are available in the nlme library as the groupedData object Phenobarb and are described in more detail in Appendix A.23.

The mechanistic model postulated for the phenobarbital kinetics is a *one-compartment open model with intravenous administration and first-order elimination* (Grasela and Donn, 1985) . The corresponding model for the expected phenobarbital concentration c_t at time t for an infant receiving a single dose D_d administered at time t_d is

$$c_t = \frac{D_d}{V} \exp\left[-\frac{Cl}{V}\left(t - t_d\right)\right], \quad t_d < t,$$

where V and Cl denote, respectively, the *volume of distribution* and the *clearance*. The model for an individual who has received several doses by

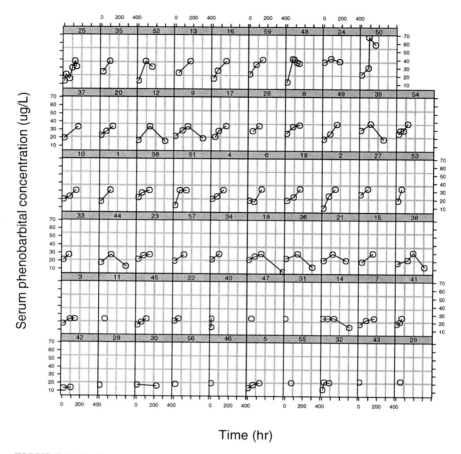

Time (hr)

FIGURE 6.15. Serum concentrations of phenobarbital in 59 newborn infants under varying dosage regimens versus time after birth.

time t is given by the sum of the individual contributions of each dose.

$$c_t = \sum_{d:t_d < t} \frac{D_d}{V} \exp\left[-\frac{Cl}{V}(t - t_d)\right]. \tag{6.8}$$

Model (6.8) can also be expressed in recursive form (Grasela and Donn, 1985). To ensure that the estimates of V and Cl are positive, we reparameterize 6.8) using the logarithm of these parameters: $lV = \log V$ and $lCl = \log Cl$. The function phenoModel in the nlme library implements the reparameterized version of model (6.8) in S. Because phenoModel is *not* self-starting, initial values need to be provided for the parameters when this function is used for estimation in S.

Because of the small number of concentrations recorded for each individual, the usual model-building approach of beginning the analysis with an `nlsList` fit cannot be used with the phenobarbital data and we must go directly to an NLME fit. The nonlinear mixed-effects model corresponding to (6.8) representing the phenobarbital concentration y_{ij} measured at time t_{ij} on individual i, following intravenous injections of dose D_{id} at times t_{id}, is expressed as

$$y_{ij} = \sum_{d:t_{id}<t_{ij}} \frac{D_{id}}{\exp(lV_i)} \exp\left[-\exp\left(lCl_i - lV_i\right)\left(t_{ij} - t_{id}\right)\right] + \epsilon_{ij},$$

$$\begin{bmatrix} lCl_i \\ lV_i \end{bmatrix} = \begin{bmatrix} \beta_1 \\ \beta_2 \end{bmatrix} + \begin{bmatrix} b_{1i} \\ b_{2i} \end{bmatrix} = \boldsymbol{\beta} + \boldsymbol{b}_i, \quad \boldsymbol{b}_i \sim \mathcal{N}\left(\boldsymbol{0}, \boldsymbol{\Psi}\right), \quad \epsilon_{ij} \sim \mathcal{N}\left(0, \sigma^2\right).$$

$$(6.9)$$

The fixed effects, $\boldsymbol{\beta}$, represent the average log-clearance and log-volume of distribution in the infant population, and the random effects, $\boldsymbol{b}_i$, account for individual differences from the population average. As usual, the random effects are assumed to be independent for different individuals and the within-group errors ϵ_{ij} are assumed to be independent for different i, j and to be independent of the random effects.

To avoid convergence problems with the optimization algorithm used in `nlme` due to the sparsity of the phenobarbital concentrations in the data, a diagonal $\boldsymbol{\Psi}$ is assumed in model (6.9), which is then fitted with

```
> fm1Pheno.nlme <-
+   nlme( conc ~ phenoModel(Subject, time, dose, 1Cl, 1V),
+      data = Phenobarb, fixed = 1Cl + 1V ~ 1,
+      random = pdDiag(1Cl + 1V ~ 1), start = c(-5, 0),
+      na.action = na.include, naPattern = ~ !is.na(conc) )
> fm1Pheno.nlme
Nonlinear mixed-effects model fit by maximum likelihood
  Model: conc ~ phenoModel(Subject, time, dose, 1Cl, 1V)
  Data: Phenobarb
  Log-likelihood: -505.41
  Fixed: 1Cl + 1V ~ 1
     1Cl       1V
 -5.0935 0.34259

Random effects:
 Formula: list(1Cl ~ 1, 1V ~ 1)
 Level: Subject
 Structure: Diagonal
             1Cl       1V Residual
StdDev: 0.43989 0.45048   2.7935

Number of Observations: 155
Number of Groups: 59
```

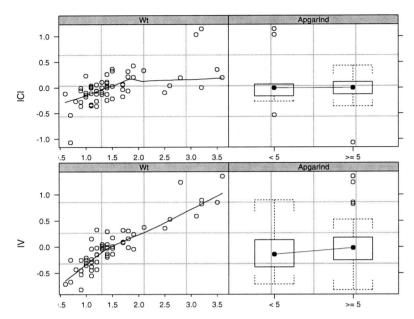

FIGURE 6.16. Estimated log-clearance and log-volume of distribution random effects from model **fm1Pheno.nlme** versus birth weight (**Wt**) and Apgar score indicator (**ApgarInd**) in the phenobarbital data. A **loess** smoother is included in the scatter plots of the continuous covariates to aid in visualizing possible trends.

Starting values for the fixed effects are obtained from Davidian and Giltinan (1995, §6.6). The **na.action** argument in the **nlme** call is used to preserve those rows with dose information. (These rows contain NA for the concentration.) The **naPattern** argument is used to remove these rows from the calculation of the objective function in the optimization algorithm.

One of the questions of interest for the phenobarbital data is the possible relationship between the pharmacokinetic parameters in (6.9) and the additional covariates available on the infants, birth weight and 5-minute Apgar score. For the purposes of modeling the pharmacokinetic parameters, the 5-minute Apgar score is converted to a binary indicator of whether the score is < 5 or ≥ 5, represented by the column **ApgarInd** in the **Phenobarb** data frame.

Figure 6.16 contains plots of the estimated random effects from **fm1Pheno.nlme** versus birth weight (**Wt**) and 5-minute Apgar score indicator (**ApgarInd**). It is produced by

```
> fm1Pheno.ranef <- ranef( fm1Pheno.nlme, augFrame = T )
> plot( fm1Pheno.ranef, form = lCl ~ Wt + ApgarInd )
> plot( fm1Pheno.ranef, form = lV ~ Wt + ApgarInd )
```

The plots in Figure 6.16 clearly indicate that both clearance and volume of distribution increase with birth weight. A linear model seems adequate to represent the increase in lV with birth weight. For birth weights less than 2.5 kg, the increase in lCl seems linear. Because there are few infants with birth weights greater than 2.5 kg in this data set, it is unclear whether the linear relationship between lCl and Wt extends beyond this limit, but we will assume it does. The Apgar score does not seem to have any relationship with clearance and is not included in the model for the lCl fixed effect. It is unclear whether the Apgar score and the volume of distribution are related, so we include ApgarInd in the model for lV to test for a possible relationship.

The updated fit with covariates included in the fixed-effects model is then obtained with

```
> options( contrasts = c("contr.treatment", "contr.poly") )
> fm2Pheno.nlme <- update( fm1Pheno.nlme,
+    fixed = list(lCl ~ Wt, lV ~ Wt + ApgarInd),
+    start = c(-5.0935, 0, 0.34259, 0, 0),
+    control = list(pnlsTol = 1e-6) )
> #pnlsTol reduced to prevent convergence problems in PNLS step
> summary( fm2Pheno.nlme )
. . .
Random effects:
 Formula: list(lCl ~ 1, lV ~ 1)
 Level: Subject
 Structure: Diagonal
         lCl.(Intercept) lV.(Intercept) Residual
StdDev:         0.21599        0.17206   2.7374

Fixed effects: list(lCl ~ Wt, lV ~ Wt + ApgarInd)
                  Value Std.Error DF t-value p-value
lCl.(Intercept) -5.9574   0.12425 92 -47.947  <.0001
        lCl.Wt  0.6197   0.07569 92   8.187  <.0001
 lV.(Intercept) -0.4744   0.07258 92  -6.537  <.0001
        lV.Wt   0.5325   0.04141 92  12.859  <.0001
   lV.ApgarInd -0.0228   0.05131 92  -0.444  0.6577
. . .
```

As expected, the fixed effects corresponding to birth weight, lCl.Wt and lV.Wt, are highly significant. The large p-value for the lV.ApgarInd fixed effect indicates that the volume of distribution is not related to the Apgar scores. The estimated standard deviations for the random effects are about half of the corresponding values in the fm1Pheno.nlme fit, indicating that a substantial part of the between-individual variability in the pharmacokinetic parameters is explained by birth weight.

The ApgarInd variable is dropped from the lV fixed effect model to give the final NLME model for the phenobarbital data considered here.

```
> fm3Pheno.nlme <- update( fm2Pheno.nlme,
+    fixed = 1Cl + 1V ~ Wt, start = fixef(fm2Pheno.nlme)[-5] )
> fm3Pheno.nlme
Nonlinear mixed-effects model fit by maximum likelihood
  Model: conc ~ phenoModel(Subject, time, dose, 1Cl, 1V)
  Data: Phenobarb
  Log-likelihood: -437.7
  Fixed: 1Cl + 1V ~ Wt
  1Cl.(Intercept)  1Cl.Wt 1V.(Intercept)    1V.Wt
          -5.9577 0.61968       -0.48452 0.53205

Random effects:
  Formula: list(1Cl ~ 1, 1V ~ 1)
  Level: Subject
  Structure: Diagonal
          1Cl.(Intercept) 1V.(Intercept) Residual
StdDev:           0.21584        0.17316   2.7326
  . . .
```

The likelihood ratio tests for dropping either of the random effects in fm3Pheno.nlme have very significant p-values (< 0.0001), indicating that both random effects are needed in the model to account for individual effects.

A plot of the augmented predictions is not meaningful for the pheno-barbital data, due to the small number of observations per individual, but we can still assess the adequacy of the fit with the plot of the observed concentrations against the within-group fitted values, produced with

```
> plot( fm3Pheno.nlme, conc ~ fitted(.), abline = c(0,1) )
```

and displayed in Figure 6.17. The good agreement between the observations and the predictions attests the adequacy of the fm3Pheno.nlme model.

6.5 Chapter Summary

This chapter gives an introductory overview of the nonlinear mixed-effects model, describing its basic concepts and assumptions and relating it to the linear mixed-effects model described in the first part of the book. Real-life examples from pharmacokinetics studies and an agricultural experiment are used to illustrate the use of the nlme function in S, and its associated methods, for fitting and analyzing NLME models.

The many similarities between NLME and LME models allow most of the lme methods defined in the first part of the book to also be used with the nlme objects introduced in this section. There are, however, important differences between the two models, and the methods used to fit them, which translate into more complex estimation algorithms and less accurate inference for NLME models.

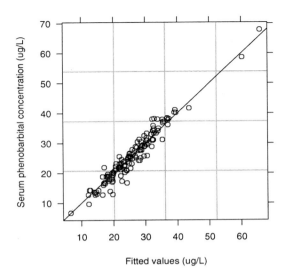

FIGURE 6.17. Observed phenobarbital concentrations versus within-group fitted values corresponding to the **fm3Pheno.nlme** model.

The purpose of this chapter is to present the motivation for using NLME models with grouped data and to set the stage for the following two chapters in the book, dealing with the theory and computational methods for NLME models (Chapter 7) and the nonlinear modeling facilities in the nlme library (Chapter 8).

Exercises

1. The **Loblolly** data described in Appendix A.13 consist of the heights of 14 Loblolly pine trees planted in the southern United States, measured at 3, 5, 10, 15, 20, and 25 years of age. An asymptotic regression model, available in nlme as the self-starting function **SSasymp** (Appendix C.1), seems adequate to explain the observed growth pattern.

   (a) Plot the data (using **plot(Loblolly)**) and verify that the same growth pattern is observed for all trees. The similarity can be emphasized by putting all the curves into a single panel using **plot(Loblolly, outer = ~1)**. What is the most noticeable difference among the curves?

   (b) Fit a separate asymptotic regression model to each tree using **nlsList** and **SSasymp**. Notice that no starting values are needed for the parameters, as they are automatically produced by **SSasymp**.

(c) Plot the confidence intervals on the individual parameters from the `nlsList` fit. Can you identify the ones that vary significantly with tree?

(d) Use the `nlsList` object to fit an NLME model with random effects for all parameters (you can use `nlme(object)`, with `object` replaced with the name of the `nlsList` object). Can you evaluate the confidence intervals on the NLME parameters (using `intervals(object)`)? The error message indicates that the maximum likelihood estimates do not correspond to a numerically stable solution (this generally occurs when the random-effects model is overparameterized).

(e) Update the previous `nlme` fit using a diagonal structure for Ψ (use `random = pdDiag(Asym+R0+lrc ~ 1)` in the call to `update`). Print the results and examine the estimated standard errors of the random effects. Which random effect can be dropped from the model?

(f) Update the `nlme` fit eliminating the random effect you identified in the previous item (continue to use a diagonal Ψ). Compare this fit to the one obtained in the previous item using `anova`. What do you conclude? Can you drop further random effects from the model?

(g) Plot the augmented predictions (using `plot(augPred(object))`) for the final model obtained in the previous item. Does the model seem adequate?

2. Davidian and Giltinan (1995, §1.1, p. 2) describe a pharmacokinetic study of the drug cefamandole in which plasma concentrations of cefamandole at 14 time points following intravenous infusion were collected on 6 healthy volunteers. These data are available in nlme as the object `Cefamandole` and are described in greater detail in Appendix A.4.

(a) Plot the cefamandole plasma concentrations versus time by subject (use `plot(Cefamandole)`. The concentration-time profiles are well described by the biexponential model (6.2) used with the `Indometh` data in §6.2.

(b) Use the self-starting function `SSbiexp` (Appendix C.4) in conjunction with `nlsList` to produce separate fits per subject. Plot the individual confidence intervals and identify the coefficients that seem to vary significantly with subject.

(c) Fit an NLME model to the data with random effects for all coefficients, using the `nlsList` object produced in the previous item. Because of the small number of subjects, convergence is not attained for a model with a general Ψ (try it). Use a diagonal structure for Ψ (random = pdDiag(A1+lrc1+A2+lrc2~1)) and examine the estimated standard errors for the random effects. Can any of the random effects be dropped from the model? Does this agree with your conclusions from the plot of the confidence intervals for the nlsList fit?

(d) Update the nlme fit in the previous item according to your previous conclusions for the random-effects model. Use anova to compare the two models. Produce the confidence intervals for the parameters in the updated model. Do you think any further random effects can be dropped from the model? If so, update the fit and compare it to the previous model using anova.

(e) Plot the standardized residuals versus the fitted values for the final model obtained in the previous item. Do you observe any patterns that contradict the model's assumptions? Plot the observed concentrations versus the fitted values by subject. Does the fitted model produce sensible predictions?

3. Data on the intensity of 23 large earthquakes in western North America between 1940 and 1980 were reported by Joyner and Boore (1981). The data, included in the object Earthquake , are described in more detail in Appendix A.8. The objective of the study was to predict the maximum horizontal acceleration (accel) at a given location during a large earthquake based on the magnitude of the quake (Richter) and its distance from the epicenter (distance). These data are analyzed in Davidian and Giltinan (1995, §11.4, pp. 319–326). The model proposed by Joyner and Boore (1981) can be written as

$$\log_{10}(\texttt{accel}) = \phi_1 + \phi_2 \texttt{Richter}$$
$$- \log_{10} \sqrt{\texttt{distance}^2 + \exp(\phi_3)} - \phi_4 \sqrt{\texttt{distance}^2 + \exp(\phi_3)}.$$

(a) Plot the data and verify that acceleration measurements are sparse or noisy for most of the quakes. No common attenuation pattern is evident from the plot.

(b) No self-starting function is available in nlme for the Earthquake data model. The estimates reported in Davidian and Giltinan (1995) ($\hat{\phi}_1 = -0.764$, $\hat{\phi}_2 = 0.218$, $\hat{\phi}_3 = 1.845$, and $\hat{\phi}_4 = 5.657 \times 10^{-3}$) can be used as initial estimates for an nlsList fit (use start = c(phi1 = -0.764, phi2 = 0.218, phi3 = 1.845, phi4 = 0.005657). However, due to sparse and noisy nature of the data, convergence is not attained for any of the quakes in the nlsList fit (verify it).

(c) Fit an NLME model to the data with random effects for all coefficients and a diagonal Ψ, using as starting values for the fixed effects the estimates reported in Davidian and Giltinan (1995) (listed in the previous item). Note that, even though the model did not converge for any of the individual quakes in the `nlsList` fit, it converged for the combined data in the `nlme` fit. The individual quakes "borrow strength" from each other in the `nlme` fit.

(d) Examine the estimated standard errors for the random effects in the `nlme` fit relative to the absolute value of the estimated fixed effects. Can any of the random effects be dropped from the model? If so, refit the model with fewer random effects an compare it to the previous model using `anova`. Repeat the procedure until no further random effects can be removed from the model. (You may have to use `nls` to fit a model with no random effects, which can also be compared to an `nlme` fit using `anova`, provided the `nlme` object comes first in the argument list.)

(e) Examine the plot of the standard residuals versus the fitted values and the normal plot of the standardized residuals for the final model obtained in the previous item. Are there any apparent departures from the model's assumptions? Plot the observed `log10(accel)` measurements versus the fitted values. How well does the model predict the accelerations?

(f) One of the questions of interest in this study was the possible effect of soil type (represented in `Earthquake` by the indicator variable `soil`, taking values 0 for "rock" and 1 for "soil") on acceleration. Update the final `nlme` model obtained before to include a "soil effect" for the fixed effects of coefficients with an associated random effect. For example, if `phi4` is the only coefficient with a random effect, you can use `fixed = list(phi1+phi2+phi3 ~ 1, phi4 ~ soil)`. You will also need to provide initial estimates for the fixed effects, using for example `start = c(fixef(object), 0)`, where `object` should be replaced with the name of the `nlme` object being updated. Use `summary` to test the significance of soil type.

7
Theory and Computational Methods for Nonlinear Mixed-Effects Models

This chapter presents the theory for the nonlinear mixed-effects model introduced in Chapter 6. A general formulation of NLME models is presented and illustrated with examples. Estimation methods for fitting NLME models, based on approximations to the likelihood function, are described and discussed. The computational methods used in the `nlme` function to fit NLME models are also described. An extended class of nonlinear regression models with heteroscedastic, correlated errors, but with no random effects, is presented.

The objective of this chapter is to give an overall description of the theoretical and computational aspects of NLME models so as to allow one to evaluate the strengths and limitations of such models in practice. It is not the purpose of this chapter to present a thorough theoretical description of NLME models. Such a comprehensive treatment of the theory of nonlinear mixed-effects models can be found, for example, in Davidian and Giltinan (1995) and in Vonesh and Chinchilli (1997).

Readers who are more interested in the applications of NLME models and the use of the functions and methods in the nlme library to fit such models can, without loss of continuity, skip this chapter and go straight to Chapter 8. If you decide to skip this chapter at a first reading, it is recommended that you return to it (especially §7.1) at a later time to get a good understanding of the NLME model formulation and its assumptions and limitations.

7.1 The NLME Model Formulation

Nonlinear mixed-effects models are mixed-effects models in which some, or all, of the fixed and random effects occur nonlinearly in the model function. They can be regarded either as an extension of linear mixed-effects models in which the conditional expectation of the response given the random effects is allowed to be a nonlinear function of the coefficients, or as an extension of nonlinear regression models for independent data (Bates and Watts, 1988) in which random effects are incorporated in the coefficients to allow then to vary by group, thus inducing correlation within the groups.

This section presents a general formulation for NLME models proposed by Lindstrom and Bates (1990). The NLME model for single-level grouped data, which includes repeated measures and longitudinal data, is presented in §7.1.1. The multilevel NLME model is described in §7.1.2.

7.1.1 Single-Level of Grouping

By far the most common application of NLME models is for repeated measures data—in particular, longitudinal data. The nonlinear mixed-effects model for repeated measures proposed by Lindstrom and Bates (1990) can be thought of as a hierarchical model. At one level the jth observation on the ith group is modeled as

$$y_{ij} = f(\boldsymbol{\phi}_{ij}, \boldsymbol{v}_{ij}) + \epsilon_{ij}, \qquad i = 1, \dots, M, \; j = 1, \dots, n_i, \qquad (7.1)$$

where M is the number of groups, n_i is the number of observations on the ith group, f is a general, real-valued, differentiable function of a group-specific parameter vector $\boldsymbol{\phi}_{ij}$ and a covariate vector $\boldsymbol{v}_{ij}$, and ϵ_{ij} is a normally distributed within-group error term. The function f is nonlinear in at least one component of the group-specific parameter vector $\boldsymbol{\phi}_{ij}$, which is modeled as

$$\boldsymbol{\phi}_{ij} = \boldsymbol{A}_{ij}\boldsymbol{\beta} + \boldsymbol{B}_{ij}\boldsymbol{b}_i, \quad \boldsymbol{b}_i \sim \mathcal{N}(\boldsymbol{0}, \boldsymbol{\Psi}), \qquad (7.2)$$

where $\boldsymbol{\beta}$ is a p-dimensional vector of *fixed effects* and $\boldsymbol{b}_i$ is a q-dimensional *random effects* vector associated with the ith group (not varying with j) with variance–covariance matrix $\boldsymbol{\Psi}$. The matrices $\boldsymbol{A}_{ij}$ and $\boldsymbol{B}_{ij}$ are of appropriate dimensions and depend on the group and possibly on the values of some covariates at the jth observation. This model is a slight generalization of that described in Lindstrom and Bates (1990) in that $\boldsymbol{A}_{ij}$ and $\boldsymbol{B}_{ij}$ can depend on j. This generalization allows the incorporation of "time-varying" covariates in the fixed effects or the random effects for the model. It is assumed that observations corresponding to different groups are independent and that the within-group errors ϵ_{ij} are independently distributed as $\mathcal{N}(0, \sigma^2)$ and independent of the $\boldsymbol{b}_i$. The assumption of independence and homoscedasticity for the within-group errors can be relaxed, as shown in §7.4.

Because f can be any nonlinear function of ϕ_{ij}, the representation of the group-specific coefficients ϕ_{ij} could be chosen so that $\boldsymbol{A}_{ij}$ and $\boldsymbol{B}_{ij}$ are always simple incidence matrices. However, it is desirable to encapsulate as much modeling of the ϕ_{ij} as possible in this second stage, as this simplifies the calculation of the derivatives of the model function with respect to $\boldsymbol{\beta}$ and $\boldsymbol{b}_i$, used in the optimization algorithm. In a call to `nlme` the arguments `fixed` and `random` are used to specify the $\boldsymbol{A}_{ij}$ and $\boldsymbol{B}_{ij}$ matrices, respectively.

We can write (7.1) and (7.2) in matrix form as

$$
\begin{aligned}
\boldsymbol{y}_i &= \boldsymbol{f}_i\left(\boldsymbol{\phi}_i, \boldsymbol{v}_i\right) + \boldsymbol{\epsilon}_i, \\
\boldsymbol{\phi}_i &= \boldsymbol{A}_i \boldsymbol{\beta} + \boldsymbol{B}_i \boldsymbol{b}_i,
\end{aligned} \tag{7.3}
$$

for $i = 1, \ldots, M$, where

$$
\boldsymbol{y}_i = \begin{bmatrix} y_{i1} \\ \vdots \\ y_{in_i} \end{bmatrix}, \quad
\boldsymbol{\phi}_i = \begin{bmatrix} \phi_{i1} \\ \vdots \\ \phi_{in_i} \end{bmatrix}, \quad
\boldsymbol{\epsilon}_i = \begin{bmatrix} \epsilon_{i1} \\ \vdots \\ \epsilon_{in_i} \end{bmatrix}, \quad
\boldsymbol{f}_i\left(\boldsymbol{\phi}_i, \boldsymbol{v}_i\right) = \begin{bmatrix} f(\phi_{i1}, v_{i1}) \\ \vdots \\ f(\phi_{in_i}, v_{in_i}) \end{bmatrix},
$$

$$
\boldsymbol{v}_i = \begin{bmatrix} v_{i1} \\ \vdots \\ v_{in_i} \end{bmatrix}, \quad
\boldsymbol{A}_i = \begin{bmatrix} \boldsymbol{A}_{i1} \\ \vdots \\ \boldsymbol{A}_{in_i} \end{bmatrix}, \quad
\boldsymbol{B}_i = \begin{bmatrix} \boldsymbol{B}_{i1} \\ \vdots \\ \boldsymbol{B}_{in_i} \end{bmatrix}. \tag{7.4}
$$

We use the examples of Chapter 6 to illustrate the general NLME model formulation.

Indomethicin Kinetics

The final model obtained in §6.2 for the indomethicin data, represented by the object `fm4Indom.nlme`, expresses the concentration measurement y_{ij} for the ith subject at time t_j as

$$
y_{ij} = \phi_{1i} \exp\left[-\exp\left(\phi_{2i}'\right) t_j\right] + \phi_{3i} \exp\left[-\exp\left(\phi_{4i}'\right) t_j\right] + \epsilon_{ij},
$$

$$
\underbrace{\begin{bmatrix} \phi_{1i} \\ \phi_{2i} \\ \phi_{3i} \\ \phi_{4i} \end{bmatrix}}_{\phi_{ij}} = \underbrace{\begin{bmatrix} 1 & 0 & 0 & 0 \\ 0 & 1 & 0 & 0 \\ 0 & 0 & 1 & 0 \\ 0 & 0 & 0 & 1 \end{bmatrix}}_{A_{ij}} \underbrace{\begin{bmatrix} \beta_1 \\ \beta_2 \\ \beta_3 \\ \beta_4 \end{bmatrix}}_{\beta} + \underbrace{\begin{bmatrix} 1 & 0 & 0 \\ 0 & 1 & 0 \\ 0 & 0 & 1 \\ 0 & 0 & 0 \end{bmatrix}}_{B_{ij}} \underbrace{\begin{bmatrix} b_{1i} \\ b_{2i} \\ b_{3i} \end{bmatrix}}_{b_i},
$$

$$
\boldsymbol{b}_i \sim \mathcal{N}\left(\boldsymbol{0}, \begin{bmatrix} \psi_{11} & \psi_{12} & 0 \\ \psi_{12} & \psi_{22} & 0 \\ 0 & 0 & \psi_{33} \end{bmatrix}\right), \quad \epsilon_{ij} \sim \mathcal{N}\left(0, \sigma^2\right).
$$

In this case, the individual coefficients ϕ_{ij} and the design matrices $\boldsymbol{A}_{ij} = \boldsymbol{I}$ and $\boldsymbol{B}_{ij}$ do not vary with time. The $\boldsymbol{\Psi}$ matrix for the random effects is block-diagonal.

Growth of Soybean Plants

The fitted object `fm4Soy.nlme` represents the final model for the soybean data obtained in §6.3. We use plot `1990P8`, for which `Year = 1990` and `Variety = P`, to illustrate the general NLME model representation, expressing the average leaf weight y_{ij} for plot i at t_{ij} days after planting as

$$y_{ij} = \frac{\phi_{1i}}{1 + \exp\left[-\left(t_{ij} - \phi_{2i}\right)/\phi_{3i}\right]} + \epsilon_{ij},$$

$$\underbrace{\begin{bmatrix} \phi_{1i} \\ \phi_{2i} \\ \phi_{3i} \end{bmatrix}}_{\phi_{ij}} = \underbrace{\begin{bmatrix} 1 & 0 & 1 & 1 & 0 & 1 & 0 & 0 & 0 & 0 & 0 & 0 & 0 \\ 0 & 0 & 0 & 0 & 0 & 0 & 1 & 0 & 1 & 1 & 0 & 0 & 0 \\ 0 & 0 & 0 & 0 & 0 & 0 & 0 & 0 & 0 & 0 & 1 & 0 & 1 \end{bmatrix}}_{A_{ij}} \underbrace{\begin{bmatrix} \beta_1 \\ \beta_2 \\ \beta_3 \\ \beta_4 \\ \beta_5 \\ \beta_6 \\ \beta_7 \\ \beta_8 \\ \beta_9 \\ \beta_{10} \\ \beta_{11} \\ \beta_{12} \\ \beta_{13} \end{bmatrix}}_{\beta} + \underbrace{\begin{bmatrix} 1 \\ 0 \\ 0 \end{bmatrix}}_{B_{ij}} \underbrace{\begin{bmatrix} b_{1i} \end{bmatrix}}_{b_i},$$

$$b_i \sim \mathcal{N}\left(0, \psi\right), \quad \epsilon_{ij}|\phi_i \sim \mathcal{N}\left(0, \sigma^2 \left[\mathrm{E}\left(y_{ij}|\phi_i\right)\right]^\theta\right).$$

The correspondence between the fixed effects, β, and the coefficient names used in `fm4Soy.nlme` is: $\beta_1 =$ `Asym.(Intercept)`, $\beta_2 =$ `Asym.Year1989`, $\beta_3 =$ `Asym.Year1990`, $\beta_4 =$ `Asym.Variety`, $\beta_5 =$ `Asym.Year1989Variety`, $\beta_6 =$ `Asym.Year1990Variety`, $\beta_7 =$ `xmid.(Intercept)`, $\beta_8 =$ `xmid.Year1989`, $\beta_9 =$ `xmid.Year1990`, $\beta_{10} =$ `xmid.Variety`, $\beta_{11} =$ `scal.(Intercept)`, $\beta_{12} =$ `scal.-Year1989`, and $\beta_{13} =$ `scal.Year1990`. The design matrices A_{ij} and B_{ij} do not vary with j in this example and, as a result, neither do the coefficients ϕ_{ij}. The use of variance functions for the within-group errors is discussed in §7.4, when we present extensions to the basic NLME model.

Clinical Study of Phenobarbital Kinetics

The final model for the phenobarbital data, represented in §6.4 by the object `fm3Pheno.nlme`, includes only the infant's birth weight w_i as a covariate for the fixed effects. The phenobarbital concentration y_{ij} for individual i measured at time t_{ij}, following intravenous injections of dose D_{id} at times

t_{id}, is expressed as

$$y_{ij} = \sum_{d:t_{id}<t_{ij}} \frac{D_{id}}{\exp(lV_i)} \exp\left[-\exp(lCl_i - lV_i)(t_{ij} - t_{id})\right] + \epsilon_{ij},$$

$$\underbrace{\begin{bmatrix} lCl_i \\ lV_i \end{bmatrix}}_{\phi_{ij}} = \underbrace{\begin{bmatrix} 1 & w_i & 0 & 0 \\ 0 & 0 & 1 & w_i \end{bmatrix}}_{A_{ij}} \underbrace{\begin{bmatrix} \beta_1 \\ \beta_2 \\ \beta_3 \\ \beta_4 \end{bmatrix}}_{\beta} + \underbrace{\begin{bmatrix} 1 & 0 \\ 0 & 1 \end{bmatrix}}_{B_{ij}} \underbrace{\begin{bmatrix} b_{1i} \\ b_{2i} \end{bmatrix}}_{b_i},$$

$$b_i \sim \mathcal{N}\left(0, \begin{bmatrix} \psi_{11} & 0 \\ 0 & \psi_{22} \end{bmatrix}\right), \qquad \epsilon_{ij} \sim \mathcal{N}\left(0, \sigma^2\right).$$

The correspondence between the fixed effects, β, and the coefficient names in the `fm3Pheno.nlme` object is: $\beta_1 = $ `1Cl.(Intercept)`, $\beta_2 = $ `1Cl.Wt`, $\beta_3 = $ `1V.(Intercept)`, and $\beta_4 = $ `1V.Wt`. A diagonal Ψ matrix is used to represent the independence between the random effects.

7.1.2 Multilevel NLME Models

The single-level NLME model (7.1) can be extended to data grouped according to multiple, nested factors by modifying the model for the random effects in (7.2). For example, the multilevel version of the Lindstrom and Bates (1990) model for two levels of nesting is written as a two-stage model in which the first stage expresses the response y_{ijk} for the kth observation on the jth second-level group of the ith first-level group as

$$y_{ijk} = f(\phi_{ijk}, v_{ijk}) + \epsilon_{ijk},$$
$$i = 1, \ldots, M, \quad j = 1, \ldots, M_i, \quad k = 1, \ldots, n_{ij}, \quad (7.5)$$

where M is the number of first-level groups, M_i is the number of second-level groups within the ith first-level group, n_{ij} is the number of observations on the jth second-level group of the ith first-level group, and ϵ_{ijk} is a normally distributed within-group error term. As in the single-level model, f is a general, real-valued, differentiable function of a group-specific parameter vector ϕ_{ijk} and a covariate vector v_{ijk}. It is nonlinear in at least one component of ϕ_{ij}. The second stage of the model expresses ϕ_{ij} as

$$\phi_{ijk} = A_{ijk}\beta + B_{i,jk}b_i + B_{ijk}b_{ij},$$
$$b_i \sim \mathcal{N}(0, \Psi_1), \quad b_{ij} \sim \mathcal{N}(0, \Psi_2). \quad (7.6)$$

As in the single-level model (7.2), β is a p-dimensional vector of fixed effects, with design matrix A_{ijk}, which may incorporate time-varying covariates. The first-level random effects b_i are independently distributed

q_1-dimensional vectors with variance–covariance matrix $\boldsymbol{\Psi}_1$. The second-level random effects $\boldsymbol{b}_{ij}$ are q_2-dimensional independently distributed vectors with variance–covariance matrix $\boldsymbol{\Psi}_2$, assumed to be independent of the first-level random effects. The random effects design matrices $\boldsymbol{B}_{i,jk}$ and $\boldsymbol{B}_{ijk}$ depend on first- and second-level groups and possibly on the values of some covariates at the kth observation. The within-group errors ϵ_{ijk} are independently distributed as $\mathcal{N}(0, \sigma^2)$ and are independent of the random effects. The assumption of independence and homoscedasticity for the within-group errors can be relaxed, as shown in §7.4.

We can express (7.5) and (7.6) in matrix form as

$$\begin{aligned} \boldsymbol{y}_{ij} &= \boldsymbol{f}_{ij}\left(\boldsymbol{\phi}_{ij}, \boldsymbol{v}_{ij}\right) + \boldsymbol{\epsilon}_{ij}, \\ \boldsymbol{\phi}_{ij} &= \boldsymbol{A}_{ij}\boldsymbol{\beta} + \boldsymbol{B}_{i,j}\boldsymbol{b}_i + \boldsymbol{B}_{ij}\boldsymbol{b}_{ij}, \end{aligned} \tag{7.7}$$

for $i = 1, \dots, M$, $j = 1, \dots, M_i$, where

$$\boldsymbol{y}_{ij} = \begin{bmatrix} y_{ij1} \\ \vdots \\ y_{ijn_{ij}} \end{bmatrix}, \quad \boldsymbol{\phi}_{ij} = \begin{bmatrix} \phi_{ij1} \\ \vdots \\ \phi_{ijn_{ij}} \end{bmatrix}, \quad \boldsymbol{\epsilon}_{ij} = \begin{bmatrix} \epsilon_{ij1} \\ \vdots \\ \epsilon_{ijn_{ij}} \end{bmatrix},$$

$$\boldsymbol{f}_{ij}\left(\boldsymbol{\phi}_{ij}, \boldsymbol{v}_{ij}\right) = \begin{bmatrix} f(\phi_{ij1}, v_{ij1}) \\ \vdots \\ f(\phi_{ijn_{ij}}, v_{ijn_{ij}}) \end{bmatrix}, \quad \boldsymbol{v}_{ij} = \begin{bmatrix} v_{ij1} \\ \vdots \\ v_{ijn_{ij}} \end{bmatrix},$$

$$\boldsymbol{A}_{ij} = \begin{bmatrix} A_{ij1} \\ \vdots \\ A_{ijn_{ij}} \end{bmatrix}, \quad \boldsymbol{B}_{i,j} = \begin{bmatrix} B_{i,j1} \\ \vdots \\ B_{i,jn_{ij}} \end{bmatrix}, \quad \boldsymbol{B}_{ij} = \begin{bmatrix} B_{ij1} \\ \vdots \\ B_{ijn_{ij}} \end{bmatrix}.$$

Extensions of the NLME model to more than two levels of nesting are straightforward. For example, with three levels of nesting the second-stage model for the group-specific coefficients is

$$\begin{aligned} \boldsymbol{\phi}_{ijk\ell} &= \boldsymbol{A}_{ijk\ell}\boldsymbol{\beta} + \boldsymbol{B}_{i,jk\ell}\boldsymbol{b}_i + \boldsymbol{B}_{ij,k\ell}\boldsymbol{b}_{ij} + \boldsymbol{B}_{ijk\ell}\boldsymbol{b}_{ijk}, \\ \boldsymbol{b}_i &\sim \mathcal{N}(\boldsymbol{0}, \boldsymbol{\Psi}_1), \quad \boldsymbol{b}_{ij} \sim \mathcal{N}(\boldsymbol{0}, \boldsymbol{\Psi}_2), \quad \boldsymbol{b}_{ijk} \sim \mathcal{N}(\boldsymbol{0}, \boldsymbol{\Psi}_3). \end{aligned}$$

7.1.3 Other NLME Models

The first developments of nonlinear mixed-effects models appear in Sheiner and Beal (1980). Their models and estimation methods are incorporated in the NONMEM program (Beal and Sheiner, 1980), which is widely used in pharmacokinetics. They introduced a model similar to (7.1) and developed a maximum likelihood estimation method based on a first-order Taylor expansion of the model function around $\boldsymbol{0}$, the expected value of the random effects vector $\boldsymbol{b}$.

A nonparametric maximum likelihood method for nonlinear mixed-effects models was proposed by Mallet, Mentre, Steimer and Lokiek (1988). They use a model similar to (7.1), but make no assumptions about the distribution of the random effects. The conditional distribution of the response given the random effects is assumed to be known. The objective of the estimation procedure is to get the probability distribution of the group-specific coefficients (ϕ_{ij}) that maximizes the likelihood of the data. Mallet (1986) showed that the maximum likelihood solution is a discrete distribution with the number of discontinuity points less than or equal to the number of groups in the sample. Inference for this model is based on the maximum likelihood distribution from which summary statistics (e.g., means and variance–covariance matrices) and plots are obtained.

Davidian and Gallant (1992) introduced a smooth, nonparametric maximum likelihood estimation method for nonlinear mixed effects. Their model is similar to (7.1), but with a more general definition for the group-specific coefficients, $\phi_{ij} = g(\beta, b_i, v_{ij})$, where g is a general, possibly nonlinear function. As in Mallet et al. (1988), Davidian and Gallant assume that the conditional distribution of the response vector given the random effects is known (up to the parameters that define it), but the distribution of the random effects is free to vary within a class of smooth densities defined in Gallant and Nychka (1987).

A Bayesian approach using hierarchical models for nonlinear mixed effects is described in Bennett and Wakefield (1993) and Wakefield (1996). The first stage model is similar to (7.1) and the distributions of both the random effects and the errors ϵ_{ij} are assumed known up to population parameters. Prior distributions for the population parameters must be provided then Markov-chain Monte Carlo methods, such as the Gibbs sampler (Geman and Geman, 1984) or the Metropolis algorithm (Hastings, 1970), are used to approximate the posterior density of the random effects.

Vonesh and Carter (1992) developed a mixed-effects model that is nonlinear in the fixed effects, but linear in the random effects. Their model is

$$y_i = f(\beta, v_i) + Z_i(\beta)b_i + \epsilon_i,$$

where, as before, β, b_i, and ϵ_i denote, respectively, the fixed effects, the random effects, and the within-group error term, v_i is a matrix of covariates, and Z_i is a full-rank matrix of known functions of the fixed effects β. It is further assumed that $b_i \sim \mathcal{N}(0, \Psi)$, $\epsilon_i \sim \mathcal{N}(0, \sigma^2 I)$, and the two vectors are independent. In some sense, Vonesh and Carter incorporate in the model the approximations suggested by Sheiner and Beal (1980) and Lindstrom and Bates (1990). Their approach concentrates more on inferences about the fixed effects, and less on the variance–covariance components of the random effects.

7.2   Estimation and Inference in NLME Models

Different methods have been proposed to estimate the parameters in the NLME model described in §7.1.1 and §7.1.2. In this book we restrict ourselves to methods based on the likelihood function. Descriptions and comparisons of other estimation methods proposed for NLME models can be found, for example, in Ramos and Pantula (1995), Davidian and Giltinan (1995), and Vonesh and Chinchilli (1997).

7.2.1   Likelihood Estimation

Because the random effects are unobserved quantities, maximum likelihood estimation in mixed-effects models is based on the marginal density of the responses y, which, for a model with Q levels of nesting, is calculated as

$$p\left(y|\beta,\sigma^2,\Psi_1,\ldots,\Psi_Q\right) = \int p\left(y|b,\beta,\sigma^2\right) p\left(b|\Psi_1,\ldots,\Psi_Q\right) db, \quad (7.8)$$

where $p\left(y|\beta,\sigma^2,\Psi_1,\ldots,\Psi_Q\right)$ is the marginal density of y, $p(y|b,\beta,\sigma^2)$ is the conditional density of y given the random effects b, and the marginal distribution of b is $p\left(b|\Psi_1,\ldots,\Psi_Q\right)$. For the NLME model (7.1), expressing the random effects variance–covariance matrix in terms of the precision factor Δ, so that $\Psi^{-1} = \sigma^{-2}\Delta^T\Delta$ as described in §2.1.1, provides the marginal density of y as

$$p\left(y|\beta,\sigma^2,\Delta\right) =$$

$$\frac{|\Delta|^M}{(2\pi\sigma^2)^{(N+Mq)/2}} \prod_{i=1}^{M} \int \exp\left\{\frac{\|y_i - f_i\left(\beta,b_i\right)\|^2 + \|\Delta b_i\|^2}{-2\sigma^2}\right\} db_i, \quad (7.9)$$

where $f_i\left(\beta,b_i\right) = f_i\left[\phi_i\left(\beta,b_i\right),v_i\right]$.

Because the model function f can be nonlinear in the random effects, the integral in (7.8) generally does not have a closed-form expression. To make the numerical optimization of the likelihood function a tractable problem, different approximations to (7.8) have been proposed. Some of these methods consist of taking a first-order Taylor expansion of the model function f around the expected value of the random effects (Sheiner and Beal, 1980; Vonesh and Carter, 1992), or around the conditional (on Δ) modes of the random effects (Lindstrom and Bates, 1990). Gaussian quadrature rules have also been used (Davidian and Gallant, 1992).

We describe three different methods for approximating the likelihood function in the NLME model. The first, proposed by Lindstrom and Bates (1990), approximates   (7.8) by the likelihood of a linear mixed-effects model. We call this the LME approximation. It is the basis of the estimation algorithm currently implemented in the `nlme` function. The second method uses a Laplacian approximation to the likelihood function, and the last method uses an adaptive Gaussian quadrature rule to improve the Laplacian approximation. The LME, Laplacian, and adaptive Gaussian approximations have increasing degrees of accuracy, at the cost of increasing computational complexity. The three approximations to the NLME likelihood are discussed and compared in Pinheiro and Bates (1995).

Lindstrom and Bates Algorithm

The estimation algorithm described by Lindstrom and Bates (1990) alternates between two steps, a penalized nonlinear least squares (PNLS) step, and a linear mixed effects (LME) step, as described below. We initially consider the alternating algorithm for the single-level NLME model (7.1).

In the PNLS step, the current estimate of $\boldsymbol{\Delta}$ (the precision factor) is held fixed, and the conditional modes of the random effects $\boldsymbol{b}_i$ and the conditional estimates of the fixed effects $\boldsymbol{\beta}$ are obtained by minimizing a penalized nonlinear least squares objective function

$$\sum_{i=1}^{M} \left[\|\boldsymbol{y}_i - \boldsymbol{f}_i\left(\boldsymbol{\beta}, \boldsymbol{b}_i\right)\|^2 + \|\boldsymbol{\Delta b}_i\|^2 \right]. \tag{7.10}$$

The LME step updates the estimate of $\boldsymbol{\Delta}$ based on a first-order Taylor expansion of the model function f around the current estimates of $\boldsymbol{\beta}$ and the conditional modes of the random effects $\boldsymbol{b}_i$, which we will denote by $\widehat{\boldsymbol{\beta}}^{(w)}$ and $\widehat{\boldsymbol{b}}_i^{(w)}$, respectively. Letting

$$\widehat{\boldsymbol{X}}_i^{(w)} = \left.\frac{\partial \boldsymbol{f}_i}{\partial \boldsymbol{\beta}^T}\right|_{\widehat{\boldsymbol{\beta}}^{(w)}, \widehat{\boldsymbol{b}}_i^{(w)}}, \qquad \widehat{\boldsymbol{Z}}_i^{(w)} = \left.\frac{\partial \boldsymbol{f}_i}{\partial \boldsymbol{b}_i^T}\right|_{\widehat{\boldsymbol{\beta}}^{(w)}, \widehat{\boldsymbol{b}}_i^{(w)}},$$
$$\widehat{\boldsymbol{w}}_i^{(w)} = \boldsymbol{y}_i - \boldsymbol{f}_i(\widehat{\boldsymbol{\beta}}^{(w)}, \widehat{\boldsymbol{b}}_i^{(w)}) + \widehat{\boldsymbol{X}}_i^{(w)}\widehat{\boldsymbol{\beta}}^{(w)} + \widehat{\boldsymbol{Z}}_i^{(w)}\widehat{\boldsymbol{b}}_i^{(w)}, \tag{7.11}$$

the approximate log-likelihood function used to estimate $\boldsymbol{\Delta}$ is

$$\ell_{\text{LME}}\left(\boldsymbol{\beta}, \sigma^2, \boldsymbol{\Delta} \mid \boldsymbol{y}\right) = -\frac{N}{2}\log\left(2\pi\sigma^2\right) - \frac{1}{2}\sum_{i=1}^{M}\left\{\log\left|\boldsymbol{\Sigma}_i(\boldsymbol{\Delta})\right|\right.$$
$$\left. + \sigma^{-2}\left[\widehat{\boldsymbol{w}}_i^{(w)} - \widehat{\boldsymbol{X}}_i^{(w)}\boldsymbol{\beta}\right]^T \boldsymbol{\Sigma}_i^{-1}(\boldsymbol{\Delta})\left[\widehat{\boldsymbol{w}}_i^{(w)} - \widehat{\boldsymbol{X}}_i^{(w)}\boldsymbol{\beta}\right]\right\}, \tag{7.12}$$

where $\boldsymbol{\Sigma}_i(\boldsymbol{\Delta}) = \boldsymbol{I} + \widehat{\boldsymbol{Z}}_i^{(w)}\boldsymbol{\Delta}^{-1}\boldsymbol{\Delta}^{-T}\widehat{\boldsymbol{Z}}_i^{(w)T}$. This log-likelihood is identical to that of a linear mixed-effects model in which the response vector is given by $\widehat{\boldsymbol{w}}^{(w)}$ and the fixed- and random-effects design matrices are given by $\widehat{\boldsymbol{X}}^{(w)}$ and $\widehat{\boldsymbol{Z}}^{(w)}$, respectively. Using the results in §2.2, one can express the optimal values of $\boldsymbol{\beta}$ and σ^2 as functions of $\boldsymbol{\Delta}$ and work with the profiled log-likelihood of $\boldsymbol{\Delta}$, greatly simplifying the optimization problem.

Lindstrom and Bates (1990) also proposed a restricted maximum likelihood estimation method for $\mathbf{\Delta}$, which consists of replacing the log-likelihood in the LME step of the alternating algorithm by the log-restricted-likelihood

$$\ell_{\text{LME}}^R \left(\sigma^2, \mathbf{\Delta} \mid \boldsymbol{y} \right) =$$

$$\ell_{\text{LME}} \left(\widehat{\boldsymbol{\beta}} \left(\mathbf{\Delta} \right), \sigma^2, \mathbf{\Delta} \mid \boldsymbol{y} \right) - \frac{1}{2} \sum_{i=1}^M \log \left| \sigma^{-2} \widehat{\boldsymbol{X}}_i^{(w)^T} \boldsymbol{\Sigma}_i^{-1}(\mathbf{\Delta}) \widehat{\boldsymbol{X}}_i^{(w)} \right|. \quad (7.13)$$

Note that, because $\widehat{\boldsymbol{X}}_i^{(w)}$ depends on both $\widehat{\boldsymbol{\beta}}^{(w)}$ and $\widehat{\boldsymbol{b}}_i^{(w)}$, changes in either the fixed effects model or the random effects model imply changes in the penalty factor for the log-restricted-likelihood (7.13). Therefore, log-restricted-likelihoods from NLME models with different fixed or random effects models are not comparable.

The algorithm alternates between the PNLS and LME steps until a convergence criterion is met. Such alternating algorithms tend to be more efficient when the estimates of the variance–covariance components ($\mathbf{\Delta}$ and σ^2) are not highly correlated with the estimates of the fixed effects ($\boldsymbol{\beta}$). Pinheiro (1994) has shown that, in the linear mixed-effects model, the maximum likelihood estimates of $\mathbf{\Delta}$ and σ^2 are asymptotically independent of the maximum likelihood estimates of $\boldsymbol{\beta}$. These results have not yet been extended to the nonlinear mixed-effects model (7.1).

Lindstrom and Bates (1990) only use the LME step to update the estimate of $\mathbf{\Delta}$. However, the LME step also produces updated estimates of $\boldsymbol{\beta}$ and the conditional modes of $\boldsymbol{b}_i$. Thus, one can iterate LME steps by re-evaluating (7.11) and (7.12) (or (7.13) for the log-restricted-likelihood) at the updated estimates of $\boldsymbol{\beta}$ and $\boldsymbol{b}_i$, as described in Wolfinger (1993). Because the updated estimates correspond to the values obtained in the first iteration of a Gauss–Newton algorithm for the PNLS step, iterated LME steps will converge to the same values as the alternating algorithm, though possibly not as quickly.

Wolfinger (1993) also shows that, when a flat prior is assumed for $\boldsymbol{\beta}$, the LME approximation to the log-restricted-likelihood (7.13) is equivalent to a Laplacian approximation (Tierney and Kadane, 1986) to the integral (7.9).

The alternating algorithm and the LME approximation to the NLME log-likelihood can be extended to multilevel models. For example, for an NLME model with two levels of nesting, the PNLS step consists of minimizing the penalized nonlinear least-squares function

$$\sum_{i=1}^M \left\{ \sum_{j=1}^{M_i} \left[\| \boldsymbol{y}_{ij} - \boldsymbol{f}_{ij}(\boldsymbol{\beta}, \boldsymbol{b}_i, \boldsymbol{b}_{ij}) \|^2 + \| \mathbf{\Delta}_2 \boldsymbol{b}_{ij} \|^2 \right] + \| \mathbf{\Delta}_1 \boldsymbol{b}_i \|^2 \right\} \quad (7.14)$$

to obtain estimates for the fixed effects $\boldsymbol{\beta}$ and the conditional (on $\mathbf{\Delta}_1$ and $\mathbf{\Delta}_2$) modes of the random effects $\boldsymbol{b}_i$ and $\boldsymbol{b}_{ij}$.

Letting

$$\widehat{X}_{ij}^{(w)} = \frac{\partial f_{ij}}{\partial \beta^T}\Big|_{\widehat{\beta}^{(w)}, \widehat{b}_i^{(w)}, \widehat{b}_{ij}^{(w)}}, \qquad \widehat{Z}_{i,j}^{(w)} = \frac{\partial f_{ij}}{\partial b_i^T}\Big|_{\widehat{\beta}^{(w)}, \widehat{b}_i^{(w)}, \widehat{b}_{ij}^{(w)}},$$

$$\widehat{Z}_{ij}^{(w)} = \frac{\partial f_{ij}}{\partial b_{ij}^T}\Big|_{\widehat{\beta}^{(w)}, \widehat{b}_i^{(w)}, \widehat{b}_{ij}^{(w)}},$$

$$\widehat{w}_{ij}^{(w)} = y_{ij} - f_{ij}(\widehat{\beta}^{(w)}, \widehat{b}_i^{(w)}, \widehat{b}_{ij}^{(w)}) + \widehat{X}_{ij}^{(w)} \widehat{\beta}^{(w)} + \widehat{Z}_{i,j}^{(w)} \widehat{b}_i^{(w)} + \widehat{Z}_{ij}^{(w)} \widehat{b}_{ij}^{(w)},$$

$$\widehat{X}_i^{(w)} = \begin{bmatrix} \widehat{X}_{i1}^{(w)} \\ \vdots \\ \widehat{X}_{iM_i}^{(w)} \end{bmatrix}, \qquad \widehat{Z}_i^{(w)} = \begin{bmatrix} \widehat{Z}_{i,1}^{(w)} \\ \vdots \\ \widehat{Z}_{i,M_i}^{(w)} \end{bmatrix}, \qquad \widehat{w}_i^{(w)} = \begin{bmatrix} \widehat{w}_{i1}^{(w)} \\ \vdots \\ \widehat{w}_{iM_i}^{(w)} \end{bmatrix},$$

$$(7.15)$$

the approximate log-likelihood function used to estimate Δ_1 and Δ_2 in the two-level NLME models is

$$\ell_{\text{LME}}\left(\beta, \sigma^2, \Delta_1, \Delta_2 \mid y\right) = -\frac{N}{2}\log\left(2\pi\sigma^2\right) - \frac{1}{2}\sum_{i=1}^{M}\left\{\log|\Sigma_i(\Delta_1, \Delta_2)|\right.$$

$$\left. +\sigma^{-2}\left[\widehat{w}_i^{(w)} - \widehat{X}_i^{(w)}\beta\right]^T \Sigma_i^{-1}(\Delta_1, \Delta_2)\left[\widehat{w}_i^{(w)} - \widehat{X}_i^{(w)}\beta\right]\right\},$$

where $\Sigma_i(\Delta_1, \Delta_2) = I + \widehat{Z}_i^{(w)}\Delta_1^{-1}\Delta_1^{-T}\widehat{Z}_i^{(w)^T} + \bigoplus_{j=1}^{M_i}\widehat{Z}_{ij}^{(w)}\Delta_2^{-1}\Delta_2^{-T}\widehat{Z}_{ij}^{(w)^T}$ and $\oplus$ denotes the direct sum operator. The corresponding log-restricted-likelihood is

$$\ell_{\text{LME}}^R\left(\sigma^2, \Delta_1, \Delta_2 \mid y\right) = \ell_{\text{LME}}\left(\widehat{\beta}(\Delta_1, \Delta_2), \sigma^2, \Delta_1, \Delta_2 \mid y\right)$$

$$-\frac{1}{2}\sum_{i=1}^{M}\log\left|\sigma^{-2}\widehat{X}_i^{(w)^T}\Sigma_i^{-1}(\Delta_1, \Delta_2)\widehat{X}_i^{(w)}\right|.$$

This formulation can be extended to multilevel NLME models with an arbitrary number of levels.

The alternating algorithm is the only estimation algorithm used in the nlme function. It is implemented for maximum likelihood and restricted maximum likelihood estimation in single and multilevel NLME models.

Laplacian Approximation

Laplacian approximations are used frequently in Bayesian inference to estimate marginal posterior densities and predictive distributions (Tierney and Kadane, 1986; Leonard, Hsu and Tsui, 1989). These techniques can also be used for approximating the likelihood function in NLME models.

We consider initially the single-level NLME models. The integral that we want to estimate to obtain the marginal distribution of $\boldsymbol{y}_i$ in (7.9) can be written as

$$p(\boldsymbol{y}_i \mid \boldsymbol{\beta}, \sigma^2, \boldsymbol{\Delta}) = \int (2\pi\sigma^2)^{-(n_i+q)/2} |\boldsymbol{\Delta}| \exp\left[-g(\boldsymbol{\beta}, \boldsymbol{\Delta}, \boldsymbol{y}_i, \boldsymbol{b}_i)/2\sigma^2\right] d\boldsymbol{b}_i,$$

where $g(\boldsymbol{\beta}, \boldsymbol{\Delta}, \boldsymbol{y}_i, \boldsymbol{b}_i) = \|\boldsymbol{y}_i - \boldsymbol{f}_i(\boldsymbol{\beta}, \boldsymbol{b}_i)\|^2 + \|\boldsymbol{\Delta}\boldsymbol{b}_i\|^2$, the sum of which is the objective function for the PNLS step of the alternating algorithm defined in (7.10). Let

$$\widehat{\boldsymbol{b}}_i = \widehat{\boldsymbol{b}}_i(\boldsymbol{\beta}, \boldsymbol{\Delta}, \boldsymbol{y}_i) = \arg\min_{\boldsymbol{b}_i} g(\boldsymbol{\beta}, \boldsymbol{\Delta}, \boldsymbol{y}_i, \boldsymbol{b}_i),$$

$$g'(\boldsymbol{\beta}, \boldsymbol{\Delta}, \boldsymbol{y}_i, \boldsymbol{b}_i) = \frac{\partial g(\boldsymbol{\beta}, \boldsymbol{\Delta}, \boldsymbol{y}_i, \boldsymbol{b}_i)}{\partial \boldsymbol{b}_i}, \tag{7.16}$$

$$g''(\boldsymbol{\beta}, \boldsymbol{\Delta}, \boldsymbol{y}_i, \boldsymbol{b}_i) = \frac{\partial^2 g(\boldsymbol{\beta}, \boldsymbol{\Delta}, \boldsymbol{y}_i, \boldsymbol{b}_i)}{\partial \boldsymbol{b}_i \partial \boldsymbol{b}_i^T},$$

and consider a second-order Taylor expansion of g around $\widehat{\boldsymbol{b}}_i$

$$g(\boldsymbol{\beta}, \boldsymbol{\Delta}, \boldsymbol{y}_i, \boldsymbol{b}_i) \simeq g\left(\boldsymbol{\beta}, \boldsymbol{\Delta}, \boldsymbol{y}_i, \widehat{\boldsymbol{b}}_i\right) + \frac{1}{2}\left[\boldsymbol{b}_i - \widehat{\boldsymbol{b}}_i\right]^T g''\left(\boldsymbol{\beta}, \boldsymbol{\Delta}, \boldsymbol{y}_i, \widehat{\boldsymbol{b}}_i\right)\left[\boldsymbol{b}_i - \widehat{\boldsymbol{b}}_i\right].$$
$$\tag{7.17}$$

(The linear term in the expansion vanishes because $g'(\boldsymbol{\beta}, \boldsymbol{\Delta}, \boldsymbol{y}_i, \widehat{\boldsymbol{b}}_i) = \boldsymbol{0}$.) The Laplacian approximation is defined as

$$p\left(\boldsymbol{y} \mid \boldsymbol{\beta}, \sigma^2, \boldsymbol{\Delta}\right)$$

$$\simeq (2\pi\sigma^2)^{-\frac{N}{2}} |\boldsymbol{\Delta}|^M \exp\left[-\frac{1}{2\sigma^2}\sum_{i=1}^M g(\boldsymbol{\beta}, \boldsymbol{\Delta}, \boldsymbol{y}_i, \widehat{\boldsymbol{b}}_i)\right]$$

$$\times \prod_{i=1}^M \int (2\pi\sigma^2)^{\frac{q}{2}} \exp\left\{-\frac{1}{2\sigma^2}\left[\boldsymbol{b}_i - \widehat{\boldsymbol{b}}_i\right]^T g''(\boldsymbol{\beta}, \boldsymbol{\Delta}, \boldsymbol{y}_i, \widehat{\boldsymbol{b}}_i)\left[\boldsymbol{b}_i - \widehat{\boldsymbol{b}}_i\right]\right\} d\boldsymbol{b}_i$$

$$= (2\pi\sigma^2)^{-\frac{N}{2}} |\boldsymbol{\Delta}|^M \exp\left[-\frac{1}{2\sigma^2}\sum_{i=1}^M g(\boldsymbol{\beta}, \boldsymbol{\Delta}, \boldsymbol{y}_i, \widehat{\boldsymbol{b}}_i)\right] \prod_{i=1}^M \left|g''(\boldsymbol{\beta}, \boldsymbol{\Delta}, \boldsymbol{y}_i, \widehat{\boldsymbol{b}}_i)\right|^{-\frac{1}{2}}.$$

The Hessian

$$g''\left(\boldsymbol{\beta}, \boldsymbol{\Delta}, \boldsymbol{y}_i, \widehat{\boldsymbol{b}}_i\right) = -\left.\frac{\partial^2 \boldsymbol{f}_i(\boldsymbol{\beta}, \boldsymbol{b}_i)}{\partial \boldsymbol{b}_i \partial \boldsymbol{b}_i^T}\right|_{\widehat{\boldsymbol{b}}_i}\left[\boldsymbol{y}_i - \boldsymbol{f}_i(\boldsymbol{\beta}, \widehat{\boldsymbol{b}}_i)\right]$$

$$+ \left.\frac{\partial \boldsymbol{f}_i(\boldsymbol{\beta}, \boldsymbol{b}_i)}{\partial \boldsymbol{b}_i}\right|_{\widehat{\boldsymbol{b}}_i} \left.\frac{\partial \boldsymbol{f}_i(\boldsymbol{\beta}, \boldsymbol{b}_i)}{\partial \boldsymbol{b}_i^T}\right|_{\widehat{\boldsymbol{b}}_i} + \boldsymbol{\Delta}^T \boldsymbol{\Delta}$$

involves second derivatives of f but, at $\widehat{\boldsymbol{b}}_i$, the contribution of

$$\left.\frac{\partial^2 \boldsymbol{f}_i(\boldsymbol{\beta}, \boldsymbol{b}_i)}{\partial \boldsymbol{b}_i \partial \boldsymbol{b}_i^T}\right|_{\widehat{\boldsymbol{b}}_i}\left[\boldsymbol{y}_i - \boldsymbol{f}_i(\boldsymbol{\beta}, \widehat{\boldsymbol{b}}_i)\right]$$

is usually negligible compared to that of $\partial \boldsymbol{f}_i(\boldsymbol{\beta}, \boldsymbol{b}_i)/\partial \boldsymbol{b}_i|_{\widehat{\boldsymbol{b}}_i} \, \partial \boldsymbol{f}_i(\boldsymbol{\beta}, \boldsymbol{b}_i)/\partial \boldsymbol{b}_i^T|_{\widehat{\boldsymbol{b}}_i}$ (Bates and Watts, 1980). Therefore, we use the approximation

$$
\begin{aligned}
g''\left(\boldsymbol{\beta}, \boldsymbol{\Delta}, \boldsymbol{y}_i, \widehat{\boldsymbol{b}}_i\right) &\simeq \boldsymbol{G}\left(\boldsymbol{\beta}, \boldsymbol{\Delta}, \boldsymbol{y}_i\right) \\
&= \left.\frac{\partial \boldsymbol{f}_i(\boldsymbol{\beta}, \boldsymbol{b}_i)}{\partial \boldsymbol{b}_i}\right|_{\widehat{\boldsymbol{b}}_i} \left.\frac{\partial \boldsymbol{f}_i(\boldsymbol{\beta}, \boldsymbol{b}_i)}{\partial \boldsymbol{b}_i^T}\right|_{\widehat{\boldsymbol{b}}_i} + \boldsymbol{\Delta}^T \boldsymbol{\Delta}.
\end{aligned}
\tag{7.18}
$$

This approximation is similar to that used in the Gauss–Newton algorithm for nonlinear least squares and has the advantage of requiring only the first-order partial derivatives of f with respect to the random effects. These are usually available as a by-product of the estimation of $\widehat{\boldsymbol{b}}_i$, which is a penalized least squares problem, for which standard and reliable code is available.

The modified Laplacian approximation to the log-likelihood of the single-level NLME model (7.1) is then given by

$$
\begin{aligned}
\ell_{\mathrm{LA}}\left(\boldsymbol{\beta}, \sigma^2, \boldsymbol{\Delta}, \mid \boldsymbol{y}\right) = &-\frac{N}{2}\log\left(2\pi\sigma^2\right) + M\log|\boldsymbol{\Delta}| \\
&- \frac{1}{2}\left\{\sum_{i=1}^{M}\log|\boldsymbol{G}\left(\boldsymbol{\beta}, \boldsymbol{\Delta}, \boldsymbol{y}_i\right)| + \sigma^{-2}\sum_{i=1}^{M}g\left(\boldsymbol{\beta}, \boldsymbol{\Delta}, \boldsymbol{y}_i, \widehat{\boldsymbol{b}}_i\right)\right\}.
\end{aligned}
\tag{7.19}
$$

Because $\widehat{\boldsymbol{b}}_i$ does not depend on σ^2, for given $\boldsymbol{\beta}$ and $\boldsymbol{\Delta}$ the maximum likelihood estimate of σ^2 (based upon ℓ_{LA}) is

$$
\widehat{\sigma}^2 = \widehat{\sigma}^2\left(\boldsymbol{\beta}, \boldsymbol{\Delta}, \boldsymbol{y}\right) = \sum_{i=1}^{M}g\left(\boldsymbol{\beta}, \boldsymbol{\Delta}, \boldsymbol{y}_i, \widehat{\boldsymbol{b}}_i\right)/N.
$$

We can profile ℓ_{LA} on σ^2 to reduce the dimension of the optimization problem, obtaining

$$
\begin{aligned}
\ell_{\mathrm{LAp}}\left(\boldsymbol{\beta}, \boldsymbol{\Delta}\right) = \\
-\frac{N}{2}\left[1 + \log\left(2\pi\right) + \log\left(\widehat{\sigma}^2\right)\right] + M\log|\boldsymbol{\Delta}| - \frac{1}{2}\sum_{i=1}^{M}\log|\boldsymbol{G}\left(\boldsymbol{\beta}, \boldsymbol{\Delta}, \boldsymbol{y}_i\right)|.
\end{aligned}
$$

If the model function f is linear in the random effects, then the modified Laplacian approximation is exact because the second-order Taylor expansion in (7.17) is exact when $\boldsymbol{f}_i\left(\boldsymbol{\beta}, \boldsymbol{b}_i\right) = \boldsymbol{f}_i\left(\boldsymbol{\beta}\right) + \boldsymbol{Z}_i\left(\boldsymbol{\beta}\right)\boldsymbol{b}_i$.

There does not yet seem to be a straightforward generalization of the concept of restricted maximum likelihood to NLME models. The difficulty is that restricted maximum likelihood depends heavily upon the linearity of the fixed effects in the model function, which does not occur in nonlinear models. Lindstrom and Bates (1990) circumvented that problem by using an approximation to the model function $\boldsymbol{f}$ in which the fixed effects, $\boldsymbol{\beta}$, occur linearly. This cannot be done for the Laplacian approximation, unless we consider yet another Taylor expansion of the model function, what would lead us back to something very similar to Lindstrom and Bates's approach.

The Laplacian approximation (7.19) can be extended to multilevel NLME models. For example, in a two-level NLME model, let

$$
\boldsymbol{b}_i^{\mathrm{aug}} = \begin{bmatrix} \boldsymbol{b}_i \\ \boldsymbol{b}_{i1} \\ \vdots \\ \boldsymbol{b}_{iM_i} \end{bmatrix}, \qquad i = 1, \dots, M
$$

denote the *augmented* random effects vector for the ith first-level group, containing the first-level random effects $\boldsymbol{b}_i$ and all the second-level random effects $\boldsymbol{b}_{ij}$ pertaining to first-level i. The two-level NLME likelihood can then be expressed as

$$
p(\boldsymbol{y}_i \mid \boldsymbol{\beta}, \sigma^2, \boldsymbol{\Delta}_1, \boldsymbol{\Delta}_2) = \int \left(2\pi\sigma^2\right)^{-(n_i+q_1+M_i q_2)/2} |\boldsymbol{\Delta}_1| |\boldsymbol{\Delta}_2|^{M_i}
$$
$$
\times \exp\left[-g(\boldsymbol{\beta}, \boldsymbol{\Delta}_1, \boldsymbol{\Delta}_2, \boldsymbol{y}_i, \boldsymbol{b}_i^{\mathrm{aug}})/2\sigma^2\right] d\boldsymbol{b}_i^{\mathrm{aug}},
$$

where $g(\boldsymbol{\beta}, \boldsymbol{\Delta}_1, \boldsymbol{\Delta}_2, \boldsymbol{y}_i, \boldsymbol{b}_i^{\mathrm{aug}})$ is the objective function for the PNLS step in the alternating algorithm for two-level NLME models, defined in (7.14). We proceed as in the single-level case and define

$$
\widehat{\boldsymbol{b}}_i^{\mathrm{aug}} = \widehat{\boldsymbol{b}}_i^{\mathrm{aug}}(\boldsymbol{\beta}, \boldsymbol{\Delta}_1, \boldsymbol{\Delta}_2, \boldsymbol{y}_i) = \arg\min_{\boldsymbol{b}_i^{\mathrm{aug}}} g(\boldsymbol{\beta}, \boldsymbol{\Delta}_1, \boldsymbol{\Delta}_2, \boldsymbol{y}_i, \boldsymbol{b}_i^{\mathrm{aug}}),
$$
$$
g'(\boldsymbol{\beta}, \boldsymbol{\Delta}_1, \boldsymbol{\Delta}_2, \boldsymbol{y}_i, \boldsymbol{b}_i^{\mathrm{aug}}) = \frac{\partial g(\boldsymbol{\beta}, \boldsymbol{\Delta}_1, \boldsymbol{\Delta}_2, \boldsymbol{y}_i, \boldsymbol{b}_i^{\mathrm{aug}})}{\partial \boldsymbol{b}_i^{\mathrm{aug}}},
$$
$$
g''(\boldsymbol{\beta}, \boldsymbol{\Delta}_1, \boldsymbol{\Delta}_2, \boldsymbol{y}_i, \boldsymbol{b}_i^{\mathrm{aug}}) = \frac{\partial^2 g(\boldsymbol{\beta}, \boldsymbol{\Delta}_1, \boldsymbol{\Delta}_2, \boldsymbol{y}_i, \boldsymbol{b}_i^{\mathrm{aug}})}{\partial \boldsymbol{b}_i^{\mathrm{aug}} \partial (\boldsymbol{b}_i^{\mathrm{aug}})^T},
$$

to obtain the second-order approximation

$$
g(\boldsymbol{\beta}, \boldsymbol{\Delta}_1, \boldsymbol{\Delta}_2, \boldsymbol{y}_i, \boldsymbol{b}_i^{\mathrm{aug}}) \simeq g\left(\boldsymbol{\beta}, \boldsymbol{\Delta}_1, \boldsymbol{\Delta}_2, \boldsymbol{y}_i, \widehat{\boldsymbol{b}}_i^{\mathrm{aug}}\right)
$$
$$
+ \frac{1}{2}\left[\boldsymbol{b}_i^{\mathrm{aug}} - \widehat{\boldsymbol{b}}_i^{\mathrm{aug}}\right]^T g''\left(\boldsymbol{\beta}, \boldsymbol{\Delta}_1, \boldsymbol{\Delta}_2, \boldsymbol{y}_i, \widehat{\boldsymbol{b}}_i^{\mathrm{aug}}\right)\left[\boldsymbol{b}_i^{\mathrm{aug}} - \widehat{\boldsymbol{b}}_i^{\mathrm{aug}}\right].
$$

We note that $\partial^2 g(\boldsymbol{\beta}, \boldsymbol{\Delta}_1, \boldsymbol{\Delta}_2, \boldsymbol{y}_i, \boldsymbol{b}_i^{\mathrm{aug}})/\partial \boldsymbol{b}_{ij} \partial \boldsymbol{b}_{ik}^T = \boldsymbol{0}$ for any $j \neq k$ and use the same reasoning as in (7.18), to approximate the matrix $g''(\boldsymbol{\beta}, \boldsymbol{\Delta}, \boldsymbol{y}_i, \widehat{\boldsymbol{b}}_i)$

by

$$g''\left(\boldsymbol{\beta},\boldsymbol{\Delta}_1,\boldsymbol{\Delta}_2\boldsymbol{y}_i,\widehat{\boldsymbol{b}}_i\right) \simeq \boldsymbol{G}\left(\boldsymbol{\beta},\boldsymbol{\Delta}_1,\boldsymbol{\Delta}_2,\boldsymbol{y}_i\right) = \begin{bmatrix} \boldsymbol{G}_1 & \boldsymbol{G}_2 \\ \boldsymbol{G}_2^T & \boldsymbol{G}_3 \end{bmatrix}, \quad \text{where}$$

$$\boldsymbol{G}_1 = \left.\frac{\partial \boldsymbol{f}_i(\boldsymbol{\beta},\boldsymbol{b}_i^{\text{aug}})}{\partial \boldsymbol{b}_i}\right|_{\widehat{\boldsymbol{b}}_i} \left.\frac{\partial \boldsymbol{f}_i(\boldsymbol{\beta},\boldsymbol{b}_i^{\text{aug}})}{\partial \boldsymbol{b}_i^T}\right|_{\widehat{\boldsymbol{b}}_i^{\text{aug}}} + \boldsymbol{\Delta}_1^T \boldsymbol{\Delta}_1,$$

$$\boldsymbol{G}_2 = \left.\frac{\partial \boldsymbol{f}_i(\boldsymbol{\beta},\boldsymbol{b}_i^{\text{aug}})}{\partial \boldsymbol{b}_i}\right|_{\widehat{\boldsymbol{b}}_i} \left.\left[\frac{\partial \boldsymbol{f}_{i1}(\boldsymbol{\beta},\boldsymbol{b}_i,\boldsymbol{b}_{i1})}{\partial \boldsymbol{b}_{i1}^T} \quad \cdots \quad \frac{\partial \boldsymbol{f}_{iM_i}(\boldsymbol{\beta},\boldsymbol{b}_i,\boldsymbol{b}_{iM_i})}{\partial \boldsymbol{b}_{iM_i}^T}\right]\right|_{\widehat{\boldsymbol{b}}_i^{\text{aug}}},$$

$$\boldsymbol{G}_3 = \bigoplus_{j=1}^{M_i} \left\{ \left.\frac{\partial \boldsymbol{f}_{ij}(\boldsymbol{\beta},\boldsymbol{b}_i,\boldsymbol{b}_{ij})}{\partial \boldsymbol{b}_{ij}}\right|_{\widehat{\boldsymbol{b}}_i,\widehat{\boldsymbol{b}}_{ij}} \left.\frac{\partial \boldsymbol{f}_{ij}(\boldsymbol{\beta},\boldsymbol{b}_i,\boldsymbol{b}_{ij})}{\partial \boldsymbol{b}_{ij}^T}\right|_{\widehat{\boldsymbol{b}}_i,\widehat{\boldsymbol{b}}_{ij}} + \boldsymbol{\Delta}_2^T \boldsymbol{\Delta}_2 \right\},$$

$$\boldsymbol{f}_i(\boldsymbol{\beta},\boldsymbol{b}_i^{\text{aug}}) = \begin{bmatrix} \boldsymbol{f}_{i1}(\boldsymbol{\beta},\boldsymbol{b}_i,\boldsymbol{b}_{i1}) \\ \vdots \\ \boldsymbol{f}_{iM_i}(\boldsymbol{\beta},\boldsymbol{b}_i,\boldsymbol{b}_{iM_i}) \end{bmatrix}.$$

The modified, profiled Laplacian approximation to the log-likelihood of the two-level NLME model is then given by

$$\ell_{\text{LAp}}(\boldsymbol{\beta},\boldsymbol{\Delta}_1,\boldsymbol{\Delta}_2) = -\frac{N}{2}\left[1 + \log(2\pi) + \log(\widehat{\sigma}^2)\right] + M\log|\boldsymbol{\Delta}_1|$$
$$+ \sum_{i=1}^{M} M_i \log|\boldsymbol{\Delta}_2| - \frac{1}{2}\sum_{i=1}^{M}\log|\boldsymbol{G}(\boldsymbol{\beta},\boldsymbol{\Delta}_1,\boldsymbol{\Delta}_2,\boldsymbol{y}_i)|,$$

where $\widehat{\sigma}^2 = \sum_{i=1}^{M} g\left(\boldsymbol{\beta},\boldsymbol{\Delta}_1,\boldsymbol{\Delta}_2,\boldsymbol{y}_i,\widehat{\boldsymbol{b}}_i^{\text{aug}}\right)/N$. This formulation can be extended to multilevel NLME models with an arbitrary number of levels.

The Laplacian approximation generally gives more accurate estimates than the alternating algorithm, as it uses an expansion around the estimated random effects only, while the LME approximation in the alternating algorithm uses an expansion around the estimated fixed and random effects. Because it requires solving a different penalized nonlinear least-squares problem for each group in the data and its objective function cannot be profiled on the fixed effects, the Laplacian approximation is more computationally intensive than the alternating algorithm. The algorithm for calculating the Laplacian approximation can be easily parallelized, because the individual PNLS problems are independently optimized.

Adaptive Gaussian Approximation

Gaussian quadrature rules are used to approximate integrals of functions with respect to a given kernel by a weighted average of the integrand evaluated at predetermined abscissas. The weights and abscissas used in Gaussian quadrature rules for the most common kernels can be obtained from the tables of Abramowitz and Stegun (1964) or by using an algorithm proposed by Golub (1973) (see also Golub and Welsch (1969)). Gaussian

quadrature rules for multiple integrals are known to be numerically complex
(Davis and Rabinowitz, 1984), but using the structure of the integrand in
the nonlinear mixed-effects model we can transform the problem into suc-
cessive applications of simple one-dimensional Gaussian quadrature rules.
We consider initially the single-level NLME model.

A natural candidate for the kernel function for the quadrature rule in the
single-level NLME model is the marginal distribution of the random effects,
that is, the $\mathcal{N}(\mathbf{0}, \mathbf{\Psi})$ density. The Gaussian quadrature rule in this case can
be viewed as a deterministic version of a Monte Carlo integration algorithm
in which random samples of the random effects, $\mathbf{b}_i$, are generated from
the $\mathcal{N}(\mathbf{0}, \mathbf{\Psi})$ distribution. The samples and the weights in the Gaussian
quadrature rule are fixed beforehand, while in Monte Carlo integration
they are left to random choice. Because importance sampling tends to be
much more efficient than simple Monte Carlo integration (Geweke, 1989),
we consider an importance sample version of the Gaussian quadrature rule,
which we denote by *adaptive Gaussian* quadrature.

The critical step for the success of importance sampling is the choice
of an importance distribution that approximates the integrand. For the
single-level NLME model the integrand is proportional to

$$\exp\left[-g\left(\boldsymbol{\beta}, \boldsymbol{\Delta}, \boldsymbol{y}_i, \boldsymbol{b}_i\right)/2\sigma^2\right],$$

which is approximated by a $\mathcal{N}(\widehat{\boldsymbol{b}}_i, \sigma^2 \boldsymbol{G}^{-1}(\boldsymbol{\beta}, \boldsymbol{\Delta}, \boldsymbol{y}_i))$ density, with $\widehat{\boldsymbol{b}}_i$ defined
as in (7.16) and $\boldsymbol{G}(\boldsymbol{\beta}, \boldsymbol{\Delta}, \boldsymbol{y}_i)$ defined as in (7.18). This is the importance
distribution used in the adaptive Gaussian quadrature, so that the grid
of abscissas in the $\boldsymbol{b}_i$ scale is centered around the conditional modes $\widehat{\boldsymbol{b}}_i$
and $\boldsymbol{G}(\boldsymbol{\beta}, \boldsymbol{D}, \boldsymbol{y}_i)$ is used for scaling. Letting $z_j, w_j\ j = 1, \ldots, N_{GQ}$ denote,
respectively, the abscissas and the weights for the (one-dimensional) Gaus-
sian quadrature rule with N_{GQ} points based on the $\mathcal{N}(0, 1)$ kernel, the
adaptive Gaussian quadrature rule is given by

$$
\begin{aligned}
\int \exp\left[-g\left(\boldsymbol{\beta}, \boldsymbol{\Delta}, \boldsymbol{y}_i, \boldsymbol{b}_i\right)/2\sigma^2\right]d\boldsymbol{b}_i &= \int \sigma^q \left|\boldsymbol{G}\left(\boldsymbol{\beta}, \boldsymbol{\Delta}, \boldsymbol{y}_i\right)\right|^{-1/2} \\
&\times \exp\left\{-g\left[\boldsymbol{\beta}, \boldsymbol{\Delta}, \boldsymbol{y}_i, \widehat{\boldsymbol{b}}_i + \sigma \boldsymbol{G}^{-1/2}\left(\boldsymbol{\beta}, \boldsymbol{\Delta}, \boldsymbol{y}_i\right)\boldsymbol{z}\right]/2\sigma^2 + \|\boldsymbol{z}\|^2/2\right\} \\
&\times \exp\left(-\|\boldsymbol{z}\|^2/2\right)d\boldsymbol{z} \\
&\simeq \sigma^q \left|\boldsymbol{G}\left(\boldsymbol{\beta}, \boldsymbol{\Delta}, \boldsymbol{y}_i\right)\right|^{-1/2} \\
&\times \sum_{j_1=1}^{N_{GQ}}\cdots\sum_{j_q=1}^{N_{GQ}} \exp\left(-g\left\{\boldsymbol{\beta}, \boldsymbol{\Delta}, \boldsymbol{y}_i, \widehat{\boldsymbol{b}}_i + \sigma \boldsymbol{G}^{-1/2}\left(\boldsymbol{\beta}, \boldsymbol{\Delta}, \boldsymbol{y}_i\right)\boldsymbol{z}_j\right\}/2\sigma^2 \right. \\
&\left. + \|\boldsymbol{z}_j\|^2/2\right) \prod_{k=1}^{q} w_{j_k},
\end{aligned}
$$

where $[G(\boldsymbol{\beta}, \boldsymbol{\Delta}, \boldsymbol{y}_i)]^{1/2}$ denotes a square root of $G(\boldsymbol{\beta}, \boldsymbol{\Delta}, \boldsymbol{y}_i)$ and $\boldsymbol{z}_j = (z_{j_1}, \dots, z_{j_q})^T$.

The adaptive Gaussian approximation to the log-likelihood function in the single-level NLME model is then

$$\ell_{AGQ}(\boldsymbol{\beta}, \sigma^2, \boldsymbol{\Delta} \mid \boldsymbol{y}) = -\tfrac{N}{2}\log(2\pi\sigma^2) + M\log|\boldsymbol{\Delta}| - \tfrac{1}{2}\sum_{i=1}^{M}\log|G(\boldsymbol{\beta}, \boldsymbol{\Delta}, \boldsymbol{y}_i)|$$

$$+ \sum_{i=1}^{M}\log\left(\sum_{j}^{N_{GQ}}\exp\left\{-g\left[\boldsymbol{\beta}, \boldsymbol{\Delta}, \boldsymbol{y}_i, \widehat{\boldsymbol{b}}_i + \sigma G^{-1/2}(\boldsymbol{\beta}, \boldsymbol{\Delta}, \boldsymbol{y}_i)\boldsymbol{z}_j\right]/2\sigma^2\right.$$

$$+ \left. \|\boldsymbol{z}_j\|^2/2\right\}\prod_{k=1}^{q}w_{j_k}\right).$$

The one point (i.e., $N_{GQ} = 1$) adaptive Gaussian quadrature approximation is simply the modified Laplacian approximation (7.19), because in this case $z_1 = 0$ and $w_1 = 1$. The adaptive Gaussian quadrature also gives the exact log-likelihood when the model function f is linear in the random effects $\boldsymbol{b}$.

The adaptive Gaussian approximation can be made arbitrarily accurate by increasing the number of abscissas, N_{GQ}. However, because N_{GQ}^q grid points are used to calculate the adaptive Gaussian quadrature for each group, it quickly becomes prohibitively computationally intensive as the number of abscissas increases. In practice $N_{GQ} \leq 7$ generally suffices and $N_{GQ} = 1$ often provides a reasonable approximation (Pinheiro and Bates, 1995).

The adaptive Gaussian approximation can be generalized to multilevel NLME models, using the same steps as in the multilevel Laplacian approximation. For example, the adaptive Gaussian approximation to the log-likelihood of a two-level NLME model is

$$\ell_{AGQ}(\boldsymbol{\beta}, \sigma^2, \boldsymbol{\Delta}_1, \boldsymbol{\Delta}_2 \mid \boldsymbol{y}) = -\frac{N}{2}\log(2\pi\sigma^2) + M\log|\boldsymbol{\Delta}_1| + \sum_{i=1}^{M}M_i\log|\boldsymbol{\Delta}_2|$$

$$-\frac{1}{2}\sum_{i=1}^{M}\log|G(\boldsymbol{\beta}, \boldsymbol{\Delta}_1, \boldsymbol{\Delta}_2, \boldsymbol{y}_i)| + \sum_{i=1}^{M}\log\left[\sum_{j}^{N_{GQ}}\left(\exp\left\{-g\left[\boldsymbol{\beta}, \boldsymbol{\Delta}_1, \boldsymbol{\Delta}_2, \boldsymbol{y}_i,\right.\right.\right.\right.$$

$$\widehat{\boldsymbol{b}}_i^{\text{aug}} + \sigma G^{-1/2}(\boldsymbol{\beta}, \boldsymbol{\Delta}_1, \boldsymbol{\Delta}_2, \boldsymbol{y}_i)\boldsymbol{z}_j\right]/2\sigma^2 + \|\boldsymbol{z}_j\|^2/2\Big\}\prod_{k=1}^{q_1+M_iq_2}w_{j_k}\Bigg)\Bigg],$$

where $\boldsymbol{z}_j = (z_{j_1}, \dots, z_{j_{q_1+Mq_2}})^T$. In this case, the number of grid points for the ith first-level group is $N_{GQ}^{q_1+M_iq_2}$, so that the computational complexity of the calculations increases exponentially with the number of second-level groups. This formulation can be extended to multilevel NLME models with arbitrary number of levels.

7.2.2   Inference and Predictions

Because the alternating algorithm of Lindstrom and Bates (1990) is the only estimation algorithm implemented in the `nlme` function, we restrict ourselves for the rest of this section to inference and predictions using this estimation algorithm. The results described here can be easily extended to other likelihood estimation approaches, such as the Laplacian and adaptive Gaussian approximations described in §7.2.1.

Inference

Inference on the parameters of an NLME model estimated via the alternating algorithm is based on the LME approximation to the log-likelihood function, defined in §7.2.1. Using this approximation at the estimated values of the parameters and the asymptotic results for LME models described in §2.3 we obtain standard errors, confidence intervals, and hypothesis tests for the parameters in the NLME model. We use the single-level NLME model of §7.1.1 to illustrate the inference results derived from the LME approximation to the log-likelihood. Extensions to multilevel NLME models are straightforward.

Under the LME approximation, the distribution of the (restricted) maximum likelihood estimators $\widehat{\boldsymbol{\beta}}$ of the fixed effects is

$$\widehat{\boldsymbol{\beta}} \overset{\cdot}{\sim} \mathcal{N}\left(\boldsymbol{\beta}, \sigma^2 \left[\sum_{i=1}^{M} \widehat{\boldsymbol{X}}_i^T \boldsymbol{\Sigma}_i^{-1} \widehat{\boldsymbol{X}}_i\right]^{-1}\right), \tag{7.20}$$

where $\boldsymbol{\Sigma}_i = \boldsymbol{I} + \widehat{\boldsymbol{Z}}_i \boldsymbol{\Delta}^{-1} \boldsymbol{\Delta}^{-T} \widehat{\boldsymbol{Z}}_i^T$, with $\widehat{\boldsymbol{X}}_i$ and $\widehat{\boldsymbol{Z}}_i$ are defined as in (7.11). The standard errors included in the `summary` method for nlme objects are obtained from the approximate variance–covariance matrix in (7.20). The t and F tests reported in the `summary` method and in the `anova` method with a single argument are also based on (7.20). The degrees-of-freedom for t and F tests are calculated as described in §2.4.2.

Now let $\boldsymbol{\theta}$ denote an unconstrained set of parameters that determine the precision factor $\boldsymbol{\Delta}$. The LME approximation is also used to provide an approximate distribution for the (RE)ML estimators $(\widehat{\boldsymbol{\theta}}, \log \widehat{\sigma})^T$. We use $\log \sigma$ in place of σ^2 to give an unrestricted parameterization for which the normal approximation tends to be more accurate.

$$\begin{bmatrix} \widehat{\boldsymbol{\theta}} \\ \log \widehat{\sigma} \end{bmatrix} \,\dot{\sim}\, \mathcal{N}\left(\begin{bmatrix} \boldsymbol{\theta} \\ \log \sigma \end{bmatrix}, \, \mathcal{I}^{-1}(\boldsymbol{\theta}, \sigma) \right),$$

$$\mathcal{I}(\boldsymbol{\theta}, \sigma) = - \begin{bmatrix} \partial^2 \ell_{\mathrm{LMEp}}/\partial\boldsymbol{\theta}\partial\boldsymbol{\theta}^T & \partial^2 \ell_{\mathrm{LMEp}}/\partial \log \sigma \partial\boldsymbol{\theta}^T \\ \partial^2 \ell_{\mathrm{LMEp}}/\partial\boldsymbol{\theta}\partial \log \sigma & \partial^2 \ell_{\mathrm{LMEp}}/\partial^2 \log \sigma \end{bmatrix}, \tag{7.21}$$

where $\ell_{\mathrm{LMEp}} = \ell_{\mathrm{LMEp}}(\boldsymbol{\Delta}, \sigma)$ denotes the LME approximation to the log-likelihood, profiled on the fixed effects, and $\mathcal{I}$ denotes the empirical information matrix. The same approximate distribution is valid for the REML estimators with ℓ_{LMEp} replaced by the log-restricted-likelihood ℓ^R_{LME} defined in (7.13).

In practice, $\boldsymbol{\Delta}$ and σ^2 are replaced by their respective (RE)ML estimates in the expressions for the approximate variance–covariance matrices in (7.20) and (7.21). The approximate distributions for the (RE)ML estimators are used to produce the confidence intervals reported in the `intervals` method for nlme objects.

The LME approximate log-likelihood is also used to compare nested NLME models through likelihood ratio tests, as described in §2.4.1. In the case of REML estimation, only models with identical fixed and random-effects structures can be compared, because the $\widehat{\boldsymbol{X}}_i$ matrices depend on both $\widehat{\boldsymbol{\beta}}$ and the $\widehat{\boldsymbol{b}}_i$. The same recommendations stated in §2.4.1, on the use of likelihood ratio tests for comparing LME models, remain valid for likelihood ratio tests (based on the LME approximate log-likelihood) for comparing NLME models. Hypotheses on the fixed effects should be tested using t and F tests, because likelihood ratio tests tend to be "anticonservative." Likelihood ratio tests for variance–covariance parameters tend to be somewhat conservative, but are generally used to compare NLME models with nested random effects structures. Information criterion statistics, for example, AIC and BIC, based on the LME approximate log-likelihood are also used to compare NLME models.

The inference results for NLME models based on the LME approximation to the log-likelihood are "approximately asymptotic," making them less reliable than the asymptotic inference results for LME models described in §2.3.

Predictions

As with LME models, fitted values and predictions for NLME models may be obtained at different levels of nesting, or at the population level. Population-level predictions estimate the expected response when the random effects are equal to their mean value, $\boldsymbol{0}$. For example, letting $\boldsymbol{x}_h$ represent a vector of fixed-effects covariates and $\boldsymbol{v}_h$ a vector of other model covariates, the population prediction for the corresponding response y_h estimates $f(\boldsymbol{x}_h^T\boldsymbol{\beta}, \boldsymbol{v}_h)$.

Predicted values at the kth level of nesting estimate the conditional expectation of the response given the random effects at levels $\leq k$ and with the random effects at higher levels of nesting set to zero. For example, letting $z_h(i)$ denote a vector of covariates corresponding to random effects associated with the ith group at the first level of nesting, the *level-1* predictions estimate $f(x_h^T\beta + z_h(i)^T b_i, v_h)$. Similarly, letting $z_h(i, j)$ denote a covariate vector associated with the jth level-2 group within the ith level-1 group, the *level-2* predicted values estimate $f(x_h^T\beta + z_h(i)^T b_i + z_h(i, j)^T b_{ij}, v_h)$. This extends naturally to an arbitrary level of nesting.

The (RE)ML estimates of the fixed effects and the conditional modes of the random effects, which are estimated Best Linear Unbiased Estimates (*BLUPs*) of the random effects in the LME approximate log-likelihood, are used to obtain predicted values for the response. For example, the population, level-1, and level-2 predictions for y_h are

$$\widehat{y}_h = f(x_h^T\widehat{\beta}, v_h),$$
$$\widehat{y}_h(i) = f(x_h^T\widehat{\beta} + z_h(i)^T\widehat{b}_i, v_h),$$
$$\widehat{y}_h(i, j) = f(x_h^T\widehat{\beta} + z_h(i)^T\widehat{b}_i + z_h(i, j)^T\widehat{b}_{i,j}, v_h).$$

7.3   Computational Methods

In this section we describe efficient computational methods for estimating the parameters in an NLME model using the alternating algorithm presented in §7.2.1. The LME step of the alternating algorithm consists of optimizing a linear mixed-effects log-likelihood, or log-restricted-likelihood, for which efficient computational algorithms are discussed in §2.2.8. Therefore, we concentrate here on computational methods for the PNLS step of the alternating algorithm, focusing initially on the single-level NLME model.

The objective function optimized in the PNLS step of the alternating algorithm for a single-level NLME model is the penalized sum of squares

$$\sum_{i=1}^{M} \left[\|y_i - f_i(\beta, b_i)\|^2 + \|\Delta b_i\|^2 \right]. \tag{7.22}$$

By adding "pseudo" observations to the data, we can convert (7.22) into a simple nonlinear sum of squares. Define the augmented response and model function vectors

$$\tilde{y}_i = \begin{bmatrix} y_i \\ 0 \end{bmatrix}, \qquad \tilde{f}_i(\beta, b_i) = \begin{bmatrix} f_i(\beta, b_i) \\ \Delta b_i \end{bmatrix}.$$

The penalized sum of squares (7.22) can then be re-expressed as

$$\sum_{i=1}^{M} \|\tilde{\boldsymbol{y}}_i - \tilde{\boldsymbol{f}}_i(\boldsymbol{\beta}, \boldsymbol{b}_i)\|^2. \qquad (7.23)$$

It follows from (7.23) that, conditional on $\boldsymbol{\Delta}$, the estimation of $\boldsymbol{\beta}$ and $\boldsymbol{b}_i$ in the PNLS step can be regarded as a standard nonlinear least-squares problem. A common iterative estimation method for standard nonlinear least-squares problems is the Gauss–Newton method (Bates and Watts, 1988, §2.2) wherein a nonlinear model $\boldsymbol{f}(\boldsymbol{\alpha})$ is replaced by a first-order Taylor series approximation about current estimates $\widehat{\boldsymbol{\alpha}}^{(w)}$ as

$$\boldsymbol{f}(\boldsymbol{\alpha}) \approx \boldsymbol{f}(\widehat{\boldsymbol{\alpha}}^{(w)}) + \left.\frac{\partial \boldsymbol{f}}{\partial \boldsymbol{\alpha}^T}\right|_{\widehat{\boldsymbol{\alpha}}^{(w)}} \left(\boldsymbol{\alpha} - \widehat{\boldsymbol{\alpha}}^{(w)}\right).$$

The parameter increment $\widehat{\boldsymbol{\delta}}^{(w+1)} = \widehat{\boldsymbol{\alpha}}^{(w+1)} - \widehat{\boldsymbol{\alpha}}^{(w)}$ for the wth iteration is calculated as the solution of the least-squares problem

$$\left\|\left[\boldsymbol{y} - \boldsymbol{f}(\widehat{\boldsymbol{\alpha}}^{(w)})\right] - \left.\frac{\partial \boldsymbol{f}}{\partial \boldsymbol{\alpha}^T}\right|_{\widehat{\boldsymbol{\alpha}}^{(w)}} \left(\boldsymbol{\alpha} - \widehat{\boldsymbol{\alpha}}^{(w)}\right)\right\|^2.$$

Step-halving is used at each Gauss–Newton iteration to ensure that the updated parameter estimates result in a decrease of the objective function. That is, the new estimate is set to $\widehat{\boldsymbol{\alpha}}^{(w)} + \widehat{\boldsymbol{\delta}}^{(w+1)}$ and the corresponding value of the objective function is calculated. If it is less than the value at $\widehat{\boldsymbol{\alpha}}^{(w)}$, the value is retained and the algorithm proceed to the next step, or declares convergence. Otherwise, the new estimate is set to $\widehat{\boldsymbol{\alpha}}^{(w)} + \widehat{\boldsymbol{\delta}}^{(w+1)}/2$ and the procedure is repeated, with the increment being halved until a decrease in the objective function is observed or some predetermined minimum step size is reached.

The Gauss–Newton algorithm is used to estimate $\boldsymbol{\beta}$ and the $\boldsymbol{b}_i$ in the PNLS step of the alternating algorithm. Because of the "loosely coupled" structure of the PNLS problem (Soo and Bates, 1992), efficient nonlinear least-squares algorithms can be employed.

The derivative matrices for the Gauss–Newton optimization of (7.23) are, for $i = 1, \ldots, M$,

$$\left.\frac{\partial \tilde{\boldsymbol{f}}_i(\boldsymbol{\beta}, \boldsymbol{b}_i|\boldsymbol{\Delta})}{\partial \boldsymbol{\beta}^T}\right|_{\widehat{\boldsymbol{\beta}}^{(w)}, \widehat{\boldsymbol{b}}_i^{(w)}} = \tilde{\boldsymbol{X}}_i^{(w)} = \begin{bmatrix} \widehat{\boldsymbol{X}}_i^{(w)} \\ \boldsymbol{0} \end{bmatrix},$$

$$\left.\frac{\partial \tilde{\boldsymbol{f}}_i(\boldsymbol{\beta}, \boldsymbol{b}_i|\boldsymbol{\Delta})}{\partial \boldsymbol{b}_i^T}\right|_{\widehat{\boldsymbol{\beta}}^{(w)}, \widehat{\boldsymbol{b}}_i^{(w)}} = \tilde{\boldsymbol{Z}}_i^{(w)} = \begin{bmatrix} \widehat{\boldsymbol{Z}}_i^{(w)} \\ \boldsymbol{\Delta} \end{bmatrix},$$

with $\widehat{\boldsymbol{X}}_i^{(w)}$ and $\widehat{\boldsymbol{Z}}_i^{(w)}$ defined as in (7.11). The least-squares problem to be solved at each Gauss–Newton iteration is

$$\sum_{i=1}^{M} \left\| \left[\tilde{\boldsymbol{y}}_i - \tilde{\boldsymbol{f}}_i \left(\widehat{\boldsymbol{\beta}}^{(w)}, \widehat{\boldsymbol{b}}_i^{(w)} \right) \right] - \tilde{\boldsymbol{X}}_i^{(w)} \left(\boldsymbol{\beta} - \widehat{\boldsymbol{\beta}}^{(w)} \right) - \tilde{\boldsymbol{Z}}_i^{(w)} \left(\boldsymbol{b}_i - \widehat{\boldsymbol{b}}_i^{(w)} \right) \right\|^2$$

or, equivalently,

$$\sum_{i=1}^{M} \left\| \tilde{\boldsymbol{w}}_i^{(w)} - \tilde{\boldsymbol{X}}_i^{(w)} \boldsymbol{\beta} - \tilde{\boldsymbol{Z}}_i^{(w)} \boldsymbol{b}_i \right\|^2, \quad \text{where} \quad \tilde{\boldsymbol{w}}_i^{(w)} = \begin{bmatrix} \widehat{\boldsymbol{w}}_i^{(w)} \\ \boldsymbol{0} \end{bmatrix}, \qquad (7.24)$$

with $\widehat{\boldsymbol{w}}^{(w)}$ defined as in (7.11).

We use the same matrix decomposition methods as in §2.2.3 to obtain an efficient algorithm for solving (7.24). Consider first the orthogonal-triangular decomposition

$$\begin{bmatrix} \widehat{\boldsymbol{Z}}_i^{(w)} & \widehat{\boldsymbol{X}}_i^{(w)} & \widehat{\boldsymbol{w}}_i^{(w)} \\ \boldsymbol{\Delta} & \boldsymbol{0} & \boldsymbol{0} \end{bmatrix} = \boldsymbol{Q}_{1(i)} \begin{bmatrix} \boldsymbol{R}_{11(i)} & \boldsymbol{R}_{10(i)} & \boldsymbol{c}_{1(i)} \\ \boldsymbol{0} & \boldsymbol{R}_{00(i)} & \boldsymbol{c}_{0(i)} \end{bmatrix}, \qquad (7.25)$$

where the reduction to triangular form is halted after the first q columns. The numbering scheme used for the components in (7.25) is the same introduced for the LME model in §2.2.3. Because $\boldsymbol{\Delta}$ is assumed to be of full rank, so is the upper-triangular matrix $\boldsymbol{R}_{11(i)}$ in (7.25).

Forming another orthogonal-triangular decomposition

$$\begin{bmatrix} \boldsymbol{R}_{00(1)} & \boldsymbol{c}_{0(1)} \\ \vdots & \vdots \\ \boldsymbol{R}_{00(M)} & \boldsymbol{c}_{0(M)} \end{bmatrix} = \boldsymbol{Q}_0 \begin{bmatrix} \boldsymbol{R}_{00} & \boldsymbol{c}_0 \\ \boldsymbol{0} & \boldsymbol{c}_{-1} \end{bmatrix} \qquad (7.26)$$

and noticing that the $\boldsymbol{Q}_{1(i)}$ and $\boldsymbol{Q}_0$ are orthogonal matrices, we can rewrite (7.24) as

$$\sum_{i=1}^{M} \left\| \boldsymbol{c}_{1(i)} - \boldsymbol{R}_{11(i)} \boldsymbol{b}_i - \boldsymbol{R}_{10(i)} \boldsymbol{\beta} \right\|^2 + \left\| \boldsymbol{c}_0 - \boldsymbol{R}_{00} \boldsymbol{\beta} \right\|^2 + \left\| \boldsymbol{c}_{-1} \right\|^2. \qquad (7.27)$$

We assume that the $\boldsymbol{R}_{00}$ is of full-rank, in which case (7.27) is uniquely minimized by the least-squares estimates

$$\begin{aligned} \widehat{\boldsymbol{\beta}} &= \boldsymbol{R}_{00}^{-1} \boldsymbol{c}_0, \\ \widehat{\boldsymbol{b}}_i &= \boldsymbol{R}_{11(i)}^{-1} \left(\boldsymbol{c}_{1(i)} - \boldsymbol{R}_{10(i)} \widehat{\boldsymbol{\beta}} \right), \quad i = 1, \dots, M. \end{aligned} \qquad (7.28)$$

The Gauss–Newton increments are then obtained as the difference between the least-squares estimates (7.28) and the current estimates, $\widehat{\boldsymbol{\beta}}^{(w)}$ and $\widehat{\boldsymbol{b}}_i^{(w)}$.

Step-halving is used to ensure that the new estimates result in a decrease of the objective function (7.23).

The efficient Gauss–Newton algorithm described above can be extended to multilevel PNLS optimization problems. For example, in the two-level NLME model, the PNLS step consists in optimizing

$$
\sum_{i=1}^{M}\left\{\sum_{j=1}^{M_i}\left[\|\boldsymbol{y}_{ij}-\boldsymbol{f}_{ij}(\boldsymbol{\beta},\boldsymbol{b}_i,\boldsymbol{b}_{ij})\|^2+\|\boldsymbol{\Delta}_2\boldsymbol{b}_{ij}\|^2\right]+\|\boldsymbol{\Delta}_1\boldsymbol{b}_i\|^2\right\} \qquad (7.29)
$$

over $\boldsymbol{\beta}$, $\boldsymbol{b}_i$, and $\boldsymbol{b}_{ij}$.

The Gauss–Newton iteration is implemented by solving the least-squares problem

$$
\sum_{i=1}^{M}\left\{\sum_{j=1}^{M_i}[\|\widehat{\boldsymbol{w}}_{ij}^{(w)}-\widehat{\boldsymbol{X}}_{ij}^{(w)}\boldsymbol{\beta}-\widehat{\boldsymbol{Z}}_{i,j}^{(w)}\boldsymbol{b}_i-\widehat{\boldsymbol{Z}}_{ij}^{(w)}\boldsymbol{b}_{ij}\|^2+\|\boldsymbol{\Delta}_2\boldsymbol{b}_{ij}\|^2]+\|\boldsymbol{\Delta}_1\boldsymbol{b}_i\|^2\right\},
$$
$$(7.30)$$

with $\widehat{\boldsymbol{w}}_{ij}^{(w)}$, $\widehat{\boldsymbol{X}}_{ij}^{(w)}$, $\widehat{\boldsymbol{Z}}_{i,j}^{(w)}$, and $\widehat{\boldsymbol{Z}}_{ij}^{(w)}$ defined as in (7.15). To solve it efficiently, we first consider the orthogonal-triangular decomposition

$$
\begin{bmatrix}\widehat{\boldsymbol{Z}}_{ij}^{(w)} & \widehat{\boldsymbol{Z}}_{i,j}^{(w)} & \widehat{\boldsymbol{X}}_{ij}^{(w)} & \widehat{\boldsymbol{w}}_{ij}^{(w)} \\ \boldsymbol{\Delta}_2 & \boldsymbol{0} & \boldsymbol{0} & \boldsymbol{0}\end{bmatrix}=\boldsymbol{Q}_{2(ij)}\begin{bmatrix}\boldsymbol{R}_{22(ij)} & \boldsymbol{R}_{21(ij)} & \boldsymbol{R}_{20(ij)} & \boldsymbol{c}_{2(ij)} \\ \boldsymbol{0} & \boldsymbol{R}_{11(i)} & \boldsymbol{R}_{10(i)} & \boldsymbol{c}_{1(i)}\end{bmatrix},
$$

where the reduction to triangular form is halted after the first q_2 columns. Because $\boldsymbol{\Delta}_2$ is assumed of full rank, so is $\boldsymbol{R}_{22(ij)}$. We then form a second orthogonal-triangular decomposition for each first-level group

$$
\begin{bmatrix}\boldsymbol{R}_{11(i1)} & \boldsymbol{R}_{10(i1)} & \boldsymbol{c}_{1(i1)} \\ \vdots & \vdots & \vdots \\ \boldsymbol{R}_{11(iM_i)} & \boldsymbol{R}_{10(iM_i)} & \boldsymbol{c}_{1(iM_i)} \\ \boldsymbol{\Delta}_1 & \boldsymbol{0} & \boldsymbol{0}\end{bmatrix}=\boldsymbol{Q}_{1(i)}\begin{bmatrix}\boldsymbol{R}_{11(i)} & \boldsymbol{R}_{10(i)} & \boldsymbol{c}_{1(i)} \\ \boldsymbol{0} & \boldsymbol{R}_{00(i)} & \boldsymbol{c}_{0(i)}\end{bmatrix},
$$

where the reduction to triangular form is stopped after the first q_1 columns. The $\boldsymbol{\Delta}_1$ matrix is assumed of full rank and, as a result, so is $\boldsymbol{R}_{11(i)}$. A final orthogonal-decomposition, identical to (7.26), is then formed.

Because the matrices $\boldsymbol{Q}_{2(ij)}$, $\boldsymbol{Q}_{1(i)}$, and $\boldsymbol{Q}_0$ are orthogonal, (7.30) can be re-expressed as

$$
\sum_{i=1}^{M}\left\{\sum_{j=1}^{M_i}\left[\|\boldsymbol{c}_{2(ij)}-\boldsymbol{R}_{20(ij)}\boldsymbol{\beta}-\boldsymbol{R}_{21(ij)}\boldsymbol{b}_i-\boldsymbol{R}_{22(ij)}\boldsymbol{b}_{ij}\|^2\right]\right.
$$
$$
\left.+\|\boldsymbol{c}_{1(i)}-\boldsymbol{R}_{10(i)}\boldsymbol{\beta}-\boldsymbol{R}_{11(i)}\boldsymbol{b}_i\|^2\right\}+\|\boldsymbol{c}_0-\boldsymbol{R}_{00}\boldsymbol{\beta}\|^2,
$$

which is uniquely minimized by the least-squares estimates

$$\widehat{\beta} = R_{00}^{-1} c_0,$$

$$\widehat{b}_i = R_{11(i)}^{-1} \left(c_{1(i)} - R_{10(i)} \widehat{\beta} \right), \quad i = 1, \dots, M,$$

$$\widehat{b}_{ij} = R_{22(ij)}^{-1} (c_{2(ij)} - R_{21(ij)} \widehat{b}_i - R_{20(i)} \widehat{\beta}), \ i = 1, \dots, M, \ j = 1, \dots, M_i.$$

$$(7.31)$$

The Gauss–Newton increments are then obtained as the difference between the least-squares estimates (7.31) and the current estimates $\widehat{\beta}^{(w)}$, $\widehat{b}_i^{(w)}$, and $\widehat{b}_{ij}^{(w)}$, with step-halving used to ensure that the objective function (7.29) decreases. This extends naturally to multilevel NLME models with arbitrary number of levels.

The efficiency of the Gauss–Newton algorithm described in this section derives from the fact that, at each iteration, the orthogonal-triangular decompositions are performed separately for each group and then once for the fixed effects. This allows efficient memory allocation for storing intermediate results and reduces the numerical complexity of the decompositions. Also, the matrix inversions required to calculate the Gauss–Newton increments involve only upper-triangular matrices of small dimension, which are easy to invert. (In fact, although (7.28) and (7.31) are written in terms of matrix inverses, such as R_{00}^{-1}, the actual calculation performed is the solution of the triangular system of equations $R_{00}\widehat{\beta} = c_0$, which is even simpler.)

7.4   Extending the Basic NLME Model

The nonlinear mixed-effects model formulation used so far in this chapter allows considerable flexibility in the specification of the random-effects structure, but restricts the within-group errors to be independent and to have constant variance. This *basic* NLME model provides an adequate model for a broad range of applications, but there are many cases in which the within-group errors are *heteroscedastic* (i.e., have unequal variances) or are *correlated* or are both heteroscedastic and correlated.

This section extends the basic NLME model to allow heteroscedastic, correlated within-group errors. We show how the estimation methods of §7.2 and the computational methods of §7.3 can be adapted to the extended model formulation.

7.4.1   General Formulation of the Extended NLME Model

As described in §7.1.1, the basic single-level NLME model (7.3) assumes that the within-group errors ϵ_i are independent $\mathcal{N}\left(0, \sigma^2 I\right)$ random vectors.

The extended single-level NLME model relaxes this assumption by allowing heteroscedastic and correlated within-group errors, being expressed for $i = 1, \ldots, M$ as

$$\boldsymbol{y}_i = \boldsymbol{f}_i \left(\boldsymbol{\phi}_i, \boldsymbol{v}_i \right) + \boldsymbol{\epsilon}_i, \quad \boldsymbol{\phi}_i = \boldsymbol{A}_i \boldsymbol{\beta} + \boldsymbol{B}_i \boldsymbol{b}_i,$$
$$\boldsymbol{b}_i \sim \mathcal{N}(\boldsymbol{0}, \boldsymbol{\Psi}), \quad \boldsymbol{\epsilon}_i \sim \mathcal{N} \left(\boldsymbol{0}, \sigma^2 \boldsymbol{\Lambda}_i \right). \tag{7.32}$$

The $\boldsymbol{\Lambda}_i$ are positive-definite matrices parametrized by a fixed, generally small, set of parameters $\boldsymbol{\lambda}$. As in the basic NLME model, the within-group errors $\boldsymbol{\epsilon}_i$ are assumed to be independent for different i and to be independent of the random effects $\boldsymbol{b}_i$. The σ^2 is factored out of the $\boldsymbol{\Lambda}_i$ for computational reasons (it can then be eliminated from the profiled likelihood function).

Similarly, the extended two-level NLME model generalizes the basic two-level NLME model (7.7) described in §7.1.2 by letting

$$\boldsymbol{\epsilon}_{ij} \sim \mathcal{N} \left(\boldsymbol{0}, \sigma^2 \boldsymbol{\Lambda}_{ij} \right), \quad i = 1, \ldots, M \quad j = 1, \ldots, M_i,$$

where the $\boldsymbol{\Lambda}_{ij}$ are positive-definite matrices parametrized by a fixed $\boldsymbol{\lambda}$ vector. This readily generalizes to a multilevel model with Q levels of random effects. For simplicity, we concentrate for the remainder of this section on the extended single-level NLME model (7.32), but the results we obtain are easily generalizable to multilevel models with an arbitrary number of levels of random effects.

As described in §5.1.3, the variance–covariance structure of the within-group errors can be decomposed into two independent components: a *variance* structure and a *correlation* structure. Variance function models to represent the variance structure component of the within-group errors are described and illustrated in §5.2. Correlation models to represent the correlation structure of the within-group errors are presented and have their use illustrated in §5.3. The use of the `nlme` function to fit the extended NLME model is described in §8.3.

7.4.2 Estimation and Computational Methods

Because $\boldsymbol{\Lambda}_i$ is positive-definite, it admits an invertible square-root $\boldsymbol{\Lambda}_i^{1/2}$ (Thisted, 1988, §3), with inverse $\boldsymbol{\Lambda}_i^{-1/2}$, such that

$$\boldsymbol{\Lambda}_i = \boldsymbol{\Lambda}_i^{T/2} \boldsymbol{\Lambda}_i^{1/2} \quad \text{and} \quad \boldsymbol{\Lambda}_i^{-1} = \boldsymbol{\Lambda}_i^{-1/2} \boldsymbol{\Lambda}_i^{-T/2}.$$

Letting

$$\boldsymbol{y}_i^* = \boldsymbol{\Lambda}_i^{-T/2} \boldsymbol{y}_i,$$
$$\boldsymbol{f}_i^* \left(\boldsymbol{\phi}_i, \boldsymbol{v}_i \right) = \boldsymbol{\Lambda}_i^{-T/2} \boldsymbol{f}_i \left(\boldsymbol{\phi}_i, \boldsymbol{v}_i \right), \tag{7.33}$$
$$\boldsymbol{\epsilon}_i^* = \boldsymbol{\Lambda}_i^{-T/2} \boldsymbol{\epsilon}_i,$$

and noting that $\epsilon_i^* \sim \mathcal{N}\left[\mathbf{\Lambda}_i^{-T/2}\mathbf{0}, \sigma^2\mathbf{\Lambda}_i^{-T/2}\mathbf{\Lambda}_i\mathbf{\Lambda}_i^{-1/2}\right] = \mathcal{N}\left(\mathbf{0}, \sigma^2 \boldsymbol{I}\right)$, we can rewrite (7.32) as

$$y_i^* = f_i^*(\boldsymbol{\phi}_i, \boldsymbol{v}_i) + \boldsymbol{\epsilon}_i^*,$$
$$\boldsymbol{\phi}_i = \boldsymbol{A}_i\boldsymbol{\beta} + \boldsymbol{B}_i\boldsymbol{b}_i,$$
$$\boldsymbol{b}_i \sim \mathcal{N}(\mathbf{0}, \boldsymbol{\Psi}), \quad \boldsymbol{\epsilon}_i^* \sim \mathcal{N}\left(\mathbf{0}, \sigma^2 \boldsymbol{I}\right).$$

That is, y_i^* is described by a basic NLME model.

Because the differential of the linear transformation $y_i^* = \mathbf{\Lambda}_i^{-T/2}\boldsymbol{y}_i$ is simply $d\boldsymbol{y}_i^* = |\mathbf{\Lambda}_i|^{-1/2}\,d\boldsymbol{y}_i$, the log-likelihood function corresponding to the extended NLME model (7.32) is expressed as

$$\begin{aligned}
\ell\left(\boldsymbol{\beta}, \sigma^2, \boldsymbol{\Delta}, \boldsymbol{\lambda}|\boldsymbol{y}\right) &= \sum_{i=1}^{M} \log p\left(\boldsymbol{y}_i|\boldsymbol{\beta}, \sigma^2, \boldsymbol{\Delta}, \boldsymbol{\lambda}\right) \\
&= \sum_{i=1}^{M} \log p\left(\boldsymbol{y}_i^*|\boldsymbol{\beta}, \sigma^2, \boldsymbol{\Delta}, \boldsymbol{\lambda}\right) - \frac{1}{2}\sum_{i=1}^{M} \log |\mathbf{\Lambda}_i| \\
&= \ell\left(\boldsymbol{\beta}, \sigma^2, \boldsymbol{\Delta}, \boldsymbol{\lambda}|\boldsymbol{y}^*\right) - \frac{1}{2}\sum_{i=1}^{M} \log |\mathbf{\Lambda}_i|.
\end{aligned}$$

The log-likelihood function $\ell\left(\boldsymbol{\beta}, \sigma^2, \boldsymbol{\Delta}, \boldsymbol{\lambda}|\boldsymbol{y}^*\right)$ corresponds to a basic NLME model with model function f_i^* and, therefore, the approximations presented in §7.2.1 can be applied to it. The inference results described in §7.2.2 also remain valid.

Alternating Algorithm

The PNLS step of the alternating algorithm for the extended NLME model consists of minimizing, over $\boldsymbol{\beta}$ and $\boldsymbol{b}_i$, $i = 1, \ldots, M$, the penalized nonlinear least-squares function

$$\sum_{i=1}^{M}\left[\|\boldsymbol{y}_i^* - \boldsymbol{f}_i^*(\boldsymbol{\beta}, \boldsymbol{b}_i)\|^2 + \|\boldsymbol{\Delta}\boldsymbol{b}_i\|^2\right] =$$
$$\sum_{i=1}^{M}\left\{\left\|\mathbf{\Lambda}_i^{-T/2}\left[\boldsymbol{y}_i - \boldsymbol{f}_i(\boldsymbol{\beta}, \boldsymbol{b}_i)\right]\right\|^2 + \|\boldsymbol{\Delta}\boldsymbol{b}_i\|^2\right\}.$$

The derivative matrices and working vector used in the Gauss–Newton algorithm for the PNLS step and also in the LME step are defined as

$$\widehat{\boldsymbol{X}}_i^{*(w)} = \left.\frac{\partial \boldsymbol{f}_i^*}{\partial \boldsymbol{\beta}^T}\right|_{\widehat{\boldsymbol{\beta}}^{(w)}, \widehat{\boldsymbol{b}}_i^{(w)}} = \mathbf{\Lambda}_i^{-T/2}\widehat{\boldsymbol{X}}_i^{(w)},$$

$$\widehat{\boldsymbol{Z}}_i^{*(w)} = \left.\frac{\partial \boldsymbol{f}_i^*}{\partial \boldsymbol{b}_i^T}\right|_{\widehat{\boldsymbol{\beta}}^{(w)}, \widehat{\boldsymbol{b}}_i^{(w)}} = \mathbf{\Lambda}_i^{-T/2}\widehat{\boldsymbol{Z}}_i^{(w)},$$

$$\widehat{\boldsymbol{w}}_i^{*(w)} = \boldsymbol{y}_i^* - \boldsymbol{f}_i^*(\widehat{\boldsymbol{\beta}}^{(w)}, \widehat{\boldsymbol{b}}_i^{(w)}) + \widehat{\boldsymbol{X}}_i^{*(w)}\widehat{\boldsymbol{\beta}}^{(w)} + \widehat{\boldsymbol{Z}}_i^{*(w)}\widehat{\boldsymbol{b}}_i^{(w)} = \mathbf{\Lambda}_i^{-T/2}\widehat{\boldsymbol{w}}_i^{(w)},$$

with $\widehat{\boldsymbol{X}}_i^{(w)}$, $\widehat{\boldsymbol{Z}}_i^{(w)}$, and $\widehat{\boldsymbol{w}}_i^{(w)}$ defined as in (7.11).

The Gauss–Newton algorithm for the PNLS step is identical to the algorithm described in §7.3, with $\widehat{\boldsymbol{X}}_i^{(w)}$, $\widehat{\boldsymbol{Z}}_i^{(w)}$, and $\widehat{\boldsymbol{w}}_i^{(w)}$ replaced, respectively, by $\widehat{\boldsymbol{X}}_i^{*(w)}$, $\widehat{\boldsymbol{Z}}_i^{*(w)}$, and $\widehat{\boldsymbol{w}}_i^{*(w)}$. The LME approximation to the log-likelihood function of the extended single-level NLME model is

$$\ell_{\text{LME}}^* \left(\boldsymbol{\beta}, \sigma^2, \boldsymbol{\Delta}, \boldsymbol{\lambda} \mid \boldsymbol{y}\right) = \ell_{\text{LME}} \left(\boldsymbol{\beta}, \sigma^2, \boldsymbol{\Delta}, \boldsymbol{\lambda} \mid \boldsymbol{y}^*\right) - \frac{1}{2} \sum_{i=1}^{M} \log |\boldsymbol{\Delta}_i|,$$

which has the same form as the log-likelihood of the extended single-level LME model described in §5.1. The log-restricted-likelihood for the extended NLME model is similarly defined.

Laplacian and Adaptive Gaussian Approximations

For the extended single-level NLME model, the objective function which is minimized to produce the conditional modes $\widehat{\boldsymbol{b}}_i$ used in the Laplacian and adaptive Gaussian approximations is

$$g^* \left(\boldsymbol{\beta}, \boldsymbol{\Delta}, \boldsymbol{\lambda}, \boldsymbol{y}_i, \boldsymbol{b}_i\right) = \left\| \boldsymbol{\Lambda}_i^{-T/2} \left[\boldsymbol{y}_i - \boldsymbol{f}_i \left(\boldsymbol{\beta}, \boldsymbol{b}_i\right)\right] \right\|^2 + \|\boldsymbol{\Delta} \boldsymbol{b}_i\|^2.$$

The corresponding approximation to the second-derivative matrix of g^* with respect $\boldsymbol{b}_i$ evaluated at $\widehat{\boldsymbol{b}}_i$ is

$$\left. \frac{\partial \partial^2 g^* \left(\boldsymbol{\beta}, \boldsymbol{\Delta}, \boldsymbol{\lambda}, \boldsymbol{y}_i, \boldsymbol{b}_i\right)}{\partial \boldsymbol{b}_i \partial \boldsymbol{b}_i^T} \right|_{\widehat{\boldsymbol{b}}_i} \simeq \boldsymbol{G}^* \left(\boldsymbol{\beta}, \boldsymbol{\Delta}, \boldsymbol{\lambda}, \boldsymbol{y}_i\right) =$$

$$\left. \frac{\partial \boldsymbol{f}_i(\boldsymbol{\beta}, \boldsymbol{b}_i)}{\partial \boldsymbol{b}_i} \right|_{\widehat{\boldsymbol{b}}_i} \boldsymbol{\Lambda}_i^{-1} \left. \frac{\partial \boldsymbol{f}_i(\boldsymbol{\beta}, \boldsymbol{b}_i)}{\partial \boldsymbol{b}_i^T} \right|_{\widehat{\boldsymbol{b}}_i} + \boldsymbol{\Delta}^T \boldsymbol{\Delta}.$$

The modified Laplacian approximation to the log-likelihood of the extended single-level NLME model is then given by

$$\ell_{\text{LA}}^* \left(\boldsymbol{\beta}, \sigma^2, \boldsymbol{\Delta}, \boldsymbol{\lambda}, \mid \boldsymbol{y}\right) = -\frac{N}{2} \log \left(2\pi\sigma^2\right) + M \log |\boldsymbol{\Delta}|$$

$$-\frac{1}{2} \left\{ \sum_{i=1}^{M} \log |\boldsymbol{G}^* \left(\boldsymbol{\beta}, \boldsymbol{\Delta}, \boldsymbol{\lambda}, \boldsymbol{y}_i\right)| + \sigma^{-2} \sum_{i=1}^{M} g^* \left(\boldsymbol{\beta}, \boldsymbol{\Delta}, \boldsymbol{\lambda}, \boldsymbol{y}_i, \widehat{\boldsymbol{b}}_i\right) \right\} - \frac{1}{2} \sum_{i=1}^{M} \log |\boldsymbol{\Lambda}_i|$$

and the adaptive Gaussian approximation is given by

$$
\ell^*_{\mathrm{AGQ}}\left(\boldsymbol{\beta}, \sigma^2, \boldsymbol{\Delta}, \boldsymbol{\lambda}, \mid \boldsymbol{y}\right) =
$$

$$
-\ \tfrac{N}{2}\log\left(2\pi\sigma^2\right) + M\log|\boldsymbol{\Delta}| - \tfrac{1}{2}\sum_{i=1}^{M}\log|\boldsymbol{G}^*\left(\boldsymbol{\beta}, \boldsymbol{\Delta}, \boldsymbol{\lambda}, \boldsymbol{y}_i\right)|
$$

$$
+\ \sum_{i=1}^{M}\log(\sum_{j}^{N_{GQ}}\exp\{-g^*[\boldsymbol{\beta}, \boldsymbol{\Delta}, \boldsymbol{\lambda}, \boldsymbol{y}_i, \widehat{\boldsymbol{b}}_i + \sigma\left(\boldsymbol{G}^*\right)^{-\frac{1}{2}}\left(\boldsymbol{\beta}, \boldsymbol{\Delta}, \boldsymbol{\lambda}, \boldsymbol{y}_i\right)\boldsymbol{z}_j]/2\sigma^2
$$

$$
+\ \|\boldsymbol{z}_j\|^2/2\}\prod_{k=1}^{q}w_{j_k}) - \frac{1}{2}\sum_{i=1}^{M}\log|\boldsymbol{\Lambda}_i|.
$$

The same comments and conclusions presented in §5.2 for the case when the within-group variance function depends on the fixed effects and/or the random effects also apply to the extended NLME model. As in the LME case, to keep the optimization problem feasible, an "iteratively reweighted" scheme is used to approximate the variance function. The fixed and random effects used in the variance function are replaced by their current estimates and held fixed during the log-likelihood optimization. New estimates for the fixed and random effects are then produced and the procedure is repeated until convergence. In the case of the alternating algorithm, the estimates for the fixed and random effects obtained in the PNLS step are used to calculate the variance function weights in the LME step. If the variance function does not depend on either the fixed effects or the random effects, then no approximation is necessary.

7.5   An Extended Nonlinear Regression Model

In many applications of nonlinear regression models to grouped data, one wishes to represent the within-group variance–covariance structure through the $\boldsymbol{\Lambda}_i$ matrices only, avoiding the use of random effects to account for within-group dependence. This results in a simplified version of the extended single-level NLME model (7.32), which we call the *extended nonlinear regression* model. In this section, we present the general formulation of the extended nonlinear regression model, describe methods for estimating its parameters, and present computational algorithms for implementing such estimation methods.

The modeling function `gnls` in the nlme library fits the extended nonlinear regression model using maximum likelihood. The use of this function is described and illustrated in §8.3.3.

7.5.1 General Model Formulation

The extended nonlinear regression model for the jth observation on the ith group, y_{ij}, is

$$y_{ij} = f\left(\phi_{ij}, v_{ij}\right) + \epsilon_{ij}, \qquad i = 1, \ldots, M, \ j = 1, \ldots, n_i,$$
$$\phi_{ij} = A_{ij}\beta. \tag{7.34}$$

The real-valued function f depends on a group-specific parameter vector ϕ_{ij} and a known covariate vector v_{ij}. It is nonlinear in at least one component of ϕ_{ij} and differentiable with respect to the group-specific parameters. M is the number of groups, n_i is the number of observations on the ith group, and ϵ_{ij} is a normally distributed error term.

The extended nonlinear regression model (7.34) is a two-stage model in which the second stage expresses the group-specific parameters ϕ_{ij} as a linear function of a fixed set of parameters β. The design matrices A_{ij} are known. We note that the coefficients β could be incorporated directly into the model function f, thus eliminating the need of a second stage in the model. However, there are advantages in having the second stage in (7.34), some of which are (i) group-specific parameters generally have a more natural interpretation in the model, (ii) inclusion and elimination of covariates in the model can be done at the second-stage model only, facilitating the understanding of the model building process, and (iii) because the ϕ_{ij} are linear functions of the β, derivatives with respect to ϕ_{ij} are easily obtained from derivatives with respect to β.

Using the same definitions of vectors and matrices given in (7.4), we can express the extended nonlinear regression in matrix form as

$$y_i = f_i\left(\phi_i, v_i\right) + \epsilon_i,$$
$$\phi_i = A_i\beta, \quad \epsilon_i \sim \mathcal{N}\left(0, \sigma^2 \Lambda_i\right).$$

As in the extended NLME model of §7.4, the Λ_i matrices are determined by a fixed, generally small, set of parameters λ.

Estimation and inference under this model has been studied extensively in the nonlinear regression literature (Carroll and Ruppert, 1988; Seber and Wild, 1989), usually assuming that the Λ_i matrices are known, being referred to as the generalized least-squares model (Seber and Wild, 1989, §2.1.4). We refer to it as the *generalized nonlinear least-squares* (GNLS) model to differentiate from the extended linear model described in §5.1.2.

Using the same transformations described in (7.33), the GNLS model (7.34) can be re-expressed as a "classic" nonlinear regression model:

$$y_i^* = f_i^*\left(\phi_i, v_i\right) + \epsilon_i^*,$$
$$\phi_i = A_i\beta, \quad \epsilon_i^* \sim \mathcal{N}\left(0, \sigma^2 I\right).$$

7.5.2   Estimation and Computational Methods

Different estimation methods have been proposed for the parameters in the GNLS model (Davidian and Giltinan, 1995, §2.5). We concentrate here on maximum likelihood estimation.

The log-likelihood function for the GNLS model is

$$\ell\left(\boldsymbol{\beta}, \sigma^2, \boldsymbol{\delta} \middle| \, \boldsymbol{y}\right) = -\frac{1}{2} \left\{ N \log\left(2\pi\sigma^2\right) + \sum_{i=1}^{M} \left[\frac{\|\boldsymbol{y}_i^* - \boldsymbol{f}_i^*(\boldsymbol{\beta})\|^2}{\sigma^2} + \log|\boldsymbol{\Lambda}_i| \right] \right\},$$

$$(7.35)$$

where N represents the total number of observations and, for simplicity, we use $\boldsymbol{f}_i^*(\boldsymbol{\beta}) = \boldsymbol{f}_i^*(\boldsymbol{\phi}_i, \boldsymbol{v}_i)$.

For fixed $\boldsymbol{\beta}$ and $\boldsymbol{\lambda}$, the maximum likelihood estimator of σ^2 is

$$\widehat{\sigma}^2(\boldsymbol{\beta}, \boldsymbol{\lambda}) = \sum_{i=1}^{M} \|\boldsymbol{y}_i^* - \boldsymbol{f}_i^*(\boldsymbol{\beta})\|^2 / N, \tag{7.36}$$

so that the profiled log-likelihood, obtained by replacing σ^2 with $\widehat{\sigma}^2(\boldsymbol{\beta}, \boldsymbol{\lambda})$ in (7.35), is

$$\ell\left(\boldsymbol{\beta}, \boldsymbol{\lambda} \middle| \, \boldsymbol{y}\right) = -\frac{1}{2} \left\{ N\left[\log\left(2\pi/N\right) + 1\right] + \log\left(\sum_{i=1}^{M} \|\boldsymbol{y}_i^* - \boldsymbol{f}_i^*(\boldsymbol{\beta})\|^2\right) \right.$$
$$\left. + \sum_{i=1}^{M} \log|\boldsymbol{\Lambda}_i| \right\}. \tag{7.37}$$

A Gauss–Seidel algorithm (Thisted, 1988, §3.11.2) is used with the profiled log-likelihood (7.37) to obtain the maximum likelihood estimates of $\boldsymbol{\beta}$ and $\boldsymbol{\lambda}$. Given the current estimate $\widehat{\boldsymbol{\lambda}}^{(w)}$ of $\boldsymbol{\lambda}$, a new estimate $\widehat{\boldsymbol{\beta}}^{(w)}$ for $\boldsymbol{\beta}$ is produced by maximizing $\ell(\boldsymbol{\beta}, \widehat{\boldsymbol{\lambda}}^{(w)})$. The roles are then reversed and a new estimate $\boldsymbol{\lambda}^{(k+1)}$ is produced by maximizing $\ell(\widehat{\boldsymbol{\beta}}^{(w)}, \boldsymbol{\lambda})$. The procedure iterates between the two optimizations until a convergence criterion is met.

It follows from (7.37) that, conditional on $\boldsymbol{\lambda}$, the maximum likelihood estimator of $\boldsymbol{\beta}$ is obtained by solving an ordinary nonlinear least-squares problem

$$\widehat{\boldsymbol{\beta}}(\boldsymbol{\lambda}) = \arg\min_{\boldsymbol{\beta}} \sum_{i=1}^{M} \|\boldsymbol{y}_i^* - \boldsymbol{f}_i^*(\boldsymbol{\beta})\|^2,$$

for which we can use a standard Gauss–Newton algorithm. If k is the iteration counter for this algorithm and $\widehat{\boldsymbol{\beta}}^{(k)} = \widehat{\boldsymbol{\beta}}^{(k)}(\boldsymbol{\lambda}^{(w)})$ is the current estimate of $\boldsymbol{\beta}$, then the derivative matrices

$$\widehat{\boldsymbol{X}}_i^{(k)} = \left.\frac{\partial \boldsymbol{f}_i(\boldsymbol{\beta})}{\partial \boldsymbol{\beta}^T}\right|_{\widehat{\boldsymbol{\beta}}^{(k)}}, \qquad \widehat{\boldsymbol{X}}_i^{*(k)} = \boldsymbol{\Lambda}_i^{-T/2} \widehat{\boldsymbol{X}}_i^{(k)},$$

provide the Gauss–Newton increment $\widehat{\boldsymbol{\delta}}^{(k+1)}$ for $\widehat{\boldsymbol{\beta}}$ as the (ordinary) least-squares solution of

$$\sum_{i=1}^{M} \left\| \widehat{\boldsymbol{w}}_i^{*(k)} - \widehat{\boldsymbol{X}}_i^{*(k)} \boldsymbol{\delta} \right\|^2,$$

where $\widehat{\boldsymbol{w}}_i^{*(k)} = \boldsymbol{y}_i^* - \boldsymbol{f}_i^*(\widehat{\boldsymbol{\beta}}^{(k)})$. Orthogonal-triangular decomposition methods similar to the ones described in §7.3 can be used to obtain a compact and numerically efficient implementation of the Gauss–Newton algorithm for estimating $\boldsymbol{\beta}$. The derivation is left to the reader as an exercise.

Inference on the parameters of the GNLS model generally relies on "classical" asymptotic theory for maximum likelihood estimation (Cox and Hinkley, 1974, §9.2), which states that, for large N, the MLEs are approximately normally distributed with mean equal to the true parameter values and variance–covariance matrix given by the inverse of the information matrix. Because $\mathrm{E}[\partial^2 \ell(\boldsymbol{\beta}, \sigma^2, \boldsymbol{\lambda})/\partial\boldsymbol{\beta}\partial\boldsymbol{\lambda}^T] = \mathbf{0}$ and $\mathrm{E}[\partial^2 \ell(\boldsymbol{\beta}, \sigma^2, \boldsymbol{\lambda})/\partial\boldsymbol{\beta}\partial\sigma^2] = \mathbf{0}$, the expected information matrix for the GNLS likelihood is block-diagonal and the MLE of $\boldsymbol{\beta}$ is asymptotically uncorrelated with the MLEs of $\boldsymbol{\lambda}$ and σ^2.

The approximate distributions for the MLEs in the GNLS model which are used for constructing confidence intervals and hypothesis tests are

$$\widehat{\boldsymbol{\beta}} \stackrel{.}{\sim} \mathcal{N}\left(\boldsymbol{\beta}, \sigma^2 \left[\sum_{i=1}^{M} \widehat{\boldsymbol{X}}_i^T \boldsymbol{\Lambda}_i^{-1} \widehat{\boldsymbol{X}}_i\right]^{-1}\right),$$

$$\begin{bmatrix} \widehat{\boldsymbol{\lambda}} \\ \log \widehat{\sigma} \end{bmatrix} \stackrel{.}{\sim} \mathcal{N}\left(\begin{bmatrix} \boldsymbol{\lambda} \\ \log \sigma \end{bmatrix}, \mathcal{I}^{-1}(\boldsymbol{\lambda}, \sigma)\right), \qquad (7.38)$$

$$\mathcal{I}(\boldsymbol{\lambda}, \sigma) = -\begin{bmatrix} \partial^2 \ell/\partial\boldsymbol{\lambda}\partial\boldsymbol{\lambda}^T & \partial^2 \ell/\partial\log\sigma\partial\boldsymbol{\lambda}^T \\ \partial^2 \ell/\partial\boldsymbol{\lambda}\partial\log\sigma & \partial^2 \ell/\partial^2\log\sigma \end{bmatrix},$$

where $\widehat{\boldsymbol{X}}_i$ is the derivative matrix evaluated at the true parameter values. As in §7.2.2, $\log\sigma$ is used in place of σ^2 to give an unrestricted parameterization for which the normal approximation tends to be more accurate. In practice, the parameters in the approximate variance–covariance matrices in (7.38) are replaced by their respective MLEs.

To reduce the bias associated with the maximum likelihood estimation of σ^2, the following modified version of (7.36) is used,

$$\tilde{\sigma}^2 = \sum_{i=1}^{M} \left\| \widehat{\boldsymbol{\Lambda}}_i^{-T/2} \left[\boldsymbol{y}_i - \boldsymbol{f}_i\left(\widehat{\boldsymbol{\beta}}\right)\right] \right\|^2 / (N - p),$$

with p denoting the length of $\boldsymbol{\beta}$. $(N - p)\tilde{\sigma}^2$ is approximately distributed as a $\sigma^2 \chi^2_{N-p}$ random variable and is asymptotically independent of $\widehat{\boldsymbol{\beta}}$. This is used to produce approximate t and F tests for the coefficients β. These tests tend to have better small sample properties than the tests obtained from the normal approximation (7.38) alone.

7.6 Chapter Summary

This chapter presents the theoretical foundations of the nonlinear mixed-effects model for single- and multilevel grouped data, including the general model formulation and its underlying distributional assumptions. Efficient computational methods for maximum likelihood estimation in the NLME model are described and discussed. Different approximations to the NLME model log-likelihood with varying degrees of accuracy and computational complexity are derived.

The basic NLME model with independent, homoscedastic within-group errors is extended to allow correlated, heteroscedastic within-group errors and efficient computational methods are described for maximum likelihood estimation of its parameters.

An extended class of nonlinear regression models, with correlated and heteroscedastic errors, but with no random effects, is presented. An efficient maximum likelihood estimation algorithm is described and approximate inference results for the parameters in this extended nonlinear regression are presented.

8
Fitting Nonlinear Mixed-Effects Models

As shown in the examples in Chapter 6, nonlinear mixed-effects models offer a flexible tool for analyzing grouped data with models that depend nonlinearly upon their parameters. As nonlinear models are usually based on a *mechanistic* model of the relationship between the response and the covariates, their parameters can have a theoretical interpretation and are often of interest in their own right. In this chapter, we describe in detail the facilities in the nlme library for fitting nonlinear mixed-effects models.

The first section presents a brief review of the standard nonlinear regression function in S, `nls`, to illustrate the use of nonlinear formulas and the derivation of starting estimates for the model parameters. We describe the selfStart class of nonlinear model functions that can calculate initial values for their parameters from the data. The `nlsList` function, which produces separate `nls` fits for each level of a grouping variable, is described and illustrated through examples.

Section 8.2 describes the `nlme` function for fitting nonlinear mixed-effects models with single or multiple levels of nesting. Method functions for displaying, plotting, updating, making predictions, and obtaining confidence intervals from a fitted nlme object are described and illustrated with examples.

The use of correlation structures and variance functions to extend the basic nonlinear mixed-effects model is discussed in Section 8.3. The `gnls` function for fitting the extended nonlinear regression model without random effects, presented in §7.5.1, is described, together with its associated methods.

8.1 Fitting Nonlinear Models in S with nls and nlsList

In this section we describe the use of the nonlinear least squares (nls) function for fitting a single nonlinear regression model and the nlsList function for fitting a set of nonlinear regression models to grouped data. Especially with nlsList, it is very helpful to have the model function itself defined as a *self-starting* model, as described in §8.1.2.

8.1.1 *Using the* nls *Function*

The S function nls uses a Gauss–Newton algorithm, described in §7.5.2, to determine the *nonlinear least squares* estimates of the parameters in a nonlinear regression model. A typical call to nls is of the form

```
nls( formula, data, start )
```

where formula is a two-sided nonlinear formula specifying the model, data is a data frame with the variables used in formula, and start is a named vector or list containing the starting estimates for the model parameters. Several other arguments to nls are available, as described in Bates and Chambers (1992) and Venables and Ripley (1999, Chapter 8).

We illustrate the use of nls to fit nonlinear regression models with the Orange data from a study of the growth of orange trees, reported in Draper and Smith (1998, Exercise 24.N, p. 559). The data, shown in Figure 8.1 and described in Appendix A.16, are the trunk circumferences (in millimeters) of each of five trees at each of seven occasions.

```
> ## outer = ~1 is used to display all five curves in one panel
> plot( Orange, outer = ~1 )                        # Figure 8.1
```

Because all trees are measured on the same occasions these are balanced, longitudinal data. It is clear from Figure 8.1 that a "tree effect" is present, and we will take this into account when fitting nlme or nlsList models. To illustrate some of the details of fitting nls models, we will temporarily ignore the tree effect and fit a single logistic model to all the data. Recall from §6.1 that this model expresses the trunk circumference y_{ij} of tree i at age x_{ij} for $i = 1, \ldots, 5 \; j = 1, \ldots, 7$ as

$$y_{ij} = \frac{\phi_1}{1 + \exp\left[-\left(t_{ij} - \phi_2\right)/\phi_3\right]} + \epsilon_{ij}, \tag{8.1}$$

where the error terms ϵ_{ij} are assumed to be distributed independently as $\mathcal{N}(0, \sigma^2)$. As explained in §6.1, the model parameters are the asymptotic trunk circumference ϕ_1, the age at which the tree attains half of its asymptotic trunk circumference ϕ_2, and the growth scale ϕ_3. This function is nonlinear in ϕ_2 and ϕ_3.

Model (8.1) can be represented in S by the nonlinear formula

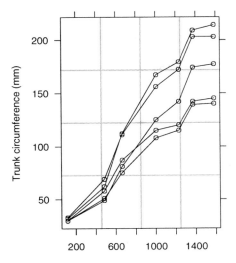

Time since December 31, 1968 (days)

FIGURE 8.1. Circumference of five orange trees from a grove in southern California over time. The measurement is probably the "circumference at breast height" commonly used by foresters. Points corresponding to the same tree are connected by lines.

```
circumference ~ Asym/(1 + exp(-(age - xmid)/scal)),
```

where `Asym` $= \phi_1$, `xmid` $= \phi_2$, and `scal` $= \phi_3$. Unlike in the linear case, the parameters must be declared explicitly in a nonlinear model formula and an intercept is not assumed by default.

An alternative approach is to write an S function representing the logistic model as, say,

```
> logist <-
+     function(x, Asym, xmid, scal) Asym/(1 + exp(-(x - xmid)/scal))
```

and then use it in the nonlinear formula

```
circumference ~ logist(age, Asym, xmid, scal)
```

An advantage of this latter approach is that we can modify our `logist` function to include a `gradient` attribute with its returned value. This would then be used as the gradient matrix in the Gauss–Newton nonlinear least-squares optimization, increasing the numerical stability and the rate of convergence of the algorithm, compared to the default use of numerical derivatives. The `deriv` function can be used to produce a function that returns a `gradient` attribute with its value.

```
> logist <- deriv( ~Asym/(1+exp(-(x-xmid)/scal)),
```

```
+    c("Asym", "xmid", "scal"), function(x, Asym, xmid, scal){} )
> Asym <- 180; xmid <- 700; scal <- 300
> logist( Orange$age[1:7], Asym, xmid, scal )
[1]   22.617  58.931  84.606 132.061 153.802 162.681 170.962
attr(, "gradient"):
        Asym       xmid       scal
[1,] 0.12565 -0.065916  0.127878
[2,] 0.32739 -0.132124  0.095129
[3,] 0.47004 -0.149461  0.017935
[4,] 0.73367 -0.117238 -0.118802
[5,] 0.85446 -0.074616 -0.132070
[6,] 0.90378 -0.052175 -0.116872
[7,] 0.94979 -0.028614 -0.084125
```

As mentioned in §6.1, one important difference between linear and non-linear regression is that the nonlinear models require starting estimates for the parameters. Determining reasonable starting estimates for a nonlinear regression problem is something of an art, but some general recommendations are available (Bates and Watts, 1988, §3.2). We return to this issue in §8.1.2, where we describe the selfStart class of model functions.

Because the parameters in the logistic model (8.1) have a graphical interpretation, we can determine initial estimates from a plot of the data. In Figure 8.1 it appears that the mean asymptotic trunk circumference is around 170 mm and that the trees attain half of their asymptotic trunk circumference at about 700 days of age. Therefore, we use the inital estimates of $\tilde{\phi}_1 = 170$ for the asymptotic trunk circumference and $\tilde{\phi}_2 = 700$ for the location of the inflection point. To obtain an initial estimate for ϕ_3, we note that the logistic curve reaches approximately 3/4 of its asymptotic value when $x = \phi_2 + \phi_3$. Inspection of Figure 8.1 suggests that the trees attain 3/4 of their final trunk circumference at about 1200 days, giving an inital estimate of $\tilde{\phi}_3 = 500$.

We combine all this information in the following call to nls

```
> fm1Oran.nls <- nls(circumference ~ logist(age, Asym, xmid, scal),
+    data = Orange, start = c(Asym = 170, xmid = 700, scal = 500) )
```

Our initial estimates are reasonable and the call converges. Following the usual framework for modeling functions in S, the object fm1Oran.nls produced by the call to nls is of class nls, for which several methods are available to display, plot, update, and extract components from a fitted object. For example, the summary method provides information about the estimated parameters.

```
> summary( fm1Oran.nls )
Formula: circumference ~ logist(age, Asym, xmid, scal)
Parameters:
       Value Std. Error t value
Asym 192.68     20.239  9.5203
```

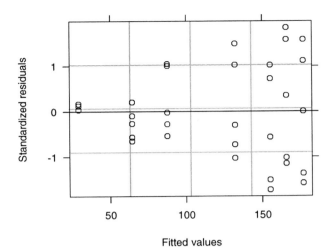

FIGURE 8.2. Scatter plots of standardized residuals versus fitted values for fm1Oran.nls, a nonlinear least squares fit of the logistic growth model to the entire orange tree data set.

```
xmid 728.71    107.272  6.7931
scal 353.49     81.460  4.3395

Residual standard error: 23.3721 on 32 degrees of freedom

Correlation of Parameter Estimates:
      Asym   xmid
xmid 0.922
scal 0.869 0.770
```

The final estimates are close to the initial values derived from Figure 8.1. The standard errors for the parameter estimates are relatively large, suggesting that there is considerable variability in the data.

The plot method for nls objects (which is included with the nlme library), has a syntax similar to the lme and gls plot methods described in §4.3.1 and §5.4. By default, the plot of the standardized residuals versus fitted values, shown in Figure 8.2, is produced.

```
> plot( fm1Oran.nls )                          # Figure 8.2
```

The variability in the residuals increases with the fitted values, but, in this case, the wedge-shaped pattern is due to the correlation among observations in the same tree and not to heteroscedastic errors. We can get a better understanding of the problem by looking at the plot of the residuals by tree presented in Figure 8.3.

```
> plot(fm1Oran.nls, Tree ~ resid(.), abline = 0)     # Figure 8.3
```

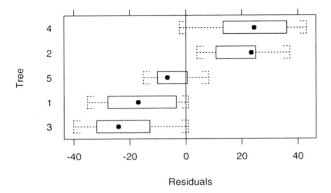

FIGURE 8.3. Boxplots of residuals by tree for `fm1Oran.nls`, a nonlinear least squares fit of the logistic growth model to the entire orange tree data set.

The residuals are mostly negative for trees 1 and 3 and mostly positive for trees 2 and 4, giving strong evidence that a "tree effect" should be included in the model.

A simple approach to account for a tree effect is to allow different parameters for each tree, resulting in separate `nls` fits. This is the approach used in the `nlsList` function, described in §8.1.3, which provides a valuable tool for model building, but usually produces overparametrized models. As illustrated in Chapter 6, nonlinear mixed-effects models strike a balance between the simple `nls` model and the overparametrized `nlsList` model, by allowing random effects to account for among-group variation in some of the parameters, while preserving a moderate number of parameters in the model.

8.1.2 Self-Starting Nonlinear Model Functions

Bates and Watts (1988, §3.2) describe several techniques for determining starting estimates in a nonlinear regression. Some of the more effective techniques are:

- Take advantage of partially linear models, as described in Bates and Chambers (1992, §10.2.5), so that initial estimates are needed only for those parameters that enter the model nonlinearly.

- Choose parameters that have meaningful graphical interpretations, as we did for the logistic model.

- Refine the estimates of some of the parameters by iterating on them while holding all the other parameters fixed at their current values.

The application of these techniques to a particular nonlinear regression model applied to a given set of data can be straightforward but tedious.

Especially when the same model will be applied to several similar sets of data, as is the case in many of the examples we consider here, we do not want to have to manually repeat the same series of steps in determining starting estimates. A more sensible approach is to encapsulate the steps used to derive initial estimates for a given nonlinear model into a function that can be used to generate intial estimates from any dataset. Self-starting nonlinear regression models are S functions that contain an auxillary function to calculate the initial parameter estimates. They are represented in S as selfStart objects.

The S objects of the selfStart class are composed of two functions; one that evaluates the nonlinear regression model itself and an auxillary function, called the initial attribute, that determines starting estimates for the model's parameters from a set of data. When a selfStart object for a model is available, there is no need to determine starting values for the parameters. The user can simply specify the formula for the model and the data to which it should be applied. From a user's point of view, fitting self-starting nonlinear regression models is nearly as easy as fitting linear regression models.

We illustrate the construction and use of self-starting models by building function one for the logistic model. The basic steps in the calculation of initial estimates for the logistic model (8.1) from the **Orange** dataset are:

1. *Sort/average*: sort the data according to the x variable and obtain the average response y for each unique x.

2. *Asymptote*: use the maximum y as an initial value $\tilde{\phi}_1$ for the asymptote.

3. *Inflection point*: use the x corresponding to $0.5\tilde{\phi}_1$ as an initial value $\tilde{\phi}_2$ for the inflection point.

4. *Scale*: use the difference between the x corresponding to $0.75\tilde{\phi}_1$ and $\tilde{\phi}_2$ as an initial value $\tilde{\phi}_3$ for the growth scale

Step 1, *sort/average*, was carried out implicitly in our graphical derivation of initial values for the orange trees example, but is now explicitly incorporated in the algorithm.

Two auxiliary functions, **sortedXyData** and **NLSstClosestX**, included in the nlme library, are particularly useful for constructing self-starting models. The **sortedXyData** function performs the *sort/average* step. It takes the arguments x, y, and data and returns a data.frame with two columns: y, the average y for each unique value of x, and x, the unique values of x, sorted in ascending order. The arguments x and y can be numeric vectors or they can be expressions or strings to be evaluated in data. The returned object is of class sortedXyData. For example, the pointwise averages of the growth curves of the orange trees are obtained with

```
> Orange.sortAvg <- sortedXyData( "age", "circumference", Orange )
> Orange.sortAvg
     x     y
1  118  31.0
2  484  57.8
3  664  93.2
4 1004 134.2
5 1231 145.6
6 1372 173.4
7 1582 175.8
```

The NLSstClosestX function estimates the value of x corresponding to a given y, using linear interpolation. It takes two arguments: xy, a sortedXy-Data object, and yval, the desired value of y, and returns a numeric value. For example, the estimated age at which the average growth curve reaches 130 mm is

```
> NLSstClosestX(Orange.sortAvg, 130)
[1] 969.17
```

A function to serve as an initial attribute of a selfStart function must be defined with the arguments mCall, LHS, and data, in that order, and must return the initial values as a vector, or list, with elements named according to the actual parameters in the model function's call. The first argument, mCall, is a *matched call* to the selfStart model. It contains the arguments of the function call (as expressions), matched with the formal parameters, which are the argument names used in the original definition of the model. In the case of the logist function, defined before as

```
function(x, Asym, xmid, scal)
```

the formal parameters are x, Asym, xmid, and scal. If the function is called as

```
logist(age, A, xmid, scal)
```

then mCall will have components x = age, Asym = A, xmid = xmid and scal = scal. As described above, the names of these components are the formal parameters in the model function definition. The values of these components are the names (or, more generally, the expressions) that are the actual arguments in the call to the model function.

The LHS argument is the expression on the left-hand side of the model formula in the call to nls. It determines the response vector. The data argument gives a data.frame in which to find the variables named in the other two arguments. The function logistInit below implements a slightly more general version of the algorithm described above for calculating initial estimates in the logistic model, using the required syntax.

```
> logistInit
function(mCall, LHS, data)
{
  xy <- sortedXyData(mCall[["x"]], LHS, data)
  if(nrow(xy) < 3) {
    stop("Too few distinct input values to fit a logistic")
  }
  Asym <- max(abs(xy[,"y"]))
  if (Asym != max(xy[,"y"])) Asym <- -Asym  # negative asymptote
  xmid <- NLSstClosestX(xy, 0.5 * Asym)
  scal <- NLSstClosestX(xy, 0.75 * Asym) - xmid
  value <- c(Asym, xmid, scal)
  names(value) <- mCall[c("Asym", "xmid", "scal")]
  value
}
```

The algorithm in logistInit includes a check for an adequate number of distinct observations to fit a logistic model and allows negative as well as positive asymptotes.

Before the starting values are returned, we ensure that the components are named according to the actual parameters in the model function call. As described above, the object mCall provides the correspondence between the formal parameter names and the actual parameter names, so we assign the names indirectly through mCall. This step is important. All functions to be used as the initial attribute of a selfStart model should indirectly assign the names to the result in this way.

The selfStart constructor is used to create a self-starting model. It can be called with two functions, the model function itself and the initial attribute, as in

```
> logist <- selfStart( logist, initial = logistInit )
> class( logist )
[1] "selfStart"
```

Alternatively, it can be called with a one-sided formula defining the nonlinear model, the function for the initial attribute, and a character vector giving the parameter names.

```
> logist <- selfStart( ~ Asym/(1 + exp(-(x - xmid)/scal)),
+     initial = logistInit, parameters = c("Asym", "xmid", "scal"))
```

When selfStart is called like this, the model function is produced by applying deriv to the right-hand side of the model formula.

The getInitial function is used to extract the initial parameter estimates from a given dataset when using a selfStart model function. It takes two arguments: a two-sided model formula, which must include a selfStart function on its right-hand side, and a data.frame in which to evaluate the variables in the model formula.

TABLE 8.1. Standard selfStart functions in the NLME 3.0 distribution.

SSasymp	asymptotic regression
SSasympOff	asymptotic regression with an offset
SSasympOrig	asymptotic regression through the origin
SSbiexp	biexponential
SSfol	first-order compartment
SSfpl	four-parameter logistic
SSlogis	logistic
SSmicmen	Michaelis–Menten

```
> getInitial(circumference ~ logist(age, Asym, xmid, scal), Orange)
  Asym   xmid   scal
175.8 637.05 347.46
```

As expected, the initial values produced by logist are similar to those obtained previously using the graphical interpretation of the parameters and Figure 8.1.

When nls is called without initial values for the parameters and a self-Start model function is provided, nls calls getInitial to provide the initial values. In this case, only the model formula and the data are required to fit the model, making the call nearly as simple as an lm call. For example, the logistic model can be fit to the orange tree data with

```
> nls( circumference ~ logist(age, Asym, xmid, scal), Orange )
Residual sum of squares : 17480
parameters:
   Asym   xmid   scal
192.69 728.77 353.54
formula: circumference ~ logist(age, Asym, xmid, scal)
35 observations
```

The nlme library includes several self-starting model functions that can be used to fit nonlinear regression models in S without specifying starting estimates for the parameters. They are listed in Table 8.1 and are described in detail in Appendix B. The SSlogis function is a more sophisticated version of our simple logist self-starting model, but with the same argument sequence. It uses several techniques, such as the algorithm for partially linear models, to refine the starting estimates so the returned values are actually the converged estimates.

```
> getInitial(circumference ~ SSlogis(age,Asym,xmid,scal), Orange)
   Asym   xmid   scal
 192.68 728.72 353.5
> nls( circumference ~ SSlogis(age, Asym, xmid, scal), Orange )
Residual sum of squares : 17480
```

```
parameters:
   Asym   xmid  scal
 192.68 728.72 353.5
formula: circumference ~ SSlogis(age, Asym, xmid, scal)
35 observations
```

We can see that selfStart model objects relieve the user of much of the effort required for a nonlinear regression analysis. If a versatile, effective strategy for determining starting estimates is represented carefully in the object, it can make the use of nonlinear models nearly as simple as the use of linear models. If you frequently use a nonlinear model not included in Table 8.1, you should consider writing your own self-starting model to represent it. The selfStart functions in Table 8.1 and the logist function described in this section can be used as templates.

8.1.3 Separate Nonlinear Fits by Group: The nlsList Function

In the indomethicin and soybean examples of Chapter 6 we saw how the nlsList function can be used to produce separate fits of a nonlinear model for each group in a groupedData object. These separate fits by group are a powerful tool for model building with nonlinear mixed-effects models because the individual estimates can suggest the type of random-effects structure to use. Also, these estimates provide starting values for the parameters in the mixed-effects model. In this section we provide more detail on the use of nlsList function itself and the methods for the nlsList objects that it creates.

The nls fits performed within nlsList will require starting estimates for the parameters. Although it is sometimes possible to use a single set of starting estimates for all the groups, we recommend using a selfStart function, as described in §8.1.2, to automatically generate individual initial estimates for each group.

A typical call to nlsList is

    nlsList(model, data)

where model is a two-sided formula whose right-hand side consists of two parts separated by the | operator. The first part defines the nonlinear model, generally involving a selfStart function to be fitted to each subset of data, and the second part specifies the grouping factor. Any nonlinear model formula allowed in nls can also be used with nlsList. The data argument gives a data frame in which the variables in model can be evaluated. The grouping factor can be omitted from the model formula when data is a groupedData object.

To illustrate the use of nlsList, let us continue with the analysis of the orange trees data, fitting a separate logistic curve to each tree and using the selfStart function SSlogis to produce initial estimates.

```
> fm1Oran.lis <-
+     nlsList(circumference ~ SSlogis(age, Asym, xmid, scal) | Tree,
+             data = Orange)
```

Because Orange is a groupedData object, the grouping factor Tree could have been omitted from the model formula. When a selfStart function depends on only one covariate, as does SSlogis, and data is a groupedData object, the entire form of the model formula can be inferred from the display formula stored with the groupedData object. In this case, only the selfStart function and the groupedData object need to be passed to the nlsList function. For example, we could use

```
> fm1Oran.lis <- nlsList( SSlogis, Orange )
```

to obtain the same nlsList fit as before.

The nlsList function can also be used with regular, non-self-starting nonlinear functions, but in this case the same set of starting values, specified in the start argument, will be used for every group. The use of common starting estimates may not be a sensible choice in many applications, so we strongly encourage the use of selfStart functions with nlsList. In the orange trees example, the individual growth patterns are similar enough that a common set of starting estimates can be used. For example, using the same initial estimates as used to fit fm1Oran.nls in §8.1.1, we have

```
> fm1Oran.lis.noSS <-
+     nlsList( circumference ~ Asym/(1+exp(-(age-xmid)/scal)),
+              data = Orange,
+              start = c(Asym = 170, xmid = 700, scal = 500) )
```

Because the model fits fm1Oran.lis and fm1Oran.lis.noSS are derived from different starting values, the parameter estimates will differ slightly.

Objects returned by nlsList are of class nlsList which inherits from class lmList. Therefore, all summary and display methods, as well as methods for extracting components from the fitted object, for class lmList can also be applied to an nlsList object. Table 8.2 lists the most commonly used methods for nlsList objects. We illustrate the use of some of these methods below.

The print method gives minimal information about the individual nls fits.

```
> fm1Oran.lis
 Call:
  Model: circumference ~ SSlogis(age, Asym, xmid, scal) | Tree
   Data: Orange

Coefficients:
    Asym   xmid   scal
3 158.83 734.85 400.95
```

TABLE 8.2. Main nlsList methods.

augPred	predictions augmented with observed values
coef	coefficients from individual nls fits
fitted	fitted values from individual nls fits
fixef	average of individual nls coefficients
intervals	confidence intervals on coefficients
nlme	nonlinear mixed-effects model from nlsList fit
logLik	sum of individual nls log-likelihoods
pairs	scatter-plot matrix of coefficients or random effects
plot	diagnostic Trellis plots
predict	predictions for individual nls fits
print	brief information about the lm fits
qqnorm	normal probability plots
ranef	deviations of coefficients from average
resid	residuals from individual nls fits
summary	more detailed information about nls fits
update	update the individual nls fits

```
1 154.15 627.12 362.50
5 207.27 861.35 379.99
2 218.99 700.32 332.47
4 225.30 710.69 303.13

Degrees of freedom: 35 total; 20 residual
Residual standard error: 7.98
```

More detailed information on the coefficients is obtained with summary

```
> summary( fm1Oran.lis )
 . . .
Coefficients:
  Asym
  Value Std. Error t value
3 158.83      19.235  8.2574
1 154.15      13.588 11.3446
5 207.27      22.028  9.4092
2 218.99      13.357 16.3959
4 225.30      11.838 19.0318
  xmid
  Value Std. Error t value
3 734.85     130.807  5.6178
1 627.12      92.854  6.7538
5 861.35     108.017  7.9742
2 700.32      61.342 11.4167
4 710.69      51.166 13.8899
```

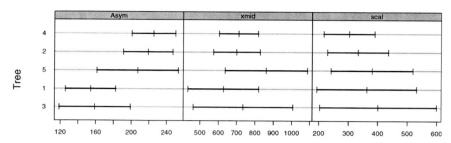

FIGURE 8.4. Ninety-five percent confidence intervals on the logistic model parameters for each tree in the orange trees data.

```
    scal
    Value Std. Error  t value
3  400.95     94.776   4.2306
1  362.50     81.185   4.4652
5  379.99     66.761   5.6917
2  332.47     49.381   6.7327
4  303.13     41.608   7.2853
.  .  .
```

Although the estimates for all the parameters vary with tree, there appears to be relatively more variability in the Asym estimates. We can investigate this better with the intervals method.

```
> plot( intervals( fm1Oran.lis ), layout = c(3,1) )    # Figure 8.4
```

As mentioned in §6.2, confidence intervals for the same parameter in different groups within an nlsList fit do not necessarily have the same length, even with balanced data. This is evident in Figure 8.4.

The only parameter for which all the confidence intervals do not overlap in Figure 8.4 is Asym, suggesting that it is the only parameter for which random effects are needed to account for variation among trees.

The same plot method used for lmList objects is used to obtain diagnostic plots for an nlsList object. The boxplots of the residuals by tree, obtained with

```
> plot( fm1Oran.lis, Tree ~ resid(.), abline = 0 )    # Figure 8.5
```

and displayed in Figure 8.5, no longer indicate the "tree effect" observed in Figure 8.3. The basic drawback of the nlsList model is that uses 15 parameters to account for the individual tree effects. A more parsimonious representation is provided by the nonlinear mixed-effects model discussed in §8.2.

As a second example to illustrate the use of the nlsList function, we consider the theophylline data, which we used in §3.4. Recall that these data, displayed in Figure 8.6, are the serum concentrations of theophylline

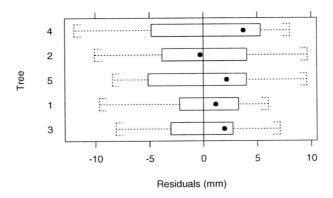

FIGURE 8.5. Boxplots of residuals by tree for fm1Oran.lis.

measured on twelve subjects at eleven times up to 25 hours after receiving an oral dose of the drug.

```
> Theoph[1:4,]
Grouped Data: conc ~ Time | Subject
  Subject   Wt Dose Time  conc
1       1 79.6 4.02 0.00  0.74
2       1 79.6 4.02 0.25  2.84
3       1 79.6 4.02 0.57  6.57
4       1 79.6 4.02 1.12 10.50
```

The column Wt in Theoph gives the Subject's weight (in kilograms).

As when modeling the indomethicin data in §6.2 and the phenobarbital data in §6.4, we use a compartment model for these data. A first-order open-compartment model expresses the theophylline concentration c_t at time t after an initial dose D as

$$c_t = \frac{D k_e k_a}{Cl\,(k_a - k_e)}\left[\exp\left(-k_e t\right) - \exp\left(-k_a t\right)\right]. \qquad (8.2)$$

The parameters in the model are the *elimination* rate constant k_e, the *absorption* rate constant k_a, and the *clearance* Cl. For the model to be meaningful, all three parameters must be positive. To ensure ourselves of positive estimates while keeping the optimization problem unconstrained, we reparameterize model (8.2) in terms of the logarithm of the clearance and the rate constants.

$$c_t = \frac{D\exp\left(lKe + lKa - lCl\right)}{\exp\left(lKa\right) - \exp\left(lKe\right)}\left\{\exp\left[-\exp\left(lKe\right)t\right] - \exp\left[-\exp\left(lKa\right)t\right]\right\},$$
$$\qquad (8.3)$$

where $lKe = \log(k_e)$, $lKa = \log(k_a)$, and $lCl = \log(Cl)$.

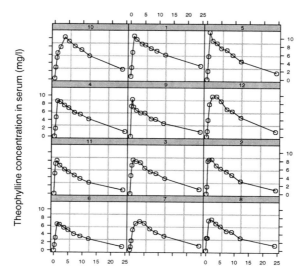

Time since drug administration (hr)

FIGURE 8.6. Serum concentrations of theophylline versus time since oral administration of the drug in twelve subjects.

The selfStart function SSfol, described in Appendix C.5, provides a self-starting implementation of (8.3). Because two covariates, *dose* and *time*, are present in (8.3), and hence also in the argument sequence of SSfol, we must specify the full model formula when calling nlsList.

```
> fm1Theo.lis <- nlsList( conc ~ SSfol(Dose, Time, lKe, lKa, lCl),
+    data = Theoph )
> fm1Theo.lis
Call:
  Model: conc ~ SSfol(Dose, Time, lKe, lKa, lCl) | Subject
   Data: Theoph

Coefficients:
        lKe       lKa       lCl
 6  -2.3074   0.15171  -2.9733
 7  -2.2803  -0.38617  -2.9643
 8  -2.3863   0.31862  -3.0691
11  -2.3215   1.34779  -2.8604
 3  -2.5081   0.89755  -3.2300
 2  -2.2862   0.66417  -3.1063
 4  -2.4365   0.15834  -3.2861
 9  -2.4461   2.18201  -3.4208
12  -2.2483  -0.18292  -3.1701
10  -2.6042  -0.36309  -3.4283
 1  -2.9196   0.57509  -3.9158
```

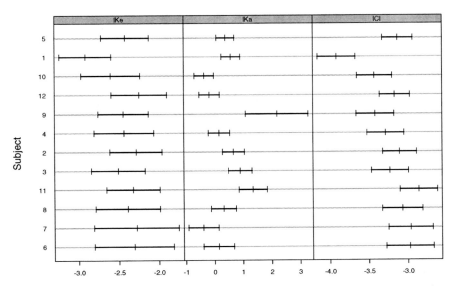

FIGURE 8.7. Ninety-five percent confidence intervals on the parameters in the first-order open-compartment model (8.3) for each subject in the theophylline data.

```
5 -2.4254  0.38616 -3.1326

Degrees of freedom: 132 total; 96 residual
Residual standard error: 0.70019
```

The individual estimates suggest that the absorption rate constant is more variable among subjects than either the elimination rate constant or the clearance. As usual, we recommend using the plot of the confidence intervals on the individual parameters to analyze their between-group variation.

```
> plot( intervals( fm1Theo.lis ), layout = c(3,1) )     # Figure 8.7
```

The individual confidence intervals in Figure 8.7 indicate that there is substantial subject-to-subject variation in the absorption rate constant and moderate variation in the clearance. The elimination rate constant does not seem to vary significantly with subject.

The main purpose of the preliminary analysis provided by nlsList is to suggest a structure for the random effects to be used in a nonlinear mixed-effects model. We must decide which random effects to include in the model (intervals and its associated plot method are often useful for that) and what covariance structure these random effects should have. The pairs method, which is the same for lmList and nlsList objects, provides one view of the random effects covariance structure.

```
> pairs( fm1Theo.lis, id = 0.1 )                        # Figure 8.8
```

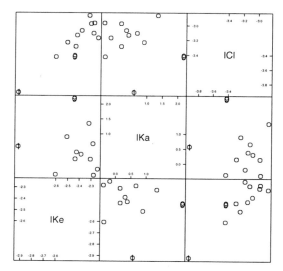

FIGURE 8.8. Pairs plot for the random effects estimates corresponding to fm1Theo.lis.

The scatter plots in Figure 8.8 suggest that Subject 1 has an unusually low elimination rate constant and clearance and that Subject 9 has an unusually high absorption rate constant. Overall, there do not appear to be significant correlations between the individual parameter estimates.

8.2   Fitting Nonlinear Mixed-Effects Models with nlme

The general formulation of a nonlinear mixed-effects model, and the estimation methods used to fit it, are described in §7.1. This section concentrates on the capabilities available in the nlme library for fitting nonlinear mixed-effects models with independent, homoscedastic within-group errors. Fitting nlme models to single-level grouped data is described in §8.2.1. The use of covariates to explain between-group variability and to reduce the number of random effects in an nlme model is presented in §8.2.2. In §8.2.3, we describe how to fit multilevel nonlinear mixed-effects models with nlme.

8.2.1   Fitting Single-Level nlme Models

Several optional arguments can be used with the nlme function, but a typical call looks like

```
nlme( model, data, fixed, random, groups, start )
```

The `model` argument is required and consists of either a two-sided formula specifying the nonlinear model to be fitted, or an `nlsList` object. Any S nonlinear formula can be used, giving the function considerable flexibility. For example, in the orange trees example one could either write down the logistic model explicitly

```
circumference ~ Asym/(1 + exp(-(age - xmid)/scal))
```

or use the selfStart function SSlogis

```
circumference ~ SSlogis(age, Asym, xmid, scal)
```

As with `nls` fits, there is an advantage of encapsulating the model expression in an S function when fitting an `nlme` model, in that it allows analytic derivatives of the model function to be passed to `nlme` and used in the optimization algorithm. The S function `deriv` can be used to create model functions that return the value of the model and its derivatives as a `gradient` attribute. If the value returned by the model function does not have a `gradient` attribute, numerical derivatives are used in the optimization.

The arguments `fixed` and `random` are formulas, or lists of formulas, defining the structures of the fixed and random effects in the model. In these formulas a 1 on the right-hand side of a formula indicates that a single parameter is associated with the effect, but any linear formula in S can be used. This gives considerable flexibility to the model, as time-dependent parameters can be incorporated easily (e.g., when a formula in the `fixed` list involves a covariate that changes with time). Each parameter in the model will usually have an associated fixed effect, but it may, or may not, have an associated random effect. Because the `nlme` model assumes that all random effects have expected value zero, the inclusion of a random effect without a corresponding fixed effect would be unusual. Any covariates defined in the `fixed` and `random` formulas can, alternatively, be directly incorporated in the `model` formula. However, declaring the covariates in `fixed` and `random` allows for more efficient calculation of derivatives and is useful for `update` methods. `Fixed` is required when `model` is declared as a formula. By default, when `random` is omitted, all fixed effects in the model are assumed to have an associated random effect.

In the theophylline example, to have random effects only for the log of the absorption rate constant, `1Ka`, and the log-clearance, `1Cl`, we use

```
fixed = list(1Ke ~ 1, 1Ka ~ 1, 1Cl ~ 1),
random = list(1Ka ~ 1, 1Cl ~ 1)
```

Model parameters with common right-hand side expressions in the `fixed` or `random` formulas can be collapsed into a single formula, in which the left-hand side lists the parameters separated by the + operator. For example, `fixed` and `random` in the theophylline example can be expressed more compactly as

```
fixed = 1Ke + 1Ka + 1Cl ~ 1
random = 1Ka + 1Cl ~ 1
```

If we wanted to allow the log-clearance fixed effect to depend linearly on the subject's weight while retaining a single component for the other fixed effects, we would use

```
fixed = list(1Ke + 1Ka ~ 1, 1Cl ~ Wt)
```

Note that there would be two fixed effects associated with 1Cl in this case: an intercept and a slope with respect to Wt.

Data names a data frame in which any variables used in model, fixed, random, and groups are to be evaluated. By default, data is set to the environment from which nlme was called.

The groups argument is a one-sided formula, or an S expression, which, when evaluated in data, returns a factor with the group label of each observation. This argument does not need to be specified when object is an nlsList object, or when data is a groupedData object, or when random is a named list (in which case the name is used as the grouping factor).

The start argument provides a list, or a vector, of starting values for the iterative algorithm. When given as a vector, it is used as starting estimates for the fixed effects only. It is only required when model is given as a formula and the model function is not a selfStart object. In this case, starting values for the fixed effects must be specified. Starting estimates for the remaining parameters are generated automatically. By default, the random effects are initialized to zero.

Objects returned by nlme are of class nlme, which inherits from class lme. As a result, all the methods available for lme objects can also be applied to an nlme object. In fact, most methods are common to both classes. Table 8.3 lists the most important methods for class nlme. We illustrate the use of these methods through the examples in the next sections.

Growth of Orange Trees

The nonlinear mixed-effects model corresponding to the logistic model 8.1, with random effects for all parameters, is

$$
y_{ij} = \frac{\phi_{1i}}{1 + \exp\left[-\left(t_{ij} - \phi_{2i}\right)/\phi_{3i}\right]} + \epsilon_{ij},
$$
$$
\boldsymbol{\phi}_i = \begin{bmatrix} \phi_{1i} \\ \phi_{2i} \\ \phi_{3i} \end{bmatrix} = \begin{bmatrix} \beta_1 \\ \beta_2 \\ \beta_3 \end{bmatrix} + \begin{bmatrix} b_{1i} \\ b_{2i} \\ b_{3i} \end{bmatrix} = \boldsymbol{\beta} + \boldsymbol{b}_i,
$$
$$
\boldsymbol{b}_i \sim \mathcal{N}\left(\boldsymbol{0}, \boldsymbol{\Psi}\right), \qquad \epsilon_{ij} \sim \mathcal{N}\left(0, \sigma^2\right). \tag{8.4}
$$

The model parameters $\boldsymbol{\phi}_i$ have the same interpretation as in (8.1), but are now allowed to vary with tree. The fixed effects, $\boldsymbol{\beta}$, represent the average value of the individual parameters, $\boldsymbol{\phi}_i$, in the population of orange

TABLE 8.3. Main nlme methods.

ACF	empirical autocorrelation function of within-group residuals
anova	likelihood ratio or Wald-type tests
augPred	predictions augmented with observed values
coef	estimated coefficients for different levels of grouping
fitted	fitted values for different levels of grouping
fixef	fixed-effects estimates
intervals	confidence intervals on model parameters
logLik	log-likelihood at convergence
pairs	scatter-plot matrix of coefficients or random effects
plot	diagnostic Trellis plots
predict	predictions for different levels of grouping
print	brief information about the fit
qqnorm	normal probability plots
ranef	random-effects estimates
resid	residuals for different levels of grouping
summary	more detailed information about the fit
update	update the nlme fit
Variogram	semivariogram of within-group residuals

trees and the random effects, b_i, represent the deviations of the ϕ_i from their population average. The random effects are assumed to be independent for different trees and the within-group errors ϵ_{ij} are assumed to be independent for different i, j and to be independent of the random effects.

The nonlinear mixed-effects model (8.4) uses ten parameters to represent the fixed effects (three parameters), the random-effects variance–covariance structure (six parameters), and the within-group variance (one parameter). These numbers remain unchanged if we increase the number of trees being analyzed. In comparison, the number of parameters in the corresponding `nlsList` model, described in §8.1.3, is equal to three times the number of trees (15, in the orange trees example).

A nonlinear mixed-effects fit of model (8.4) can be obtained with

```
> ## no need to specify groups, as Orange is a groupedData object
> ## random is omitted - by default it is equal to fixed
> fm1Oran.nlme <-
+   nlme( circumference ~ SSlogis(age, Asym, xmid, scal),
+         data = Orange,
+         fixed = Asym + xmid + scal ~ 1,
+         start = fixef(fm1Oran.lis) )
```

but a much simpler, equivalent form is

```
> fm1Oran.nlme <- nlme( fm1Oran.lis )
```

The `fm1Oran.lis` object stores information about the model function, the parameters in the model, the groups formula, and the data used to fit the model. These are retrieved by `nlme`, allowing a more compact call. Another important advantage of using an `nlsList` object as the first argument to `nlme` is that it automatically provides initial estimates for the fixed effects, the random effects, and the random-effects covariance matrix.

We can now use the `nlme` methods to display the results and to assess the quality of the fit. As with `lme` objects, the `print` method gives some brief information about the fitted object.

```
> fm1Oran.nlme
Nonlinear mixed-effects model fit by maximum likelihood
  Model: circumference ~ SSlogis(age, Asym, xmid, scal)
  Data: Orange
  Log-likelihood: -129.99
  Fixed: list(Asym ~ 1, xmid ~ 1, scal ~ 1)
    Asym    xmid    scal
  192.12  727.74  356.73

Random effects:
 Formula: list(Asym ~ 1, xmid ~ 1, scal ~ 1)
 Level: Tree
 Structure: General positive-definite
          StdDev   Corr
     Asym 27.0302 Asym   xmid
     xmid 24.3761 -0.331
     scal 36.7363 -0.992  0.447
 Residual  7.3208

Number of Observations: 35
Number of Groups: 5
```

Note that the default estimation method in `nlme` is maximum likelihood (ML), while the default in `lme` is restricted maximum likelihood (REML). The reason for this, as described in §7.2.2, is because nested `nlme` models which differ in either their fixed effects, or their random effects, cannot be compared using their REML likelihoods, thus invalidating most REML likelihood ratio tests of practical interest. In the linear case, REML likelihoods of nested mixed-effects models with common fixed-effects structure are comparable.

More detailed information about the fit is provided by the `summary` method.

```
> summary(fm1Oran.nlme)
 Nonlinear mixed-effects model fit by maximum likelihood
  Model: circumference ~ SSlogis(age, Asym, xmid, scal)
  Data: Orange
      AIC     BIC   logLik
   279.98  295.53  -129.99
```

```
Random effects:
 Formula: list(Asym ~ 1, xmid ~ 1, scal ~ 1)
 Level: Tree
 Structure: General positive-definite
          StdDev   Corr
    Asym 27.0302 Asym   xmid
    xmid 24.3761 -0.331
    scal 36.7363 -0.992  0.447
Residual  7.3208

Fixed effects: list(Asym ~ 1, xmid ~ 1, scal ~ 1)
      Value Std.Error DF t-value p-value
Asym 192.12   14.045 28  13.679  <.0001
xmid 727.74   34.618 28  21.022  <.0001
scal 356.73   30.537 28  11.682  <.0001
 Correlation:
        Asym    xmid
xmid  0.275
scal -0.194   0.666
. . .
```

It is interesting, at this point, to compare the `nlme` fit with the simple `nls` fit that ignores the grouped structure of the data, obtained in §8.1.1.

```
> summary(fm1Oran.nls)
. . .
Parameters:
      Value Std. Error t value
Asym 192.68    20.239  9.5203
xmid 728.71   107.272  6.7931
scal 353.49    81.460  4.3395

Residual standard error: 23.3721 on 32 degrees of freedom
. . .
```

The fixed-effects estimates are similar, but the standard errors are much smaller in the `nlme` fit. The estimated within-group standard error is also considerably smaller in the `nlme` fit. This is because the between-group variability is not incorporated in the `nls` model, being absorbed in the standard error. This pattern is generally observed when comparing mixed-effects versus fixed-effects fits.

In the `fm1Oran.nlme` fit the estimated correlation of -0.992 between `Asym` and `xmid` suggests that the estimated variance–covariance matrix is ill-conditioned and that the random-effects structure may be over-parameterized. The scatter-plot matrix of the estimated random effects, produced by the `pairs` method, provides a useful diagnostic plot for assessing over-parameterization problems.

```
> pairs( fm1Oran.nlme )                          # Figure 8.9
```

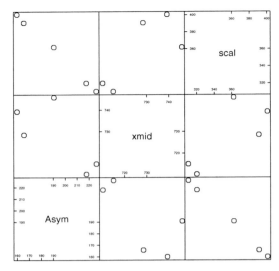

FIGURE 8.9. Pairs plot for the random-effects estimates corresponding to fm1Oran.nlme.

The nearly perfect alignment between the `Asym` random effects and the `scal` random effects further indicates that the model is over-parameterized.

The individual confidence intervals for the parameters in the `nlsList` model described by `fm1Oran.lis`, displayed in Figure 8.4 and discussed in §8.1.3, suggested that `Asym` was the only parameter requiring a random effect to account for its variation among trees. We can fit the corresponding model using the `update` method and compare it to the *full* model `fm1Oran.nlme` using the `anova` method.

```
> fm2Oran.nlme <- update( fm1Oran.nlme, random = Asym ~ 1 )
> anova( fm1Oran.nlme, fm2Oran.nlme )
            Model df    AIC    BIC  logLik    Test L.Ratio p-value
fm1Oran.nlme    1 10 279.98 295.53 -129.99
fm2Oran.nlme    2  5 273.17 280.95 -131.58 1 vs 2  3.1896  0.6708
```

The large p-value for the likelihood ratio test confirms that the `xmid` and `scal` random effects are not needed in the `nlme` model and that the simpler model `fm2Oran.nlme` is to be preferred.

As in the linear case, we must check if the assumptions underlying the nonlinear mixed-effects model appear valid for the model fitted to the data. The two basic distributional assumptions are the same as for the `lme` model:

Assumption 1: the within-group errors are independent and identically normally distributed, with mean zero and variance σ^2, and they are independent of the random effects.

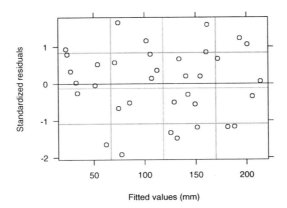

FIGURE 8.10. Scatter plot of standardized residuals versus fitted values for `fm1Oran.nlme`.

Assumption 2: the random effects are normally distributed, with mean zero and covariance matrix $\boldsymbol{\Psi}$ (not depending on the group) and are independent for different groups;

The `plot` and `qqnorm` methods provide the most useful tools for assessing these assumptions. For example, a plot of the standardized residuals versus the fitted values in Figure 8.10

```
> plot( fm1Oran.nlme )                        # Figure 8.10
```

shows that the residuals are distributed symmetrically around zero, with an approximately constant variance. It does not indicate any violations of the assumptions for the within-group error:

The adequacy of the fitted model is better visualized by displaying the fitted and observed values in the same plot. The `augPred` method, which produces the *augmented predictions*, and its associated `plot` method are used for that. For comparison, both the population predictions (obtained by setting the random effects to zero) and the within-group predictions (using the estimated random effects) are displayed in Figure 8.11, produced with

```
> ## level = 0:1 requests fixed (0) and within-group (1) predictions
> plot( augPred(fm2Oran.nlme, level = 0:1),         # Figure 8.11
+    layout = c(5,1) )
```

The tree-specific predictions follow the observed values closely, indicating that the logistic mixed-effects model (8.4) explains the trunk circumference growth in the orange trees data well.

The normal plot of the standardized residuals, shown in Figure 8.12, does not indicate any violations of the normality assumption for the within-group errors.

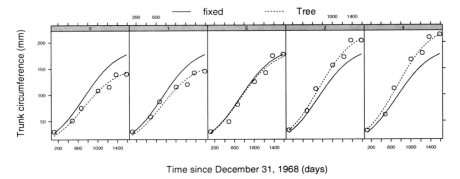

Time since December 31, 1968 (days)

FIGURE 8.11. Population predictions (**fixed**), within-group predictions (**Tree**), and observed trunk circumferences (circles) versus age of tree, for the **fm2Oran.nlme** model .

```
> qqnorm( fm2Oran.nlme, abline = c(0,1) )            #Figure 8.12
```

Because there are only five trees in the data, with just seven observations each, we cannot reliably test assumptions about the random-effects distribution and the independence of the within-group errors.

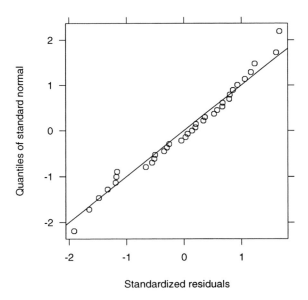

FIGURE 8.12. Normal probability plot of the standardized residuals from a nonlinear mixed-effects model fit **fm1Oran.nlme** to the orange data.

Theophylline Kinetics

The nonlinear mixed-effects version of the first-order open-compartment model (8.3), with all parameters as mixed-effects, is

$$c_{ij} = \frac{D_i \exp\left(lKe_i + lKa_i - lCl_i\right)}{\exp\left(lKa_i\right) - \exp\left(lKe_i\right)}$$
$$\times \left\{ \exp\left[-\exp\left(lKe_i\right) t_{ij}\right] - \exp\left[-\exp\left(lKa_i\right) t_{ij}\right] \right\} + \epsilon_{ij},$$
$$\phi_i = \begin{bmatrix} lKe_i \\ lKa_i \\ lCl_i \end{bmatrix} = \begin{bmatrix} \beta_1 \\ \beta_2 \\ \beta_3 \end{bmatrix} + \begin{bmatrix} b_{1i} \\ b_{2i} \\ b_{3i} \end{bmatrix} = \boldsymbol{\beta} + \boldsymbol{b}_i, \tag{8.5}$$
$$\boldsymbol{b}_i \sim \mathcal{N}\left(\boldsymbol{0}, \boldsymbol{\Psi}\right), \quad \epsilon_{ij} \sim \mathcal{N}\left(0, \sigma^2\right),$$

where c_{ij} is the theophylline concentration for patient i measured at time t_{ij}, with an initial dose D_i. The pharmacokinetic parameters ϕ_i are interpreted as in (8.3), but are now allowed to vary with subject. The fixed effects, $\boldsymbol{\beta}$, represent the population average of the ϕ_i and the random effects, $\boldsymbol{b}_i$, their deviations from the population average.

Fitting the nonlinear mixed-effects model (8.5) to the theophylline data is done with the simple call

```
> fm1Theo.nlme <- nlme( fm1Theo.lis )
> fm1Theo.nlme
Nonlinear mixed-effects model fit by maximum likelihood
  Model: conc ~ SSfol(Dose, Time, lKe, lKa, lCl)
  Data: Theoph
  Log-likelihood: -173.32
  Fixed: list(lKe ~ 1, lKa ~ 1, lCl ~ 1)
     lKe     lKa     lCl
 -2.4327 0.45146 -3.2145

Random effects:
 Formula: list(lKe ~ 1, lKa ~ 1, lCl ~ 1)
 Level: Subject
 Structure: General positive-definite
         StdDev    Corr
     lKe 0.13104 lKe     lKa
     lKa 0.63783  0.012
     lCl 0.25118  0.995 -0.089
Residual 0.68183

Number of Observations: 132
Number of Groups: 12
```

The estimated correlation between the `lKe` and `lCl` random effects is near one, indicating that the estimated random-effects variance–covariance matrix is ill-conditioned. We can investigate the precision of the correlation estimates with the `intervals` method.

```
> intervals( fm1Theo.nlme, which = "var-cov" )
Approximate 95% confidence intervals

Random Effects:
  Level: Subject
                   lower       est.    upper
        sd(1Ke)  0.058102  0.131142  0.29600
        sd(1Ka)  0.397740  0.637838  1.02287
        sd(1Cl)  0.156811  0.251216  0.40246
    cor(1Ke,1Ka) -0.399511  0.011723  0.41903
    cor(1Ke,1Cl) -0.995248  0.994868  1.00000
    cor(1Ka,1Cl) -0.520140 -0.089480  0.37746

Within-group standard error:
  lower     est.     upper
 0.59598  0.68183  0.78005
```

All three confidence intervals for the correlations include zero. The interval on the correlation between 1Ke and 1Cl shows this quantity is not estimated with any precision whatsoever. It must lie in the interval $[-1, 1]$ and the confidence interval is essentially that complete range.

As a first attempt at simplifying model (8.5), we investigate the assumption that the random effects are independent, that is, the matrix Ψ is diagonal. Structured random-effects variance—covariance matrices are specified in nlme in the same way as in lme: by using a pdMat constructor to specify the desired class of positive-definite matrix. The pdMat classes and methods are described in §4.2.2 and the standard pdMat classes available in the nlme library are listed in Table 4.3. By default, when no pdMat class is specified, a general positive-definite matrix (pdSymm class) is used to represent the random-effects variance–covariance structure. Alternative pdMat classes are specified by calling the corresponding constructor with the random effects formula, or list of formulas, as its first argument. For example, we specify a model with independent random effects for the theophylline data with either

```
> fm2Theo.nlme <- update( fm1Theo.nlme,
+   random = pdDiag(list(1Ke ~ 1, 1Ka ~ 1, 1Cl ~ 1)) )
```

or

```
> fm2Theo.nlme <- update( fm1Theo.nlme,
+   random = pdDiag(1Ke + 1Ka + 1Cl ~ 1) )
> fm2Theo.nlme
  . . .
  Log-likelihood: -177.02
  Fixed: list(1Ke ~ 1, 1Ka ~ 1, 1Cl ~ 1)
     1Ke     1Ka      1Cl
  -2.4547  0.46554  -3.2272
```

```
Random effects:
 Formula: list(lKe ~ 1, lKa ~ 1, lCl ~ 1)
 Level: Subject
 Structure: Diagonal
                lKe      lKa      lCl  Residual
StdDev: 1.9858e-05 0.64382 0.16692   0.70923
. . .
```

The very small estimated standard deviation for `lKe` suggests that the corresponding random effect could be omitted from the model.

```
> fm3Theo.nlme <-
+   update( fm2Theo.nlme, random = pdDiag(lKa + lCl ~ 1) )
```

We use the `anova` method to test the equivalence of the different `nlme` models used so far for the theophylline data. By including `fm3Theo.nlme` as the second model, we obtain the p-values comparing this model with each of the other two.

```
> anova( fm1Theo.nlme, fm3Theo.nlme, fm2Theo.nlme )
             Model df    AIC    BIC  logLik   Test L.Ratio p-value
fm1Theo.nlme     1 10 366.64 395.47 -173.32
fm3Theo.nlme     2  6 366.04 383.34 -177.02 1 vs 2  7.4046  0.1160
fm2Theo.nlme     3  7 368.05 388.23 -177.02 2 vs 3  0.0024  0.9611
```

The simpler `fm3Theo.nlme` model, with two independent random effects for `lKa` and `lCl`, has the smallest AIC and BIC. Also, the large p-values for the likelihood ratio tests comparing it to the other two models indicate that it should be preferred.

The plot of the standardized residuals versus the fitted values in Figure 8.13 gives some indication that the within-group variability increases with the drug concentration.

```
> plot( fm3Theo.nlme )                        # Figure 8.13
```

The use of variance functions in `nlme` to model within-group heteroscedasticity is described in §8.3.1. We postpone further investigation of the within-group error assumptions in the theophylline until that section.

The `qqnorm` method is used to investigate the normality of the random effects.

```
> qqnorm( fm3Theo.nlme, ~ ranef(.) )          # Figure 8.14
```

Figure 8.14 does not indicate any violations of the assumption of normality for the random effects.

8.2.2   Using Covariates with *nlme*

The random effects in a mixed-effects model represent deviations of the individual parameters from the fixed effects. In some applications, these

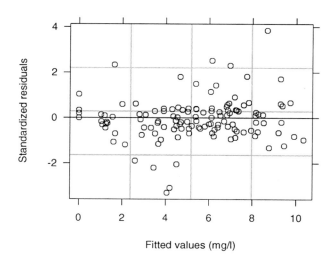

FIGURE 8.13. Scatter plot of standardized residuals versus fitted values for fm3Theo.nlme.

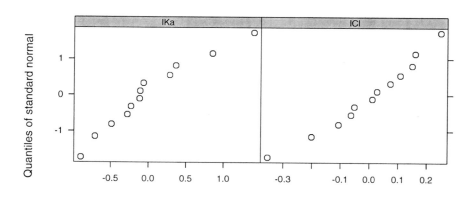

FIGURE 8.14. Normal plot of the estimated random effects corresponding to fm3Theo.nlme fit.

deviations arise from unexplained intergroup variation but, frequently, they can be at least partially explained by differences in covariate values among groups. For example, differences in the absorption rate constant among the subjects in the theophylline example of §8.2.1 may be attributed to differences in age, weight, blood pressure, etc.

Including covariates in the model to explain intergroup variation generally reduces the number of random effects in the model and leads to a better understanding of the mechanism producing the response. We have seen this in the soybean and the phenobarbital examples in Chapter 6.

Some of the questions that need to be addressed in the covariate modeling process are the following:

1. Among the candidate covariates, which are potentially useful in explaining the random-effects variation?

2. Which random effects have their variation best explained by covariates?

3. How should the potentially useful covariates be tested for inclusion in the model?

4. Should random effects be included in, or eliminated from, the modified model?

This section describes a model-building strategy for addressing these questions in the context of nonlinear mixed-effects models and using the capabilities of the nlme library.

Our general approach to address questions 1 and 2 above is to start with an `nlme` model with no covariates to explain random-effects variation, and use plots of the estimated random effects versus the candidate covariates to identify interesting patterns. Because the random effects accommodate individual departures from the population mean, plotting the estimated random effects against the candidate covariates provides useful information for the model-building process. A systematic pattern in a given random effect with respect to a covariate would indicate that the covariate should be included in the model.

If no interesting patterns are observed, we keep the current model. Otherwise, we choose the covariate-coefficient pair with the most promising pattern and test for the inclusion of the covariate in the model. After a covariate has been included in the model, we repeat the procedure for the estimated random effects in the updated model and the remaining candidate covariates, searching for further useful covariates.

The number of additional parameters to be estimated tends to grow considerably with the inclusion of covariates and their associated random effects in the model. If the number of covariate-coefficient combinations is large, we suggest using a *forward stepwise* approach in which covariate-coefficient pairs are included in the model one at a time and the potential importance of the remaining covariates is graphically assessed at each step. The significance of the fixed-effects associated with a covariate included in the model is assessed using the Wald-type tests that are described in §7.2.2 and are included in the output of the summary and anova methods for nlme objects.

The inclusion of new random effects in the model when a covariate is added is rare, but should be investigated. The more common situation is that random effects can be eliminated from the model after covariates are included to account for intergroup variation. In both cases we proceed by comparing nested models using either likelihood ratio tests, or information criterion statistics (AIC and BIC).

We illustrate the use of the proposed model-building strategy with two examples, one from an experiment to evaluate the cold tolerance of grass species and the other from a clinical study of the antiarrhytimic drug quinidine.

Carbon Dioxide Uptake

Potvin, Lechowicz and Tardif (1990) report data from a study of the cold tolerance of a C_4 grass species, *Echinochloa crus-galli*. A total of 12 four-week-old plants, 6 from Québec and 6 from Mississippi, were divided into two groups: control plants that were kept at $26°C$ and chilled plants that were subject to 14 h of chilling at $7°C$. After 10 h of recovery at $20°C$, carbon dioxide (CO_2) uptake rates (in $\mu mol/m^2 s$) were measured for each plant at seven concentrations of ambient CO_2 ($\mu L/L$). The objective of the experiment was to evaluate the effect of plant type and chilling treatment on the CO_2 uptake. The CO_2 data, displayed in Figure 8.15 (and in Figure 3.6, p. 113), are available in the nlme library as the groupedData object CO2. More details about these data are given in Appendix A.5.

```
> CO2
Grouped Data: uptake ~ conc | Plant
    Plant           Type  Treatment conc uptake
 1    Qn1        Quebec nonchilled   95   16.0
 2    Qn1        Quebec nonchilled  175   30.4
 . . .
83   Mc3 Mississippi    chilled  675   18.9
84   Mc3 Mississippi    chilled 1000   19.9
> plot(CO2, outer = ~Treatment*Type, layout = c(4,1)) # Figure 8.15
```

It is clear from Figure 8.15 that the CO_2 uptake rates of Québec plants are greater than those of Mississippi plants and that chilling the plants reduces their CO_2 uptake rates.

An asymptotic regression model with an offset is used in Potvin et al. (1990) to represent the expected CO_2 uptake rate $U(c)$ as a function of the ambient CO_2 concentration c:

$$U(c) = \phi_1 \left\{ 1 - \exp\left[-\exp\left(\phi_2\right)\left(c - \phi_3\right)\right]\right\}, \tag{8.6}$$

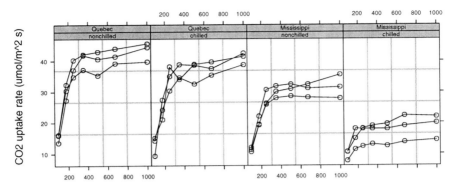

FIGURE 8.15. CO_2 uptake versus ambient CO_2 by treatment and type for *Echinochloa crus-galli* plants, 6 from Quebec and 6 from Mississippi. Half the plants of each type were chilled overnight before the measurements were taken.

where ϕ_1 is the asymptotic CO_2 uptake rate, ϕ_2 is the logarithm of the rate constant, and ϕ_3 is the maximum ambient concentration of CO_2 at which there is no uptake. The logarithm of the rate constant is used to enforce the positivity of the estimated rate constant, while keeping the optimization problem unconstrained.

The selfStart function SSasympOff gives a self-starting implementation of model (8.6), which is used to automatically generate starting estimates for the parameters in an nlsList fit.

```
> fm1CO2.lis <- nlsList( SSasympOff, CO2)
> fm1CO2.lis
Call:
  Model: uptake ~ SSasympOff(conc, Asym, lrc, c0) | Plant
   Data: CO2

Coefficients:
       Asym     lrc      c0
Qn1 38.140 -4.3807  51.221
Qn2 42.872 -4.6658  55.856
 . . .
Mc3 18.535 -3.4654  67.843
Mc1 21.787 -5.1422 -20.395

Degrees of freedom: 84 total; 48 residual
Residual standard error: 1.7982
```

with $\phi_1 = $ Asym, $\phi_2 = $ lrc, and $\phi_3 = $ c0.

The plot of the individual confidence intervals from fm1CO2.lis (not shown) indicates that there is substantial between-plant variation in the asymptote `Asym` and only moderate variation in the log-rate `lrc` and the offset `c0`. We initially consider a nonlinear mixed-effects version of the CO_2 uptake model (8.6) with all parameters as mixed effects and no treatment covariates. The corresponding model for the CO_2 uptake u_{ij} of plant i at ambient CO_2 concentration c_{ij} is

$$u_{ij} = \phi_{1i} \left\{ 1 - \exp\left[-\exp\left(\phi_{2i}\right) \left(c_{ij} - \phi_{3i}\right)\right]\right\} + \epsilon_{ij},$$

$$\boldsymbol{\phi}_i = \begin{bmatrix} \phi_{1i} \\ \phi_{2i} \\ \phi_{3i} \end{bmatrix} = \begin{bmatrix} \beta_1 \\ \beta_2 \\ \beta_3 \end{bmatrix} + \begin{bmatrix} b_{1i} \\ b_{2i} \\ b_{3i} \end{bmatrix} = \boldsymbol{\beta} + \boldsymbol{b}_i, \qquad (8.7)$$

$$\boldsymbol{b}_i \sim \mathcal{N}\left(\boldsymbol{0}, \boldsymbol{\Psi}\right), \quad \epsilon_{ij} \sim \mathcal{N}\left(0, \sigma^2\right),$$

where ϕ_{1i}, ϕ_{2i}, and ϕ_{3i} have the same interpretation as in model (8.6), but are now allowed to vary with plant. The fixed effects, $\boldsymbol{\beta}$, represent the population average of the individual parameters, $\boldsymbol{\phi}_i$, and the random effects, $\boldsymbol{b}_i$, represent the deviations of the $\boldsymbol{\phi}_i$ from their population average. The random effects are assumed to be independent for different plots and the within-group errors ϵ_{ij} are assumed to be independent for different i, j and to be independent of the random effects.

The `nlsList` object `fm1CO2.lis` is used to produce starting estimates for the `nlme` fit of model (8.7).

```
> fm1CO2.nlme <- nlme( fm1CO2.lis )
> fm1CO2.nlme
Nonlinear mixed-effects model fit by maximum likelihood
  Model: uptake ~ SSasympOff(conc, Asym, lrc, c0)
  Data: CO2
  Log-likelihood: -201.31
  Fixed: list(Asym ~ 1, lrc ~ 1, c0 ~ 1)
    Asym      lrc       c0
  32.474  -4.6362  43.543

Random effects:
 Formula: list(Asym ~ 1, lrc ~ 1, c0 ~ 1)
 Level: Plant
 Structure: General positive-definite
          StdDev    Corr
    Asym  9.50999  Asym   lrc
     lrc  0.12828  -0.160
      c0 10.40519   0.999  -0.139
 Residual  1.76641

Number of Observations: 84
Number of Groups: 12
```

The very high correlation between `Asym` and `c0` suggests that the random-effects model is over-parameterized. The scatter plot matrix of the estimated random effects (not shown) confirms that `Asym` and `c0` are in almost

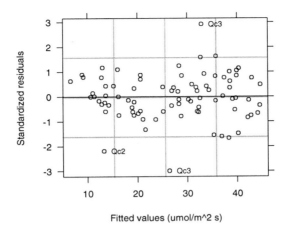

FIGURE 8.16. Scatter plot of standardized residuals versus fitted values for `fm2CO2.nlme`.

perfect linear alignment. A simpler model with just `Asym` and `lrc` as random effects gives an equivalent fit of the data.

```
> fm2CO2.nlme <- update( fm1CO2.nlme, random = Asym + lrc ~ 1 )
> fm2CO2.nlme
 . . .
Random effects:
 Formula: list(Asym ~ 1, lrc ~ 1)
 Level: Plant
 Structure: General positive-definite
         StdDev   Corr
    Asym 9.65939 Asym
     lrc 0.19951 -0.777
Residual 1.80792
 . . .
> anova( fm1CO2.nlme, fm2CO2.nlme )
            Model df    AIC    BIC  logLik    Test L.Ratio p-value
fm1CO2.nlme     1 10 422.62 446.93 -201.31
fm2CO2.nlme     2  7 419.52 436.53 -202.76 1 vs 2   2.8961  0.4079
```

The plot of the standardized residuals versus the fitted values in Figure 8.16 does not indicate any violations from the assumptions on the within-group error. The residuals are distributed symmetrically around zero, with uniform variance. Two large standardized residuals are observed for plant Qc3 and one for plant Qc2.

```
> plot( fm2CO2.nlme,id = 0.05,cex = 0.8,adj = -0.5 )   # Figure 8.16
```

The normal plot of the within-group residuals, not shown, does not indicate violations in the normality of the within-group errors.

The primary question of interest for the CO_2 data is the effect of plant type and chilling treatment on the individual model parameters ϕ_i. The random effects accommodate individual deviations from the fixed effects. Plotting the estimated random effects against the candidate covariates provides useful information for choosing covariates to include in the model. First, we need to extract the estimated random effects from the fitted model and combine them with the covariates. The `ranef` method accomplishes that.

```
> fm2CO2.nlmeRE <- ranef( fm2CO2.nlme, augFrame = T )
> fm2CO2.nlmeRE
            Asym        lrc         Type  Treatment conc uptake
Qn1      6.17160   0.0483563      Quebec nonchilled  435 33.229
Qn2     10.53264  -0.1728531      Quebec nonchilled  435 35.157
Qn3     12.21810  -0.0579930      Quebec nonchilled  435 37.614
Qc1      3.35213  -0.0755880      Quebec    chilled  435 29.971
Qc3      7.47431  -0.1924203      Quebec    chilled  435 32.586
Qc2      7.92855  -0.1803391      Quebec    chilled  435 32.700
Mn3     -4.07333   0.0334485 Mississippi nonchilled  435 24.114
Mn2     -0.14198   0.0056463 Mississippi nonchilled  435 27.343
Mn1      0.24056  -0.1938500 Mississippi nonchilled  435 26.400
Mc2    -18.79914   0.3193732 Mississippi    chilled  435 12.143
Mc3    -13.11688   0.2994393 Mississippi    chilled  435 17.300
Mc1    -11.78655   0.1667798 Mississippi    chilled  435 18.000
> class( fm2CO2.nlmeRE )
[1] "ranef.lme" "data.frame"
```

The `augFrame` argument, when `TRUE`, indicates that summary values for all the variables in the data frame should be returned along with the random effects. The summary values are calculated as in the `gsummary` function (§3.4). When a covariate is constant within a group, such as `Treatment` and `Type` in the `CO2` data, its unique values per group are returned. Otherwise, if the covariate varies within the group and is numeric, such as `conc` and `uptake` in `CO2`, the group means are returned; if it is a categorical variable (`factor` or `ordered`), the most frequent values (modes) within each group are returned.

The `plot` method for the ranef.lme class is the most useful tool for identifying relationships between individual parameters and covariates. The `form` argument is used to specify the desired covariates and plot type. A one-sided formula on the right-hand side, with covariates separated by the `*` operator, results in a `dotplot` of the estimated random effects versus all combinations of the unique values of the variables named in the formula. This plot is particularly useful for a moderate number of categorical covariates (`factor` or `ordered` variables) with a relatively small number of levels, as in the `CO2` example.

```
> plot( fm2CO2.nlmeRE, form = ~ Type * Treatment )    # Figure 8.17
```

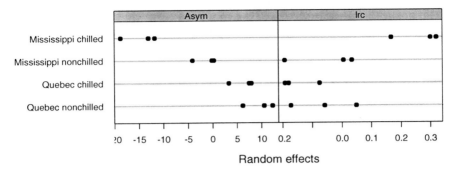

FIGURE 8.17. Dotplots of estimated random effects corresponding to `fm2CO2.nlme` versus all combinations of plant type and chilling treatment.

Figure 8.17 shows a strong relationship between the asymptotic update rate and the covariates: `Asym` decreases when the plants are chilled and is higher among Québec plants than Mississippi plants. The increase in `Asym` from chilled to nonchilled plants is larger among Mississippi plants than Québec plants, suggesting an interaction between `Type` and `Treatment`. There is also some evidence of a `Type:Treatment` interaction on the log-rate `lrc`, but it is less striking than in the case of `Asym`. We include both covariates in the model to explain the `Asym` plant-to-plant variation. The only change required in model (8.7) is in the formulation of ϕ_{1i}.

$$\phi_{1i} = \beta_1 + \gamma_1 x_{1i} + \gamma_2 x_{2i} + \gamma_3 x_{1i} x_{2i} + b_{1i},$$

$$x_{1i} = \begin{cases} -1, & \text{Type of Plant } i = \text{Québec}, \\ 1, & \text{Type of Plant } i = \text{Mississippi}, \end{cases} \qquad (8.8)$$

$$x_{2i} = \begin{cases} -1, & \text{Treatment of Plant } i = \text{nonchilled}, \\ 1, & \text{Treatment of Plant } i = \text{chilled}, \end{cases}$$

where β_1 represents the average asymptotic uptake rate, γ_1 and γ_2 represent, respectively, the plant type and chilling treatment main effects, and γ_3 represents the plant type–chilling treatment interaction. The parameterization used for x_{1i} and x_{2i} in (8.8) is consistent with the default parameterization for factors in S.

```
> contrasts(CO2$Type)
            [,1]
    Quebec   -1
Mississippi   1
> contrasts(CO2$Treatment)
            [,1]
nonchilled   -1
   chilled    1
```

The `update` method is used to fit the model with the covariate terms, which are specified through the `fixed` argument.

```
> fm3CO2.nlme <- update( fm2CO2.nlme,
+    fixed = list(Asym ~ Type * Treatment, lrc + c0 ~ 1),
+    start = c(32.412, 0, 0, 0, -4.5603, 49.344) )
```

Because the fixed-effects model has been reformulated, new starting values must be provided. We use the previous estimates for β_1, β_2 and β_3 and set the initial values for γ_1, γ_2 and γ_3 to zero. The fixed effects are represented internally in nlme in the same order they appear in fixed.

The summary method gives information about the significance of the individual fixed effects.

```
> summary( fm3CO2.nlme )
 .  .  .
    AIC    BIC   logLik
 393.68 417.98 -186.84


Random effects:
 Formula: list(Asym ~ 1, lrc ~ 1)
 Level: Plant
 Structure: General positive-definite
                  StdDev   Corr
Asym.(Intercept) 2.92980 Asym.(
            lrc 0.16373 -0.906
      Residual 1.84957


Fixed effects: list(Asym ~ Type * Treatment, lrc + c0 ~ 1)
                    Value Std.Error DF t-value p-value
   Asym.(Intercept) 32.447   0.9359 67  34.670  <.0001
        Asym.Type  -7.108   0.5981 67 -11.885  <.0001
   Asym.Treatment  -3.815   0.5884 67  -6.483  <.0001
Asym.Type:Treatment -1.197   0.5884 67  -2.033   0.046
              lrc  -4.589   0.0848 67 -54.108  <.0001
               c0  49.479   4.4569 67  11.102  <.0001
 .  .  .
```

The names of the fixed-effects terms include the parameter name. All fixed effects introduced in the model to explain the variability in Asym are significantly different from zero at the 5% level, confirming the previous conclusions from Figure 8.17. The *joint* significance of the fixed effects introduced in the model can be tested with the anova method.

```
> anova( fm3CO2.nlme, Terms = 2:4 )
F-test for: Asym.Type, Asym.Treatment, Asym.Type:Treatment
   numDF denDF F-value p-value
1     3    67  54.835  <.0001
```

As expected, the approximate F-test indicates that the added terms are highly significant.

The inclusion of the experimental factors in the model resulted in a reduction in the estimated standard deviation for the Asym random effects

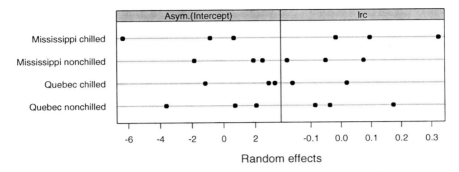

FIGURE 8.18. Dotplots of estimated random effects corresponding to `fm3CO2.nlme` versus all combinations of plant type and chilling treatment.

from 9.66 to 2.93, indicating that a substantial part of the plant-to-plant variation in the asymptotic uptake rate is explained by differences in plant type and chilling treatment. The standard deviations for the `lrc` random effects and the within-group error remained about the same.

We now investigate if any covariates should be included to account for the variability in the `lrc` random effects.

```
> fm3CO2.nlmeRE <- ranef( fm3CO2.nlme, aug = T )
> plot( fm3CO2.nlmeRE, form = ~ Type * Treatment )     # Figure 8.18
```

No systematic pattern can be observed for the estimated `Asym.(Intercept)` random effects in Figure 8.18, but it appears that the chilling treatment has opposite effects on Québec and Mississippi plants, suggesting an interaction. As before, we fit the augmented model with `update`, setting the initial values for the new fixed effects in the model to zero, and test the significance of the new terms using `summary`.

```
> fm3CO2.fix <- fixef( fm3CO2.nlme )          # for starting values
> fm4CO2.nlme <- update( fm3CO2.nlme,
+    fixed = list(Asym + lrc ~ Type * Treatment, c0 ~ 1),
+    start = c(fm3CO2.fix[1:5], 0, 0, 0, fm3CO2.fix[6]) )
> summary( fm4CO2.nlme )
. . .
Random effects:
 Formula: list(Asym ~ 1, lrc ~ 1)
 Level: Plant
 Structure: General positive-definite
                 StdDev   Corr
Asym.(Intercept) 2.349663 Asym.(
 lrc.(Intercept) 0.079608 -0.92
        Residual 1.791950

Fixed effects: list(Asym + lrc ~ Type * Treatment, c0 ~ 1)
```

	Value	Std.Error	DF	t-value	p-value
Asym.(Intercept)	32.342	0.7849	64	41.208	<.0001
Asym.Type	-7.990	0.7785	64	-10.264	<.0001
Asym.Treatment	-4.210	0.7781	64	-5.410	<.0001
Asym.Type:Treatment	-2.725	0.7781	64	-3.502	0.0008
lrc.(Intercept)	-4.509	0.0809	64	-55.743	<.0001
lrc.Type	0.133	0.0552	64	2.417	0.0185
lrc.Treatment	0.100	0.0551	64	1.812	0.0747
lrc.Type:Treatment	0.185	0.0554	64	3.345	0.0014
c0	50.512	4.3646	64	11.573	<.0001

. . .

The `lrc.Type:Treatment` coefficient is highly significant and the `lrc.Type` coefficient is moderately significant. Even though `lrc.Treatment` is not significant (at a 5% level) we keep it in the model because the `Treatment` effect on `lrc` is involved in a highly significant interaction. The estimated standard deviation for the `lrc.(Intercept)` random effect is about 50% of the corresponding estimate in the `fm3CO2.nlme` fit. The remaining standard deviations are about the same as in the previous fit.

After covariates have been introduced in the model to account for intergroup variation, a natural question is which random effects, if any, are still needed. The ratio between a random-effects standard deviation and the absolute value of the corresponding fixed effect gives an idea of the relative intergroup variability for the coefficient, which is often useful in deciding which random effects should be tested for deletion from the model. For the `fm4CO2.nlme` fit these ratios are 7.3% for `Asym.(Intercept)` and 1.8% for `lrc.(Intercept)`, suggesting that the latter should be tested for exclusion first.

```
> fm5CO2.nlme <- update( fm4CO2.nlme, random = Asym ~ 1)
> anova( fm4CO2.nlme, fm5CO2.nlme )
            Model df    AIC    BIC  logLik   Test L.Ratio p-value
fm4CO2.nlme     1 13 388.42 420.02 -181.21
fm5CO2.nlme     2 11 387.06 413.79 -182.53 1 vs 2  2.6369  0.2675
```

The large p-value for the likelihood ratio test and the smaller AIC and BIC values for `fm5CO2.nlme` indicate that no random effects are needed for `lrc`.

To test if a random effect is needed for the asymptotic uptake rate, we need to fit a nonlinear fixed-effects model to the `CO2` data. The `nls` function can be used for that, though it is not designed to efficiently handle parameters that are expressed as linear combinations of covariates. (The `gnls` function, described in §8.3.3, is better suited for this type of model.) To use `nls`, we must first create variables representing the contrasts of interest

```
> CO2$type <- 2 * (as.integer(CO2$Type) - 1.5)
> CO2$treatment <- 2 * (as.integer(CO2$Treatment) - 1.5)
```

and then define all coefficients explicitly in the model formula.

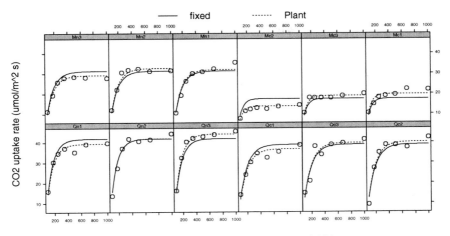

FIGURE 8.19. Plant-specific (`Plant`) and population average (`fixed`) predicted CO_2 uptake rates obtained from `fm5CO2.nlme`.

```
> fm1CO2.nls <- nls(uptake ~ SSasympOff(conc, Asym.Intercept +
+    Asym.Type * type + Asym.Treatment * treatment +
+    Asym.TypeTreatment * type * treatment, lrc.Intercept +
+    lrc.Type * type + lrc.Treatment * treatment +
+    lrc.TypeTreatment * type * treatment, c0), data = CO2,
+    start = c(Asym.Intercept = 32.371, Asym.Type = -8.0086,
+      Asym.Treatment = -4.2001, Asym.TypeTreatment = -2.7253,
+      lrc.Intercept = -4.5267, lrc.Type =  0.13112,
+      lrc.Treatment = 0.093928, lrc.TypeTreatment = 0.17941,
+      c0 = 50.126) )
```

The `anova` method can then be used to compare the models. (Note that the nlme object must appear first in the calling sequence to `anova`, so that the correct method is invoked.)

```
> anova( fm5CO2.nlme, fm1CO2.nls )
            Model df    AIC    BIC  logLik   Test L.Ratio p-value
fm5CO2.nlme     1 11 387.06 413.79 -182.53
fm1CO2.nls      2 10 418.34 442.65 -199.17 1 vs 2  33.289  <.0001
```

The very significant p-value for the likelihood ratio test indicates that the `Asym.(Intercept)` random effect is still needed in the model.

A final assessment of the quality of the fitted model is provided by the plot of the augmented predictions included in Figure 8.19.

```
> plot( augPred(fm5CO2.nlme, level = 0:1),          # Figure 8.19
+    layout = c(6,2) )
```

The plant-specific predictions are in good agreement with the observed CO_2 uptake rates, attesting to the adequacy of the asymptotic regression model. Note that the population average predictions vary with plant type and chilling treatment.

Clinical Study of Quinidine

The `Quinidine` data were described in §3.4 where we explored the structure of the data through summaries. Like the phenobarbital data analyzed in §6.4, the quinidine data are routine clinical pharmacokinetic data characterized by extensive dosage histories for each patient, but relatively sparse information on concentration. Recall that a total of 361 quinidine concentration measurements were made on 136 hospitalized patients under varying dosage regimens. The times since hospitalization at which the quinidine concentrations were measured varied between 0.13 and 8095.5 hours. Most patients have only a few concentration measurements: 34% have only one and 80% have three or fewer. Only 5% of the patients have seven or more observations.

Additional demographic and physiological data were collected for each subject. The additional available covariates are described in Table 8.4. Some of these covariates, such as age, body weight, and creatinine clearance, were "time-varying." That is, their value for a particular patient could change during the course of the study. Others, such as race, remained constant. One of the main objectives of the study was to investigate relationships between the individual pharmacokinetic parameters and the covariates. Statistical analyses of these data using alternative modeling approaches are given in Davidian and Gallant (1992) and in Wakefield (1996).

The model that has been suggested for the quinidine data is the one-compartment open model with first-order absorption. This model can be defined recursively as follows. Suppose that, at time t, a patient receives a dose d_t and prior to that time the last dose was given at time t'. The expected concentration in the serum compartment, C_t, and in the absorption

---

TABLE 8.4. Demographic and physiological covariates in the quinidine data.

Age (yr)	42–92
Glycoprotein concentration (mg/100 dL)	0.39–3.16
Body weight (kg)	41–119
Congestive heart failure	no/mild, moderate, severe
Creatinine clearance (ml/min)	$< 50, \geq 50$
Ethanol abuse	none, current, former
Height (in.)	60–79
Race	Caucasian, Latin, Black
Smoking status	no, yes

compartment, Ca_t, are given by

$$
\begin{aligned}
C_t =\ & C_{t'} \exp\left[-k_e\left(t-t'\right)\right] + \frac{Ca_{t'}\,k_a}{k_a-k_e} \\
& \times\left\{\exp\left[-k_e\left(t-t'\right)\right]-\exp\left[-k_a\left(t-t'\right)\right]\right\},
\end{aligned}
\qquad (8.9)
$$

$$
Ca_t = Ca_{t'}\exp\left[-k_a\left(t-t'\right)\right] + \frac{d_t}{V},
$$

where V is the apparent volume of distribution, k_a is the absorption rate constant, and k_e is the elimination rate constant.

When a patient receives the same dose d at regular time intervals Δ, model (8.9) converges to the steady state model

$$
C_t = \frac{d\,k_a}{V\,(k_a-k_e)}\left[\frac{1}{1-\exp\left(-k_e\Delta\right)}-\frac{1}{1-\exp\left(-k_a\Delta\right)}\right],
$$
$$
Ca_t = \frac{d}{V\left[1-\exp\left(-k_a\Delta\right)\right]}.
\qquad (8.10)
$$

Finally, for a between-dosages time t, the model for the expected concentration C_t, given that the last dose was received at time t', is identical to (8.9).

Using the fact that the elimination rate constant k_e is equal to the ratio between the clearance (Cl) and the volume of distribution (V), we can reparameterize models (8.9) and (8.10) in terms of V, k_a, and Cl.

To ensure that the estimates of V, k_a, and Cl are positive, we can rewrite models (8.9) and (8.10) in terms of $lV = \log(V)$, $lKa = \log(k_a)$ and $lCl = \log(Cl)$.

The initial conditions for the recursive models (8.9) and (8.10) are $C_0 = 0$ and $Ca_0 = d_0/V$, with d_0 denoting the initial dose received by the patient. It is assumed in the model's definition that the bioavailability of the drug—the percentage of the administered dose that reaches the measurement compartment—is equal to one.

The function `quinModel` in the nlme library implements the recursive models (8.9) and (8.10) in S, parameterized in terms of lV, lKa and lCl. This is *not* a self-starting model, so initial values for the fixed effects need to be provided when calling `nlme`. We used values reported in the literature as starting estimates for the fixed effects.

Preliminary analyses of the data, without using any covariates to explain intersubject variation, indicates that only lCl and lV need random effects to account for their variability in the patient population, and that the corresponding random effects can be assumed to be independent. The corresponding model for the fixed and random effects is

$$
lCl_i = \beta_1 + b_{1i}, \quad lV_i = \beta_2 + b_{2i}, \quad lKa_i = \beta_3,
$$
$$
b_i = \begin{bmatrix} b_{1i} \\ b_{2i} \end{bmatrix} \sim \mathcal{N}\left(\mathbf{0}, \begin{bmatrix} \psi_1 & 0 \\ 0 & \psi_2 \end{bmatrix}\right),
\qquad (8.11)
$$

which is fitted in S with

```
> fm1Quin.nlme <-
+   nlme(conc ~ quinModel(Subject, time, conc, dose, interval,
+                         1V, 1Ka, 1Cl),
+        data = Quinidine, fixed = 1V + 1Ka + 1Cl ~ 1,
+        random = pdDiag(1V + 1Cl ~ 1), groups =  ~ Subject,
+        start = list(fixed = c(5, -0.3, 2)),
+        na.action = na.include, naPattern =  ~ !is.na(conc) )
> fm1Quin.nlme
Nonlinear mixed-effects model fit by maximum likelihood
  Model: conc ~ quinModel(Subject, time, conc, dose, interval, ...
  Data: Quinidine
  Log-likelihood: -495.77
  Fixed: 1V + 1Ka + 1Cl ~ 1
      1V      1Ka      1Cl
  5.3796 -0.20535 2.4687

Random effects:
  Formula: list(1V ~ 1, 1Cl ~ 1)
  Level: Subject
  Structure: Diagonal
                1V      1Cl Residual
StdDev: 0.31173 0.32276  0.73871

Number of Observations: 361
Number of Groups: 136
```

The **na.action** and **naPattern** arguments in this call to **nlme** are described in §6.4.

To investigate which covariates may account for patient-to-patient variation in the pharmacokinetic parameters, we first extract the estimated random effects, augmented with summary values for the available covariates (the modal value is used for time-varying factors and the mean for time-varying numeric variables).

```
> fm1Quin.nlmeRE <- ranef( fm1Quin.nlme, aug = T )
> fm1Quin.nlmeRE[1:3,]
            1V         1Cl  time    conc dose interval Age Height
109  0.0005212 -0.0028369 61.58 0.50000   NA       NA  70     67
70   0.0362214  0.3227614  1.50 0.60000   NA       NA  68     69
23  -0.0254211  0.4402551 91.14 0.56667   NA       NA  75     72
      Weight      Race Smoke Ethanol   Heart Creatinine   glyco
109       58 Caucasian    no    none No/Mild      >= 50 0.46000
70        75 Caucasian    no  former No/Mild      >= 50 1.15000
23       108 Caucasian   yes    none No/Mild      >= 50 0.83667
```

The **dotplot** displays used to visualize the relationships between the estimated random effects and the covariates in the CO_2 example do not scale up well when there are a large number of groups, or a large number of covariates in the data, as in the quinidine study. Also, they cannot be used with numeric covariates, like `Weight` and `Age`. The `plot` method for class `ranef` allows a more flexible type of trellis display for these situations. Relationships between estimated random effects and factors are displayed using boxplots, while scatter plots are used for displaying the relationships between the estimated random effects and numeric covariates. Specifying a two-sided formula in the `form` argument, with the random effect on left-hand side and the desired covariates, separated by the + operator, on the right-hand side, indicates to the `plot` method that the more general trellis display should be used. For example, to plot the estimated `lCl` random effects against the available covariates we use

```
> plot( fm1Quin.nlmeRE, form = lCl ~ Age + Smoke + Ethanol +
+           Weight + Race + Height + glyco + Creatinine + Heart,
+           control = list(cex.axis = 0.7) )          # Figure 8.20
```

The resulting plot, shown in Figure 8.20, indicates that clearance decreases with glycoprotein concentration and age, and increases with creatinine clearance and weight. There is also evidence that clearance decreases with severity of congestive heart failure and is smaller in Blacks than in both Caucasians and Latins. The glycoprotein concentration is clearly the most important covariate for explaining the `lCl` interindividual variation. A straight line seems adequate to model the observed relationship.

Figure 8.21 presents the plots of the estimated `lV` random effects versus the available covariates. None of the covariates seems helpful in explaining the variability of this random effect and we do not pursue the modeling of its variability any further.

Initially, only the glycoprotein concentration is included in the model to explain the `lCl` random-effect variation according to a linear model. This modification of model (8.11) is accomplished by writing

$$lCl_{ij} = (\beta_1 + b_{1i}) + \beta_2 \text{glyco}_{ij}. \tag{8.12}$$

Because the glycoprotein concentration may change with time on the same patient, the random effects for `lCl` need to be indexed by both patient i and time j. We fit the mixed-effects model corresponding to (8.12) with

```
> fm1Quin.fix <- fixef( fm1Quin.nlme)          # for initial values
> fm2Quin.nlme <- update( fm1Quin.nlme,
+     fixed = list(lCl ~ glyco, lKa + lV ~ 1),
+     start = c(fm1Quin.fix[3], 0, fm1Quin.fix[2:1]) )
> summary( fm2Quin.nlme )
  . . .
  AIC    BIC   logLik
 891.3 918.52 -438.65
```

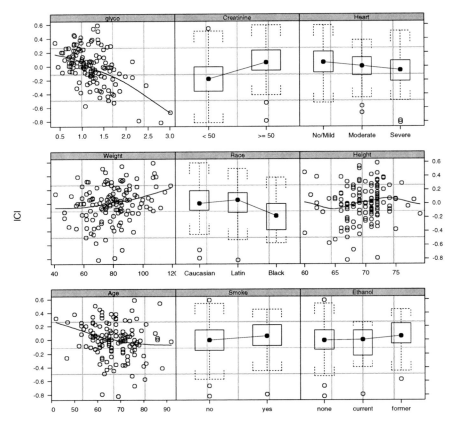

FIGURE 8.20. Estimated log-clearance random effects from model `fm1Quin.nlme` versus demographic and physiological covariates in the quinidine data. A `loess` smoother is included in the scatter plots of the continuous covariates to aid in visualizing possible trends.

```
Random effects:
 Formula: list(lV ~ 1, lCl ~ 1)
 Level: Subject
 Structure: Diagonal
           lV lCl.(Intercept) Residual
StdDev: 0.26764        0.27037  0.63746

Fixed effects: list(lCl ~ glyco, lKa + lV ~ 1)
                Value Std.Error  DF t-value p-value
lCl.(Intercept) 3.1067   0.06473 222  47.997  <.0001
    lCl.glyco -0.4914   0.04263 222 -11.527  <.0001
          lKa -0.6662   0.30251 222  -2.202  0.0287
           lV  5.3085   0.10244 222  51.818  <.0001
```

. . .

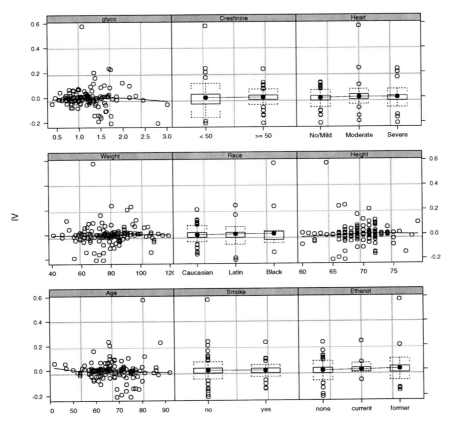

FIGURE 8.21. Estimated log-volume random effects from model `fm1Quin.nlme` versus demographic and physiological covariates in the quinidine data. A `loess` smoother is included in the scatter plots of the continuous covariates to aid in visualizing possible trends.

The estimated `lCl.glyco` fixed effect is very significant, indicating that the glycoprotein concentration should be kept in the model.

To search for further variables to include in the model, we consider the plots of the estimated `lCl.(Intercept)` random effects from the `fm2Quin.nlme` fit versus the covariates, presented in Figure 8.22.

These plots indicate that the estimated `lCl.(Intercept)` random effects increase with creatinine clearance, weight, and height, decrease with age and severity of congestive heart failure, and are smaller in Blacks than in Caucasians and Latins. The most relevant variable appears to be the creatinine clearance, which is included in the model as a binary variable taking value 0 when creatinine is < 50 and 1 when creatinine is ≥ 50.

```
> options( contrasts = c("contr.treatment", "contr.poly") )
```

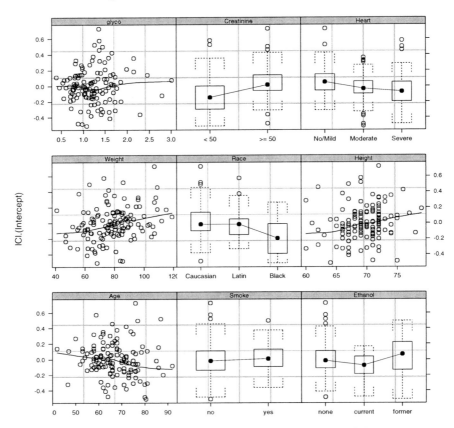

FIGURE 8.22. Estimated log-clearance random effects from model `fm2Quin.nlme` versus demographic and physiological covariates in the quinidine data. A `loess` smoother is included in the scatter plots of the continuous covariates to aid in visualizing possible trends.

```
> fm2Quin.fix <- fixef( fm2Quin.nlme )
> fm3Quin.nlme <- update( fm2Quin.nlme,
+    fixed = list(lCl ~ glyco + Creatinine, lKa + lV ~ 1),
+    start = c(fm2Quin.fix[1:2], 0.2, fm2Quin.fix[3:4]) )
> summary( fm3Quin.nlme )
 . . .
Fixed effects: list(lCl ~ glyco + Creatinine, lKa + lV ~ 1)
                  Value Std.Error  DF t-value p-value
lCl.(Intercept)  3.0291   0.06387 221  47.426  <.0001
     lCl.glyco  -0.4631   0.04117 221 -11.249  <.0001
lCl.Creatinine   0.1503   0.03175 221   4.732  <.0001
           lKa  -0.7458   0.29619 221  -2.518  0.0125
            lV   5.2893   0.10625 221  49.784  <.0001
 . . .
```

The final model produced by this stepwise model-building approach includes an extra term for the patient's body weight to explain the clearance variation. The corresponding model for the log-clearance is expressed as

$$lCl_{ij} = (\beta_1 + b_{1i}) + \beta_4 \text{glyco}_{ij} + \beta_5 \text{Creatinine}_{ij} + \beta_6 \text{Weight}_{ij} \qquad (8.13)$$

and is fit in S with

```
> fm3Quin.fix <- fixef( fm3Quin.nlme )
> fm4Quin.nlme <- update( fm3Quin.nlme,
+    fixed = list(lCl ~ glyco + Creatinine + Weight, lKa + lV ~ 1),
+    start = c(fm3Quin.fix[1:3], 0, fm3Quin.fix[4:5]) )
> summary( fm4Quin.nlme )
. . .
    AIC    BIC   logLik
 870.94 905.94 -426.47

Random effects:
 Formula: list(lV ~ 1, lCl ~ 1)
 Level: Subject
 Structure: Diagonal
              lV lCl.(Intercept) Residual
 StdDev: 0.28154         0.24128  0.63083

Fixed effects: list(lCl~glyco + Creatinine + Weight, lKa+lV ~ 1)
                 Value Std.Error  DF t-value p-value
lCl.(Intercept)  2.7883   0.15167 220  18.384  <.0001
     lCl.glyco  -0.4645   0.04100 220 -11.328  <.0001
 lCl.Creatinine  0.1373   0.03264 220   4.207  <.0001
     lCl.Weight  0.0031   0.00180 220   1.749  0.0816
            lKa -0.7974   0.29959 220  -2.662  0.0083
             lV  5.2833   0.10655 220  49.587  <.0001
. . .
```

The `lCl.Weight` coefficient is not significant at a 5% level, but it is significant at a less conservative 10% level. Given the high level of noise and the small number of observations per patient in the quinidine data, we considered a p-value of 8.2% to be small enough to justify the inclusion of `Weight` in the model.

As reported in previous analyses of the quinidine data (Davidian and Giltinan, 1995, §9.3), there is evidence that the variability in the concentration measurements increases with the quinidine concentration. We postpone the investigation of heteroscedasticity in the quinidine data until §8.3.1, when we describe the use of variance functions in `nlme`.

8.2.3 Fitting Multilevel *nlme* Models

Nonlinear mixed-effects models with nested grouping factors, called *multilevel nonlinear mixed-effects* models, are described in §7.1.2. In this section we describe how to fit and analyze them with the nlme library. The Wafer example of §4.2.3 is used to illustrate the various methods available in nlme for multilevel NLME models.

As in the linear case, the only difference between a single-level and a multilevel fit in nlme is in the specification of the random argument. In the multilevel case, random must provide information about the nested grouping structure and the random-effects model at each level of grouping. This is generally accomplished by specifying random as a *named* list, with names corresponding to the grouping factors. The order of nesting is assumed to be the same as the order of the elements in the list, with the outermost grouping factor appearing first and the innermost grouping factor appearing last. Each element of the random list has the same structure as random in a single-level call: it can be a formula, a list of formulas, a pdMat object, or a list of pdMat objects. Most of the nlme extractors, such as resid and ranef, include a level argument indicating the desired level(s) of grouping.

Manufacturing of Analog MOS Circuits

A multilevel *linear* mixed-effects analysis of the Wafer data is presented in §4.2.3 to illustrate the multilevel capabilities of lme. The final multilevel model obtained in that section to represent the intensity of current y_{ijk} at the kth level of voltage v_k in the jth site within the ith wafer is expressed, for $i = 1, \ldots, 10$, $j = 1, \ldots, 8$, and $k = 1, \ldots, 5$, as

$$y_{ijk} = (\beta_0 + b_{0i} + b_{0i,j}) + (\beta_1 + b_{1i} + b_{1i,j}) v_k + (\beta_2 + b_{2i} + b_{2i,j}) v_k^2$$
$$+ \beta_3 \cos(\omega v_k) + \beta_4 \sin(\omega v_k) + \epsilon_{ijk}$$

$$\boldsymbol{b}_i = \begin{bmatrix} b_{0i} \\ b_{1i} \\ b_{2i} \end{bmatrix} \sim \mathcal{N}(0, \boldsymbol{\Psi}_1), \qquad \boldsymbol{b}_{i,j} = \begin{bmatrix} b_{0i,j} \\ b_{1i,j} \\ b_{2i,j} \end{bmatrix} \sim \mathcal{N}(\boldsymbol{0}, \boldsymbol{\Psi}_2),$$

$$\epsilon_{ijk} \sim \mathcal{N}(0, \sigma^2).$$

$$(8.14)$$

where β_0, β_1, and β_2 are the fixed effects in the quadratic model, β_3 and β_4 are the fixed effects for the cosine wave of frequency ω, $\boldsymbol{b}_i$ is the *wafer-level* random effects vector, $\boldsymbol{b}_{i,j}$ is the *site within wafer-level* random-effects vector, and ϵ_{ijk} is the within-group error. The $\boldsymbol{b}_i$ are assumed to be independent for different i, the $\boldsymbol{b}_{i,j}$ are assumed to be independent for different i, j and independent of the $\boldsymbol{b}_i$, and the ϵ_{ijk} are assumed to be independent for different i, j, k and independent of the random effects. The wafer-level variance–covariance matrix $\boldsymbol{\Psi}_1$ is general positive-definite and the site-within-wafer-level matrix $\boldsymbol{\Psi}_2$ is block-diagonal, with a 1×1 block corresponding to the variance of $b_{0i,j}$ and a 2×2 block corresponding to variance–covariance matrix of $[b_{1i,j}, b_{2i,j}]^T$.

Because its model function is nonlinear in the frequency ω, model (8.14) is actually an example of a multilevel nonlinear mixed-effects model. It is treated as a linear model in §4.2.3 by holding ω fixed at a previously estimated value. Some of the disadvantages of this approach are that no precision can be attached to the estimate of ω and we cannot test if random effects are needed to account for variation in ω, either at the wafer level, or at the site-within-wafer level. In this section we treat (8.14) as a multilevel NLME model, allowing ω to be estimated with the other parameters and testing if random effects are needed to account for its variation in the data.

We can rewrite (8.14) in the usual two-stage NLME model formulation.

$$y_{ijk} = \phi_{1ij} + \phi_{2ij}\cos(\omega v_k) + \phi_{3ij}\sin(\omega v_k) + \epsilon_{ijk},$$
$$\phi_{1ij} = (\beta_0 + b_{0i} + b_{0i,j}) + (\beta_1 + b_{1i} + b_{1i,j})v_k + (\beta_2 + b_{2i} + b_{2i,j})v_k^2,$$
$$\phi_{2ij} = \beta_3, \qquad \phi_{3ij} = \beta_4,$$

$$(8.15)$$

where the fixed effects, $\boldsymbol{\beta} = (\beta_1, \dots, \beta_4)$, the random effects, $\boldsymbol{b}_i$, $\boldsymbol{b}_{i,j}$, and the within-group errors, ϵ_{ijk}, are defined as in (8.14). To illustrate some of the multilevel capabilities in `nlme`, we initially fit model (8.15) with fixed $\omega = 4.5679$ as in §4.2.3. To get results that are comparable to the `fm5Wafer` fit in §4.2.3, we need to set the estimation method in `nlme` to REML.

```
> fm1Wafer.nlmeR <- nlme( current ~ A + B * cos(4.5679 * voltage) +
+    C * sin(4.5679 * voltage), data = Wafer,
+    fixed = list(A ~ voltage + voltage^2, B + C ~ 1),
+    random = list(Wafer = A ~ voltage + voltage^2,
+                  Site = pdBlocked(list(A~1, A~voltage+voltage^2-1))),
+    start = fixef(fm4Wafer), method = "REML")
> fm1Wafer.nlmeR
 . . .
  Fixed: list(A ~ voltage + voltage^2, B + C ~ 1)
  A.(Intercept) A.voltage A.I(voltage^2)        B        C
        -4.2554    5.6224         1.2585 -0.095557  0.10435

Random effects:
 Formula: A ~ voltage + voltage^2 | Wafer
 Structure: General positive-definite
                   StdDev    Corr
  A.(Intercept) 0.131805  A.(In) A.vltg
      A.voltage 0.354743  -0.967
 A.I(voltage^2) 0.049955   0.814 -0.935

 Composite Structure: Blocked

 Block 1: A.(Intercept)
 Formula: A ~ 1 | Site %in% Wafer
           A.(Intercept)
 StdDev:        0.066563
```

```
Block 2: A.voltage, A.I(voltage^2)
Formula: A ~ voltage + voltage^2 - 1 | Site %in% Wafer
Structure: General positive-definite
                    StdDev     Corr
        A.voltage 0.2674083 A.volt
A.I(voltage^2) 0.0556444 -0.973
        Residual 0.0091086
 . . .
```

The only difference in the multilevel model specification in `random` between `lme` and `nlme` is the use of one-sided formulas in the former and two-sided formulas in the latter. As expected, the estimation results for `fm1Wafer.nlme` are almost identical to the ones for `fm5Wafer`.

Because only five distinct voltages are used in the MOS circuit experiment, at most five different fixed effects can be used in a mixed-effects model fitted to the `Wafer` data. This is true in general: the number of *estimable* fixed effects in a mixed-effects model cannot exceed the number of distinct design points in the data used to fit it. Therefore, in order to allow the frequency ω to be estimated from the data, we need to drop at least one fixed effect from model (8.15).

Examining the fixed effects estimates in `fm1Wafer.nlme` we see that $\widehat{\beta}_3 \simeq -\widehat{\beta}_4$. We make the assumption that $\beta_3 = -\beta_4$ in (8.15), which, using the identity $\cos(\theta) - \sin(\theta) = \cos(\theta + \pi/4)$, gives the modified model

$$y_{ijk} = \phi_{1ij} + \phi_{2ij} \cos\left(\omega_{ij} v_k + \pi/4\right) + \epsilon_{ijk}. \tag{8.16}$$

To compensate for the restriction that $\beta_3 = -\beta_4$, we include random effects for ϕ_{2ij} at the wafer and site-within-wafer levels. Preliminary analyses of the modified model indicated that random effects for ω_{ij} are needed at the wafer level only. The corresponding model for the fixed and random effects is

$$\phi_{1ij} = (\beta_0 + b_{0i} + b_{0i,j}) + (\beta_1 + b_{1i} + b_{1i,j}) v_k + (\beta_2 + b_{2i} + b_{2i,j}) v_k^2,$$
$$\phi_{2ij} = \beta_3 + b_{3i} + b_{3i,j},$$
$$\omega_{ij} = \beta_4^* + b_{4i},$$

$$\boldsymbol{b}_i = \begin{bmatrix} b_{0i} \\ b_{1i} \\ b_{2i} \\ b_{3i} \\ b_{4i} \end{bmatrix} \sim \mathcal{N}\left(0, \boldsymbol{\Psi}_1\right), \qquad \boldsymbol{b}_{i,j} = \begin{bmatrix} b_{0i,j} \\ b_{1i,j} \\ b_{2i,j} \\ b_{3i,j} \end{bmatrix} \sim \mathcal{N}\left(\mathbf{0}, \boldsymbol{\Psi}_2\right),$$

$$\epsilon_{ijk} \sim \mathcal{N}\left(0, \sigma^2\right). \tag{8.17}$$

Because of the large number of random effects at each grouping level, to make the estimation problem more numerically stable, we make the simplifying assumption that the random effects are independent. That is, we assume that $\boldsymbol{\Psi}_1$ and $\boldsymbol{\Psi}_2$ are diagonal matrices.

A maximum likelihood fit of model (8.16), with fixed and random-effects structures given in (8.17), is obtained with

```
> fm2Wafer.nlme <- nlme( current ~ A + B * cos(w * voltage + pi/4),
+    data = Wafer, fixed = list(A ~ voltage + voltage^2, B + w ~ 1),
+    random = list(Wafer = pdDiag(list(A ~ voltage + voltage^2,
+                                      B + w ~ 1)),
+                  Site = pdDiag(list(A ~ voltage+voltage^2, B ~ 1))),
+    start = c(fixef(fm1Wafer.nlme)[-5], 4.5679) )
> fm2Wafer.nlme
Nonlinear mixed-effects model fit by maximum likelihood
  Model: current ~ A + B * cos(w * voltage + pi/4)
  Data: Wafer
  Log-likelihood: 766.44
  Fixed: list(A ~ voltage + voltage^2, B + w ~ 1)
  A.(Intercept) A.voltage A.I(voltage^2)       B       w
      -4.2653     5.6329         1.256 -0.14069  4.5937

Random effects:
 Formula: list(A ~ voltage + voltage^2, B ~ 1, w ~ 1)
 Level: Wafer
 Structure: Diagonal
         A.(Intercept) A.voltage A.I(voltage^2)        B        w
StdDev:         0.1332   0.34134       0.048243 0.0048037 0.014629

 Formula: list(A ~ voltage + voltage^2, B ~ 1)
 Level: Site %in% Wafer
 Structure: Diagonal
         A.(Intercept) A.voltage A.I(voltage^2)        B  Residual
StdDev:        0.084082   0.30872       0.067259 0.0067264 0.0008428
 . . .
```

Even though models (8.15) and (8.16) are not nested, they can compared using information criterion statistics. The `anova` method can be used for that, but we must first obtain a maximum likelihood fit of model (8.15).

```
> fm1Wafer.nlme <- update( fm1Wafer.nlmeR, method = "ML" )
> anova( fm1Wafer.nlme, fm2Wafer.nlme, test = F )
              Model df     AIC     BIC logLik
fm1Wafer.nlme     1 16 -1503.9 -1440.1 767.96
fm2Wafer.nlme     2 15 -1502.9 -1443.0 766.44
```

The more conservative BIC favors the model with fewer parameters, `fm2Wafer.nlme`, while the more liberal AIC favors the model with larger log-likelihood, `fm1Wafer.nlme`.

The `intervals` method is used to obtain confidence intervals on the fixed effects and the variance components.

```
> intervals( fm2Wafer.nlme )
Approximate 95% confidence intervals
```

```
Fixed effects:
                   lower      est.     upper
A.(Intercept) -4.35017 -4.26525 -4.18033
   A.voltage   5.40994  5.63292  5.85589
A.I(voltage^2)  1.22254  1.25601  1.28948
            B -0.14403 -0.14069 -0.13735
            w  4.58451  4.59366  4.60281

Random Effects:
 Level: Wafer
                         lower      est.     upper
   sd(A.(Intercept))  0.0693442 0.1331979 0.255849
     sd(A.voltage)    0.1715436 0.3413426 0.679214
 sd(A.I(voltage^2))   0.0222685 0.0482433 0.104516
             sd(B)    0.0022125 0.0048037 0.010430
             sd(w)    0.0078146 0.0146290 0.027385
 Level: Site
                         lower      est.     upper
   sd(A.(Intercept))  0.0664952 0.0840825 0.1063214
     sd(A.voltage)    0.2442003 0.3087244 0.3902973
 sd(A.I(voltage^2))   0.0532046 0.0672589 0.0850257
             sd(B)    0.0053128 0.0067264 0.0085162

Within-group standard error:
     lower        est.       upper
 0.00066588  0.0008428  0.0010667
```

The fixed effects and the within-group standard error are estimated with
more relative precision than the random-effects variance components. In
the random-effects variance components, the *site-within-wafer* standard
deviations are estimated with greater precision than the *wafer* standard
deviations.

The plot of the within-group residuals versus voltage by wafer, displayed
in Figure 8.23, and produced with

```
> plot( fm2Wafer.nlme, resid(.) ~ voltage | Wafer,
+        panel = function(x, y, ...) {
+                panel.grid()
+                panel.xyplot(x, y)
+                panel.loess(x, y, lty = 2)
+                panel.abline(0, 0)
+                } )                              # Figure 8.23
```

does not reveal any periodic patterns as observed, for example, in Figure 4.27, indicating that the inclusion of a random effect for ω accounted
successfully for variations in frequency among wafers.

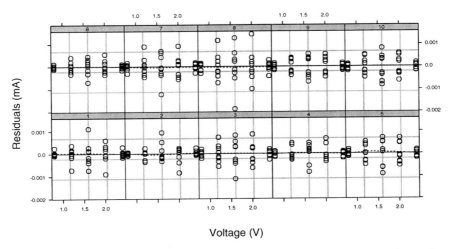

Voltage (V)

FIGURE 8.23. Scatter plots of within-group residuals versus voltage by wafer for the fm2Wafer.nlme fit. A loess smoother has been added to each panel to enhance the visualization of the residual pattern.

The normal plots of the within-group residuals and of the estimated site-within-wafer random effects, not shown here, do not indicate any violations of the NLME model assumptions.

8.3 Extending the Basic **nlme** Model

The extended nonlinear mixed-effects model with heteroscedastic, correlated within-group errors was introduced in §7.4.1. In this section, we describe the use of the **nlme** function for fitting such models. The use of variance functions for modeling heteroscedasticity in NLME models are discussed and illustrated in §8.3.1. Correlation structures for modeling within-group error dependence in NLME models are illustrated in §8.3.2. The **gnls** function to fit the extended nonlinear regression model presented in §7.5.1 is described and illustrated in §8.3.3, together with its associated class and methods.

8.3.1 Variance Functions in **nlme**

Variance functions are specified for an **nlme** model in the same way as for an lme model: through the **weights** argument. Any of the varFunc classes described in §5.2.1, and listed in Table 5.1, can also be used with **nlme**. The same diagnostic plots discussed in §5.2.1 for identifying within-group heteroscedasticity and assessing the adequacy of a variance function for lme objects, can also be used with nlme objects. We revisit the theophylline example of §8.2.1 and the quinidine example of §8.2.2 to illustrate the use of variance functions in `nlme`.

Theophylline Kinetics

The plot of the standardized residuals versus the fitted values for the fitted object `fm3Theo.nlme`, displayed in Figure 8.13, suggests that the within-group variance increases with the concentration of theophylline.

The definition of the first-order open-compartment model (8.2) implies that the fitted value for the concentration at time $t = 0$ is $\hat{c}_0 = 0$. Therefore, a *power* variance function, a natural candidate for this type of heteroscedastic pattern, cannot be used in this example, as the corresponding weights are undefined at $t = 0$. (Davidian and Giltinan (1995, §5.5, p. 145) argue that the observations at $t = 0$ do not add any information for the model and should be omitted from the data. We retain them here for illustration.) The *constant plus power* variance function, described in §5.2 and represented in the nlme library by the varConstPower class, accommodates the problem with $\hat{c}_0 = 0$ by adding a constant to the power of the fitted value. In the theophylline example, the varConstPower variance function is expressed as $g(\hat{c}_{ij}, \boldsymbol{\delta}) = \delta_1 + \hat{c}_{ij}^{\delta_2}$. We incorporate it in the nlme fit using

```
> fm4Theo.nlme <- update( fm3Theo.nlme,
+     weights = varConstPower(power = 0.1) )
> fm4Theo.nlme
Nonlinear mixed-effects model fit by maximum likelihood
  Model: conc ~ SSfol(Dose, Time, lKe, lKa, lCl)
  Data: Theoph
  Log-likelihood: -167.68
  Fixed: list(lKe ~ 1, lKa ~ 1, lCl ~ 1)
     lKe      lKa      lCl
 -2.4538  0.43348  -3.2275


Random effects:
 Formula: list(lKa ~ 1, lCl ~ 1)
 Level: Subject
 Structure: Diagonal
            lKa      lCl Residual
StdDev: 0.6387  0.16979    0.3155


Variance function:
 Structure: Constant plus power of variance covariate
 Formula:   ~ fitted(.)
 Parameter estimates:
   const    power
 0.71966  0.31408
```

An initial value for the power parameter (δ_2) is specified in the call to the varConstPower constructor to avoid convergence problems associated with the default value power=0, for this example.

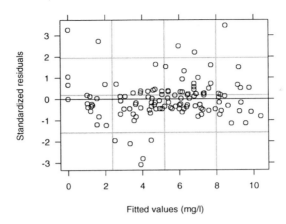

FIGURE 8.24. Scatter plot of standardized residuals versus fitted values for fm4Theo.nlme.

The anova method is used to assess the statistical significance of the variance function.

```
> anova( fm3Theo.nlme, fm4Theo.nlme )
              Model df    AIC    BIC  logLik    Test L.Ratio p-value
fm3Theo.nlme      1  6 366.04 383.34 -177.02
fm4Theo.nlme      2  8 351.35 374.41 -167.68 1 vs 2  18.694    1e-04
```

The small p-value for the likelihood ratio test indicates that the incorporation of the variance function in the model produced a significant increase in the log-likelihood. Both the AIC and the BIC also favor the fm4Theo.nlme fit. The plot of the standardized residuals versus the fitted values, shown in Figure 8.24, confirms the adequacy of the varConstPower variance function.

```
> plot( fm4Theo.nlme )                          # Figure 8.24
```

Clinical Study of Quinidine

Figure 8.25 presents the scatter plot of the standardized residuals versus the fitted values, corresponding to the final model for the quinidine data in §8.2.2, represented by the fitted object fm4Quin.nlme.

```
> ## xlim used to hide an unusually high fitted value and enhance
> ## visualization of the heteroscedastic pattern
> plot( fm4Quin.nlme, xlim = c(0, 6.2) )        # Figure 8.25
```

As reported in previous analyses of the quinidine data (Davidian and Giltinan, 1995, §9.3), and indicated in Figure 8.25, the variance of the within-group errors appears to increase with the quinidine concentration. The fitted values do not get sufficiently close to zero to cause problems in

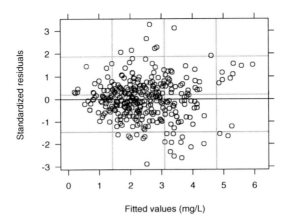

FIGURE 8.25. Scatter plot of standardized residuals versus fitted values for fm4Quin.nlme.

the calculation of weights for the power variance function and we choose the varPower class to model the within-group heteroscedasticity.

```
> fm5Quin.nlme <- update( fm4Quin.nlme, weights = varPower() )
> summary( fm5Quin.nlme )
. . .
Random effects:
 Formula: list(lV ~ 1, lCl ~ 1)
 Level: Subject
 Structure: Diagonal
            lV lCl.(Intercept) Residual
StdDev: 0.32475    0.25689 0.25548

Variance function:
 Structure: Power of variance covariate
 Formula:   ~ fitted(.)
 Parameter estimates:
   power
 0.96616
Fixed effects: list(lCl ~ glyco + Creatinine + Weight, lKa + lV ~ 1)
                 Value Std.Error  DF t-value p-value
lCl.(Intercept)  2.7076   0.15262 220  17.741  <.0001
       lCl.glyco -0.4110   0.04487 220  -9.161  <.0001
  lCl.Creatinine  0.1292   0.03494 220   3.696  0.0003
      lCl.Weight  0.0033   0.00179 220   1.828  0.0689
             lKa -0.4269   0.25518 220  -1.673  0.0958
              lV  5.3700   0.08398 220  63.941  <.0001
. . .
```

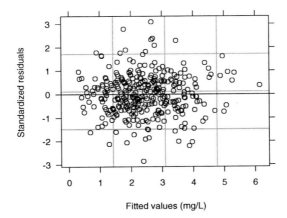

Fitted values (mg/L)

FIGURE 8.26. Scatter plot of standardized residuals versus fitted values for fm5Quin.nlme.

```
> anova( fm4Quin.nlme, fm5Quin.nlme )
             Model df    AIC    BIC  logLik    Test L.Ratio p-value
fm4Quin.nlme     1  9 870.94 905.94 -426.47
fm5Quin.nlme     2 10 812.13 851.01 -396.06 1 vs 2  60.813  <.0001
```

The incorporation of the power variance function in the NLME model for the quinidine data produced a significant increase in the log-likelihood, as evidenced by the small p-value for the likelihood ratio statistics. The plot of the standardized residuals versus fitted values, displayed in Figure 8.26, gives further evidence of the adequacy of the variance function model.

```
> plot( fm5Quin.nlme, xlim = c(0, 6.2) )          # Figure 8.26
```

8.3.2 Correlation Structures in nlme

Just as in lme and gls, correlation structures are specified in nlme through the correlation argument. All of the corStruct classes described in §5.3.3, and listed in Table 5.3, can be used with nlme. As in the linear case, we investigate the need for within-group correlation structures in the NLME model by looking at plots of the empirical autocorrelation function (ACF) and the sample semivariogram, and assess the adequacy of a particular corStruct class by examining plots of the normalized residuals. We revisit the ovary example of §5.3.4 to illustrate the use of correlation structures with nlme.

Counts of Ovarian Follicles

The ovary example was analyzed in §5.3.4 using an LME model, by assuming that the number of ovarian follicles was a periodic function of time with known frequency equal to 1. A more general model formulation assumes that the frequency is an unknown parameter to be estimated from the data, with a possible random effect associated with it. Because the frequency enters the model nonlinearly, this becomes an NLME model, represented as

$$y_{ij} = \phi_{0i} + \phi_{1i} \sin\left(2\pi\phi_{2i}t_{ij}\right) + \phi_{3i} \cos\left(2\pi\phi_{2i}t_{ij}\right) + \epsilon_{ij}, \qquad (8.18)$$

where y_{ij} represents the number of follicles observed for mare i at time t_{ij}, ϕ_{0i}, ϕ_{1i}, and ϕ_{3i} represent the intercept and the terms defining the amplitude and the phase of the cosine wave for mare i, ϕ_{2i} is the frequency of cosine wave for mare i, and ϵ_{ij} is the within-group error. We initially assume that the within-group errors are independently distributed as $\mathcal{N}(0, \sigma^2)$.

The final LME model used to fit the Ovary data in §5.3.4 used independent random effects for the ϕ_{0i} and ϕ_{1i} coefficients in model (8.18). We use this as a starting point for the random-effects model, incorporating an extra independent random effect for the frequency ϕ_{2i}. The fixed- and random-effects models corresponding to (8.18) are then expressed as

$$\phi_{0i} = \beta_0 + b_{0i}, \quad \phi_{1i} = \beta_1 + b_{1i}, \quad \phi_{2i} = \beta_2 + b_{2i}, \quad \phi_{3i} = \beta_3,$$

$$\boldsymbol{b}_i = \left[\begin{array}{c} b_{0i} \\ b_{1i} \\ b_{2i} \end{array} \right] \sim \mathcal{N}\left(\boldsymbol{0}, \boldsymbol{\Psi}\right), \qquad (8.19)$$

where $\boldsymbol{\Psi}$ is diagonal. The random effects, $\boldsymbol{b}_i$, are assumed to be independent for different i and to be independent of the within-group errors.

We fit model (8.18), with fixed- and random-effects structures given by (8.19), using

```
> fm1Ovar.nlme <- nlme(follicles ~ A + B * sin(2 * pi * w * Time) +
+    C * cos(2 * pi * w *Time), data = Ovary,
+    fixed = A + B + C + w ~ 1, random = pdDiag(A + B + w ~ 1),
+    start = c(fixef(fm5Ovar.lme), 1))
> fm1Ovar.nlme
Nonlinear mixed-effects model fit by maximum likelihood
  Model: follicles ~ A + B * sin(2 * pi * w * Time) +
         C * cos(2 * pi * w * Time)
  Data: Ovary
  Log-likelihood: -803.83
  Fixed: A + B + C + w ~ 1
      A       B        C        w
 12.184  -3.376  -1.6812  0.93605

Random effects:
 Formula: list(A ~ 1, B ~ 1, w ~ 1)
 Level: Mare
 Structure: Diagonal
              A        B         w  Residual
 StdDev: 2.9051  2.0061  0.073598    2.9387
  . . .
```

The estimated fixed effects for fm50var.lme are used as initial values for β_0, β_1, and β_3.

As mentioned in §5.3.4, the observations in the Ovary data were collected at equally spaced calendar times, and then converted to an ovulation cycle scale. Therefore, the empirical ACF can be used to investigate the correlation at different lags. The ACF method can also be used with nlme objects.

```
> ACF( fm10var.nlme )
   lag        ACF
1    0  1.0000000
2    1  0.3110027
3    2  0.0887701
4    3 -0.0668554
5    4 -0.0314934
6    5 -0.0810381
7    6 -0.0010647
8    7  0.0216463
9    8  0.0137578
10   9  0.0097497
11  10 -0.0377027
12  11 -0.0741284
13  12 -0.1504872
14  13 -0.1616297
15  14 -0.2395797
```

Because they are based on fewer residual pairs, empirical autocorrelations at larger lags are less reliable. We can control the number of lags calculated in ACF using the maxLag argument. We use it in the plot of empirical ACF, displayed in Figure 8.27 and obtained with

```
> plot( ACF(fm10var.nlme,  maxLag = 10),
+         alpha = 0.05 )                              # Figure 8.27
```

Figure 8.27 shows that only the lag-1 autocorrelation is significant at the 5% level, but the lag-2 autocorrelation, which is approximately equal to the square of the lag-1 autocorrelation, is nearly significant. This suggests two different candidate correlation structures for modeling the within-group error covariance structure: $AR(1)$ and $MA(2)$. The two correlation models are not nested, but can be compared using using the information criteria provided by the anova method, AIC and BIC. The empirical lag-1 autocorrelation is used as a starting value for the corAR1 coefficient.

```
> fm20var.nlme <- update( fm10var.nlme, corr = corAR1(0.311) )
> fm30var.nlme <- update( fm10var.nlme, corr = corARMA(p=0, q=2) )
> anova( fm20var.nlme, fm30var.nlme, test = F )
             Model df    AIC    BIC  logLik
fm20var.nlme     1  9 1568.3 1601.9 -775.15
fm30var.nlme     2 10 1572.1 1609.4 -776.07
```

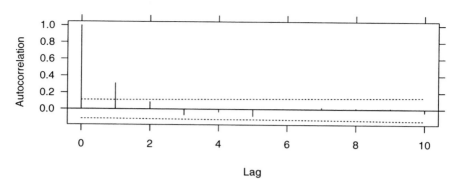

FIGURE 8.27. Empirical autocorrelation function corresponding to the standardized residuals of the `fm1Ovar.nlme` fitted object.

The $AR(1)$ model uses one fewer parameter than the $MA(2)$ model to give a larger log-likelihood and hence is the preferred model by both AIC and BIC.

The approximate 95% confidence intervals for the variance components in `fm2Ovar.nlme`, obtained with

```
> intervals( fm2Ovar.nlme )
 . . .
  Random Effects:
  Level: Mare
             lower        est.        upper
sd(A)  1.5465e+00  3.3083316  7.0772e+00
sd(B)  3.4902e-01  1.4257894  5.8245e+00
sd(w)  2.4457e-89  0.0020967  1.7974e+83
 . . .
```

indicate that there is no precision in the estimate of the standard deviation for the frequency ϕ_{2i} and little precision in the estimate of the standard deviation for ϕ_{1i}. The incorporation of the within-group autocorrelation structure into the NLME model seems to have reduced the need for random effects in the model. This is not uncommon: the random-effects model and the within-group correlation model compete with each other, in the sense that fewer random effects may be needed when within-group correlation structures are present, and viceversa. We test if the two variance components can be dropped from the model using `anova`.

```
> fm4Ovar.nlme <- update( fm2Ovar.nlme, random = A ~ 1 )
> anova( fm2Ovar.nlme, fm4Ovar.nlme )
             Model df     AIC     BIC  logLik   Test L.Ratio p-value
fm2Ovar.nlme     1  9  1568.3  1601.9 -775.15
fm4Ovar.nlme     2  7  1565.2  1591.4 -775.62 1 vs 2 0.94001   0.625
```

The high p-value for the likelihood ratio test suggests that the two models give essentially equivalent fits so the simpler model is preferred.

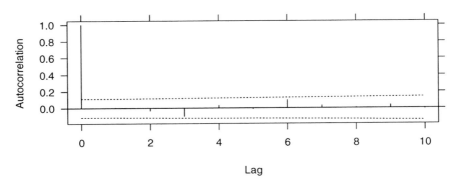

FIGURE 8.28. Empirical autocorrelation function corresponding to the normalized residuals of the `fm10var.nlme` fitted object.

An alternative, "intermediate" model between the $AR(1)$ and $MA(2)$ correlation structures is the $ARMA(1,1)$ model. This structure has an exponentially decaying ACF for lags ≥ 2, but allows greater flexibility in the lag-1 autocorrelation. Because the $AR(1)$ model is nested within the $ARMA(1,1)$ model, they can be compared via a likelihood ratio test.

```
> fm50var.nlme <- update( fm40var.nlme, corr = corARMA(p=1, q=1))
> anova( fm40var.nlme, fm50var.nlme )
             Model df    AIC    BIC  logLik   Test L.Ratio p-value
fm40var.nlme     1  7 1565.2 1591.4 -775.62
fm50var.nlme     2  8 1562.1 1592.0 -773.07 1 vs 2  5.1134  0.0237
```

The $ARMA(1,1)$ gives a significantly better representation of the within-group correlation, as indicated by the small p-value for the likelihood ratio test.

The plot of the empirical ACF of the normalized residuals, displayed in Figure 8.28, attests the the adequacy of the $ARMA(1,1)$ model for the `Ovary` data. No significant autocorrelations are detected, indicating that the normalized residuals behave like uncorrelated noise, as expected under the appropriate correlation model.

```
> plot( ACF(fm50var.nlme,  maxLag = 10, resType = "n"),
+        alpha = 0.05 )                              # Figure 8.28
```

It is illustrative, at this point, to compare the `nlme` fit represented by `fm50var.nlme` to the `lme` fit corresponding to `fm50var.lme` of §5.3.4. To have comparable log-likelihoods, we need to first obtain a maximum likelihood version of `fm50var.lme`.

```
> fm50var.lmeML <- update( fm50var.lme, method = "ML" )
> intervals( fm50var.lmeML )
 . . .
 Random Effects:
```

```
Level: Mare
                             lower      est.    upper
         sd((Intercept)) 0.988120 2.45659   6.1074
 sd(sin(2 * pi * Time)) 0.071055 0.85199 10.2158
 . . .
```

The wide confidence interval for the standard deviation of the random effect corresponding to sin(2*pi*Time) indicates that the fit is not very sensitive to the value of this coefficient and perhaps it could be eliminated from the model. We test this assumption using the likelihood ratio test.

```
> fm60var.lmeML <- update( fm50var.lmeML, random = ~1 )
> anova( fm50var.lmeML, fm60var.lmeML )
               Model df    AIC    BIC  logLik   Test L.Ratio
fm50var.lmeML      1  8 1562.7 1592.6 -773.37
fm60var.lmeML      2  7 1561.0 1587.1 -773.51 1 vs 2 0.28057
               p-value
fm50var.lmeML
fm60var.lmeML   0.5963
```

The large p-value for the test indicates that the two models are essentially equivalent so the simpler model with a single random intercept is preferred.

The LME model represented by fm60var.lmeML is nested within the NLME model represented by fm50var.nlme, corresponding to the case of $\beta_2 = 1$. Hence, we can test the assumption that the frequency of the ovulation cycle is equal to 1 using the likelihood ratio test.

```
> anova( fm60var.lmeML, fm50var.nlme )
               Model df    AIC    BIC  logLik   Test L.Ratio
fm60var.lmeML      1  7 1561.0 1587.1 -773.51
 fm50var.nlme      2  8 1562.1 1592.0 -773.07 1 vs 2 0.87881
               p-value
fm60var.lmeML
 fm50var.nlme   0.3485
```

There is no significant evidence that $\beta_2 \neq 1$. This conclusion is also supported by the approximate confidence interval for β_2, which contains 1.

```
> intervals( fm50var.nlme, which = "fixed" )
Approximate 95% confidence intervals

 Fixed effects:
       lower       est.      upper
A   10.37917   12.15566  13.932151
B   -3.91347   -2.87191  -1.830353
C   -3.07594   -1.56879  -0.061645
w    0.81565    0.93111   1.046568
```

8.3.3 Fitting Extended Nonlinear Regression Models with gnls

The general formulation of the extended nonlinear regression model, as well as the estimation methods used to fit it, have been described in §7.5.1 and §7.5.2. In this section, we present and illustrate the capabilities available in the nlme library for fitting and analyzing such models.

The gnls function fits the extended nonlinear regression model (7.34) using maximum likelihood. It can be viewed either as a version of nlme without the argument random, or as a version of nls with the arguments weights and correlation. Several arguments are available in gls, but typical calls are of the form

```
gnls(model, data, params, start, correlation)    # correl. errors
gnls(model, data, params, start, weights)        # heterosc. errors
gnls(model, data, params, start, correlation, weights)  # both
```

The first argument, model, is a two-sided nonlinear formula specifying the model for the expected value of the response. It uses the same syntax as the model argument to nlme. Correlation and weights are used as in lme, gls, and nlme to define, respectively, the correlation model and the variance function model for the error term. Data specifies a data frame in which the variables named in model, correlation, and weights can be evaluated. The parameters in the model are specified via the params argument, which can be either a two-sided linear formula, or a list of two-sided linear formulas. The syntax for params is identical to that of the fixed argument to nlme. Starting values for the model parameters are specified in start, which uses the same syntax as the argument with the same name to nlme. Starting values need not be given when the model function defined in model is a self-starting model and the the right-hand side of the parameter formulas in param do not include any covariates.

The fitted object returned by gnls is of class gnls, for which several methods are available to display, plot, update, and further explore the estimation results. Table 8.5 lists the most important gnls methods. The syntax of the gnls methods is identical to the syntax of the gls methods, described in §5.4. In fact, with the exception of coef, formula, logLik, predict, and update, all methods listed in Table 8.5 are common to both classes.

The use of the gnls function and its associated methods is described and illustrated through the re-analysis of the hemodialyzer example introduced in §5.2.2.

High-Flux Hemodialyzer Ultrafiltration Rates

The hemodialyzer ultrafiltration rates data were analyzed in §5.2.2 and §5.4 using an empirical polynomial model suggested by Littell et al. (1996). The model originally proposed for these data by Vonesh and Carter (1992) is

TABLE 8.5. Main gnls methods.

ACF	empirical autocorrelation function of residuals
anova	likelihood ratio or Wald-type tests
augPred	predictions augmented with observed values
coef	estimated coefficients for expected response model
fitted	fitted values
intervals	confidence intervals on model parameters
logLik	log-likelihood at convergence
plot	diagnostic Trellis plots
predict	predicted values
print	brief information about the fit
qqnorm	normal probability plots
resid	residuals
summary	more detailed information about the fit
update	update the gnls fit
Variogram	semivariogram of residuals

an asymptotic regression model with an offset, identical to the one used for the CO_2 uptake data in §8.2.2. The model for the expected ultrafiltration rate y at transmembrane pressure x is written as

$$E[y] = \phi_1 \left\{ 1 - \exp\left[-\exp(\phi_2)\left(x - \phi_3 \right) \right] \right\}. \qquad (8.20)$$

The parameters in model (8.20) have a physiological interpretation: ϕ_1 is the maximum ultrafiltration rate that can be attained, ϕ_2 is the logarithm of the hydraulic permeability transport rate, and ϕ_3 is the transmembrane pressure required to offset the oncotic pressure.

Vonesh and Carter (1992) suggest using different parameters in (8.20) for each blood flow rate level. We use the self-starting function SSasympOff and the nlsList function to investigate which parameters in the asymptotic regression model (8.20) depend on the blood flow rate.

```
> fm1Dial.lis <-
+     nlsList( rate ~ SSasympOff(pressure, Asym, lrc, c0) | QB,
+             data = Dialyzer )
> fm1Dial.lis
  .  .  .
Coefficients:
      Asym     lrc       c0
200 44.988 0.76493 0.22425
300 62.217 0.25282 0.22484
```

The coefficient estimates suggest that Asym (ϕ_1) and lrc ($\log \phi_2$) depend on the blood flow rate level, but c0 (ϕ_3) does not. The plot of the individual

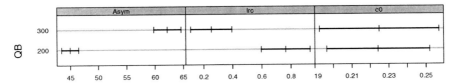

FIGURE 8.29. Ninety-five percent confidence intervals on the asymptotic regression model parameters for each level of blood flow rate (QB) in the dialyzer data.

confidence intervals in Figure 8.29 confirms that only `Asym` and `lrc` vary with blood flow level.

```
> plot( intervals(fm1Dial.lis) )                    # Figure 8.29
```

The ultrafiltration rate y_{ij} at the jth transmembrane pressure x_{ij} for the ith subject is represented by the nonlinear model

$$y_{ij} = (\phi_1 + \gamma_1 Q_i)\left\{1 - \exp\left[-\exp\left(\phi_2 + \gamma_2 Q i\right)\left(x_{ij} - \phi_3\right)\right]\right\} + \epsilon_{ij}, \quad (8.21)$$

where Q_i is a binary variable taking values -1 for 200 dl/min hemodialyzers and 1 for 300 dl/min hemodialyzers; ϕ_1, ϕ_2, and ϕ_3 are, respectively, the asymptotic ultrafiltration rate, the log-transport rate, and the transmembrane pressure offset averaged over the levels of Q; γ_i is the blood flow effect associated with the coefficient ϕ_i; and ϵ_{ij} is the error term, initially assumed to be independently distributed $\mathcal{N}(0, \sigma^2)$ random variables.

The nonlinear model (8.21) can be fitted with `nls`, but it is easier to express the dependency of the asymptote and the log-rate on the blood flow rate using `gnls`. The average of the `fm1Dial.lis` coefficients are used as the initial estimates for ϕ_1, ϕ_2, and ϕ_3, while the half differences between the first two coefficients are used as initial estimates for γ_1 and γ_2.

```
> fm1Dial.gnls <- gnls( rate ~ SSasympOff(pressure, Asym, lrc, c0),
+     data = Dialyzer, params = list(Asym + lrc ~ QB, c0 ~ 1),
+     start = c(53.6, 8.6, 0.51, -0.26, 0.225) )
> fm1Dial.gnls
Generalized nonlinear least squares fit
  Model: rate ~ SSasympOff(pressure, Asym, lrc, c0)
  Data: Dialyzer
  Log-likelihood: -382.65

Coefficients:
 Asym.(Intercept) Asym.QB lrc.(Intercept)    lrc.QB       c0
           53.606    8.62         0.50874  -0.25684  0.22449

Degrees of freedom: 140 total; 135 residual
Residual standard error: 3.7902
```

To fit the same model in `nls`, we need first to create a binary variable representing Q in (8.21) and then include all coefficients explicitly in the model formula.

```
> Dialyzer$QBcontr <- 2 * (Dialyzer$QB == 300) - 1
> fm1Dial.nls <-
+    nls( rate ~ SSasympOff(pressure, Asym.Int + Asym.QB * QBcontr,
+    lrc.Int + lrc.QB * QBcontr, c0), data = Dialyzer,
+    start = c(Asym.Int = 53.6, Asym.QB = 8.6, lrc.Int = 0.51,
+    lrc.QB = -0.26, c0 = 0.225) )
> summary( fm1Dial.nls )


Formula: rate ~ SSasympOff(pressure, Asym.Int + Asym.QB * QBcontr,
            lrc.Int + lrc.QB * QBcontr, c0)


Parameters:
              Value Std. Error  t value
Asym.Int   53.60660   0.705409  75.9937
 Asym.QB    8.61999   0.679240  12.6906
 lrc.Int    0.50872   0.055233   9.2105
  lrc.QB   -0.25683   0.045021  -5.7047
      c0    0.22448   0.010623  21.1318


Residual standard error: 3.79022 on 135 degrees of freedom
> logLik( fm1Dial.nls )
[1] -382.65
```

As expected, the results are nearly identical.

The `plot` method is the primary tool for assessing the quality of a `gnls` fit. It uses the same syntax as the other `plot` methods in the nlme library. For example, the plot of the residuals versus the transmembrane pressure, shown in Figure 8.30 and obtained with

```
> plot(fm1Dial.gnls, resid(.) ~ pressure, abline = 0) # Figure 8.30
```

indicates that the error variability increases with the transmembrane pressure. This heteroscedastic pattern is also observed in the linear model fits of the hemodialyzer data, presented in §5.2.2 and §5.4.

As in the previous analyses of the hemodialyzer data presented in §5.2.2 and §5.4, the power variance function, represented in nlme by the varPower class, is used to model the heteroscedasticity in the ultrafiltration rates.

```
> fm2Dial.gnls <- update( fm1Dial.gnls,
+                          weights = varPower(form = ~ pressure) )
> anova( fm1Dial.gnls, fm2Dial.gnls)
            Model df    AIC    BIC  logLik   Test L.Ratio p-value
fm1Dial.gnls    1  6 777.29 794.94 -382.65
fm2Dial.gnls    2  7 748.47 769.07 -367.24 1 vs 2  30.815  <.0001
```

As expected, the likelihood ratio test strongly rejects the assumption of homoscedasticity. The plot of the standard residuals for `fm2Dial.gnls`

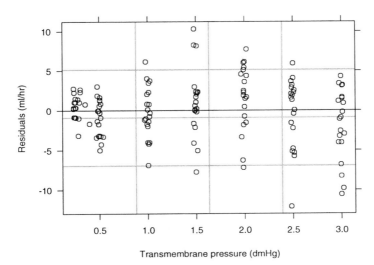

FIGURE 8.30. Plot of residuals versus transmembrane pressure for the homoscedastic fitted object fm1Dial.gls.

versus pressure, shown in Figure 8.31, indicates that the power variance function successfully models the heteroscedasticity in the data.

The hemodialyzer ultrafiltration rates measurements made sequentially on the same subject are correlated. Random effects can be used in an NLME model to account for the within-group correlation, but we choose here to model the within-subject dependence directly by incorporating a correlation structure for the error term in the gnls model. Because the measurements are equally spaced in time, the empirical autocorrelation function can be used to investigate the within-subject correlation. The ACF method is used to obtain the empirical ACF, with the time covariate and the grouping factor specified via the form argument.

```
> ACF( fm2Dial.gnls, form = ~ 1 | Subject )
    lag       ACF
1   0   1.00000
2   1   0.71567
3   2   0.50454
4   3   0.29481
5   4   0.20975
6   5   0.13857
7   6  -0.00202
```

The empirical ACF values confirm the within-group correlation and indicates that the correlation decreases with lag. As usual, it is more informative to look at a plot of the empirical ACF, displayed in Figure 8.32 and obtained with

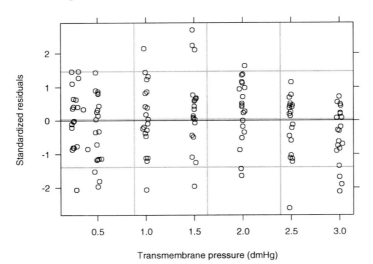

FIGURE 8.31. Plot of standardized residuals versus transmembrane pressure for the heteroscedastic fitted object `fm2Dial.gnls`.

```
> plot( ACF( fm2Dial.gnls, form = ~ 1 | Subject),
+        alpha = 0.05 )                                # Figure 8.32
```

The autocorrelation pattern in Figure 8.32 suggests that an $AR(1)$ model, represented in nlme by the corAR1 class, may be appropriate to describe the within-group correlation.

```
> fm3Dial.gnls <-
+  update(fm2Dial.gnls, corr = corAR1(0.716, form = ~ 1 | Subject))
> fm3Dial.gnls
. . .
Coefficients:
 Asym.(Intercept) Asym.QB lrc.(Intercept)    lrc.QB      c0
          55.111  8.1999         0.37193  -0.16974 0.21478

Correlation Structure: AR(1)
 Formula: ~ 1 | Subject
 Parameter estimate(s):
    Phi
 0.7444
Variance function:
 Structure: Power of variance covariate
 Formula:  ~ pressure
 Parameter estimates:
  power
 0.5723
Degrees of freedom: 140 total; 135 residual
Residual standard error: 3.1844
```

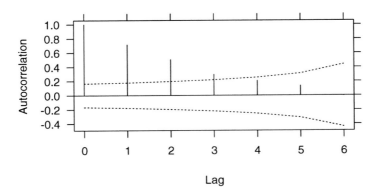

FIGURE 8.32. Empirical autocorrelation function corresponding to the standardized residuals of the **fm2Dial.gnls** fitted object.

The lag-1 empirical autocorrelation is used as initial value for the **corAR1**. The variability in the estimates is assessed with the **intervals** method.

```
> intervals( fm3Dial.gnls )
 . . .
  Correlation structure:
        lower    est.   upper
Phi 0.55913 0.7444 0.85886
 . . .
```

The confidence interval on the autocorrelation parameter is bounded away from zero, suggesting that the $AR(1)$ model provides a significantly better fit. We confirm this with the likelihood ratio test.

```
> anova( fm2Dial.gnls, fm3Dial.gnls )
             Model df    AIC    BIC  logLik    Test L.Ratio p-value
fm2Dial.gnls     1  7 748.47 769.07 -367.24
fm3Dial.gnls     2  8 661.04 684.58 -322.52 1 vs 2  89.433  <.0001
```

The plot of the empirical ACF for the normalized residuals corresponding to **fm3Dial.gnls**, displayed in Figure 8.33, does not show any significant correlations, indicating that the $AR(1)$ adequately represents the within-subject dependence in the **gnls** model for the hemodialyzer data.

The plot of the standardized residuals versus the transmembrane pressure in Figure 8.34 suggests a certain lack-of-fit for the asymptotic regression model (8.21): the residuals for the highest two transmembrane pressures are predominantly negative. This is consistent with the plot of the hemodialyzer data, shown in Figure 5.1, and also with Figure 3 in Vonesh and Carter (1992), which indicate that, for many subjects, the ultrafiltration rates decrease for the highest two transmembrane pressures. The asymptotic regression model is monotonically increasing in the transmembrane pressure and cannot properly accommodate the nonmonotonic behavior of the ultrafiltration rates.

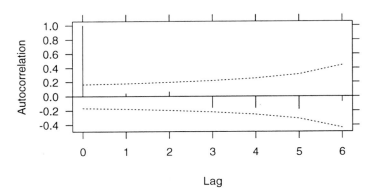

FIGURE 8.33. Empirical autocorrelation function for the normalized residuals corresponding to the `fm3Dial.gnls` fitted object.

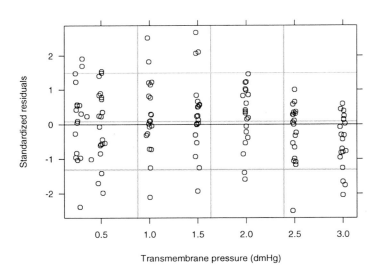

FIGURE 8.34. Plot of standardized residuals versus transmembrane pressure for the `fm3Dial.gnls` fitted object.

The gnls model corresponding to `fm3Dial.gnls` may be compared to
the best models obtained for the `Dialyzer` data in §5.2.2 and §5.4, corresponding, respectively, to the objects `fm2Dial.lme` and `fm3Dial.gls`. To
have comparable fits, we need to first obtain ML versions of `fm2Dial.lme`
and `fm3Dial.gls`.

```
> fm2Dial.lmeML <- update( fm2Dial.lme, method = "ML")
> fm3Dial.glsML <- update( fm3Dial.gls, method = "ML")
```

As the models are not nested, only the information criterion statistics can
be compared.

```
> anova( fm2Dial.lmeML, fm3Dial.glsML, fm3Dial.gnls, test = F )
                Model df    AIC    BIC  logLik
fm2Dial.lmeML       1 18 651.75 704.70 -307.87
fm3Dial.glsML       2 13 647.56 685.80 -310.78
 fm3Dial.gnls       3  8 661.04 684.58 -322.52
```

The more conservative BIC favors the `fm3Dial.gnls` model because of the
fewer number of parameters it uses; the more liberal AIC favors
`fm3Dial.glsML` because of its larger log-likelihood value. In practice, the
choice of the "best model" should take into account other factors besides
the information criteria, such as the interpretability of the parameters.

8.4   Chapter Summary

This chapter describes the nonlinear modeling capabilities available in the
nlme library. A brief review of the nonlinear least-squares function `nls` in
S is presented and self-starting models for automatically producing starting values for the coefficients in a nonlinear model are introduced and
illustrated. The `nlsList` function for fitting separate nonlinear regression
models to data partitioned according to the levels of a grouping factor is
described and its use for model building of nonlinear mixed-effects models
illustrated.

Nonlinear mixed-effects models are fitted with the `nlme` function. Data
from several real-life applications are used to illustrate the various capabilities available in `nlme` for fitting and analyzing single and multilevel NLME
models. Variance functions and correlation structures to model the within-
group variance–covariance structure are used with `nlme` in the exact same
way as with `lme`, the linear mixed-effects modeling function. Several examples are used to illustrate the use of varFunc and corStruct classes with
`nlme`.

A new modeling function, `gnls`, for fitting the extended nonlinear model
with heteroscedastic, correlated errors is introduced. The `gnls` function can
be regarded as an extended version of `nls` which allows the use of varFunc
and corStruct objects to model the error variance–covariance structure, or

as a simplified version of nlme, without random effects. The hemodialyzer example is used to illustrate the use of gnls and its associated methods.

Exercises

1. Plots of the DNase data in §3.3.2 and in Appendix A.7 suggest a sigmoidal relationship between the density and log(conc). Because there are no data points at sufficiently high DNase concentrations to define the upper part of the curve, the sigmoidal relationship is only suggested and is not definite.

   A common model for this type of assay data is the four-parameter logistic model available as SSfpl (Appendix C.6).

   (a) Create separate fits of the SSfpl model to each run using nlsList. Note that the display formula, density ~conc | Run, defines the primary covariate as conc but the model should be fit as a function of log(conc). You will either need to give explicit arguments to the SSfpl function or define a new groupedData object with a formula based on log(conc).

   (b) Examine the plot of the residuals versus the fitted values from this nlsList fit. Does the plot indicate nonconstant variance?

   (c) Fit an NLME model to these data using random effects for each model parameter and a general positive-definite Ψ. Perform the usual checks on the model using diagnostic plots, summaries, and intervals. Can the confidence intervals on the variance–covariance parameters be evaluated?

   (d) Davidian and Giltinan (1995, §5.2.4) conclude from a similar analysis that the variance of the ϵ_{ij} increases with increasing optical density. They estimate a variance function that corresponds to the varPower variance function. Update the previous fit by adding weights = varPower(). Does this updated model converge? If not, change the definition of the random effects so that Ψ is diagonal.

   (e) Compare the model fit with the varPower variance function to one fit without it. Note that if you have fit the model with the variance function by imposing a diagonal Ψ, you should refit the model without the variance function but with a diagonal Ψ before comparing. Does the addition of the varPower variance function make a significant contribution to the model?

   (f) Examine the confidence intervals on your best-fitting model to this point. Are there parameters that can be modeled as purely fixed effects? (In our analysis the standard deviation of the random effect for the xmid parameter turned out to be negligible.) If you determine that some random effects are no longer needed, remove them and refit the model.

(g) If you have modified the model to reduce the number of random effects, refit with a general positive-definite Ψ rather than a diagonal Ψ. Does this fit converge? If so, compare the fits with diagonal Ψ and general positive-definite Ψ.

(h) Write a report on the analysis, including data plots and diagnostic plots where appropriate.

2. As shown in Figure 3.4 (p. 107), the relationship between deltaBP and log(dose) in the phenybiguanide data PBG is roughly sigmoidal. There is a strong indication that the effect of the Treatment is to shift the curve to the right.

(a) Fit separate four-parameter logistic models (SSfpl, Appendix C.6) to the data from each Treatment within each Rabbit using nlsList. Recall from §3.2.1 that the primary covariate in the display formula for PBG is dose but we want the model to be fit as a function of log(dose). You will either need to specify the model explicitly or to re-define the display formula for the data. Also note that the grouping formula should be ~Rabbit/Treatment, but the grouping in the display formula is ~Rabbit.

(b) Plot the confidence intervals on the coefficients. Which parameters appear to be constant across all Rabbit/Treatment combinations? Does there appear to be a systematic shift in the xmid parameter according to Treatment?

(c) Fit a two-level NLME model with a fixed effect for Treatment on the xmid parameter and with random effects for Rabbit on the B parameter and for Treatment within Rabbit on the B and xmid parameters. Begin with a diagonal Ψ_2 matrix. If that model fit converges, update to a general positive-definite Ψ_2 matrix and compare the two fitted models with anova. Which model is preferred?

(d) Summarize your preferred model. Is the fixed effect for Treatment on xmid significant? Also check using the output from intervals.

(e) Plot the augmented predictions from your preferred model. Remember that if the display formula for the data has dose as the primary covariate then you will need to use scales = list(x = list(log = 2)) to get the symmetric shape of the logistic curve. Adjust the layout argument for the plot so the panels are aligned by Rabbit. Do these plots indicate deficiencies in the model?

(f) Examine residual plots for other possible deficiencies in the model. Also check plots of the random effects versus covariates to see if important fixed effects have been omitted.

(g) Is it necessary to use a four-parameter logistic curve? Experiment with fitting the three-parameter logistic model (SSlogis, Appendix C.7) to see if a comparable fit can be obtained. Check residual plots from your fitted models and comment.

3. The Glucose2 data described in Appendix A.10 consist of blood glucose levels measured 14 times over a period of 5 hours on 7 volunteers who took alcohol at time 0. The same experiment was repeated on a second occasion with the same subjects but with a dietary additive used for all subjects. These data are analyzed in Hand and Crowder (1996, Example 8.4, pp. 118–120), where the following empirical model relating the expected glucose level to Time is proposed.

$$\texttt{glucose} = \phi_1 + \phi_2 \texttt{Time}^3 \exp\left(-\phi_3 \texttt{Time}\right)$$

Note that there are two levels of grouping in the data: Subject and Date within Subject.

(a) Plot the data at the Subject display level (use plot(Glucose2, display = 1). Do there appear to be systematic differences between the two dates on which the experiments were conducted (which could be associated with the dietary supplement)?

(b) There is no self-starting function representing the model for the glucose level included in the nlme library. Use nlsList with starting values start = c(phi1=5, phi2=-1, phi3=1) (derived from Hand and Crowder (1996)) to fit separate models for each Subject and for each Date within Subject. Plot the individual confidence intervals for each of the two nlsList fits. Verify that phi1 and phi2 seem to vary significantly for both levels of grouping, but phi3 does not. (There is an unusual estimate of phi3 for Subject 6, Date 1 but all other confidence intervals overlap.)

(c) Fit a two-level NLME model with random effects for phi1 and phi2, using as starting values for the fixed effects the estimates from either of the nlsList fits (start = fixef(object), with object replaced with the name of the nlsList object). Examine the confidence intervals on the variance–covariance components; what can you say about the precision of the estimated correlation coefficients?

(d) Refit the NLME model using diagonal $\boldsymbol{\Psi}_q$ matrices for both grouping levels and compare the new fit to the previous one using anova. Investigate if there are random effects that can be dropped from the model using intervals. If so, refit the model with fewer random effects and compare it to the previous fit using `anova`. Plot the residuals versus `Time` and comment on the apparent adequacy of the empirical model.

(e) Plot the semivariogram of the standardized residuals corresponding to the final NLME fit obtained in the last item. Use `Time` as the covariate to define the distances in the semivariogram and consider only distances≤ 22 (that is, use `plot(Variogram(object, form = ~Time, maxDist = 22)))`. Notice the increase in the semivariogram for smaller distances, which suggests that there the within-group errors may be correlated.

(f) Hand and Crowder (1996) use a continuous AR1 model for the within-group errors, with `Time` as the covariate. Update the NLME fit adding a continuous AR1 correlation structure on `Time` (`corr = corCAR1(form = Time)`. Compare the fits using `anova`. Examine the plot of the semivariogram of the *normalized* residuals for the NLME with `corCAR1` correlation structure and comment on the adequacy of the within-group correlation model.

(g) The NLME model used by Hand and Crowder (1996) includes random effects for `phi1` and `phi2` only at the `Subject` level and uses a `corCAR1` correlation structure on `Time` for the `Date` within `Subject` errors, with errors from different `Dates` within the same `Subject` assumed to be independent. Fit such model using

```
random = B1 + B2 ~ 1 | Subject,
corr = corCAR1(form = ~Time | Subject/Date)
```

and compare it to the multilevel NLME model obtained in the previous item. Which one do you think gives a better fit?

(h) The main objective of the experiment was to determine if there were significant differences between the blood glucose level profiles over time associated with the use of the dietary additive (which is totally confounded with `Dose` in this case). Investigate the significance of the dietary additive by refitting the NLME model (with `corCAR1` correlation) incorporating a "`Date` effect" for each fixed effect (use `fixed = phi1 + phi2 + phi3    Date`). You will need to give new starting values for the fixed effects (use the previous fixed effects estimates for the `Intercept` terms and 0 for the `Date` terms). Use `summary` to assess the significance of the dietary additive effect; does it seem to make any difference on the blood glucose levels? Are your conclusions consistent with the plot of the data?

4. An NLME analysis of the `Theoph` data is presented in §8.2.1. The final `nlme` fit obtained in that section, `fm3Theo.nlme`, includes random effects for the `lKa` and `lCl` coefficients and a diagonal Ψ.

(a) Use the `gnls` function described in §8.3.3 to fit the `SSfol` model to the theophylline concentrations with no random effects. Compare this fit to `fm3Theo.nlme` using `anova`. Obtain the boxplots of the residuals by `Subject` (`plot(object, Subject~resid(.))`) and comment on the observed pattern.

(b) Print and plot the ACF of the standardized residuals for the `gnls` fit (use `form = ~1 | Subject` to specify the grouping structure). The decrease in the ACF with lag suggests that an AR1 model may be adequate.

(c) Update the `gnls` fit incorporating an AR1 correlation structure, using the lag-1 autocorrelation from the `ACF` output as an initial estimate for the correlation parameter (`corr = corAR1(0.725, form = ~1 | Subject)`). Compare this fit to the previous `gnls` fit using `anova`. Is there significant evidence of autocorrelation? Examine the plot of the ACF of the normalized residuals. Does the AR1 model seem adequate?

(d) Compare the `gnls` fit with AR1 correlation structure to the `fm3Theo.nlme` fit of §8.2.1 using `anova` with the argument `test` set to `FALSE` (why?). Which model seems better?

References

Abramowitz, M. and Stegun, I. A. (1964). *Handbook of Mathematical Functions with Formulas, Graphs, and Mathematical Tables*, Dover, New York.

Bates, D. M. and Chambers, J. M. (1992). "Nonlinear models," in Chambers and Hastie (1992), Chapter 10, pp. 421–454.

Bates, D. M. and Pinheiro, J. C. (1998). Computational methods for multilevel models, *Technical Memorandum BL0112140-980226-01TM*, Bell Labs, Lucent Technologies, Murray Hill, NJ.

Bates, D. M. and Watts, D. G. (1980). Relative curvature measures of nonlinearity, *Journal of the Royal Statistical Society, Ser. B* **42**: 1–25.

Bates, D. M. and Watts, D. G. (1988). *Nonlinear Regression Analysis and Its Applications*, Wiley, New York.

Beal, S. and Sheiner, L. (1980). The NONMEM system, *American Statistician* **34**: 118–119.

Becker, R. A., Cleveland, W. S. and Shyu, M.-J. (1996). The visual design and control of trellis graphics displays, *Journal of Computational and Graphical Statistics* **5**(2): 123–156.

Bennett, J. E. and Wakefield, J. C. (1993). Markov chain Monte Carlo for nonlinear hierarchical models, *Technical Report TR-93-11*, Statistics Section, Imperial College, London.

Boeckmann, A. J., Sheiner, L. B. and Beal, S. L. (1994). *NONMEM Users Guide: Part V*, NONMEM Project Group, University of California, San Francisco.

Box, G. E. P., Hunter, W. G. and Hunter, J. S. (1978). *Statistics for Experimenters*, Wiley, New York.

Box, G. E. P., Jenkins, G. M. and Reinsel, G. C. (1994). *Time Series Analysis: Forecasting and Control*, 3rd ed., Holden-Day, San Francisco.

Brillinger, D. (1987). Comment on a paper by C. R. Rao, *Statistical Science* **2**: 448–450.

Bryk, A. and Raudenbush, S. (1992). *Hierarchical Linear Models for Social and Behavioral Research*, Sage, Newbury Park, CA.

Carroll, R. J. and Ruppert, D. (1988). *Transformation and Weighting in Regression*, Chapman & Hall, New York.

Chambers, J. M. (1977). *Computational Methods for Data Analysis*, Wiley, New York.

Chambers, J. M. and Hastie, T. J. (eds) (1992). *Statistical Models in S*, Chapman & Hall, New York.

Cleveland, W. S. (1994). *Visualizing Data*, Hobart Press, Summit, NJ.

Cleveland, W. S., Grosse, E. and Shyu, W. M. (1992). "Local regression models," in Chambers and Hastie (1992), Chapter 8, pp. 309–376.

Cochran, W. G. and Cox, G. M. (1957). *Experimental Designs*, 2nd ed., Wiley, New York.

Cox, D. R. and Hinkley, D. V. (1974). *Theoretical Statistics*, Chapman & Hall, London.

Cressie, N. A. C. (1993). *Statistics for Spatial Data*, Wiley, New York.

Cressie, N. A. C. and Hawkins, D. M. (1980). Robust estimation of the variogram, *Journal of the International Association of Mathematical Geology* **12**: 115–125.

Crowder, M. and Hand, D. (1990). *Analysis of Repeated Measures*, Chapman & Hall, London.

Davidian, M. and Gallant, A. R. (1992). Smooth nonparametric maximum likelihood estimation for population pharmacokinetics, with application to quinidine, *Journal of Pharmacokinetics and Biopharmaceutics* **20**: 529–556.

Davidian, M. and Giltinan, D. M. (1995). *Nonlinear Models for Repeated Measurement Data*, Chapman & Hall, London.

Davis, P. J. and Rabinowitz, P. (1984). *Methods of Numerical Integration*, 2nd ed., Academic Press, New York.

Dempster, A. P., Laird, N. M. and Rubin, D. B. (1977). Maximum likelihood from incomplete data via the EM algorithm, *Journal of the Royal Statistical Society, Ser. B* **39**: 1–22.

Devore, J. L. (2000). *Probability and Statistics for Engineering and the Sciences*, 5th ed., Wadsworth, Belmont, CA.

Diggle, P. J., Liang, K.-Y. and Zeger, S. L. (1994). *Analysis of Longitudinal Data*, Oxford University Press, New York.

Dongarra, J. J., Bunch, J. R., Moler, C. B. and Stewart, G. W. (1979). *Linpack Users' Guide*, SIAM, Philadelphia.

Draper, N. R. and Smith, H. (1998). *Applied Regression Analysis*, 3rd ed., Wiley, New York.

Gallant, A. R. and Nychka, D. W. (1987). Seminonparametric maximum likelihood estimation, *Econometrica* **55**: 363–390.

Geman, S. and Geman, D. (1984). Stochastic relaxation, Gibbs distributions and the Bayesian restoration of images, *IEEE Transactions on Pattern Analysis and Machine Intelligence* **6**: 721–741.

Geweke, J. (1989). Bayesian inference in econometric models using Monte Carlo integration, *Econometrica* **57**: 1317–1339.

Gibaldi, M. and Perrier, D. (1982). *Pharmacokinetics*, Marcel Dekker, New York.

Goldstein, H. (1987). *Multilevel Models in Education and Social Research*, Oxford University Press, Oxford.

Goldstein, H. (1995). *Multilevel Statistical Models*, Halstead Press, New York.

Golub, G. H. (1973). Some modified matrix eigenvalue problems, *SIAM Review* **15**: 318–334.

Golub, G. H. and Welsch, J. H. (1969). Calculation of Gaussian quadrature rules, *Mathematical Computing* **23**: 221–230.

Grasela and Donn (1985). Neonatal population pharmacokinetics of phenobarbital derived from routine clinical data, *Developmental Pharmacology and Therapeutics* **8**: 374–0383.

Hand, D. and Crowder, M. (1996). *Practical Longitudinal Data Analysis*, Texts in Statistical Science, Chapman & Hall, London.

Harville, D. A. (1977). Maximum likelihood approaches to variance component estimation and to related problems, *Journal of the American Statistical Association* **72**: 320–340.

Hastings, W. K. (1970). Monte Carlo sampling methods using Markov chains and their applications, *Biometrika* **57**: 97–109.

Jones, R. H. (1993). *Longitudinal Data with Serial Correlation: A State-space Approach*, Chapman & Hall, London.

Joyner and Boore (1981). Peak horizontal acceleration and velocity from strong-motion records including records from the 1979 Imperial Valley, California, earthquake, *Bulletin of the Seismological Society of America* **71**: 2011–2038.

Kennedy, William J., J. and Gentle, J. E. (1980). *Statistical Computing*, Marcel Dekker, New York.

Kung, F. H. (1986). Fitting logistic growth curve with predetermined carrying capacity, *ASA Proceedings of the Statistical Computing Section* pp. 340–343.

Kwan, K. C., Breault, G. O., Umbenhauer, E. R., McMahon, F. G. and Duggan, D. E. (1976). Kinetics of indomethicin absorption, elimination, and enterohepatic circulation in man, *Journal of Pharmacokinetics and Biopharmaceutics* **4**: 255–280.

Laird, N. M. and Ware, J. H. (1982). Random-effects models for longitudinal data, *Biometrics* **38**: 963–974.

Lehmann, E. L. (1986). *Testing Statistical Hypotheses*, Wiley, New York.

Leonard, T., Hsu, J. S. J. and Tsui, K. W. (1989). Bayesian marginal inference, *Journal of the American Statistical Association* **84**: 1051–1058.

Lindley, D. and Smith, A. (1972). Bayes estimates for the linear model, *Journal of the Royal Statistical Society, Ser. B* **34**: 1–41.

Lindstrom, M. J. and Bates, D. M. (1988). Newton–Raphson and EM algorithms for linear mixed-effects models for repeated-measures data (corr: 94v89 p1572), *Journal of the American Statistical Association* **83**: 1014–1022.

Lindstrom, M. J. and Bates, D. M. (1990). Nonlinear mixed-effects models for repeated measures data, *Biometrics* **46**: 673–687.

Littell, R. C., Milliken, G. A., Stroup, W. W. and Wolfinger, R. D. (1996). *SAS System for Mixed Models*, SAS Institute Inc., Cary, NC.

Longford, N. T. (1993). *Random Coefficient Models*, Oxford University Press, New York.

Ludbrook, J. (1994). Repeated measurements and multiple comparisons in cardiovascular research, *Cardiovascular Research* **28**: 303–311.

Mallet, A. (1986). A maximum likelihood estimation method for random coefficient regression models, *Biometrika* **73**(3): 645–656.

Mallet, A., Mentre, F., Steimer, J.-L. and Lokiek, F. (1988). Nonparametric maximum likelihood estimation for population pharmacokinetics, with applications to Cyclosporine, *Journal of Pharmacokinetics and Biopharmaceutics* **16**: 311–327.

Matheron, G. (1962). *Traite de Geostatistique Appliquee*, Vol. I of *Memoires du Bureau de Recherches Geologiques et Minieres*, Editions Technip, Paris.

Milliken, G. A. and Johnson, D. E. (1992). *Analysis of Messy Data. Volume 1: Designed Experiments*, Chapman & Hall, London.

Patterson, H. D. and Thompson, R. (1971). Recovery of interblock information when block sizes are unequal, *Biometrika* **58**: 545–554.

Pierson, R. A. and Ginther, O. J. (1987). Follicular population dynamics during the estrus cycle of the mare, *Animal Reproduction Science* **14**: 219–231.

Pinheiro, J. C. (1994). *Topics in Mixed-Effects Models*, Ph.D. thesis, University of Wisconsin, Madison, WI.

Pinheiro, J. C. and Bates, D. M. (1995). Approximations to the log-likelihood function in the nonlinear mixed-effects model, *Journal of Computational and Graphical Statistics* **4**(1): 12–35.

Potthoff, R. F. and Roy, S. N. (1964). A generalized multivariate analysis of variance model useful especially for growth curve problems, *Biometrika* **51**: 313–326.

Potvin, C., Lechowicz, M. J. and Tardif, S. (1990). The statistical analysis of ecophysiological response curves obtained from experiments involving repeated measures, *Ecology* **71**: 1389–1400.

Ramos, R. Q. and Pantula, S. G. (1995). Estimation of nonlinear random coefficient models, *Statistics & Probability Letters* **24**: 49–56.

Sakamoto, Y., Ishiguro, M. and Kitagawa, G. (1986). *Akaike Information Criterion Statistics*, Reidel, Dordrecht, Holland.

Schwarz, G. (1978). Estimating the dimension of a model, *Annals of Statistics* **6**: 461–464.

Searle, S. R., Casella, G. and McCulloch, C. E. (1992). *Variance Components*, Wiley, New York.

Seber, G. A. F. and Wild, C. J. (1989). *Nonlinear Regression*, Wiley, New York.

Self, S. G. and Liang, K. Y. (1987). Asymptotic properties of maximum likelihood estimators and likelihood ratio tests under nonstandard conditions, *Journal of the American Statistical Association* **82**: 605–610.

Sheiner, L. B. and Beal, S. L. (1980). Evaluation of methods for estimating population pharmacokinetic parameters. I. Michaelis–Menten model: Routine clinical pharmacokinetic data, *Journal of Pharmacokinetics and Biopharmaceutics* **8**(6): 553–571.

Snedecor, G. W. and Cochran, W. G. (1980). *Statistical Methods*, 7th ed., Iowa State University Press, Ames, IA.

Soo, Y.-W. and Bates, D. M. (1992). Loosely coupled nonlinear least squares, *Computational Statistics and Data Analysis* **14**: 249–259.

Stram, D. O. and Lee, J. W. (1994). Variance components testing in the longitudinal mixed-effects models, *Biometrics* **50**: 1171–1177.

Stroup, W. W. and Baenziger, P. S. (1994). Removing spatial variation from wheat yield trials: a comparison of methods, *Crop Science* **34**: 62–66.

Thisted, R. A. (1988). *Elements of Statistical Computing*, Chapman & Hall, London.

Tierney, L. and Kadane, J. B. (1986). Accurate approximations for posterior moments and densities, *Journal of the American Statistical Association* **81**(393): 82–86.

Venables, W. N. and Ripley, B. D. (1999). *Modern Applied Statistics with S-PLUS*, 3rd ed., Springer-Verlag, New York.

Verme, C. N., Ludden, T. M., Clementi, W. A. and Harris, S. C. (1992). Pharmacokinetics of quinidine in male patients: A population analysis, *Clinical Pharmacokinetics* **22**: 468–480.

Vonesh, E. F. and Carter, R. L. (1992). Mixed-effects nonlinear regression for unbalanced repeated measures, *Biometrics* **48**: 1–18.

Vonesh, E. F. and Chinchilli, V. M. (1997). *Linear and Nonlinear Models for the Analysis of Repeated Measures*, Marcel Dekker, New York.

Wakefield, J. (1996). The Bayesian analysis of population pharmacokinetic models, *Journal of the American Statistical Association* **91**: 62–75.

Wilkinson, G. N. and Rogers, C. E. (1973). Symbolic description of factorial models for analysis of variance, *Applied Statistics* **22**: 392–399.

Wolfinger, R. D. (1993). Laplace's approximation for nonlinear mixed models, *Biometrika* **80**: 791–795.

Wolfinger, R. D. and Tobias, R. D. (1998). Joint estimation of location, dispersion, and random effects in robust design, *Technometrics* **40**: 62–71.

Yates, F. (1935). Complex experiments, *Journal of the Royal Statistical Society (Supplement)* **2**: 181–247.

Appendix A
Data Used in Examples and Exercises

We have used several sets of data in our examples and exercises. In this appendix we list all the data sets that are available as the NLMEDATA library included with the nlme 3.1 distribution and we describe in greater detail the data sets referenced in the text.

The title of each section in this appendix gives the name of the corresponding groupedData object from the nlme library, followed by a short description of the data. The formula stored with the data and a short description of each of the columns is also given.

We have adopted certain conventions for the ordering and naming of columns in these descriptions. The first column provides the response, the second column is the primary covariate, if present, and the next column is the primary grouping factor. Other covariates and grouping factors, if present, follow. Usually we use lowercase for the names of the response and the primary covariate. One exception to this rule is the name Time for a covariate. We try to avoid using the name time because it conflicts with a standard S function.

Table A.1 lists the groupedData objects in the NLMEDATA library that is part of the nlme distribution.

TABLE A.1: Data sets included with the nlme library distribution. The data sets whose names are shown in bold are described in this appendix.

Alfalfa	Yields of three varieties of alfalfa
Assay	Laboratory data on a biochemical assay
BodyWeight	Rat weight over time for different diets
CO2	Carbon dioxide uptake by grass plants
Cephamadole	Pharmacokinetic data
ChickWeight	Growth of chicks on different diets
Dialyzer	Performance of high-flux hemodialyzers
DNase	Assay of DNase
Earthquake	Severity of earthquakes
ergoStool	Ergometrics experiment with stool types
Fatigue	Metal fatique data
Gasoline	Gasoline yields for different crude samples
Glucose	Glucose levels over time
Glucose2	Glucose levels over time after alcohol ingestion
Gun	Naval gun firing data from Hicks (1993)
IGF	Assay data on Insulin-like Growth Factor
Indometh	Pharmacokinetic data on indomethicin
Loblolly	Growth of Loblolly pines
Machines	Productivity of workers on machines
MathAchSchool	School demographic data for MathAchieve
MathAchieve	Mathematics achievement scores
Meat	Tenderness of meat
Milk	Milk production by diet
Muscle	Muscle response by conc of $CaCl_2$
Nitrendipene	Assay of nitrendipene
Oats	Yield under different fertilizers
Orange	Growth of orange trees
Orthodont	Orthodontic measurement over time
Ovary	Number of large ovarian follicles over time
Oxboys	Heights of boys in Oxford, England
Oxide	Oxide coating on a semiconductor
PBG	Change in blood pressure vs. dose of phenylbiguanide
PBIB	A partially balanced incomplete block design
Phenobarb	Neonatal pharmacokinetics of phenobarbitol
Pixel	X-ray pixel intensities over time
Quinidine	Pharmacokinetic study of quinidine
Rail	Travel times of ultrasonic waves in railway rails
RatPupWeight	Weights of rat pups by litter
Relaxin	Assays of relaxin
Remifentanil	Pharmacokinetics of remifentanil
Soybean	Soybean growth by variety

TABLE A.1: (continued)

Spruce	Spruce tree growth
Tetracycline1	Pharmacokinetics of tetracycline
Tetracycline2	Pharmacokinetics of tetracycline
Theoph	Pharmacokinetics of theophylline
Wafer	Current vs. voltage on semiconductor wafers
Wheat	Yields by growing conditions
Wheat2	Yields from a randomized complete block design

Other data sets may be included with later versions of the library, which will be made available at `http://nlme.stat.wisc.edu`.

A.1 Alfalfa—Split-Plot Experiment on Varieties of Alfalfa

These data are described in Snedecor and Cochran (1980, §16.15) as an example of a split-plot design. The treatment structure used in the experiment was a 3×4 full factorial, with three varieties of alfalfa and four dates of third cutting in 1943. The experimental units were arranged into six blocks, each subdivided into four plots. The varieties of alfalfa (*Cossac*, *Ladak*, and *Ranger*) were assigned randomly to the blocks and the dates of third cutting (*None*, *S1*—September 1, *S20*—September 20, and *O7*—October 7) were randomly assigned to the plots. All four dates were used on each block. The data are presented in Figure A.1.

The display formula for these data is

```
Yield ~ Date | Block / Variety
```

based on the columns named:

Yield: the plot yield (T/acre).

Date: the third cutting date—None, S1, S20, or O7.

Block: a factor identifying the block—1 through 6.

Variety: alfalfa variety—Cossac, Ladak, or Ranger.

A.2 Assay—Bioassay on Cell Culture Plate

These data, courtesy of Rich Wolfe and David Lansky from Searle, Inc., come from a bioassay run on a 96-well cell culture plate. The assay is performed using a split-block design. The 8 rows on the plate are labeled A–H

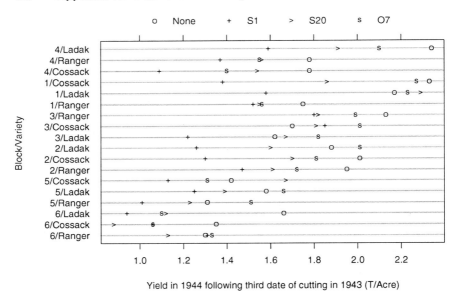

FIGURE A.1. Plot yields in a split-plot experiment on alfalfa varieties and dates of third cutting.

from top to bottom and the 12 columns on the plate are labeled 1–12 from left to right. Only the central 60 wells of the plate are used for the bioassay (the intersection of rows B–G and columns 2–11). There are two blocks in the design: Block 1 contains columns 2–6 and Block 2 contains columns 7–11. Within each block, six samples are assigned randomly to rows and five (serial) dilutions are assigned randomly to columns. The response variable is the logarithm of the optical density. The cells are treated with a compound that they metabolize to produce the stain. Only live cells can make the stain, so the optical density is a measure of the number of cells that are alive and healthy. The data are displayed in Figure 4.13 (p. 164).

Columns

The display formula for these data is

```
logDens ~ 1 | Block
```

based on the columns named:

logDens: log-optical density.

Block: a factor identifying the block where the wells are measured.

sample: a factor identifying the sample corresponding to the well, varying from "a" to "f."

dilut: a factor indicating the dilution applied to the well, varying from 1 to 5.

A.3   BodyWeight—Body Weight Growth in Rats

Hand and Crowder (1996) describe data on the body weights of rats measured over 64 days. These data also appear in Table 2.4 of Crowder and Hand (1990). The body weights of the rats (in grams) are measured on day 1 and every seven days thereafter until day 64, with an extra measurement on day 44. The experiment started several weeks before "day 1." There are three groups of rats, each on a different diet. A plot of the data is presented in Figure 3.2 (p. 104).

Columns

The display formula for these data is

    weight ~ Time | Rat

based on the columns named:

weight: body weight of the rat (grams).

Time: time at which the measurement is made (days).

Rat: a factor identifying the rat whose weight is measured.

Diet: a factor indicating the diet the rat receives.

A.4   Cefamandole—Pharmacokinetics of Cefamandole

Davidian and Giltinan (1995, §1.1, p. 2) describe data, shown in Figure A.2, obtained during a pilot study to investigate the pharmacokinetics of the drug cefamandole. Plasma concentrations of the drug were measured on six healthy volunteers at 14 time points following an intravenous dose of 15 mg/kg body weight of cefamandole.

Columns

The display formula for these data is

    conc ~ Time | Subject

based on the columns named:

conc: observed plasma concentration of cefamandole (mcg/ml).

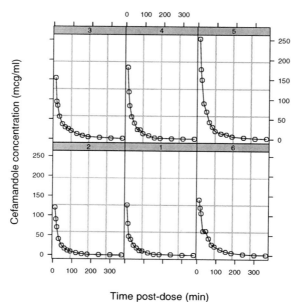

FIGURE A.2. Plasma concentration of cefamandole versus time post-injection for six healthy volunteers.

Time: time at which the sample was drawn (minutes post-injection).

Subject: a factor giving the subject from which the sample was drawn.

Models

Davidian and Giltinan (1995) use the biexponential model SSbiexp (§C.4, p. 514) with these data.

A.5   CO2—Carbon Dioxide Uptake

Potvin et al. (1990) describe an experiment on the cold tolerance of a C_4 grass species, *Echinochloa crus-galli*. The CO_2 uptake of six plants from Québec and six plants from Mississippi was measured at several levels of ambient CO_2 concentration. Half the plants of each type were chilled overnight before the experiment was conducted. The data are shown in Figure 8.15 (p. 369).

Columns

The display formula for these data is

```
uptake ~ conc | Plant
```

based on the columns named:

uptake: carbon dioxide uptake rate (μmol/m^2 sec).

conc: ambient concentration of carbon dioxide (mL/L).

Plant: a factor giving a unique identifier for each plant.

Type: origin of the plant, Québec or Mississippi.

Treatment: treatment, chilled or nonchilled.

Models

Potvin et al. (1990) suggest using a modified form of the asymptotic regression model SSasymp ($\S$C.1, p. 511), which we have coded as SSasympOff ($\S$C.2, p. 512).

A.6   Dialyzer—High-Flux Hemodialyzer

Vonesh and Carter (1992) describe data measured on high-flux hemodialyzers to assess their *in vivo* ultrafiltration characteristics. The ultrafiltration rates (in mL/hr) of 20 high-flux dialyzers were measured at seven different transmembrane pressures (in dmHg). The *in vitro* evaluation of the dialyzers used bovine blood at flow rates of either 200 dl/min or 300 dl/min. The data, shown in Figure 5.1 (p. 215), are also analyzed in Littell et al. (1996, $\S$8.2).

Columns

The display formula for these data is

    rate ~ pressure | Subject

based on the columns named:

rate: hemodialyzer ultrafiltration rate (mL/hr).

pressure: transmembrane pressure (dmHg).

Subject: a factor giving a unique identifier for each subject.

QB: bovine blood flow rate (dL/min)—200 or 300.

index: index of observation within subject—1 through 7.

A.7   DNase—Assay Data for the Protein DNase

Davidian and Giltinan (1995, $\S$5.2.4, p. 134) describe data, shown in Figure 3.8 (p. 115), obtained during the development of an ELISA assay for the recombinant protein DNase in rat serum.

Columns

The display formula for these data is

    density ˜ conc | Run

based on the columns named:

> density: the measured optical density in the assay. Duplicate optical density measurements were obtained.

> conc: the known concentration of the protein.

> Run: a factor giving the run from which the data were obtained.

Models

Davidian and Giltinan (1995) use the four-parameter logistic model, SSfpl (§C.6, p. 517) with these data, modeling the optical density as a logistic function of the logarithm of the concentration.

A.8   Earthquake—Earthquake Intensity

These data, shown in Figure A.3, are measurements recorded at available seismometer locations for 23 large earthquakes in western North America between 1940 and 1980. They were originally given in Joyner and Boore (1981); are mentioned in Brillinger (1987); and are analyzed in §11.4 of Davidian and Giltinan (1995).

Columns

The display formula for these data is

    accel ˜ distance | Quake

based on the columns named:

> accel: maximum horizontal acceleration observed (g).

> distance: the distance from the seismological measuring station to the epicenter of the earthquake (km).

> Quake: a factor indicating the earthquake on which the measurements were made.

> Richter: the intensity of the earthquake on the Richter scale.

> soil: soil condition at the measuring station—either soil or rock.

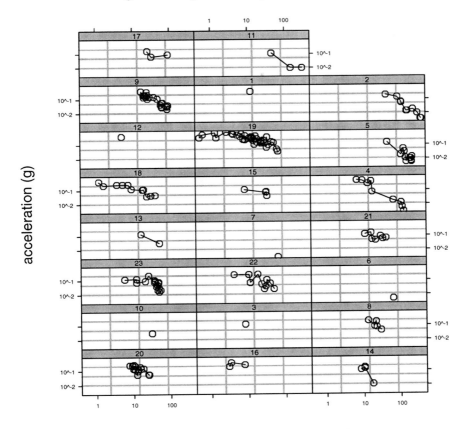

Distance from epicenter (km)

FIGURE A.3. Lateral acceleration versus distance from the epicenter for 23 large earthquakes in western North America. Both the acceleration and the distance are on a logarithmic scale. Earthquakes of greatest intensity as measured on the Richter scale are in the uppermost panels.

A.9   ergoStool—Ergometrics Experiment with Stool Types

Devore (2000, Exercise 11.9, p. 447) cites data from an article in *Ergometrics* (1993, pp. 519-535) on "The Effects of a Pneumatic Stool and a One-Legged Stool on Lower Limb Joint Load and Muscular Activity." These data are shown in Figure 1.5 (p. 13).

The display formula for these data is

```
effort ˜ Type | Subject
```

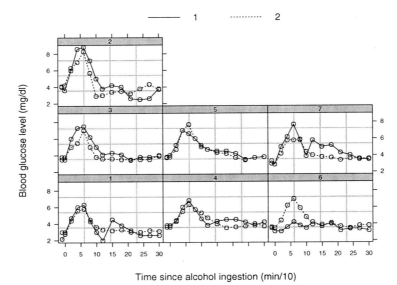

FIGURE A.4. Blood glucose levels of seven subjects measured over a period of 5 hours on two different occasions. In both dates the subjects took alcohol at time 0, but on the second occasion a dietary additive was used.

based on the columns named:

effort: effort to arise from a stool

Type: a factor giving the stool type

Subject: a factor giving a unique identifier for the subject in the experiment

A.10   Glucose2—Glucose Levels Following Alcohol Ingestion

Hand and Crowder (1996, Table A.14, pp. 180–181) describe data on the blood glucose levels measured at 14 time points over 5 hours for 7 volunteers who took alcohol at time 0. The same experiment was repeated on a second date with the same subjects but with a dietary additive used for all subjects. A plot of the data is presented in Figure A.4.

Columns

The display formula for these data is

```
glucose ~ Time | Subject/Date
```

based on the columns named:

weight: blood glucose level (in mg/dl).

Time: time since alcohol ingestion (in min/10).

Subject: a factor identifying the subject whose glucose level is measured.

Date: a factor indicating the occasion in which the experiment was conducted.

A.11   IGF—Radioimmunoassay of IGF-I Protein

Davidian and Giltinan (1995, §3.2.1, p. 65) describe data, shown in Figure 4.6 (p. 144), obtained during quality control radioimmunoassays for ten different lots of radioactive tracer used to calibrate the Insulin-like Growth Factor (IGF-I) protein concentration measurements.

Columns

The display formula for these data is

conc ˜ age | Lot

based on the columns named:

conc: the estimated concentration of IGF-I protein, in ng/ml.

age: the age (in days) of the radioactive tracer.

Lot: a factor giving the radioactive tracer lot.

A.12   Indometh—Indomethicin Kinetics

Kwan et al. (1976) present data on the plasma concentrations of indomethicin following intravenous injection. There are six different subjects in the experiment. The sampling times, ranging from 15 minutes post-injection to 8 hours post-injection, are the same for each subject. The data, presented in Figure 6.3 (p. 277), are analyzed in Davidian and Giltinan (1995, §2.1)
The display formula for these data is

conc ˜ time | Subject

based on the columns named:

conc: observed plasma concentration of indomethicin (mcg/ml).

time: time at which the sample was drawn (hours post-injection).

Subject: a factor indicating the subject from whom the sample is drawn.

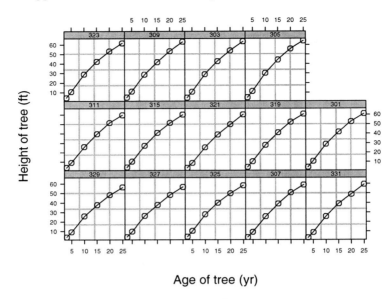

FIGURE A.5. Height of Loblolly pine trees over time

Models

Davidian and Giltinan (1995) use the biexponential model SSbiexp (§C.4, p. 514) with these data.

A.13 Loblolly—Growth of Loblolly Pine Trees

Kung (1986) presents data, shown in Figure A.5, on the growth of Loblolly pine trees.

The display formula for these data is

```
height ~ age | Seed
```

based on the columns named:

height: height of the tree (ft).

age: age of the tree (yr).

Seed: a factor indicating the seed source for the tree.

A.14   Machines—Productivity Scores for Machines and Workers

Data on an experiment to compare three brands of machines used in an industrial process are presented in Milliken and Johnson (1992, §23.1, p. 285). Six workers were chosen randomly among the employees of a factory to operate each machine three times. The response is an overall productivity score taking into account the number and quality of components produced. These data, shown in Figure 1.9 (p. 22), are analyzed in Milliken and Johnson (1992) with an ANOVA model.

The display formula for these data is

```
score ~ Machine | Worker
```

based on the columns named:

score: productivity score.

Machine: a factor identifying the machine brand—A, B, or C.

Worker: a factor giving the unique identifier for each worker.

A.15   Oats—Split-plot Experiment on Varieties of Oats

These data have been introduced by Yates (1935) as an example of a split-plot design. The treatment structure used in the experiment was a 3×4 full factorial, with three varieties of oats and four concentrations of nitrogen. The experimental units were arranged into six blocks, each with three whole-plots subdivided into four subplots. The varieties of oats were assigned randomly to the whole-plots and the concentrations of nitrogen to the subplots. All four concentrations of nitrogen were used on each whole-plot.

The data, presented in Figure 1.20 (p. 47), are analyzed in Venables and Ripley (1999, §6.11).

The display formula for these data is

```
yield ~ nitro | Block
```

based on the columns named:

yield: the subplot yield (bushels/acre).

nitro: nitrogen concentration (cwt/acre)—0.0, 0.2, 0.4, or 0.6.

Block: a factor identifying the block—I through VI.

Variety: oats variety—Golden Rain, Marvellous, or Victory.

A.16   Orange—Growth of Orange Trees

Draper and Smith (1998, Exercise 24.N, p. 559) present data on the growth of a group of orange trees. These data are plotted in Figure 8.1 (p. 339).

The display formula for these data is

```
circumference ~ age | Tree
```

based on the columns named:

circumference: circumference of the tree (mm)

age: time in days past the arbitrary origin of December 31, 1968.

Tree: a factor identifying the tree on which the measurement is made.

Models

The logistic growth model, SSlogis (§C.7, p. 519) provides a reasonable fit to these data.

A.17   Orthodont—Orthodontic Growth Data

Investigators at the University of North Carolina Dental School followed the growth of 27 children (16 males, 11 females) from age 8 until age 14. Every two years they measured the distance between the pituitary and the pterygomaxillary fissure, two points that are easily identified on x-ray exposures of the side of the head. These data are reported in Potthoff and Roy (1964) and plotted in Figure 1.11 (p. 31).

The display formula for these data is

```
distance ~ age | Subject
```

based on the columns named:

distance: the distance from the center of the pituitary to the pterygomaxillary fissure (mm).

age: the age of the subject when the measurement is made (years).

Subject: a factor identifying the subject on whom the measurement was made.

Sex: a factor indicating if the subject is male or female.

Models:

Based on the relationship shown in Figure 1.11 we begin with a simple linear relationship between distance and age

A.18   Ovary—Counts of Ovarian Follicles

Pierson and Ginther (1987) report on a study of the number of large ovarian follicles detected in different mares at several times in their estrus cycles. These data are shown in Figure 5.10 (p. 240).

The display formula for these data is

```
follicles ~ Time | Mare
```

based on the columns named:

follicles: the number of ovarian follicles greater than 10 mm in diameter.

Time: time in the estrus cycle. The data were recorded daily from 3 days before ovulation until 3 days after the next ovulation. The measurement times for each mare are scaled so that the ovulations for each mare occur at times 0 and 1.

Mare: a factor indicating the mare on which the measurement is made.

A.19   Oxboys—Heights of Boys in Oxford

These data are described in Goldstein (1987) as data on the height of a selection of boys from Oxford, England versus a standardized age. We display the data in Figure 3.1 (p. 99).

The display formula for these data is

```
height ~ age | Subject
```

based on the columns named:

height: height of the boy (cm)

age: standardized age (dimensionless)

Subject: a factor giving a unique identifier for each boy in the experiment

Occasion: an ordered factor—the result of converting age from a continuous variable to a count so these slightly unbalanced data can be analyzed as balanced.

A.20   Oxide—Variability in Semiconductor Manufacturing

These data are described in Littell et al. (1996, §4.4, p. 155) as coming "from a passive data collection study in the semiconductor industry where the objective is to estimate the variance components to determine the assignable

causes of the observed variability." The observed response is the thickness
of the oxide layer on silicon wafers, measured at three different sites of each
of three wafers selected from each of eight lots sampled from the population
of lots. We display the data in Figure 4.14 (p. 168).

The display formula for these data is

```
Thickness ~ 1 | Lot/Wafer
```

based on the columns named:

Thickness: thickness of the oxide layer.

Lot: a factor giving a unique identifier for each lot.

Wafer: a factor giving a unique identifier for each wafer within a lot.

A.21 PBG—Effect of Phenylbiguanide on Blood Pressure

Data on an experiment to examine the effect of a antagonist MDL 72222 on
the change in blood pressure experienced with increasing dosage of phenylbiguanide are described in Ludbrook (1994) and analyzed in Venables and
Ripley (1999, §8.8). Each of five rabbits was exposed to increasing doses of
phenylbiguanide after having either a placebo or the HD_5-antagonist MDL
72222 administered. The data are shown in Figure 3.4 (p. 107).

The display formula for these data is

```
deltaBP ~ dose | Rabbit
```

based on the columns named:

deltaBP: change in blood pressure (mmHg).

dose: dose of phenylbiguanide (μg).

Rabbit: a factor identifying the test animal.

Treatment: a factor identifying whether the observation was made after
administration of placebo or the HD_5-antagonist MDL 72222.

Models

The form of the response suggests a logistic model SSlogis (§C.7, p. 519)
for the change in blood pressure as function of the logarithm of the concentration of PBG.

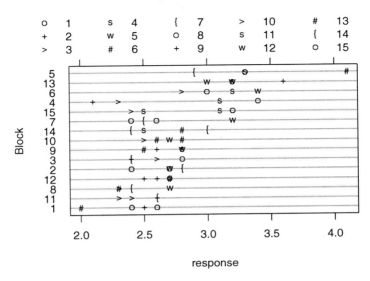

FIGURE A.6. Data on the response in an experiment conducted using fifteen treatments in fifteen blocks of size four. The responses are shown by block with different characters indicating different treatments.

A.22   PBIB—A Partially Balanced Incomplete Block Design

Data from a partially balanced incomplete block design in which there were fifteen treatments used in fifteen blocks of size four. The blocking is incomplete in that only a subset of the treatments can be used in each block. It is partially balanced in that every pair of treatments occurs together in a block the same number of times.

These data were described in Cochran and Cox (1957, p. 456). They are also used as data set 1.5.1 in Littell et al. (1996, §1.5.1). The data are shown in Figure A.6.

The display formula for these data is

```
response ~ Treatment | Block
```

based on the columns named:

response: the continuous response in the experiment

Treatment: the treatment factor

Block: the block

A.23   Phenobarb—Phenobarbitol Kinetics

Data from a pharmacokinetics study of phenobarbital in neonatal infants. During the first few days of life the infants receive multiple doses of phenobarbital for prevention of seizures. At irregular intervals blood samples are drawn and serum phenobarbital concentrations are determined. The data, displayed in Figure 6.15 (p. 296), were originally given in Grasela and Donn (1985) and are analyzed in Boeckmann et al. (1994) and in Davidian and Giltinan (1995, §6.6).

The display formula for these data is

```
conc ~ time | Subject
```

based on the columns named:

conc: phenobarbital concentration in the serum (μg/L).

time: time when the sample is drawn or drug administered (hr).

Subject: a factor identifying the infant.

Wt: birth weight of the infant (kg).

Apgar: the 5-minute Apgar score for the infant. This is an indication of health of the newborn infant. The scale is $1 - 10$.

ApgarInd: a factor indicating whether the 5-minute Apgar score is < 5 or ≥ 5.

dose: dose of drug administered (μg/kg).

Models

A one-compartments open model with intravenous administration and first-order elimination, described in §6.4, is used for these data

A.24   Pixel—Pixel Intensity in Lymphnodes

These data are from an experiment conducted by Deborah Darien, Department of Medical Sciences, School of Veterinary Medicine, University of Wisconsin, Madison. The mean pixel intensity of the right and left lymphnodes in the axillary region obtained from CT scans of 10 dogs were recorded over a period of 14 days after intravenous application of a contrast. The data are shown in Figure 1.17 (p. 42).

The display formula for these data is

```
pixel ~ day | Dog
```

based on the columns named:

`pixel`: mean pixel intensity of lymphnode in the CT scan.

`day`: number of days since contrast administration.

`Dog`: a factor giving the unique identifier for each dog.

`Side`: a factor indicating the side on which the measurement was made.

A.25   Quinidine—Quinidine Kinetics

Verme, Ludden, Clementi and Harris (1992) analyze routine clinical data on patients receiving the drug quinidine as a treatment for cardiac arrythmia (atrial fibrillation of ventricular arrythmias). All patients were receiving oral quinidine doses. At irregular intervals blood samples were drawn and serum concentrations of quinidine were determined. These data, shown in Figure A.7, are analyzed in several publications, including Davidian and Giltinan (1995, §9.3).

The display formula for these data is

`conc ~ time | Subject`

based on the columns named:

`conc`: serum quinidine concentration (mg/L).

`time`: time (hr) at which the drug was administered or the blood sample drawn. This is measured from the time the patient entered the study.

`Subject`: a factor identifying the patient on whom the data were collected.

`dose`: dose of drug administered (mg). Although there were two different forms of quinidine administered, the doses were adjusted for differences in salt content by conversion to milligrams of quinidine base.

`interval`: when the drug has been given at regular intervals for a sufficiently long period of time to assume steady state behavior, the interval is recorded.

`Age`: age of the subject on entry to the study (yr).

`Height`: height of the subject on entry to the study (in.).

`Weight`: body weight of the subject (kg).

`Race`: a factor identifying the race—Caucasian, Black, or Latin.

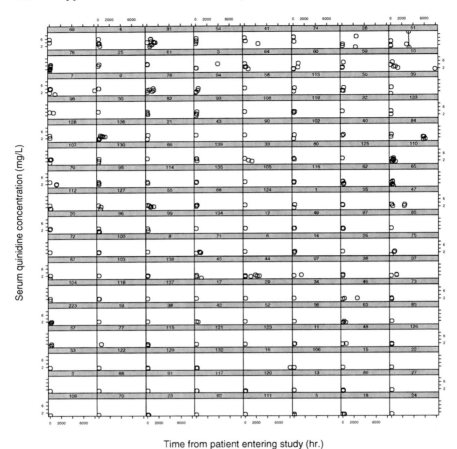

Time from patient entering study (hr.)

FIGURE A.7. Serum concentrations of quinidine in 136 hospitalized patients under varying dosage regimens versus time since entering the study.

Smoke: a factor giving smoking status at the time of the measurement—no or yes.

Ethanol: a factor giving ethanol (alcohol) abuse status at the time of the measurement—none, current, or former.

Heart: a factor indicating congestive heart failure for the subject—none/mild, moderate, or severe.

Creatinine: a factor in eight levels coding the creatinine clearance and other measurements. Creatinine clearance is divided into those greater than 50 mg/min and those less than 50 mg/min.

`glyco`: alpha-1 acid glycoprotein concentration (mg/dL). Often measured at the same time as the quinidine concentration.

Models

A model for these data is described in §8.2.2.

A.26 Rail—Evaluation of Stress in Rails

Devore (2000, Example 10.10, p. 427) cites data from an article in *Materials Evaluation* on "a study of travel time for a certain type of wave that results from longitudinal stress of rails used for railroad track." The data are displayed in Figure 1.1 (p. 4).

The display formula for these data is

```
travel ~ 1 | Rail
```

based on the columns named:

`travel`: travel time for ultrasonic head-waves in the rail (nanoseconds). The value given is the original travel time minus 36,100 nanoseconds.

`Rail`: a factor giving the number of the rail on which the measurement was made.

A.27 Soybean—Soybean Leaf Weight over Time

These data, shown in Figure 6.10 (p. 288), are described in Davidian and Giltinan (1995, §1.1.3, p. 7) as "Data from an experiment to compare growth patterns of two genotypes of soybeans: Plant Introduction #416937 (P), an experimental strain, and Forrest (F), a commercial variety."

The display formula for these data is

```
weight ~ Time | Plot
```

based on the columns named:

`weight`: average leaf weight per plant (g).

`Time`: time the sample was taken (days after planting).

`Plot`: a factor giving a unique identifier for each plot.

`Variety`: a factor indicating the variety; Forrest (F) or Plant Introduction #416937 (P)

`Year`: the year the plot was planted.

Models

The form of the response suggests a logistic model, SSlogis (§C.7, p. 519).

A.28   Spruce—Growth of Spruce Trees

Diggle et al. (1994, Example 1.3, page 5) describe data on the growth of spruce trees that have been exposed to an ozone-rich atmosphere or to a normal atmosphere. These data are plotted in Figures A.8–A.10.   The display formula for these data is

    logSize ~ days | Tree

based on the columns named:

   logSize: the logarithm of an estimate of the volume of the tree trunk

   days: number of days since the beginning of the experiment

   Tree: a factor giving a unique identifier for each tree

   Plot: a factor identifying the plot in which the tree was grown. The levels of this factor are Ozone1, Ozone2, Normal1, and Normal2.

   Treatment a factor indicating whether the tree was grown in an ozone-rich atmosphere or a normal atmosphere.

A.29   Theoph—Theophylline Kinetics

Boeckmann et al. (1994) report data from a study by Dr. Robert Upton of the kinetics of the anti-asthmatic drug theophylline. Twelve subjects were given oral doses of theophylline then serum concentrations were measured at 11 time points over the next 25 hours. Davidian and Giltinan (1995) also analyze these data, shown in Figure 8.6 (p. 352).

   The display formula for these data is

    conc ~ Time | Subject

based on the columns named:

   conc: theophylline concentration in the sample (mg/L).

   Time: time since drug administration when the sample was drawn (hr).

   Subject: a factor identifying the subject.

   Wt: weight of the subject (kg).

   Dose: dose administered to the subject (mg/kg).

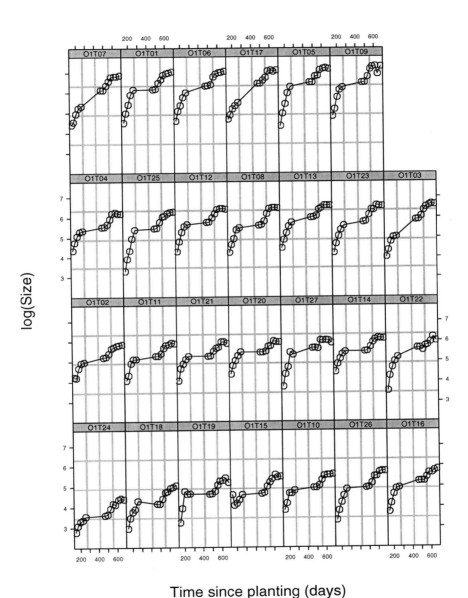

Time since planting (days)

FIGURE A.8. Growth measures in the logarithm of an estimate of the volume of the spruce tree trunk versus time. These 27 trees were in the first plot that was exposed to an ozone-rich atmosphere throughout the experiment

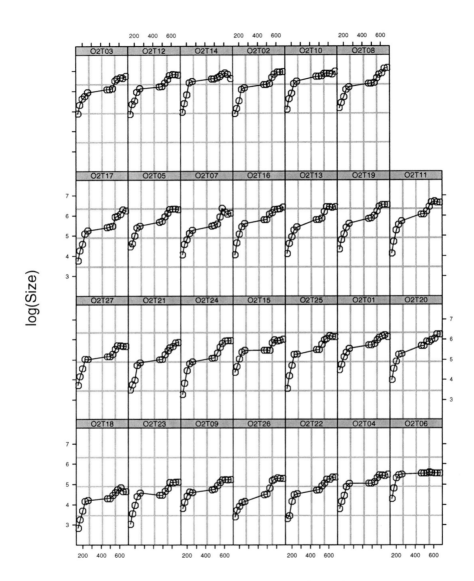

FIGURE A.9. Growth measures in the logarithm of an estimate of the volume of the spruce tree trunk versus time. These 27 trees were in the second plot that was exposed to an ozone-rich atmosphere throughout the experiment

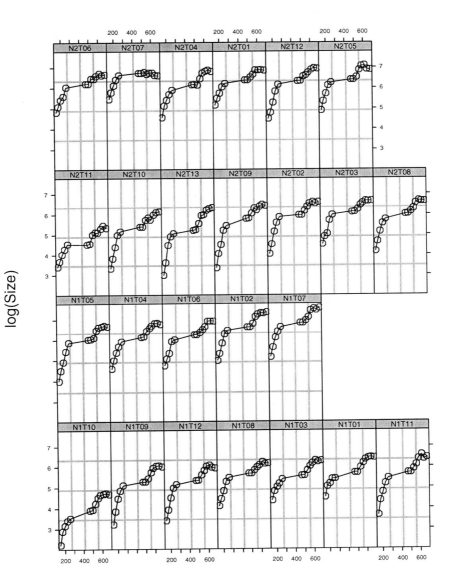

Time since planting (days)

FIGURE A.10. Growth measures in the logarithm of an estimate of the volume of the spruce tree trunk versus time. These 25 trees were in the first and second plots that were exposed to an normal atmosphere throughout the experiment

Models:

Both Boeckmann et al. (1994) and Davidian and Giltinan (1995) use a two-compartment open pharmacokinetic model, which we code as SSfol (§C.5, p. 516), for these data.

A.30   Wafer—Modeling of Analog MOS Circuits

In an experiment conducted at the Microelectronics Division of Lucent Technologies to study the variability in the manufacturing of analog MOS circuits, the intensities of the current at five ascending voltages were collected on n-channel devices. Measurements were made on eight sites of each of ten wafers. Figure 3.11 (p. 118) shows the response curves for each site, by wafer.

The display formula for these data is

```
current ~ voltage | Wafer/Site
```

based on the columns named:

current: the intensity of current (mA).

voltage: the voltage applied to the device (V).

Wafer: a factor giving a unique identifier for each wafer.

Site: a factor giving an identifier for each site within a wafer.

A.31   Wheat2—Wheat Yield Trials

Stroup and Baenziger (1994) report data on an agronomic yield trial to compare 56 different varieties of wheat. The experimental units were organized according to a randomized complete block design with four blocks. All 56 varieties of wheat were used in each block. The latitude and longitude of each experimental unit in the trial were also recorded. The data, shown in Figure 5.22 (p. 261), are also analyzed in Littell et al. (1996, §9.6.2).

Columns

The display formula for these data is

```
yield ~ variety | Block
```

based on the columns named:

yield: wheat yield.

variety: a factor giving the unique identifier for each wheat variety.

Block: a factor giving a unique identifier for each block in the experiment.

latitude: latitude of the experimental unit.

longitude: longitude of the experimental unit.

Appendix B
S Functions and Classes

There are over 300 different functions and classes defined in the nlme library. In this appendix we reproduce the on-line documentation for those functions and classes that are most frequently used in the examples in the text. The documentation for all the functions and classes in the library is available with the library.

ACF	*Autocorrelation Function*

```
ACF(object, maxLag, ...)
```

Arguments

    object   Any object from which an autocorrelation function can be obtained. Generally an object resulting from a model fit, from which residuals can be extracted.

    maxLag   Maximum lag for which the autocorrelation should be calculated.

    ...   Some methods for this generic require additional arguments.

Description

This function is generic; method functions can be written to handle specific classes of objects. Classes that already have methods for this function include gls and lme.

Value

Will depend on the method function used; see the appropriate documentation.

See Also

ACF.gls, ACF.lme

ACF.lme   *Autocorrelation Function for lme Residuals*

ACF(object, maxLag, resType)

Arguments

object   An object inheriting from class lme, representing a
                fitted linear mixed-effects model.

maxLag   An optional integer giving the maximum lag for which
                the autocorrelation should be calculated. Defaults to
                maximum lag in the within-group residuals.

resType   An optional character string specifying the type of
                residuals to be used. If "response", the "raw" resid-
                uals (observed − fitted) are used; else, if "pearson",
                the standardized residuals (raw residuals divided by
                the corresponding standard errors) are used; else, if
                "normalized", the normalized residuals (standard-
                ized residuals premultiplied by the inverse square-
                root factor of the estimated error correlation matrix)
                are used. Partial matching of arguments is used, so
                only the first character needs to be provided. Defaults
                to "pearson".

Description

This method function calculates the empirical autocorrelation function (Box et al., 1994) for the within-group residuals from an lme fit.
The autocorrelation values are calculated using pairs of residuals within
the innermost group level. The autocorrelation function is useful for investigating serial correlation models for equally spaced data.

Value

A data frame with columns lag and ACF representing, respectively, the
lag between residuals within a pair and the corresponding empirical
autocorrelation. The returned value inherits from class ACF.

See Also

ACF.gls, plot.ACF

Examples

```
fm1 <- lme(follicles ~ sin(2*pi*Time) + cos(2*pi*Time), Ovary,
           random = ~ sin(2*pi*Time) | Mare)
ACF(fm1, maxLag = 11)
```

anova.lme	*Compare Likelihoods of Fitted Objects*

```
anova(object, ..., test, type, adjustSigma, Terms, L,
      verbose)
```

Arguments

object
: A fitted model object inheriting from class lme, representing a mixed-effects model.

...
: Other optional fitted model objects inheriting from classes gls, gnls, lm, lme, lmList, nlme, nlsList, or nls.

test
: An optional logical value controlling whether likelihood ratio tests should be used to compare the fitted models represented by object and the objects in .... Defaults to TRUE.

type
: An optional character string specifying the type of sum of squares to be used in *F*-tests for the terms in the model. If "sequential", the sequential sum of squares obtained by including the terms in the order they appear in the model is used; else, if "marginal", the marginal sum of squares obtained by deleting a term from the model at a time is used. This argument is only used when a single fitted object is passed to the function. Partial matching of arguments is used, so only the first character needs to be provided. Defaults to "sequential".

adjustSigma
: An optional logical value. If TRUE and the estimation method used to obtain object was maximum likelihood, the residual standard error is multiplied by $\sqrt{n_{\mathrm{obs}}/(n_{\mathrm{obs}} - n_{\mathrm{par}})}$, converting it to a REML-like estimate. This argument is only used when a single fitted object is passed to the function. Default is TRUE.

Terms	An optional integer or character vector specifying which terms in the model should be jointly tested to be zero using a Wald F-test. If given as a character vector, its elements must correspond to term names; else, if given as an integer vector, its elements must correspond to the order in which terms are included in the model. This argument is only used when a single fitted object is passed to the function. Default is NULL.
L	An optional numeric vector or array specifying linear combinations of the coefficients in the model that should be tested to be zero. If given as an array, its rows define the linear combinations to be tested. If names are assigned to the vector elements (array columns), they must correspond to names of the coefficients and will be used to map the linear combination(s) to the coefficients; else, if no names are available, the vector elements (array columns) are assumed in the same order as the coefficients appear in the model. This argument is only used when a single fitted object is passed to the function. Default is NULL.
verbose	An optional logical value. If TRUE, the calling sequences for each fitted model object are printed with the rest of the output, being omitted if verbose = FALSE. Defaults to FALSE.

Description

When only one fitted model object is present, a data frame with the sums of squares, numerator degrees of freedom, denominator degrees of freedom, F-values, and p-values for Wald tests for the terms in the model (when Terms and L are NULL), a combination of model terms (when Terms in not NULL), or linear combinations of the model coefficients (when L is not NULL). Otherwise, when multiple fitted objects are being compared, a data frame with the degrees of freedom, the (restricted) log-likelihood, the Akaike Information Criterion (AIC), and the Bayesian Information Criterion (BIC) of each object is returned. If test=TRUE, whenever two consecutive objects have different number of degrees of freedom, a likelihood ratio statistic, with the associated p-value is included in the returned data frame.

Value

A data frame inheriting from class anova.lme.

Note

Likelihood comparisons are not meaningful for objects fit using restricted maximum likelihood and with different fixed effects.

See Also

gls, gnls, nlme, lme, AIC, BIC, print.anova.lme

Examples

```
fm1 <- lme(distance ~ age, Orthodont, random = ~ age | Subject)
anova(fm1)
fm2 <- update(fm1, random = pdDiag(~age))
anova(fm1, fm2)
```

---

coef.lme   *Extract* lme *Coefficients*

```
coef(object, augFrame, level, data, which, FUN,
     omitGroupingFactor)
```

Arguments

object	An object inheriting from class lme, representing a fitted linear mixed-effects model.
augFrame	An optional logical value. If TRUE, the returned data frame is augmented with variables defined in data; else, if FALSE, only the coefficients are returned. Defaults to FALSE.
level	An optional positive integer giving the level of grouping to be used in extracting the coefficients from an object with multiple nested grouping levels. Defaults to the highest or innermost level of grouping.
data	An optional data frame with the variables to be used for augmenting the returned data frame when augFrame = TRUE. Defaults to the data frame used to fit object.
which	An optional positive integer or character vector specifying which columns of data should be used in the augmentation of the returned data frame. Defaults to all columns in data.

FUN   An optional summary function or a list of summary
                             functions to be applied to group-varying variables,
                             when collapsing `data` by groups. Group-invariant vari-
                             ables are always summarized by the unique value that
                             they assume within that group. If FUN is a single func-
                             tion it will be applied to each noninvariant variable
                             by group to produce the summary for that variable. If
                             FUN is a list of functions, the names in the list should
                             designate classes of variables in the frame such as
                             `ordered`, `factor`, or `numeric`. The indicated func-
                             tion will be applied to any group-varying variables of
                             that class. The default functions to be used are `mean`
                             for numeric factors, and `Mode` for both `factor` and
                             `ordered`. The `Mode` function, defined internally in
                             `gsummary`, returns the modal or most popular value
                             of the variable. It is different from the `mode` function
                             that returns the S-language mode of the variable.

omitGroupingFactor
                             An optional logical value. When `TRUE` the grouping
                             factor itself will be omitted from the groupwise sum-
                             mary of `data`, but the levels of the grouping factor
                             will continue to be used as the row names for the
                             returned data frame. Defaults to `FALSE`.

Description

The estimated coefficients at level i are obtained by adding together the
fixed-effects estimates and the corresponding random-effects estimates
at grouping levels less or equal to i. The resulting estimates are returned
as a data frame, with rows corresponding to groups and columns to
coefficients. Optionally, the returned data frame may be augmented
with covariates summarized over groups.

Value

A data frame inheriting from class coef.lme with the estimated coefficients at level `level` and, optionally, other covariates summarized over
groups. The returned object also inherits from classes ranef.lme and
data.frame.

See Also

`lme`, `fixef.lme`, `ranef.lme`, `plot.ranef.lme`, `gsummary`

Examples

```
fm1 <- lme(distance ~ age, Orthodont, random = ~ age | Subject)
```

```
coef(fm1)
coef(fm1, augFrame = TRUE)
```

---

coef.lmList   *Extract lmList Coefficients*

```
coef(object, augFrame, data, which, FUN,
    omitGroupingFactor)
```

Arguments

object
: An object inheriting from class lmList, representing a list of lm objects with a common model.

augFrame
: An optional logical value. If TRUE, the returned data frame is augmented with variables defined in the data frame used to produce object; else, if FALSE, only the coefficients are returned. Defaults to FALSE.

data
: An optional data frame with the variables to be used for augmenting the returned data frame when augFrame = TRUE. Defaults to the data frame used to fit object.

which
: An optional positive integer or character vector specifying which columns of the data frame used to produce object should be used in the augmentation of the returned data frame. Defaults to all variables in the data.

FUN
: An optional summary function or a list of summary functions to be applied to group-varying variables, when collapsing the data by groups. Group-invariant variables are always summarized by the unique value that they assume within that group. If FUN is a single function it will be applied to each noninvariant variable by group to produce the summary for that variable. If FUN is a list of functions, the names in the list should designate classes of variables in the frame such as ordered, factor, or numeric. The indicated function will be applied to any group-varying variables of that class. The default functions to be used are mean for numeric factors, and Mode for both factor and ordered. The Mode function, defined internally in gsummary, returns the modal or most popular value of the variable. It is different from the mode function that returns the S-language mode of the variable.

omitGroupingFactor

An optional logical value. When TRUE the grouping factor itself will be omitted from the groupwise summary of data but the levels of the grouping factor will continue to be used as the row names for the returned data frame. Defaults to FALSE.

Description

The coefficients of each lm object in the object list are extracted and organized into a data frame, with rows corresponding to the lm components and columns corresponding to the coefficients. Optionally, the returned data frame may be augmented with covariates summarized over the groups associated with the lm components.

Value

A data frame inheriting from class coef.lmList with the estimated coefficients for each lm component of object and, optionally, other covariates summarized over the groups corresponding to the lm components. The returned object also inherits from classes ranef.lmList and data.frame.

See Also

lmList, fixed.effects.lmList, ranef.lmList,
plot.ranef.lmList, gsummary

Examples

```
fm1 <- lmList(distance ~ age|Subject, data = Orthodont)
coef(fm1)
coef(fm1, augFrame = TRUE)
```

---

fitted.lme   *Extract lme Fitted Values*

```
fitted(object, level, asList)
```

Arguments

object	An object inheriting from class lme, representing a fitted linear mixed-effects model.
level	An optional integer vector giving the level(s) of grouping to be used in extracting the fitted values from object. Level values increase from outermost to innermost grouping, with level zero corresponding to

the population fitted values. Defaults to the highest or innermost level of grouping.

aslist   An optional logical value. If TRUE and a single value is given in level, the returned object is a list with the fitted values split by groups; else the returned value is either a vector or a data frame, according to the length of level. Defaults to FALSE.

Description

The fitted values at level i are obtained by adding together the population-fitted values (based only on the fixed-effects estimates) and the estimated contributions of the random effects to the fitted values at grouping levels less or equal to i. The resulting values estimate the best linear unbiased predictions (BLUPs) at level i.

Value

If a single level of grouping is specified in level, the returned value is either a list with the fitted values split by groups (asList = TRUE) or a vector with the fitted values (asList = FALSE); else, when multiple grouping levels are specified in level, the returned object is a data frame with columns given by the fitted values at different levels and the grouping factors.

See Also

lme, residuals.lme

Examples

```
fm1 <- lme(distance ~ age + Sex, data = Orthodont, random = ~ 1)
fitted(fm1, level = 0:1)
```

---

fixef	*Extract Fixed Effects*

```
fixef(object, ...)
fixed.effects(object, ...)
```

Arguments

object   Any fitted model object from which fixed-effects estimates can be extracted.

. . .   Some methods for this generic function require additional arguments.

Description

This function is generic; method functions can be written to handle specific classes of objects. Classes that already have methods for this function include lmList and lme.

Value

Will depend on the method function used; see the appropriate documentation.

See Also

fixef.lmList, fixef.lme

---

gapply   *Apply a Function by Groups*

---

gapply(object, which, FUN, form, level, groups, ...)

Arguments

object	An object to which the function will be applied, usually a groupedData object or a data.frame. Must inherit from class data.frame.
which	An optional character or positive integer vector specifying which columns of object should be used with FUN. Defaults to all columns in object.
FUN	Function to apply to the distinct sets of rows of the data frame object defined by the values of groups.
form	An optional one-sided formula that defines the groups. When this formula is given the right-hand side is evaluated in object, converted to a factor if necessary, and the unique levels are used to define the groups. Defaults to formula(object).
level	An optional positive integer giving the level of grouping to be used in an object with multiple nested grouping levels. Defaults to the highest or innermost level of grouping.
groups	An optional factor that will be used to split the rows into groups. Defaults to getGroups(object, form, level).
...	Optional additional arguments to the summary function FUN. Often it is helpful to specify na.rm = TRUE.

Description

Applies the function to the distinct sets of rows of the data frame defined by groups.

Value

Returns a data frame with as many rows as there are levels in the groups argument.

See Also

gsummary

Examples

```
## Find number of nonmissing "conc" observations for each Subject
gapply( Quinidine, FUN = function(x) sum(!is.na(x$conc)) )
```

---

getGroups   *Extract Grouping Factors from an Object*

```
getGroups(object, form, level, data)
```

Arguments

object	Any object.
form	An optional formula with a conditioning expression on its right hand side (i.e., an expression involving the | operator). Defaults to formula(object).
level	A positive integer vector with the level(s) of grouping to be used when multiple nested levels of grouping are present. This argument is optional for most methods of this generic function and defaults to all levels of nesting.
data	A data frame in which to interpret the variables named in form. Optional for most methods.

Description

This function is generic; method functions can be written to handle specific classes of objects. Classes that already have methods for this function include corStruct, data.frame, gls, lme, lmList, and varFunc.

Value

Will depend on the method function used; see the appropriate documentation.

See Also

getGroupsFormula, getGroups.data.frame, getGroups.gls, getGroups.lmList, getGroups.lme

---

gls   *Fit Linear Model Using Generalized Least Squares*

```
gls(model, data, correlation, weights, subset, method,
    na.action, control, verbose)
```

Arguments

model
: A two-sided linear formula object describing the model, with the response on the left of a ~ operator and the terms, separated by + operators, on the right.

data
: An optional data frame containing the variables named in model, correlation, weights, and subset. By default the variables are taken from the environment from which gls is called.

correlation
: An optional corStruct object describing the within-group correlation structure. See the documentation of corClasses for a description of the available corStruct classes. If a grouping variable is to be used, it must be specified in the form argument to the corStruct constructor. Defaults to NULL, corresponding to uncorrelated errors.

weights
: An optional varFunc object or one-sided formula describing the within-group heteroscedasticity structure. If given as a formula, it is used as the argument to varFixed, corresponding to fixed variance weights. See the documentation on varClasses for a description of the available varFunc classes. Defaults to NULL, corresponding to homoscesdatic errors.

subset
: An optional expression indicating which subset of the rows of data should be used in the fit. This can be a logical vector, or a numeric vector indicating which observation numbers are to be included, or a character vector of the row names to be included. All observations are included by default.

method
: A character string. If "REML" the model is fit by maximizing the restricted log-likelihood. If "ML" the log-likelihood is maximized. Defaults to "REML".

na.action	A function that indicates what should happen when the data contain NAs. The default action (na.fail) causes gls to print an error message and terminate if there are any incomplete observations.
control	A list of control values for the estimation algorithm to replace the default values returned by the function glsControl. Defaults to an empty list.
verbose	An optional logical value. If TRUE information on the evolution of the iterative algorithm is printed. Default is FALSE.

Description

This function fits a linear model using generalized least squares. The errors are allowed to be correlated and/or have unequal variances.

Value

An object of class gls representing the linear model fit. Generic functions such as print, plot, and summary have methods to show the results of the fit. See glsObject for the components of the fit. The functions resid, coef, and fitted can be used to extract some of its components.

References

The different correlation structures available for the correlation argument are described in Box et al. (1994), Littell et al. (1996), and Venables and Ripley (1999). The use of variance functions for linear and nonlinear models is presented in detail in Carroll and Ruppert (1988) and Davidian and Giltinan (1995).

See Also

glsControl, glsObject, varFunc, corClasses, varClasses

Examples

```
# AR(1) errors within each Mare
fm1 <- gls(follicles ~ sin(2*pi*Time) + cos(2*pi*Time), Ovary,
           correlation = corAR1(form = ~ 1 | Mare))
# variance increases as a power of the absolute fitted values
fm2 <- gls(follicles ~ sin(2*pi*Time) + cos(2*pi*Time), Ovary,
           weights = varPower())
```

gnls	*Fit Nonlinear Model Using Generalized Least Squares*

```
gnls(model, data, params, start, correlation, weights,
     subset, na.action, naPattern, control, verbose)
```

Arguments

model
: A two-sided formula object describing the model, with the response on the left of a ~ operator and a non-linear expression involving parameters and covariates on the right. If data is given, all names used in the formula should be defined as parameters or variables in the data frame.

data
: An optional data frame containing the variables used in model, correlation, weights, subset, and naPattern. By default the variables are taken from the environment from which gnls is called.

params
: An optional two-sided linear formula of the form p1+···+pn~x1+···+xm, or list of two-sided formulas of the form p1~x1+···+xm, with possibly different models for each parameter. The p1,...,pn represent parameters included on the right-hand side of model and x1+···+xm define a linear model for the parameters (when the left-hand side of the formula contains several parameters, they are all assumed to follow the same linear model described by the right-hand side expression). A 1 on the right-hand side of the formula(s) indicates a single fixed effect for the corresponding parameter(s). By default, the parameters are obtained from the names of start.

start
: An optional named list, or numeric vector, with the initial values for the parameters in model. It can be omitted when a selfStarting function is used in model, in which case the starting estimates will be obtained from a single call to the nls function.

correlation
: An optional corStruct object describing the within-group correlation structure. See the documentation of corClasses for a description of the available corStruct classes. If a grouping variable is to be used, it must be specified in the form argument to the corStruct constructor. Defaults to NULL, corresponding to uncorrelated errors.

weights	An optional varFunc object or one-sided formula describing the within-group heteroscedastic structure. If given as a formula, it is used as the argument to varFixed, corresponding to fixed variance weights. See the documentation on varClasses for a description of the available varFunc classes. Defaults to NULL, corresponding to homoscesdatic errors.
subset	An optional expression indicating which subset of the rows of data should be used in the fit. This can be a logical vector, or a numeric vector indicating which observation numbers are to be included, or a character vector of the row names to be included. All observations are included by default.
na.action	A function that indicates what should happen when the data contain NAs. The default action (na.fail) causes gnls to print an error message and terminate if there are any incomplete observations.
naPattern	An expression or formula object, specifying which returned values are to be regarded as missing.
control	A list of control values for the estimation algorithm to replace the default values returned by the function gnlsControl. Defaults to an empty list.
verbose	An optional logical value. If TRUE information on the evolution of the iterative algorithm is printed. Default is FALSE.

Description

This function fits a nonlinear model using generalized least squares. The errors are allowed to be correlated and/or have unequal variances.

Value

An object of class gnls, also inheriting from class gls, representing the nonlinear model fit. Generic functions such as print, plot and summary have methods to show the results of the fit. See gnlsObject for the components of the fit. The functions resid, coef, and fitted can be used to extract some of its components.

References

The different correlation structures available for the correlation argument are described in Box et al. (1994), Littell et al. (1996), and Venables and Ripley (1999). The use of variance functions for linear and nonlinear models is presented in detail in Carroll and Ruppert (1988) and Davidian and Giltinan (1995).

See Also

`gnlsControl, gnlsObject, varFunc, corClasses, varClasses`

Examples

```
# variance increases with a power of the absolute fitted values
fm1 <- gnls(weight ~ SSlogis(Time, Asym, xmid, scal), Soybean,
            weights = varPower())
# errors follow an auto-regressive process of order 1
fm2 <- gnls(weight ~ SSlogis(Time, Asym, xmid, scal), Soybean,
            correlation = corAR1())
```

---

groupedData   *Construct a groupedData Object*

```
groupedData(formula, data, order.groups, FUN, outer, inner,
            labels, units)
```

Arguments

formula
: A formula of the form resp ~ cov | group where resp is the response, cov is the primary covariate, and group is the grouping factor. The expression 1 can be used for the primary covariate when there is no other suitable candidate. Multiple nested grouping factors can be listed separated by the / symbol as in fact1/fact2. In an expression like this the fact2 factor is nested within the fact1 factor.

data
: A data frame in which the expressions in formula can be evaluated. The resulting groupedData object will consist of the same data values in the same order, but with additional attributes.

order.groups
: An optional logical value, or list of logical values, indicating if the grouping factors should be converted to ordered factors according to the function FUN applied to the response from each group. If multiple levels of grouping are present, this argument can be either a single logical value (which will be repeated for all grouping levels) or a list of logical values. If no names are assigned to the list elements, they are

assumed in the same order as the group levels (outer-most to innermost grouping). Ordering within a level of grouping is done within the levels of the grouping factors which are outer to it. Changing the grouping factor to an ordered factor does not affect the ordering of the rows in the data frame, but it does affect the order of the panels in a trellis display of the data or models fitted to the data. Defaults to TRUE.

FUN
An optional summary function that will be applied to the values of the response for each level of the grouping factor, when order.groups = TRUE, to determine the ordering. Defaults to the max function.

outer
An optional one-sided formula, or list of one-sided formulas, indicating covariates that are outer to the grouping factor(s). If multiple levels of grouping are present, this argument can be either a single one-sided formula, or a list of one-sided formulas. If no names are assigned to the list elements, they are assumed in the same order as the group levels (out-ermost to innermost grouping). An outer covariate is invariant within the sets of rows defined by the grouping factor. Ordering of the groups is done in such a way as to preserve adjacency of groups with the same value of the outer variables. When plotting a groupedData object, the argument outer = TRUE causes the panels to be determined by the outer formula. The points within the panels are associated by level of the grouping factor. Defaults to NULL, meaning that no outer covariates are present.

inner
An optional one-sided formula, or list of one-sided formulas, indicating covariates that are inner to the grouping factor(s). If multiple levels of grouping are present, this argument can be either a single one-sided formula, or a list of one-sided formulas. If no names are assigned to the list elements, they are assumed in the same order as the group levels (outer-most to innermost grouping). An inner covariate can change within the sets of rows defined by the group-ing factor. An inner formula can be used to associate points in a plot of a groupedData object. Defaults to NULL, meaning that no inner covariates are present.

labels
An optional list of character strings giving labels for the response and the primary covariate. The label

for the primary covariate is named x and that for the response is named y. Either label can be omitted.

units   An optional list of character strings giving the units for the response and the primary covariate. The units string for the primary covariate is named x and that for the response is named y. Either units string can be omitted.

Description

An object of the groupedData class is constructed from the formula and data by attaching the formula as an attribute of the data, along with any of outer, inner, labels, and units that are given. If order.groups is TRUE the grouping factor is converted to an ordered factor with the ordering determined by FUN. Depending on the number of grouping levels and the type of primary covariate, the returned object will be of one of three classes: nfnGroupedData—numeric covariate, single level of nesting; nffGroupedData—factor covariate, single level of nesting; and nmGroupedData—multiple levels of nesting. Several modeling and plotting functions can use the formula stored with a groupedData object to construct default plots and models.

Value

An object of one of the classes nfnGroupedData, nffGroupedData, or nmGroupedData, also inheriting from classes groupedData and data.frame.

See Also

formula, gapply, gsummary, lme

Examples

```
Orth.new <-  # create a new copy of the groupedData object
  groupedData( distance ~ age | Subject,
      data = as.data.frame( Orthodont ),
      FUN = mean,
      outer = ~ Sex,
      labels = list(x = "Age",
       y = "Distance from pituitary to pterygomaxillary fissure"),
      units = list( x = "(yr)", y = "(mm)") )
plot( Orth.new )        # trellis plot by Subject
formula( Orth.new )     # extractor for the formula
gsummary( Orth.new )    # apply summary by Subject
fm1 <- lme( Orth.new )  # fixed and groups formulae extracted
                        # from object
```

| gsummary | *Summarize by Groups* |

```
gsummary(object, FUN, omitGroupingFactor, form, level,
         groups, invariantsOnly, ...)
```

Arguments

object
: An object to be summarized, usually a groupedData object or a data.frame.

FUN
: An optional summary function or a list of summary functions to be applied to each variable in the frame. The function or functions are applied only to variables in object that vary within the groups defined by groups. Invariant variables are always summarized by group using the unique value that they assume within that group. If FUN is a single function it will be applied to each noninvariant variable by group to produce the summary for that variable. If FUN is a list of functions, the names in the list should designate classes of variables in the frame such as ordered, factor, or numeric. The indicated function will be applied to any non-invariant variables of that class. The default functions to be used are mean for numeric factors, and Mode for both factor and ordered. The Mode function, defined internally in gsummary, returns the modal or most popular value of the variable. It is different from the mode function that returns the S-language mode of the variable.

omitGroupingFactor
: An optional logical value. When TRUE the grouping factor itself will be omitted from the groupwise summary, but the levels of the grouping factor will continue to be used as the row names for the data frame that is produced by the summary. Defaults to FALSE.

form
: An optional one-sided formula that defines the groups. When this formula is given, the right-hand side is evaluated in object, converted to a factor if necessary, and the unique levels are used to define the groups. Defaults to formula(object).

level
: An optional positive integer giving the level of grouping to be used in an object with multiple nested

grouping levels. Defaults to the highest or innermost level of grouping.

groups   An optional factor that will be used to split the rows into groups. Defaults to getGroups(object, form, level).

invariantsOnly

An optional logical value. When TRUE only those covariates that are invariant within each group will be summarized. The summary value for the group is always the unique value taken on by that covariate within the group. The columns in the summary are of the same class as the corresponding columns in object. By definition, the grouping factor itself must be an invariant. When combined with omitGroupingFactor = TRUE, this option can be used to discover is there are invariant covariates in the data frame. Defaults to FALSE.

...   Optional additional arguments to the summary functions that are invoked on the variables by group. Often it is helpful to specify na.rm = TRUE.

Description

Provide a summary of the variables in a data frame by groups of rows. This is most useful with a groupedData object to examine the variables by group.

Value

A data.frame with one row for each level of the grouping factor. The number of columns is at most the number of columns in object.

See Also

summary, groupedData, getGroups

Examples

```
gsummary( Orthodont )  # default summary by Subject
## gsummary with invariantsOnly = TRUE and
## omitGroupingFactor = TRUE determines whether there
## are covariates like Sex that are invariant within
## the repeated observations on the same Subject.
gsummary( Orthodont, inv = TRUE, omit = TRUE )
```

---

intervals   *Confidence Intervals on Coefficients*

```
intervals(object, level, ...)
```

Arguments

> object   A fitted model object from which parameter esti-
>   mates can be extracted.
>
> level   An optional numeric value for the interval confidence
>   level. Defaults to 0.95.
>
> ...   Some methods for the generic may require additional
>   arguments.

Description

> Confidence intervals on the parameters associated with the model rep-
> resented by object are obtained. This function is generic; method func-
> tions can be written to handle specific classes of objects. Classes which
> already have methods for this function include: gls, lme, and lmList.

Value

> Will depend on the method function used; see the appropriate docu-
> mentation.

See Also

> intervals.gls, intervals.lme, intervals.lmList

---

intervals.lme   *Confidence Intervals on lme Parameters*

```
intervals(object, level, which)
```

Arguments

> object   An object inheriting from class lme, representing a
>   fitted linear mixed-effects model.
>
> level   An optional numeric value with the confidence level
>   for the intervals. Defaults to 0.95.

which   An optional character string specifying the subset of parameters for which to construct the confidence intervals. Possible values are `"all"` for all parameters, `"var-cov"` for the variance–covariance parameters only, and `"fixed"` for the fixed effects only. Defaults to `"all"`.

Description

Approximate confidence intervals for the parameters in the linear mixed-effects model represented by `object` are obtained, using a normal approximation to the distribution of the (restricted) maximum likelihood estimators (the estimators are assumed to have a normal distribution centered at the true parameter values and with covariance matrix equal to the negative inverse Hessian matrix of the (restricted) log-likelihood evaluated at the estimated parameters). Confidence intervals are obtained in an unconstrained scale first, using the normal approximation, and, if necessary, transformed to the constrained scale. The `pdNatural` parametrization is used for general positive-definite matrices.

Value

A list with components given by data frames with rows corresponding to parameters and columns `lower`, `est.`, and `upper` representing, respectively, lower confidence limits, the estimated values, and upper confidence limits for the parameters. Possible components are:

fixed   Fixed effects, only present when `which` is not equal to `"var-cov"`.

reStruct   Random-effects variance–covariance parameters, only present when `which` is not equal to `"fixed"`.

corStruct   Within-group correlation parameters, only present when `which` is not equal to `"fixed"` and a correlation structure is used in `object`.

varFunc   Within-group variance function parameters, only present when `which` is not equal to `"fixed"` and a variance function structure is used in `object`.

sigma   Within-group standard deviation.

See Also

`lme`, `print.intervals.lme`, `pdNatural`

Examples

```
fm1 <- lme(distance ~ age, Orthodont, random = ~ age | Subject)
intervals(fm1)
```

---

intervals.lmList   *Confidence Intervals on* lmList *Coefficients*

```
intervals(object, level, pool)
```

Arguments

object   An object inheriting from class lmList, representing
            a list of lm objects with a common model.

level   An optional numeric value with the confidence level
            for the intervals. Defaults to 0.95.

pool   An optional logical value indicating whether a pooled
            estimate of the residual standard error should be
            used. Default is attr(object, "pool").

Description

Confidence intervals on the linear model coefficients are obtained for
each lm component of object and organized into a three-dimensional
array. The first dimension corresponding to the names of the object
components. The second dimension is given by lower, est., and upper
corresponding, respectively, to the lower confidence limit, estimated
coefficient, and upper confidence limit. The third dimension is given
by the coefficients names.

Value

A three-dimensional array with the confidence intervals and estimates
for the coefficients of each lm component of object.

See Also

lmList, plot.intervals.lmList

Examples

```
fm1 <- lmList(distance ~ age | Subject, Orthodont)
intervals(fm1)
```

`lme`   *Linear Mixed-Effects Models*

> `lme(fixed, data, random, correlation, weights, subset,`
> `    method, na.action, control)`

Arguments

fixed
: A two-sided linear formula object describing the fixed-effects part of the model, with the response on the left of a ˜ operator and the terms, separated by + operators, on the right, an `lmList` object, or a `grouped-Data` object. The method functions `lme.lmList` and `lme.groupedData` are documented separately.

data
: An optional data frame containing the variables named in `fixed`, `random`, `correlation`, `weights`, and `subset`. By default the variables are taken from the environment from which `lme` is called.

random
: Optionally, any of the following: (i) a one-sided formula of the form ˜x1+···+xn | g1/···/gm, with x1+···+xn specifying the model for the random effects and g1/···/gm the grouping structure (m may be equal to 1, in which case no / is required). The random-effects formula will be repeated for all levels of grouping, in the case of multiple levels of grouping; (ii) a list of one-sided formulas of the form ˜x1+···+xn | g, with possibly different random-effects models for each grouping level. The order of nesting will be assumed the same as the order of the elements in the list; (iii) a one-sided formula of the form ˜x1+···+xn, or a `pdMat` object with a formula (i.e., a non-`NULL` value for `formula(object)`), or a list of such formulas or `pdMat` objects. In this case, the grouping structure formula will be derived from the data used to fit the linear mixed-effects model, which should inherit from class `groupedData`; (iv) a named list of formulas or `pdMat` objects as in (iii), with the grouping factors as names. The order of nesting will be assumed the same as the order of the order of the elements in the list; (v) an `reStruct` object. See the documentation on `pdClasses` for a description of the available `pdMat` classes. Defaults to a formula consisting of the right-hand side of `fixed`.

correlation
: An optional `corStruct` object describing the within-group correlation structure. See the documentation of `corClasses` for a description of the available corStruct classes. Defaults to `NULL`, corresponding to no within-group correlations.

weights
: An optional `varFunc` object or one-sided formula describing the within-group heteroscedasticity structure. If given as a formula, it is used as the argument to `varFixed`, corresponding to fixed variance weights. See the documentation on `varClasses` for a description of the available varFunc classes. Defaults to `NULL`, corresponding to homocesdatic within-group errors.

subset
: An optional expression indicating the subset of the rows of `data` that should be used in the fit. This can be a logical vector, or a numeric vector indicating which observation numbers are to be included, or a character vector of the row names to be included. All observations are included by default.

method
: A character string. If `"REML"` the model is fit by maximizing the restricted log-likelihood. If `"ML"` the log-likelihood is maximized. Defaults to `"REML"`.

na.action
: A function that indicates what should happen when the data contain NAs. The default action (`na.fail`) causes `lme` to print an error message and terminate if there are any incomplete observations.

control
: A list of control values for the estimation algorithm to replace the default values returned by the function `lmeControl`. Defaults to an empty list.

Description

This generic function fits a linear mixed-effects model in the formulation described in Laird and Ware (1982), but allowing for nested random effects. The within-group errors are allowed to be correlated and/or have unequal variances.

Value

An object of class lme representing the linear mixed-effects model fit. Generic functions such as `print`, `plot` and `summary` have methods to show the results of the fit. See `lmeObject` for the components of the fit. The functions `resid`, `coef`, `fitted`, `fixef`, and `ranef` can be used to extract some of its components.

See Also

> lmeControl, lme.lmList, lme.groupedData, lmeObject, lmList,
> reStruct, reStruct, varFunc, pdClasses, corClasses, varClasses

Examples

```
fm1 <- lme(distance ~ age, data = Orthodont) # random is ~ age
fm2 <- lme(distance ~ age + Sex, data = Orthodont, random = ~ 1)
```

---

lmeControl	*Control Values for* lme *Fit*

```
lmeControl(maxIter, msMaxIter, tolerance, niterEM, msTol,
        msScale, msVerbose, returnObject, gradHess,
        apVar, .relStep, natural)
```

Arguments

maxIter
: Maximum number of iterations for the lme optimization algorithm. Default is 50.

msMaxIter
: Maximum number of iterations for the ms optimization step inside the lme optimization. Default is 50.

tolerance
: Tolerance for the convergence criterion in the lme algorithm. Default is 1e-6.

niterEM
: Number of iterations for the EM algorithm used to refine the initial estimates of the random-effects variance–covariance coefficients. Default is 25.

msTol
: Tolerance for the convergence criterion in ms, passed as the rel.tolerance argument to the function (see documentation on ms). Default is 1e-7.

msScale
: Scale function passed as the scale argument to the ms function (see documentation on that function). Default is lmeScale.

msVerbose
: A logical value passed as the trace argument to ms (see documentation on that function). Default is FALSE.

returnObject
: A logical value indicating whether the fitted object should be returned when the maximum number of iterations is reached without convergence of the algorithm. Default is FALSE.

gradHess
A logical value indicating whether numerical gradient vectors and Hessian matrices of the log-likelihood function should be used in the ms optimization. This option is only available when the correlation structure (corStruct) and the variance function structure (varFunc) have no "varying" parameters and the pdMat classes used in the random effects structure are pdSymm (general positive-definite), pdDiag (diagonal), pdIdent (multiple of the identity), or pdCompSymm (compound symmetry). Default is TRUE.

apVar
A logical value indicating whether the approximate covariance matrix of the variance–covariance parameters should be calculated. Default is TRUE.

.relStep
Relative step for numerical derivatives calculations. Default is .Machine$double.eps^(1/3).

natural
A logical value indicating whether the pdNatural parameterization should be used for general positive-definite matrices (pdSymm) in reStruct, when the approximate covariance matrix of the estimators is calculated. Default is TRUE.

Description

The values supplied in the function call replace the defaults and a list with all possible arguments is returned. The returned list is used as the control argument to the lme function.

Value

A list with components for each of the possible arguments.

See Also

lme, ms, lmeScale

Examples

```
# decrease the maximum number iterations in the ms call and
# request that information on the evolution of the ms iterations
# be printed
lmeControl(msMaxIter = 20, msVerbose = TRUE)
```

`lmList`   *List of lm Objects with a Common Model*

---

lmList(object, data, level, na.action, pool)

Arguments

object
: Either a linear formula object of the form y ~ x1+···+xn | g or a **groupedData** object. In the formula, y represents the response, x1,...,xn the covariates, and g the grouping factor specifying the partitioning of the data according to which different **lm** fits should be performed. The grouping factor g may be omitted from the formula, in which case the grouping structure will be obtained from **data**, which must inherit from class **groupedData**. The method function lmList.groupedData is documented separately.

data
: A data frame in which to interpret the variables named in object.

level
: An optional integer specifying the level of grouping to be used when multiple nested levels of grouping are present.

na.action
: A function that indicates what should happen when the data contain NAs. The default action (na.fail) causes lmList to print an error message and terminate if there are any incomplete observations.

pool
: An optional logical value that is preserved as an attribute of the returned value. This will be used as the default for pool in calculations of standard deviations or standard errors for summaries.

Description

Data is partitioned according to the levels of the grouping factor g and individual lm fits are obtained for each data partition, using the model defined in object.

Value

A list of lm objects with as many components as the number of groups defined by the grouping factor. Generic functions such as coef, fixef, lme, pairs, plot, predict, ranef, summary, and update have methods that can be applied to an lmList object.

See Also

lm, lme.lmList.

Examples

fm1 <- lmList(distance ~ age | Subject, Orthodont)

| logLik | *Extract Log-Likelihood* |

---

logLik(object, ...)

Arguments

object   Any object from which a log-likelihood, or a contribution to a log-likelihood, can be extracted.

...   Some methods for this generic function require additional arguments.

Description

This function is generic; method functions can be written to handle specific classes of objects. Classes which already have methods for this function include: corStruct, gls, lm, lme, lmList, lmeStruct, reStruct, and varFunc.

Value

Will depend on the method function used; see the appropriate documentation.

| nlme | *Nonlinear Mixed-Effects Models* |

---

nlme(model, data, fixed, random, groups, start,
     correlation, weights, subset, method, na.action,
     naPattern, control, verbose)

Arguments

model
: A nonlinear model formula, with the response on the left of a ~ operator and an expression involving parameters and covariates on the right, or an nlsList object. If data is given, all names used in the formula should be defined as parameters or variables in the data frame. The method function nlme.nlsList is documented separately.

data
: An optional data frame containing the variables named in model, fixed, random, correlation, weights, subset, and naPattern. By default the variables are taken from the environment from which nlme is called.

fixed
: A two-sided linear formula of the form f1+···+fn ~ x1+···+xm, or a list of two-sided formulas of the form f1 ~ x1+···+xm, with possibly different models for different parameters. The names of the parameters, f1,...,fn, are included on the right-hand side of model and the x1+···+xm expressions define linear models for these parameters (when the left-hand side of the formula contains several parameters, they all are assumed to follow the same linear model, described by the right-hand side expression). A 1 on the right-hand side of the formula(s) indicates a single fixed effects for the corresponding parameter(s).

random
: Optionally, any of the following: (i) a two-sided formula of the form r1+···+rn ~ x1+···+xm|g1/··· /gQ, with r1,...,rn naming parameters included on the right-hand side of model, x1+···+xm specifying the random-effects model for these parameters and g1/···/gQ the grouping structure (Q may be equal to 1, in which case no / is required). The random-effects formula will be repeated for all levels of grouping, in the case of multiple levels of grouping; (ii) a two-sided formula of the form r1+···+rn ~ x1+···+xm, a list of two-sided formulas of the form r1~x1+···+xm, with possibly different random-effects models for different parameters, a pdMat object with a two-sided formula, or list of two-sided formulas (i.e., a non-NULL value for formula(random)), or a list of pdMat objects with two-sided formulas, or lists of two-sided formulas. In this case, the grouping structure formula will be given in groups, or derived from the

data used to fit the nonlinear mixed-effects model, which should inherit from class groupedData; (iii) a named list of formulas, lists of formulas, or pdMat objects as in (ii), with the grouping factors as names. The order of nesting will be assumed the same as the order of the order of the elements in the list; (iv) an reStruct object. See the documentation on pdClasses for a description of the available pdMat classes. Defaults to fixed, resulting in all fixed effects having also random effects.

groups   An optional one-sided formula of the form ~g1 (single level of nesting) or ~g1/···/gQ (multiple levels of nesting), specifying the partitions of the data over which the random effects vary. g1,...,gQ must evaluate to factors in data. The order of nesting, when multiple levels are present, is taken from left to right (i.e., g1 is the first level, g2 the second, etc.).

start   An optional numeric vector, or list of initial estimates for the fixed effects and random effects. If declared as a numeric vector, it is converted internally to a list with a single component fixed, given by the vector. The fixed component is required, unless the model function inherits from class selfStart, in which case initial values will be derived from a call to nlsList. An optional random component is used to specify initial values for the random effects and should consist of a matrix, or a list of matrices with length equal to the number of grouping levels. Each matrix should have as many rows as the number of groups at the corresponding level and as many columns as the number of random effects in that level.

correlation   An optional corStruct object describing the within-group correlation structure. See the documentation of corClasses for a description of the available corStruct classes. Defaults to NULL, corresponding to no within-group correlations.

weights   An optional varFunc object or one-sided formula describing the within-group heteroscedasticity structure. If given as a formula, it is used as the argument to varFixed, corresponding to fixed variance weights. See the documentation on varClasses for a description of the available varFunc classes. Defaults to NULL, corresponding to homoscesdatic within-group errors.

subset
: An optional expression indicating the subset of the rows of `data` that should be used in the fit. This can be a logical vector, or a numeric vector indicating which observation numbers are to be included, or a character vector of the row names to be included. All observations are included by default.

method
: A character string. If `"REML"` the model is fit by maximizing the restricted log-likelihood. If `"ML"` the log-likelihood is maximized. Defaults to `"ML"`.

na.action
: A function that indicates what should happen when the data contain NAs. The default action (`na.fail`) causes `nlme` to print an error message and terminate if there are any incomplete observations.

naPattern
: An expression or formula object, specifying which returned values are to be regarded as missing.

control
: A list of control values for the estimation algorithm to replace the default values returned by the function `nlmeControl`. Defaults to an empty list.

verbose
: An optional logical value. If `TRUE` information on the evolution of the iterative algorithm is printed. Default is `FALSE`.

Description

This generic function fits a nonlinear mixed-effects model in the formulation described in Lindstrom and Bates (1990), but allowing for nested random effects. The within-group errors are allowed to be correlated and/or have unequal variances.

Value

An object of class nlme representing the nonlinear mixed-effects model fit. Generic functions such as `print`, `plot` and `summary` have methods to show the results of the fit. See `nlmeObject` for the components of the fit. The functions `resid`, `coef`, `fitted`, `fixef`, and `ranef` can be used to extract some of its components.

See Also

`nlmeControl, nlme.nlsList, nlmeObject, nlsList, reStruct, varFunc, pdClasses, corClasses, varClasses`

Examples

```
## all parameters as fixed and random effects
```

```
fm1 <- nlme(weight ~ SSlogis(Time, Asym, xmid, scal),
            data = Soybean, fixed = Asym + xmid + scal ~ 1,
            start = c(18, 52, 7.5))
## only Asym and xmid as random, with a diagonal covariance
fm2 <- nlme(weight ~ SSlogis(Time, Asym, xmid, scal),
            data = Soybean, fixed = Asym + xmid + scal ~ 1,
            random = pdDiag(Asym + xmid ~ 1),
            start = c(18, 52, 7.5))
```

---

nlmeControl	*Control Values for nlme Fit*

```
nlmeControl(maxIter, pnlsMaxIter, msMaxIter, minScale,
            tolerance, niterEM, pnlsTol, msTol, msScale,
            returnObject, msVerbose, gradHess, apVar,
            .relStep, natural)
```

Arguments

maxIter   Maximum number of iterations for the nlme opti-
           mization algorithm. Default is 50.

pnlsMaxIter   Maximum number of iterations for the PNLS opti-
               mization step inside the nlme optimization. Default
               is 7.

msMaxIter   Maximum number of iterations for the ms optimiza-
             tion step inside the nlme optimization. Default is 50.

minScale   Minimum factor by which to shrink the default step
            size in an attempt to decrease the sum of squares in
            the PNLS step. Default 0.001.

tolerance   Tolerance for the convergence criterion in the nlme
             algorithm. Default is 1e-6.

niterEM   Number of iterations for the EM algorithm used to
           refine the initial estimates of the random-effects var-
           iance–covariance coefficients. Default is 25.

pnlsTol   Tolerance for the convergence criterion in PNLS step.
           Default is 1e-3.

msTol   Tolerance for the convergence criterion in ms, passed
         as the rel.tolerance argument to the function (see
         documentation on ms). Default is 1e-7.

msScale   Scale function passed as the scale argument to the
           ms function (see documentation on that function).
           Default is lmeScale.
```
```

returnObject   A logical value indicating whether the fitted object should be returned when the maximum number of iterations is reached without convergence of the algorithm. Default is FALSE.

msVerbose   A logical value passed as the trace argument to ms (see documentation on that function). Default is FALSE.

gradHess   A logical value indicating whether numerical gradient vectors and Hessian matrices of the log-likelihood function should be used in the ms optimization. This option is only available when the correlation structure (corStruct) and the variance function structure (varFunc) have no "varying" parameters and the pdMat classes used in the random effects structure are pdSymm (general positive-definite), pdDiag (diagonal), pdIdent (multiple of the identity), or pdCompSymm (compound symmetry). Default is TRUE.

apVar   A logical value indicating whether the approximate covariance matrix of the variance–covariance parameters should be calculated. Default is TRUE.

.relStep   Relative step for numerical derivatives calculations. Default is .Machine$double.eps^(1/3).

natural   A logical value indicating whether the pdNatural parameterization should be used for general positive-definite matrices (pdSymm) in reStruct, when the approximate covariance matrix of the estimators is calculated. Default is TRUE.

Description

The values supplied in the function call replace the defaults and a list with all possible arguments is returned. The returned list is used as the control argument to the nlme function.

Value

A list with components for each of the possible arguments.

See Also

nlme, ms, nlmeStruct

Examples

```
# decrease the maximum number iterations in the ms call and
```

```
# request that information on the evolution of the ms iterations
# be printed
nlmeControl(msMaxIter = 20, msVerbose = TRUE)
```

---

nlsList   *List of nls Objects with a Common Model*

```
nlsList(model, data, start, control, level, na.action,
        pool)
```

Arguments

model
: Either a nonlinear model formula, with the response on the left of a ~ operator and an expression involving parameters, covariates, and a grouping factor separated by the | operator on the right, or a selfStart function. The method function nlsList.selfStart is documented separately.

data
: A data frame in which to interpret the variables named in model.

start
: An optional named list with initial values for the parameters to be estimated in model. It is passed as the start argument to each nls call and is required when the nonlinear function in model does not inherit from class selfStart.

control
: A list of control values passed as the control argument to nls. Defaults to an empty list.

level
: An optional integer specifying the level of grouping to be used when multiple nested levels of grouping are present.

na.action
: A function that indicates what should happen when the data contain NAs. The default action (na.fail) causes nlsList to print an error message and terminate if there are any incomplete observations.

pool
: An optional logical value that is preserved as an attribute of the returned value. This will be used as the default for pool in calculations of standard deviations or standard errors for summaries.

Description

Data is partitioned according to the levels of the grouping factor defined in model and individual nls fits are obtained for each data partition, using the model defined in model.

Value

A list of nls objects with as many components as the number of groups defined by the grouping factor. Generic functions such as coef, fixef, lme, pairs, plot, predict, ranef, summary, and update have methods that can be applied to an nlsList object.

See Also

nls, nlme.nlsList.

Examples

```
fm1 <- nlsList(uptake ~ SSasympOff(conc, Asym, lrc, c0),
    data = CO2, start = c(Asym = 30, lrc = -4.5, c0 = 52))
fm1
```

---

pairs.lme   *Pairs Plot of an lme Object*

```
pairs(object, form, label, id, idLabels, grid, ...)
```

Arguments

object   An object inheriting from class lme, representing a fitted linear mixed-effects model.

form   An optional one-sided formula specifying the desired type of plot. Any variable present in the original data frame used to obtain object can be referenced. In addition, object itself can be referenced in the formula using the symbol ".". Conditional expressions on the right of a | operator can be used to define separate panels in a trellis display. The expression on the right-hand side of form, and to the left of the | operator, must evaluate to a data frame with at least two columns. Default is ~ coef(.) , corresponding to a pairs plot of the coefficients evaluated at the innermost level of nesting.

id   An optional numeric value, or one-sided formula. If given as a value, it is used as a significance level for an outlier test based on the Mahalanobis distances of the estimated random effects. Groups with random effects distances greater than the $1 - value$ percentile of the appropriate chi-square distribution are identified in the plot using idLabels. If given as a one-sided formula, its right-hand side must evaluate to a logical, integer, or character vector which is used to identify points in the plot. If missing, no points are identified.

idLabels   An optional vector, or one-sided formula. If given as a vector, it is converted to character and used to label the points identified according to id. If given as a one-sided formula, its right-hand side must evaluate to a vector which is converted to character and used to label the identified points. Default is the innermost grouping factor.

grid   An optional logical value indicating whether a grid should be added to plot. Default is **FALSE**.

...   Optional arguments passed to the trellis plot function.

Description

Diagnostic plots for the linear mixed-effects fit are obtained. The form argument gives considerable flexibility in the type of plot specification. A conditioning expression (on the right side of a | operator) always implies that different panels are used for each level of the conditioning factor, according to a trellis display. The expression on the right-hand side of the formula, before a | operator, must evaluate to a data frame with at least two columns. If the data frame has two columns, a scatter plot of the two variables is displayed (the trellis function xyplot is used). Otherwise, if more than two columns are present, a scatter plot matrix with pairwise scatter plots of the columns in the data frame is displayed (the trellis function splom is used).

Value

A diagnostic trellis plot.

See Also

lme, xyplot, splom

Examples

```
fm1 <- lme(distance ~ age, Orthodont, random = ~ age | Subject)
# scatter plot of coefficients by gender, identifying
# unusual subjects
pairs(fm1, ~coef(., augFrame = T) | Sex, id = 0.1, adj = -0.5)
# scatter plot of estimated random effects
pairs(fm1, ~ranef(.))
```

---

plot.lme   *Plot an lme Object*

```
plot(object, form, abline, id, idLabels, idResType, grid,
     ...)
```

Arguments

object
> An object inheriting from class lme, representing a fitted linear mixed-effects model.

form
> An optional formula specifying the desired type of plot. Any variable present in the original data frame used to obtain object can be referenced. In addition, object itself can be referenced in the formula using the symbol ".". Conditional expressions on the right of a | operator can be used to define separate panels in a trellis display. Default is resid(., type = "p") ~ fitted(.) , corresponding to a plot of the standardized residuals versus fitted values, both evaluated at the innermost level of nesting.

abline
> An optional numeric value, or numeric vector of length two. If given as a single value, a horizontal line will be added to the plot at that coordinate; else, if given as a vector, its values are used as the intercept and slope for a line added to the plot. If missing, no lines are added to the plot.

id
> An optional numeric value, or one-sided formula. If given as a value, it is used as a significance level for a two-sided outlier test for the standardized, or normalized residuals. Observations with absolute standardized (normalized) residuals greater than the $1 - value/2$ quantile of the standard normal distribution are identified in the plot using idLabels. If given as a one-sided formula, its right-hand side must evaluate to a logical, integer, or character vector which is used to identify observations in the plot. If missing, no observations are identified.

idLabels
> An optional vector, or one-sided formula. If given as a vector, it is converted to character and used to label the observations identified according to id. If given as a one-sided formula, its right-hand side must evaluate to a vector that is converted to character and

used to label the identified observations. Default is
the innermost grouping factor.

idResType   An optional character string specifying the type of
residuals to be used in identifying outliers, when id
is a numeric value. If "pearson", the standardized
residuals (raw residuals divided by the corresponding standard errors) are used; else, if "normalized",
the normalized residuals (standardized residuals premultiplied by the inverse square-root factor of the
estimated error correlation matrix) are used. Partial
matching of arguments is used, so only the first character needs to be provided. Defaults to "pearson".

Description

Diagnostic plots for the linear mixed-effects fit are obtained. The form
argument gives considerable flexibility in the type of plot specification.
A conditioning expression (on the right side of a | operator) always
implies that different panels are used for each level of the conditioning
factor, according to a trellis display. If form is a one-sided formula,
histograms of the variable on the right-hand side of the formula, before
a | operator, are displayed (the trellis function histogram is used).
If form is two-sided and both its left- and right-hand side variables
are numeric, scatter plots are displayed (the trellis function xyplot
is used). Finally, if form is two-sided and its left-hand side variable
is a factor, boxplots of the right-hand side variable by the levels of
the left-hand side variable are displayed (the trellis function bwplot is
used).

Value

A diagnostic trellis plot.

See Also

lme, xyplot, bwplot, histogram

Examples

```
fm1 <- lme(distance ~ age, Orthodont, random = ~ age | Subject)
# standardized residuals versus fitted values by gender
plot(fm1, resid(., type = "p") ~ fitted(.) | Sex, abline = 0)
# box-plots of residuals by Subject
plot(fm1, Subject ~ resid(.))
# observed versus fitted values by Subject
plot(fm1, distance ~ fitted(.) | Subject, abline = c(0,1))
```

`plot.nfnGroupedData`   *Plot an nfnGroupedData Object*

```
plot(x, outer, inner, innerGroups, xlab, ylab, strip,
     aspect, panel, key, grid, ...)
```

Arguments

x
: An object inheriting from class nfnGroupedData, representing a groupedData object with a numeric primary covariate and a single grouping level.

outer
: An optional logical value or one-sided formula, indicating covariates that are outer to the grouping factor, which are used to determine the panels of the trellis plot. If equal to TRUE, attr(object, "outer") is used to indicate the outer covariates. An outer covariate is invariant within the sets of rows defined by the grouping factor. Ordering of the groups is done in such a way as to preserve adjacency of groups with the same value of the outer variables. Defaults to NULL, meaning that no outer covariates are to be used.

inner
: An optional logical value or one-sided formula, indicating a covariate that is inner to the grouping factor, which is used to associate points within each panel of the trellis plot. If equal to TRUE, attr(object, "inner") is used to indicate the inner covariate. An inner covariate can change within the sets of rows defined by the grouping factor. Defaults to NULL, meaning that no inner covariate is present.

innerGroups
: An optional one-sided formula specifying a factor to be used for grouping the levels of the inner covariate. Different colors, or line types, are used for each level of the innerGroups factor. Default is NULL, meaning that no innerGroups covariate is present.

xlab, ylab
: Optional character strings with the labels for the plot. Default is the corresponding elements of attr(object, "labels") and attr(object, "units") pasted together.

strip
: An optional function passed as the strip argument to the xyplot function. Default is strip.default(..., style = 1) (see trellis.args).

aspect
:   An optional character string indicating the aspect ratio for the plot passed as the `aspect` argument to the `xyplot` function. Default is `"xy"` (see `trellis.args`).

panel
:   An optional function used to generate the individual panels in the trellis display, passed as the `panel` argument to the `xyplot` function.

key
:   An optional logical function or function. If `TRUE` and `innerGroups` is non-`NULL`, a legend for the different `innerGroups` levels is included at the top of the plot. If given as a function, it is passed as the `key` argument to the `xyplot` function. Default is `TRUE` if `innerGroups` is non-`NULL` and `FALSE` otherwise.

grid
:   An optional logical value indicating whether a grid should be added to plot. Default is `TRUE`.

. . .
:   Optional arguments passed to the `xyplot` function.

Description

A trellis plot of the response versus the primary covariate is generated. If outer variables are specified, the combination of their levels are used to determine the panels of the trellis display. Otherwise, the levels of the grouping variable determine the panels. A scatter plot of the response versus the primary covariate is displayed in each panel, with observations corresponding to same inner group joined by line segments. The trellis function `xyplot` is used.

Value

A trellis plot of the response versus the primary covariate.

See Also

`groupedData`, `xyplot`

Examples

```
# different panels per Subject
plot(Orthodont)
# different panels per gender
plot(Orthodont, outer = TRUE)
```

---

plot.nmGroupedData   *Plot an nmGroupedData Object*

```
plot(x, collapseLevel, displayLevel, outer, inner,
    preserve, FUN, subset, grid, ...)
```

Arguments

x
: An object inheriting from class nmGroupedData, representing a groupedData object with multiple grouping factors.

collapseLevel
: An optional positive integer or character string indicating the grouping level to use when collapsing the data. Level values increase from outermost to innermost grouping. Default is the highest or innermost level of grouping.

displayLevel
: An optional positive integer or character string indicating the grouping level to use for determining the panels in the trellis display, when outer is missing. Default is collapseLevel.

outer
: An optional logical value or one-sided formula, indicating covariates that are outer to the displayLevel grouping factor, which are used to determine the panels of the trellis plot. If equal to TRUE, the display-Level element attr(object, "outer") is used to indicate the outer covariates. An outer covariate is invariant within the sets of rows defined by the grouping factor. Ordering of the groups is done in such a way as to preserve adjacency of groups with the same value of the outer variables. Defaults to NULL, meaning that no outer covariates are to be used.

inner
: An optional logical value or one-sided formula, indicating a covariate that is inner to the displayLevel grouping factor, which is used to associate points within each panel of the trellis plot. If equal to TRUE, attr(object, "outer") is used to indicate the inner covariate. An inner covariate can change within the sets of rows defined by the grouping factor. Defaults to NULL, meaning that no inner covariate is present.

preserve
: An optional one-sided formula indicating a covariate whose levels should be preserved when collapsing

the data according to the `collapseLevel` grouping factor. The collapsing factor is obtained by pasting together the levels of the `collapseLevel` grouping factor and the values of the covariate to be preserved. Default is `NULL`, meaning that no covariates need to be preserved.

FUN   An optional summary function or a list of summary functions to be used for collapsing the data. The function or functions are applied only to variables in `object` that vary within the groups defined by `collapseLevel`. Invariant variables are always summarized by group using the unique value that they assume within that group. If `FUN` is a single function it will be applied to each noninvariant variable by group to produce the summary for that variable. If `FUN` is a list of functions, the names in the list should designate classes of variables in the data such as `ordered`, `factor`, or `numeric`. The indicated function will be applied to any noninvariant variables of that class. The default functions to be used are `mean` for numeric factors, and `Mode` for both `factor` and `ordered`. The `Mode` function, defined internally in `gsummary`, returns the modal or most popular value of the variable. It is different from the `mode` function that returns the S-language mode of the variable.

subset   An optional named list. Names can be either positive integers representing grouping levels, or names of grouping factors. Each element in the list is a vector indicating the levels of the corresponding grouping factor to be used for plotting the data. Default is `NULL`, meaning that all levels are used.

grid   An optional logical value indicating whether a grid should be added to plot. Default is `TRUE`.

. . .   Optional arguments passed to the trellis plot function.

Description

The `groupedData` object is summarized by the values of the `displayLevel` grouping factor (or the combination of its values and the values of the covariate indicated in `preserve`, if any is present). The collapsed data is used to produce a new `groupedData` object, with grouping factor given by the `displayLevel` factor, which is plotted using the appropriate `plot` method for `groupedData` objects with single level of grouping.

Value

A trellis display of the data collapsed over the values of the collapse-
Level grouping factor and grouped according to the displayLevel
grouping factor.

See Also

groupedData, collapse.groupedData, plot.nfnGroupedData,
plot.nffGroupedData

Examples

```
# no collapsing, panels by Dog
plot(Pixel, display = "Dog", inner = ~Side)
# collapsing by Dog, preserving day
plot(Pixel, collapse = "Dog", preserve = ~day)
```

---

plot.Variogram	*Plot a **Variogram** Object*

```
plot(object, smooth, showModel, sigma, span, xlab, ylab,
     type, ylim, ...)
```

Arguments

object   An object inheriting from class Variogram, consisting
              of a data frame with two columns named variog and
              dist, representing the semivariogram values and the
              corresponding distances.

smooth   An optional logical value controlling whether a loess
              smoother should be added to the plot. Defaults to
              TRUE, when showModel is FALSE.

showModel   An optional logical value controlling whether the semi-
              variogram corresponding to an "modelVariog" at-
              tribute of object, if any is present, should be added
              to the plot. Defaults to TRUE, when the "modelVariog"
              attribute is present.

sigma   An optional numeric value used as the height of a
              horizontal line displayed in the plot. Can be used to
              represent the process standard deviation. Default is
              NULL, implying that no horizontal line is drawn.

span   An optional numeric value with the smoothing pa-
              rameter for the loess fit. Default is 0.6.

xlab,ylab	Optional character strings with the x- and y-axis labels. Default respectively to "Distance" and "Semivariogram".
type	An optional character indicating the type of plot. Defaults to "p".
ylim	An optional numeric vector with the limits for the y-axis. Defaults to c(0, max(object$variog)).
...	Optional arguments passed to the trellis xyplot function.

Description

An xyplot of the semivariogram versus the distances is produced. If smooth = TRUE, a loess smoother is added to the plot. If showModel = TRUE and object includes an "modelVariog" attribute, the corresponding semivariogram is added to the plot.

Value

An xyplot trellis plot.

See Also

Variogram, xyplot, loess

Examples

```
fm1 <- lme(follicles ~ sin(2*pi*Time) + cos(2*pi*Time), Ovary)
plot(Variogram(fm1, form = ~ Time | Mare, maxDist = 0.7))
```

| predict.lme | *Predictions from an lme Object* |

```
predict(object, newdata, level, asList, na.action)
```

Arguments

| object | An object inheriting from class lme, representing a fitted linear mixed-effects model. |
| newdata | An optional data frame to be used for obtaining the predictions. All variables used in the fixed- and random-effects models, as well as the grouping factors, must be present in the data frame. If missing, the fitted values are returned. |

level
: An optional integer vector giving the level(s) of grouping to be used in obtaining the predictions. Level values increase from outermost to innermost grouping, with level zero corresponding to the population predictions. Defaults to the highest or innermost level of grouping.

asList
: An optional logical value. If TRUE and a single value is given in level, the returned object is a list with the predictions split by groups; else the returned value is either a vector or a data frame, according to the length of level.

na.action
: A function that indicates what should happen when newdata contains NAs. The default action (na.fail) causes the function to print an error message and terminate if there are any incomplete observations.

Description

The predictions at level i are obtained by adding together the population predictions (based only on the fixed-effects estimates) and the estimated contributions of the random effects to the predictions at grouping levels less or equal to i. The resulting values estimate the best linear unbiased predictions (BLUPs) at level i. If group values not included in the original grouping factors are present in newdata, the corresponding predictions will be set to NA for levels greater or equal to the level at which the unknown groups occur.

Value

If a single level of grouping is specified in level, the returned value is either a list with the predictions split by groups (asList = TRUE) or a vector with the predictions (asList = FALSE); else, when multiple grouping levels are specified in level, the returned object is a data frame with columns given by the predictions at different levels and the grouping factors.

See Also

lme, fitted.lme

Examples

```
fm1 <- lme(distance ~ age, Orthodont, random = ~ age | Subject)
newOrth <- data.frame(Sex = c("Male","Male","Female","Female",
                             "Male","Male"),
                     age = c(15, 20, 10, 12, 2, 4),
                     Subject = c("M01","M01","F30","F30","M04",
```

"M04"))
predict(fm1, newOrth, level = 0:1)

qqnorm.lme	*Normal Plot of Residuals or Random Effects from an lme Object*

qqnorm(object, form, abline, id, idLabels, grid, ...)

Arguments

object   An object inheriting from class lme, representing a fitted linear mixed-effects model.

form   An optional one-sided formula specifying the desired type of plot. Any variable present in the original data frame used to obtain object can be referenced. In addition, object itself can be referenced in the formula using the symbol ".". Conditional expressions on the right of a | operator can be used to define separate panels in a trellis display. The expression on the right-hand side of form and to the left of a | operator must evaluate to a residuals vector, or a random effects matrix. Default is ~ resid(., type = "p"), corresponding to a normal plot of the standardized residuals evaluated at the innermost level of nesting.

abline   An optional numeric value, or numeric vector of length two. If given as a single value, a horizontal line will be added to the plot at that coordinate; else, if given as a vector, its values are used as the intercept and slope for a line added to the plot. If missing, no lines are added to the plot.

id   An optional numeric value, or one-sided formula. If given as a value, it is used as a significance level for a two-sided outlier test for the standardized residuals (random effects). Observations with absolute standardized residuals (random effects) greater than the $1 - value/2$ quantile of the standard normal distribution are identified in the plot using idLabels. If given as a one-sided formula, its right-hand side must evaluate to a logical, integer, or character vector which is used to identify observations in the plot. If missing, no observations are identified.

idLabels	An optional vector, or one-sided formula. If given as a vector, it is converted to character and used to label the observations identified according to id. If given as a one-sided formula, its right-hand side must evaluate to a vector that is converted to character and used to label the identified observations. Default is the innermost grouping factor.
grid	An optional logical value indicating whether a grid should be added to plot. Default is FALSE.
. . .	Optional arguments passed to the trellis plot function.

Description

Diagnostic plots for assessing the normality of residuals and random effects in the linear mixed-effects fit are obtained. The form argument gives considerable flexibility in the type of plot specification. A conditioning expression (on the right side of a | operator) always implies that different panels are used for each level of the conditioning factor, according to a trellis display.

Value

A diagnostic trellis plot for assessing normality of residuals or random effects.

See Also

lme, plot.lme

Examples

```
fm1 <- lme(distance ~ age, Orthodont, random = ~ age | Subject)
# normal plot of standardized residuals by gender
qqnorm(fm1, ~ resid(., type = "p") | Sex, abline = c(0, 1))
# normal plots of random effects
qqnorm(fm1, ~ranef(.))
```

---

ranef   *Extract Random Effects*

```
ranef(object, ...)
```

Arguments

| object | Any fitted model object from which random effects estimates can be extracted. |
| ... | Some methods for this generic function require additional arguments. |

Description

This function is generic; method functions can be written to handle specific classes of objects. Classes that already have methods for this function include lmList and lme.

Value

Will depend on the method function used; see the appropriate documentation.

See Also

ranef.lmList, ranef.lme

---

ranef.lme   *Extract lme Random Effects*

```
ranef(object, augFrame, level, data, which, FUN, standard,
    omitGroupingFactor)
```

Arguments

object	An object inheriting from class lme, representing a fitted linear mixed-effects model.
augFrame	An optional logical value. If TRUE, the returned data frame is augmented with variables defined in data; else, if FALSE, only the coefficients are returned. Defaults to FALSE.
level	An optional vector of positive integers giving the levels of grouping to be used in extracting the random effects from an object with multiple nested grouping levels. Defaults to all levels of grouping.
data	An optional data frame with the variables to be used for augmenting the returned data frame when augFrame = TRUE. Defaults to the data frame used to fit object.

which
: An optional positive integer vector specifying which columns of data should be used in the augmentation of the returned data frame. Defaults to all columns in data.

FUN
: An optional summary function or a list of summary functions to be applied to group-varying variables, when collapsing data by groups. Group-invariant variables are always summarized by the unique value that they assume within that group. If FUN is a single function it will be applied to each noninvariant variable by group to produce the summary for that variable. If FUN is a list of functions, the names in the list should designate classes of variables in the frame such as ordered, factor, or numeric. The indicated function will be applied to any group-varying variables of that class. The default functions to be used are mean for numeric factors, and Mode for both factor and ordered. The Mode function, defined internally in gsummary, returns the modal or most popular value of the variable. It is different from the mode function that returns the S-language mode of the variable.

standard
: An optional logical value indicating whether the estimated random effects should be "standardized" (i.e., divided by the corresponding estimated standard error). Defaults to FALSE.

omitGroupingFactor
: An optional logical value. When TRUE, the grouping factor itself will be omitted from the groupwise summary of data, but the levels of the grouping factor will continue to be used as the row names for the returned data frame. Defaults to FALSE.

Description

The estimated random effects at level i are represented as a data frame with rows given by the different groups at that level and columns given by the random effects. If a single level of grouping is specified, the returned object is a data frame; else, the returned object is a list of such data frames. Optionally, the returned data frame(s) may be augmented with covariates summarized over groups.

Value

A data frame, or list of data frames, with the estimated random effects at the grouping level(s) specified in level and, optionally, other covariates summarized over groups. The returned object inherits from classes ranef.lme and data.frame.

See Also

lme, fixed.effects.lme, coef.lme, plot.ranef.lme, gsummary

Examples

```
fm1 <- lme(distance ~ age, Orthodont, random = ~ age | Subject)
ranef(fm1)
ranef(fm1, augFrame = TRUE)
```

| ranef.lmList | *Extract lmList Random Effects* |

```
ranef(object, augFrame, data, which, FUN, standard,
    omitGroupingFactor)
```

Arguments

object
> An object inheriting from class lmList, representing a list of lm objects with a common model.

augFrame
> An optional logical value. If TRUE, the returned data frame is augmented with variables defined in the data frame used to produce object; else, if FALSE, only the random effects are returned. Defaults to FALSE.

data
> An optional data frame with the variables to be used for augmenting the returned data frame when augFrame = TRUE. Defaults to the data frame used to fit object.

which
> An optional positive integer or character vector specifying which columns of the data frame used to produce object should be used in the augmentation of the returned data frame. Defaults to all variables in the data.

FUN
> An optional summary function or a list of summary functions to be applied to group-varying variables, when collapsing the data by groups. Group-invariant variables are always summarized by the unique value that they assume within that group. If FUN is a single function it will be applied to each noninvariant variable by group to produce the summary for that variable. If FUN is a list of functions, the names in

the list should designate classes of variables in the frame such as ordered, factor, or numeric. The indicated function will be applied to any group-varying variables of that class. The default functions to be used are mean for numeric factors, and Mode for both factor and ordered. The Mode function, defined internally in gsummary, returns the modal or most popular value of the variable. It is different from the mode function that returns the S-language mode of the variable.

standard   An optional logical value indicating whether the estimated random effects should be "standardized" (i.e., divided by the corresponding estimated standard error). Defaults to FALSE.

omitGroupingFactor

An optional logical value. When TRUE, the grouping factor itself will be omitted from the groupwise summary of data, but the levels of the grouping factor will continue to be used as the row names for the returned data frame. Defaults to FALSE.

Description

A data frame containing the differences between the coefficients of the individual lm fits and the average coefficients.

Value

A data frame with the differences between the individual lm coefficients in object and their average. Optionally, the returned data frame may be augmented with covariates summarized over groups or the differences may be standardized.

See Also

lmList, fixef.lmList

Examples

```
fm1 <- lmList(distance ~ age | Subject, Orthodont)
ranef(fm1)
ranef(fm1, standard = TRUE)
ranef(fm1, augFrame = TRUE)
```

| residuals.lme | *Extract lme Residuals* |

```
residuals(object, level, type, asList)
```

Arguments

object
: An object inheriting from class lme, representing a fitted linear mixed-effects model.

level
: An optional integer vector giving the level(s) of grouping to be used in extracting the residuals from object. Level values increase from outermost to innermost grouping, with level zero corresponding to the population residuals. Defaults to the highest or innermost level of grouping.

type
: An optional character string specifying the type of residuals to be used. If "response", the "raw" residuals (observed – fitted) are used; else, if "pearson", the standardized residuals (raw residuals divided by the corresponding standard errors) are used; else, if "normalized", the normalized residuals (standardized residuals premultiplied by the inverse square-root factor of the estimated error correlation matrix) are used. Partial matching of arguments is used, so only the first character needs to be provided. Defaults to "pearson".

asList
: An optional logical value. If TRUE and a single value is given in level, the returned object is a list with the residuals split by groups; else the returned value is either a vector or a data frame, according to the length of level. Defaults to FALSE.

Description

The residuals at level i are obtained by subtracting the fitted levels at that level from the response vector (and dividing by the estimated within-group standard error, if type="pearson"). The fitted values at level i are obtained by adding together the population-fitted values (based only on the fixed-effects estimates) and the estimated contributions of the random effects to the fitted values at grouping levels less or equal to i.

Value

> If a single level of grouping is specified in `level`, the returned value is either a list with the residuals split by groups (`asList = TRUE`) or a vector with the residuals (`asList = FALSE`); else, when multiple grouping levels are specified in `level`, the returned object is a data frame with columns given by the residuals at different levels and the grouping factors.

See Also

> `lme, fitted.lme`

Examples

```
fm1 <- lme(distance ~ age + Sex, data = Orthodont, random = ~ 1)
residuals(fm1, level = 0:1)
```

`selfStart`   *Construct Self-Starting Nonlinear Models*

```
selfStart(model, initial, parameters, template)
```

Description

> This function is generic; methods functions can be written to handle specific classes of objects. Available methods include `selfStart.default` and `selfStart.formula`. See the documentation on the appropriate method function.

Value

> A function object of the selfStart class.

See Also

> `selfStart.default, selfStart.formula`

---

selfStart.default   *Construct Self-Starting Nonlinear Models*

---

        selfStart(model, initial, parameters, template)

Arguments

model
: A function object defining a nonlinear model.

initial
: A function object, taking three arguments: mCall, data, and LHS, representing, respectively, a matched call to the function model, a data frame in which to interpret the variables in mCall, and the expression from the left-hand side of the model formula in the call to nls. This function should return initial values for the parameters in model.

parameters, template
: These arguments are included for consistency with the generic function, but are not used in the default method. See the documentation on selfStart.formula.

Description

A method for the generic function selfStart for formula objects.

Value

A function object of class selfStart, corresponding to a self-starting nonlinear model function. An initial attribute (defined by the initial argument) is added to the function to calculate starting estimates for the parameters in the model automatically.

See Also

selfStart.formula

Examples

```
# 'first.order.log.model' is a function object defining a first
#  order compartment model
# 'first.order.log.initial' is a function object which calculates
# initial values for the parameters in 'first.order.log.model'
# self-starting first order compartment model
SSfol <- selfStart(first.order.log.model,
                   first.order.log.initial)
```

`selfStart.formula`   *Construct Self-Starting Nonlinear Models*

`selfStart(model, initial, parameters, template)`

Arguments

`model`
: A nonlinear formula object of the form ˜`expression`.

`initial`
: A function object, taking three arguments: `mCall`, `data`, and LHS, representing, respectively, a matched call to the function `model`, a data frame in which to interpret the variables in `mCall`, and the expression from the left-hand side of the model formula in the call to `nls`. This function should return initial values for the parameters in `model`.

`parameters`
: A character vector specifying the terms on the right-hand side of `model` for which initial estimates should be calculated. Passed as the `namevec` argument to the `deriv` function.

`template`
: An optional prototype for the calling sequence of the returned object, passed as the `function.arg` argument to the `deriv` function. By default, a template is generated with the covariates in `model` coming first and the parameters in `model` coming last in the calling sequence.

Description

A method for the generic function `selfStart` for formula objects.

Value

A function object of class selfStart, obtained by applying `deriv` to the right-hand side of the `model` formula. An `initial` attribute (defined by the `initial` argument) is added to the function to calculate starting estimates for the parameters in the model automatically.

See Also

`selfStart.default`, `deriv`

Examples

```
## self-starting logistic model
SSlogis <- selfStart(~ Asym/(1 + exp((xmid - x)/scal)),
```

```
function(mCall, data, LHS)
{
  xy <- sortedXyData(mCall[["x"]], LHS, data)
  if(nrow(xy) < 4) {
    stop("Too few distinct x values to fit a logistic")
  }
  z <- xy[["y"]]
  if (min(z) <= 0) { z <- z + 0.05 * max(z) } # avoid zeroes
  z <- z/(1.05 * max(z))          # scale to within unit height
  xy[["z"]] <- log(z/(1 - z))     # logit transformation
  aux <- coef(lm(x ~ z, xy))
  parameters(xy) <- list(xmid = aux[1], scal = aux[2])
  pars <- as.vector(coef(nls(y ~ 1/(1 + exp((xmid - x)/scal)),
                         data = xy, algorithm = "plinear")))
  value <- c(pars[3], pars[1], pars[2])
  names(value) <- mCall[c("Asym", "xmid", "scal")]
  value
}, c("Asym", "xmid", "scal"))
```

Variogram	*Calculate Semivariogram*

```
Variogram(object, distance, ...)
```

Description

This function is generic; method functions can be written to handle specific classes of objects. Classes that already have methods for this function include default, gls and lme. See the appropriate method documentation for a description of the arguments.

Value

Will depend on the method function used; see the appropriate documentation.

See Also

```
Variogram.default,Variogram.gls, Variogram.lme,
plot.Variogram
```

Variogram.lme	*Calculate Semivariogram for Residuals from an lme Object*

```
Variogram(object, distance, form, resType, data,
          na.action, maxDist, length.out, collapse, nint,
          breaks, robust, metric)
```

Arguments

object
: An object inheriting from class lme, representing a fitted linear mixed-effects model.

distance
: An optional numeric vector with the distances between residual pairs. If a grouping variable is present, only the distances between residual pairs within the same group should be given. If missing, the distances are calculated based on the values of the arguments form, data, and metric, unless object includes a corSpatial element, in which case the associated covariate (obtained with the getCovariate method) is used.

form
: An optional one-sided formula specifying the covariate(s) to be used for calculating the distances between residual pairs and, optionally, a grouping factor for partitioning the residuals (which must appear to the right of a | operator in form). Default is ~1, implying that the observation order within the groups is used to obtain the distances.

resType
: An optional character string specifying the type of residuals to be used. If "response", the "raw" residuals (observed − fitted) are used; else, if "pearson", the standardized residuals (raw residuals divided by the corresponding standard errors) are used; else, if "normalized", the normalized residuals (standardized residuals premultiplied by the inverse squareroot factor of the estimated error correlation matrix) are used. Partial matching of arguments is used, so only the first character needs to be provided. Defaults to "pearson".

data
: An optional data frame in which to interpret the variables in form. By default, the same data used to fit object is used.

na.action
: A function that indicates what should happen when the data contain NAs. The default action (na.fail)

causes an error message to be printed and the function to terminate, if there are any incomplete observations.

maxDist
An optional numeric value for the maximum distance used for calculating the semivariogram between two residuals. By default all residual pairs are included.

length.out
An optional integer value. When object includes a corSpatial element, its semivariogram values are calculated and this argument is used as the length.out argument to the corresponding Variogram method. Defaults to 50.

collapse
An optional character string specifying the type of collapsing to be applied to the individual semivariogram values. If equal to "quantiles", the semivariogram values are split according to quantiles of the distance distribution, with equal number of observations per group, with possibly varying distance interval lengths. Else, if "fixed", the semivariogram values are divided according to distance intervals of equal lengths, with possibly different number of observations per interval. Else, if "none", no collapsing is used and the individual semivariogram values are returned. Defaults to "quantiles".

nint
An optional integer with the number of intervals to be used when collapsing the semivariogram values. Defaults to 20.

robust
An optional logical value specifying if a robust semivariogram estimator should be used when collapsing the individual values. If TRUE the robust estimator is used. Defaults to FALSE.

breaks
An optional numeric vector with the breakpoints for the distance intervals to be used in collapsing the semivariogram values. If not missing, the option specified in collapse is ignored.

metric
An optional character string specifying the distance metric to be used. The currently available options are "euclidean" for the root sum-of-squares of distances; "maximum" for the maximum difference; and "manhattan" for the sum of the absolute differences. Partial matching of arguments is used, so only the first three characters need to be provided. Defaults to "euclidean".

Description

> This method function calculates the semivariogram for the within-group residuals from an lme fit. The semivariogram values are calculated for pairs of residuals within the same group. If collapse is different from "none", the individual semivariogram values are collapsed using either a robust estimator (robust = TRUE) defined in Cressie (1993), or the average of the values within the same distance interval. The semivariogram is useful for modeling the error term correlation structure.

Value

> A data frame with columns variog and dist representing, respectively, the semivariogram values and the corresponding distances. If the semivariogram values are collapsed, an extra column, n.pairs, with the number of residual pairs used in each semivariogram calculation, is included in the returned data frame. If object includes a corSpatial element, a data frame with its corresponding semivariogram is included in the returned value, as an attribute "modelVariog". The returned value inherits from class Variogram.

See Also

> lme, Variogram.default, Variogram.gls, plot.Variogram

Examples

```
fm1 <- lme(weight ~ Time * Diet, BodyWeight, ~ Time | Rat)
Variogram(fm1, form = ~ Time | Rat, nint = 10, robust = TRUE)
```

Appendix C
A Collection of Self-Starting Nonlinear Regression Models

We have mentioned several self-starting nonlinear regression models in the text. In this appendix we describe each of the self-starting models included with the nlme library. For each model we give the model formula, a description of the parameters, and the strategy used to obtain starting estimates.

C.1 SSasymp—The Asymptotic Regression Model

The asymptotic regression model is used to model a response y that approaches a horizontal asymptote as $x \to \infty$. We write it as

$$y(x) = \phi_1 + (\phi_2 - \phi_1) \exp[-\exp(\phi_3)x], \tag{C.1}$$

so that ϕ_1 is the asymptote as $x \to \infty$ and ϕ_2 is $y(0)$. These parameters are shown in Figure C.1. The parameter ϕ_3 is the logarithm of the rate constant. We use the logarithm to enforce positivity of the rate constant so the model does approach an asymptote. The corresponding half-life $t_{0.5} = \log 2/\exp(\phi_3)$ is illustrated in Figure C.1.

C.1.1 Starting Estimates for SSasymp

Starting values for the asymptotic regression model are obtained by:

1. Using NLSstRtAsymptote to get an estimate $\phi_1^{(0)}$ of the asymptote.

2. Regressing $\log(|y - \phi_1^{(0)}|)$ on t. The estimated slope is $-\exp(\phi_3^{(0)})$.

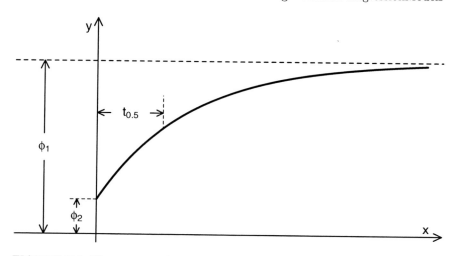

FIGURE C.1. The asymptotic regression model showing the parameters ϕ_1, the asymptotic response as $x \to \infty$, ϕ_2, the response at $x = 0$, and $t_{0.5}$, the half-life.

3. Using an algorithm for partially linear models (Bates and Chambers, 1992, §10.2.5) to refine estimates of ϕ_1, ϕ_2, and ϕ_3 in

$$y(x) = \phi_1 + (\phi_2 - \phi_1) \exp[\exp(\phi_3)x].$$

Because ϕ_1 and ϕ_2 occur linearly in the model expression, the least squares fit iterates over a single parameter.

These estimates are the final nonlinear regression estimates.

C.2   SSasympOff—Asymptotic Regression with an Offset

This is an alternative form of the asymptotic regression model that provides a more stable parameterization for the CO2 data. It is written

$$y(x) = \phi_1\{1 - \exp[-\exp(\phi_2) \times (x - \phi_3)]\}. \tag{C.2}$$

As in SSasymp, ϕ_1 is the asymptote as $x \to \infty$. In this formulation ϕ_2 is the logarithm of the rate constant, corresponding to a half-life of $t_{0.5} = \log 2 / \exp(\phi_2)$, and ϕ_3 is the value of x at which $y = 0$. The parameters ϕ_1, $t_{0.5}$, and ϕ_3 are shown in Figure C.2.

C.2.1   Starting Estimates for SSasympOff

First we fit SSasymp then we transform the parameters to the formulation used in SSasympOff. If $omega$ is the vector of parameters from SSasymp and

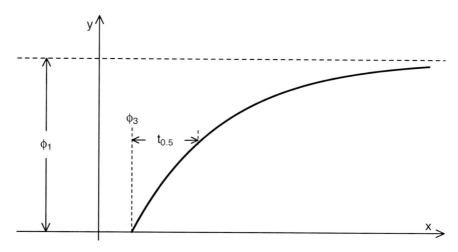

FIGURE C.2. The asymptotic regression model with an offset showing the parameters ϕ_1, the asymptote as $x \to \infty$, $t_{0.5}$, the half-life, and ϕ_3, the value of x for which $y = 0$.

ϕ is the vector of parameters for SSasympOff, the correspondence is

$$\phi_1 = \omega_1,$$
$$\phi_2 = \omega_3,$$
$$\phi_3 = \exp(-\omega_3) \log[-(\omega_2 - \omega_1)/\omega_1].$$

These estimates are the final nonlinear regression estimates.

C.3   SSasympOrig—Asymptotic Regression Through the Origin

This form of the asymptotic regression model is constrained to pass through the origin. It is called the BOD model in Bates and Watts (1988) where it is used to model Biochemical Oxygen Demand curves. The model is written

$$y(x) = \phi_1[1 - \exp(-\exp(\phi_2)x].\tag{C.3}$$

As in SSasympOff, ϕ_1 is the asymptote as $x \to \infty$ and ϕ_2 is the logarithm of the rate constant, corresponding to a half-life of $t_{0.5} = \log 2/\exp(\phi_2)$. The parameters ϕ_1 and $t_{0.5}$ are shown in Figure C.3.

C.3.1   Starting Estimates for SSasympOrig

Starting values for this regression model are obtained by:

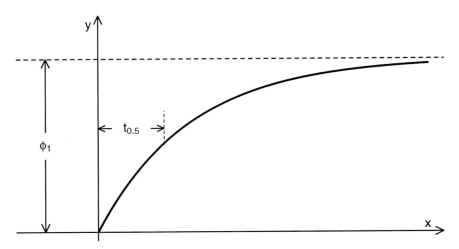

FIGURE C.3. The asymptotic regression model through the origin showing the parameters ϕ_1, the asymptote as $x \to \infty$ and $t_{0.5}$, the half-life.

1. Using `NLSstRtAsymptote` to get an estimate $\phi_1^{(0)}$ of the asymptote.

2. Obtaining an initial estimate of ϕ_2 as

$$\phi_2^{(0)} = \log \operatorname{abs} \sum_{i=1}^{n} \left[\log(1 - y_i/\phi_1^{(0)})/x_i \right] / n.$$

3. Using an algorithm for partially linear models to refine the estimates of ϕ_1 and ϕ_2. Because ϕ_1 occurs linearly in the model expression, the least squares fit iterates over a single parameter.

These estimates are the final nonlinear regression estimates.

C.4   SSbiexp—Biexponential Model

The biexponential model is a linear combination of two negative exponential terms

$$y(x) = \phi_1 \exp\left[-\exp(\phi_2)x\right] + \phi_3 \exp\left[-\exp(\phi_4)x\right]. \qquad (C.4)$$

The parameters ϕ_1 and ϕ_3 are the coefficients of the linear combination, and the parameters ϕ_2 and ϕ_4 are the logarithms of the rate constants. The two sets of parameters (ϕ_1, ϕ_2) and (ϕ_3, ϕ_4) are *exchangeable*, meaning that the values of the pairs can be exchanged without changing the value of $y(x)$. We create an identifiable parameterization by requiring that $\phi_2 > \phi_4$.

A representative biexponential model, along with its constituent exponential curves, is shown in Figure C.4.

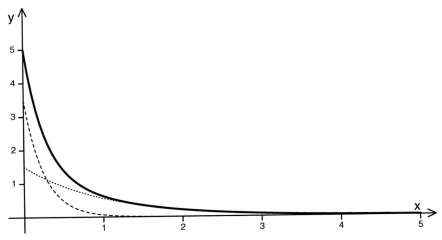

FIGURE C.4. A biexponential model showing the linear combination of the exponentials (solid line) and its constituent exponential curves (dashed line and dotted line). The dashed line is $3.5\exp(-4x)$ and the dotted line is $1.5\exp(-x)$.

C.4.1   Starting Estimates for SSbiexp

The starting estimates for the biexponential model are determined by *curve peeling*, which involves:

1. Choosing half the data with the largest x values and fitting the simple linear regression model

$$\log\text{abs}(y) = a + bx.$$

2. Setting $\phi_3^{(0)} = \exp a$ and $\phi_4^{(0)} = \log\text{abs}(b)$ and calculating the residuals $r_i = y_i - \phi_3^{(0)}\exp[-\exp(\phi_4^{(0)})x_i]$ for the half of the data with the smallest x values. Fit the simple linear regression model

$$\log\text{abs}(r) = a + bx.$$

3. Setting $\phi_2^{(0)} = \log\text{abs}(b)$ and using an algorithm for partially linear models to refine the estimates of ϕ_1, ϕ_2, ϕ_3, and ϕ_4. Because the model is linear in ϕ_1 and ϕ_3, the only starting estimates used in this step are those for ϕ_2 and ϕ_4 and the iterations are with respect to these two parameters.

The estimates obtained this way are the final nonlinear regression estimates.

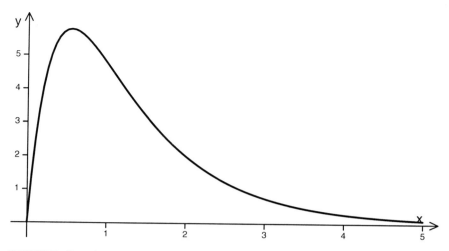

FIGURE C.5. A sample response curve from a first-order open-compartment model. The parameters correspond to an elimination rate constant of 1, an absorption rate constant of 3, and a clearance of 0.1. The dose is 1.

C.5   SSfol—First-Order Compartment Model

This model is derived from a compartment model in pharmacokinetics describing the concentration of a drug in the serum following a single oral dose. The model is based on first-order kinetics for the absorption of the drug from the digestive system and for the elimination of the drug from the circulatory system. Because the drug is eliminated from the circulatory system, the system of compartments is called an open system, and the model is a first-order open compartment model. It is written

$$y(x) = \frac{D \exp(\phi_1) \exp(\phi_2)}{\exp(\phi_3) \left[\exp(\phi_2) - \exp(\phi_1)\right]} \left\{\exp\left[-\exp(\phi_1)x\right] - \exp\left[-\exp(\phi_2)x\right]\right\},$$

$$(C.5)$$

where D is the dose, ϕ_1 is the logarithm of the elimination rate constant, ϕ_2 is the logarithm of the absorption rate constant, and ϕ_3 is the logarithm of the clearance.

A sample response curve from a first-order open compartment model is shown in Figure C.5

C.5.1   Starting Estimates for SSfol

The starting estimates for the SSfol model are also determined by curve peeling. The steps are:

1. Determine the position of the maximum response. Fit the simple linear regression model

$$\log(y) = a + bx$$

to the data with x values greater than or equal to the position of the maximum response. Set $\phi_1^{(0)} = \log \mathrm{abs}(b)$ and $\phi_2^{(0)} = \phi_1^{(0)} + 1$.

2. Use an algorithm for partially linear models to fit the nonlinear regression model

$$y(x) = k\{\exp[-\exp(\phi_1)x] - \exp[-\exp(\phi_2)x]\}$$

refining the estimates of ϕ_1 and ϕ_2.

3. Use the current estimates of ϕ_1 and ϕ_2 and an algorithm for partially linear models to fit

$$y(x) = kD\frac{\exp[-\exp(\phi_1)x] - \exp[\exp(\phi_2)x)]}{\exp(\phi_1) - \exp(\phi_2)}.$$

Set $\phi_3 = \phi_1 + \phi_2 - \log k$.

These estimates are the final nonlinear regression estimates.

C.6 SSfpl—Four-Parameter Logistic Model

The four-parameter logistic model relates a response y to an input x via a sigmoidal or "S-shaped" function. We write it as

$$y(x) = \phi_1 + \frac{\phi_2 - \phi_1}{1 + \exp\left[(\phi_3 - x)/\phi_4\right]}. \tag{C.6}$$

We require that $\phi_4 > 0$ so the parameters are:

- ϕ_1 the horizontal asymptote as $x \to \infty$

- ϕ_2 the horizontal asymptote as $x \to -\infty$

- ϕ_3 the x value at the inflection point. At this value of x the response is midway between the asymptotes.

- ϕ_4 a scale parameter on the x-axis. When $x = \phi_3 + \phi_4$ the response is $\phi_1 + (\phi_2 - \phi_1)/(1 + e^{-1})$ or roughly three-quarters of the distance from ϕ_1 to ϕ_2.

These parameters are shown in Figure C.6

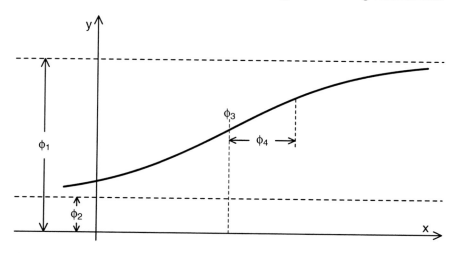

FIGURE C.6. The four-parameter logistic model. The parameters are the horizontal asymptote ϕ_1 as $x \to -\infty$, the horizontal asymptote ϕ_2 as $x \to \infty$, the x value at the inflection point (ϕ_3), and a scale parameter ϕ_4.

C.6.1 Starting Estimates for SSfpl

The steps in determining starting estimates for the SSfpl model are:

1. Use NLSstClosestX to determine $\phi_3^{(0)}$ as the x value corresponding a response at the midpoint of the range of the responses.

2. Use an algorithm for partially linear models to fit A, B, and ℓ while holding ϕ_3 fixed in the nonlinear regression model

$$y(x) = A + \frac{B}{1 + \exp[(\phi_3 - x)/\exp\ell]}.$$

The purpose of this fit is to refine the estimate of ℓ, the logarithm of the scale parameter ϕ_4. We start ℓ at zero.

3. Use the refined estimate of ℓ and an algorithm for partially linear models to fit

$$y(x) = A + \frac{B}{1 + \exp[(\phi_3 - x)/\exp\ell]}$$

with respect to A, B, ϕ_3 and ℓ. The estimates are then $\phi_1 = A$, $\phi_2 = A + B$, $\phi_4 = \exp\ell$ and ϕ_3.

These estimates are the final nonlinear regression estimates.

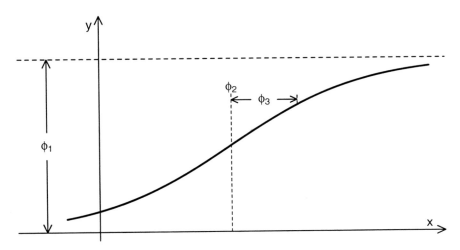

FIGURE C.7. The simple logistic model showing the parameters ϕ_1, the horizontal asymptote as $x \to \infty$, ϕ_2, the value of x for which $y = \phi_1/2$, and ϕ_3, a scale parameter on the x-axis. If $\phi_3 < 0$ the curve will be monotone decreasing instead of monotone increasing and ϕ_1 will be the horizontal asymptote as $x \to -\infty$.

C.7 SSlogis—Simple Logistic Model

The simple logistic model is a special case of the four-parameter logistic model in which one of the horizontal asymptotes is zero. We write it as

$$y(x) = \frac{\phi_1}{1 + \exp\left[(\phi_2 - x)/\phi_3\right]}. \tag{C.7}$$

For this model we do not require that the scale parameter ϕ_3 be positive. If $\phi_3 > 0$ then ϕ_1 is the horizontal asymptote as $x \to \infty$ and 0 is the horizontal asymptote as $x \to -\infty$. If $\phi_3 < 0$, these roles are reversed. The parameter ϕ_2 is the x value at which the response is $\phi_1/2$. It is the inflection point of the curve. The scale parameter ϕ_3 represents the distance on the x-axis between this inflection point and the point where the response is $\phi_1/\left(1 + e^{-1}\right) \approx 0.73\phi_1$. These parameters are shown in Figure C.7.

C.7.1 Starting Estimates for SSlogis

The starting estimates are determined by:

1. Scaling and, if necessary, shifting the responses y so the transformed responses y' are strictly within the interval $(0, 1)$.

2. Taking the logistic transformation

$$z = \log[y'/(1 - y')]$$

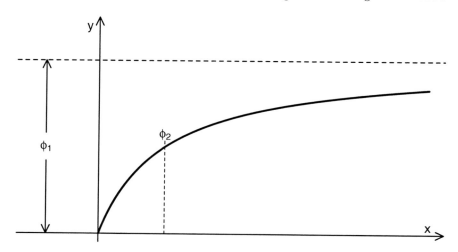

FIGURE C.8. The Michaelis–Menten model used in enzyme kinetics. The parameters are ϕ_1, the horizontal asymptote as $x \to \infty$ and ϕ_2, the value of x at which the response is $\phi_1/2$.

and fitting the simple linear regression model

$$x = a + bz.$$

3. Use $\phi_2^{(0)} = a$ and $\phi_3^{(0)} = b$ and an algorithm for partially linear models to fit

$$y = \frac{\phi_1}{1 + \exp[(\phi_2 - x)/\phi_3]}.$$

The resulting estimates are the final nonlinear regression estimates.

C.8   SSmicmen—Michaelis–Menten Model

The Michaelis–Menten model is used in enzyme kinetics to relate the initial rate of an enzymatic reaction to the concentration of the substrate. It is written

$$y(x) = \frac{\phi_1 x}{\phi_2 + x}, \tag{C.8}$$

where ϕ_1 is the horizontal asymptote as $x \to \infty$ and ϕ_2, the Michaelis parameter, is the value of x at which the response is $\phi_1/2$.

These parameters are shown in Figure C.8

C.8.1   Starting Estimates for SSmicmen

The starting estimates are obtained by:

1. Fitting a simple linear regression model

$$\frac{1}{y} = a + b\frac{1}{x}$$

   for the inverse response as a function of the inverse of x.

2. Setting $\phi_2^{(0)} = \text{abs}(b/a)$ and using an algorithm for partially linear models to fit

$$y = \frac{\phi_1 x}{\phi_2 + x}.$$

The resulting estimates are the final nonlinear regression estimates.

Index

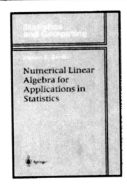